The Geology of Stratigraphic Sequences

Springer

Berlin
Heidelberg
New York
Barcelona
Budapest
Hong Kong
London
Milan
Paris
Santa Clara
Singapore
Tokyo

Andrew D. Miall

The Geology
of Stratigraphic Sequences

With 337 Figures and 11 Tables

 Springer

Prof. Dr. Andrew D. Miall
Department of Geology
University of Toronto
22 Russell Street
Toronto, Ontario M5S 3B1
Canada

ISBN 3-540-59348-9 Springer-Verlag Berlin Heidelberg New York

Library of Congress Cataloging-in-Publication Data. Miall, Andrew D. The geology of stratigraphic sequences/Andrew D. Miall.   p.   cm. Includes bibliographical references (p. 397–421) and indexes. ISBN 3-540-59348-9 (hardcover). – ISBN 0-387-59348-9 (hardcover) 1. Geology, Stratigraphic. I. Title. QE651.M46 1996, 551.7–dc20 96-26786

© Springer-Verlag Berlin Heidelberg 1997
Printed in Germany

The use of general descriptive names, registered, trademarks, etc. in this publication does not imply, even in the absence of a specific statement, that such names are exempt from the relevant protective laws and regulations and therefore free for general use.

Cover design: Springer-Verlag, E. Kirchner

Typesetter: Scientific Publishing Services (P) Ltd, Madras

SPIN: 10502355   32/3136/SPS – 5 4 3 2 1 0 – Printed on acid-free paper

For Charlene,
Christopher,
and Sarah

Preface

Sequence stratigraphy represents a new paradigm in geology. The principal hypothesis is that stratigraphic successions may be subdivided into discrete sequences bounded by widespread unconformities. There are two parts to this hypothesis.

First, it suggests that the driving forces which generate sequences and their bounding unconformities also generate predictable three-dimensional stratigraphies. In recent years stratigraphic research guided by sequence models has brought about fundamental improvements in our understanding of stratigraphic processes and the controls of basin architecture. Sequence models have provided a powerful framework for mapping and numerical modeling, enabling the science of stratigraphy to advance with rapid strides. This research has demonstrated the importance of a wide range of processes for the generation of cyclic sequences, including eustasy, tectonics, and orbital forcing of climate change. The main objective of this book is to document the sequence record and to discuss our current state of knowledge about sequence-generating processes.

To a considerable extent, however, these significant developments have been obscured by a controversy emerging from the second part of the sequence stratigraphy paradigm. It has been proposed that stratigraphic sequences are superior as chronostratigraphic indicators to all other forms of stratigraphic data, and that they therefore comprise the ideal basis for a superior standard of geological time. The main theoretical basis for the paradigm is the supposition that global stratigraphic successions are controlled primarily by eustatic sea-level changes. However, as this book demonstrates, the basic premise of the paradigm remains unproven. There is no convincing, independent evidence that a suite of globally correlatable cycles exists. Current chronostratigraphic dating techniques do not permit the level of accuracy and precision in sequence correlation claimed for the global cycle charts that have been published by Peter Vail and his former Exxon colleagues and coworkers. These charts rest on a series of assertions unsupported by published data. There has been considerable controversy on this point. A second objective of this book is to review these developments in detail, drawing on all available published literature. Examples are given from around the world, and the controversies and advances associated with sequence stratigraphy are fully described.

I may be known by some readers as a sceptic in the area of sequence stratigraphy, on the basis of a series of critical papers published in this area, plus numerous lectures given around the world. It needs to be emphasized that I find many of the developments of sequence stratigraphy to be exciting and important. There is no doubt that the emergence of sequence stratigraphy constitutes an important revolution in sedimentary geology, to add to those that have already been brought about by developments in plate tectonics, geophysical basin modeling and the concept of the process-response sedimentary model.

Part of this book is devoted to a discussion of the uses to which sequence stratigraphy can be put in the field of petroleum geology, and how the exploration and production processes may be facilitated by the employment of sequence stratigraphic techniques. It is to be hoped that the balanced approach of this book – scepticism, plus an equal focus on the positive aspects of the subject – represents the best form of scientific synthesis, and lends the book credibility and usefulness.

Andrew D. Miall

Acknowledgements

My own interest in sequence stratigraphy began slowly, as my work on regional basin analysis for the Geological Survey of Canada matured in the late 1970s, and I am grateful to this organization for introducing me to the scope and sweep of large-scale regional analysis. My developing knowledge of basin analysis provided me with a practical view of the subject that induced scepticism. In particular, my work in the Canadian Arctic included attempts to adjudicate debates between various biostratigraphic specialists who could not agree on the dating of certain subsurface sections that I was trying to correlate. My critique of sequence stratigraphy as a chronostratigraphic tool developed from this starting point. A few individuals in GSC discussed the early concepts with me and helped me to realize that something important was going on. Among these Ashton Embry stands out. Later, Jim Dixon's work provided food for thought.

Discussions with the main protagonists of sequence stratigraphy have met with mixed success. I would, however, like to acknowledge these colleagues for contributing to the development of my ideas: Phil Allen, Bert Bally, Chris Barnes, Sierd Cloetingh, Jim Coleman, Bill Galloway, Jake Hancock, Makoto Ito, David James, Alan Kendall, Dale Leckie, Peter McCabe, Dag Nummedal, Henry Posamentier, Brian Pratt, Larry Sloss, John Suter, Peter Vail, John Van Wagoner, Roger Walker, Tony Watts. Discussions with my wife, Charlene Miall, a social scientist, regarding the nature of science as a human endeavor, have been most helpful.

This book began life as an in-house report prepared for the exclusive use of the Japan National Petroleum Corporation in 1993. I am grateful to the Corporation for permission to publish their report, and to my employer, the University of Toronto, for providing the time for me both to write the original report and to prepare the material for the revisions incorporated into the final book.

Much of the material in Chapter 13 appeared in a contribution to the PaleoScene series in Geoscience Canada. I am grateful to Darrel Long for stimulating the writing of the paper, and to series editor Godfrey Nowlan and critical reviewers John Armentrout and Terry Poulton for their invaluable comments.

The entire manuscript was critically read by Brian Pratt and Phil Allen. I am most grateful to them for undertaking this onerous task, for their painstaking efforts in completing it, and for their numerous thoughtful and helpful comments. Remaining errors and omissions are, of course, my responsibility.

And once again, I must thank my wife Charlene and my children Christopher and Sarah for their encouragement, love and support.

Contents

I Review of Current Concepts

Modern sequence stratigraphy began with the work of L.L. Sloss, although it was founded on observations and ideas that emerged during the nineteenth century. The subject did not move to the center of the stratigraphic stage until modern developments in seismic stratigraphy were published by Peter Vail and his Exxon colleagues. The purpose of this first part of the book is therefore to set out the basic framework of the Exxon models. Supporting and parallel work by other authors is referred to, some of the problems with the methods and concepts are touched upon briefly, and some of the major current areas of research are listed. However, the main critical analysis of the methods and results is contained in Parts III and IV.

- In the late decades of the 18th century, geologists were striving toward a stratigraphic taxonomy within which their observations could find organization and structure. Some of the early schemes of classification were largely descriptive and relatively free of the taint of genetic implication.... By the middle of the 19th century, the gross elements of geochronology and chronostratigraphy, the periods and corresponding systems (Cambrian, Cretaceous, and so on) were widely recognized and accepted ... transplanting classical chronostratigraphic units to the New World, whether defined by unconformities and other physical changes or by paleontological changes, was not a simple or wholly satisfying operation.... As the 20th century advanced, stratigraphers were made increasingly aware of the necessity of distinguishing between what are now termed "lithostratigraphic" and "chronostratigraphic" units.... In the same decades, the three-dimensional view of stratified rocks provided by subsurface exploration and the practical requirements of subsurface stratigraphic nomenclature in the service of industry and government produced an environment within which nonclassical approaches were fostered and developed. (Sloss 1988a, pp. 1661–1662)

- The interpretation of the stratigraphic record has been greatly stimulated over the past few years by rapid conceptual advances in "sequence stratigraphy," i.e., the attempt to analyze stratigraphic successions in terms of genetically related packages of strata. The value of the concept of a "depositional sequence" lies both in the recognition of a consistent three-dimensional arrangement of facies within the sequence, the facies architecture, and the regional (and inter-regional) correlation of the sequence boundaries. It has also been argued that many sequence boundaries are correlatable globally, and that they reflect periods of sea-level lowstand, i.e., sequence-boundaries are subaerial erosion surfaces. (Nummedal 198, p. iii)

1 Introduction

1.1 Sequence Stratigraphy: A New Paradigm?

Most of the concepts encompassed by the study of sequence stratigraphy are not new. As eloquently summarized by Dott (1992), many observations and hypotheses regarding regional and global changes in sea level, and the ordering of stratigraphic successions into predictable packages, have been around since early in the nineteenth century. The word eustatic was first proposed for global changes in sea level by Suess (1885). He recognized that sea-level change could be determined by plotting the exent of marine transgression over continental areas, and by studying the changes in water depths indicated by successions of sediments and faunas (the English tranlation of this work appeared in 1906). Dott (1992) summarized the contributions of others, including Lyell, Grabau, Ulrich, Barrell, and those who studied the North American Midcontinent cyclothems, including Wanless, Weller, Moore and Shepard. The reader is referred to Dott (1992) and the memoir of which this paper is a part, for fascinating descriptions of the early controversies, many of them having a very modern flavour.

Modern work in the area of sequence stratigraphy evolved from the research of L.L. Sloss, W.C. Krumbein, and E.C. Dapples in the 1940s and 1950s, beginning with an important address they made to a symposium on "Sedimentary Facies in Geological History" in 1949 (Sloss et al. 1949). H.E. Wheeler also made notable contributions during this period, particularly in the study of time as preserved in stratigraphic sequences.

Ross (1991) pointed out that all the essential ideas that form the basis for modern sequence stratigraphy were in place by the 1960s. The concept of repetitive episodes of deposition separated by regional unconformities was developed by Wheeler and Sloss in the 1940s and 1950s. The concept of the "ideal" or "model" sequence had been developed for the Midcontinent cyclothems in the 1930s. The hypothesis of glacioeustasy was also widely dis-

cussed at that time. Van Siclen (1958) provided a diagram of the stratigraphic response of a continental margin to sea-level change and variations in sediment supply that is very similar to present-day sequence models. An important symposium on cyclic sedimentation convened by the Kansas Geological Survey marks a major milestone in the progress of research in this area (Merriam 1964); yet the subject did not "catch on." There are probably two main reasons for this. Firstly, during the 1960s and 1970s sedimentologists were preoccupied mainly by autogenic processes and the process-response model, and by the implications of plate tectonics for large-scale basin architecture. Secondly, geologists lacked the right kind of data. It was not until the advent of high-quality seismic-reflection data in the 1970s, and the development of the interpretive skills required to exploit these data, that the value and importance of sequence concepts became widely appreciated.

1.2 From Sloss to Vail

A useful history of the development of modern sequence stratigraphy has been provided by the main protagonist, Sloss (1988a), who described the evolution of the field based on the studies by himself, his colleagues, and his students, of the cratonic sedimentary cover of North America. His most important early paper (Sloss 1963) established the existence of six unconformity-bounded sequences (Fig. 1.1) which he named using Indian tribal names to distinguish them from the conventional litho- and chronostratigraphic subdivisions, the latter having been mainly imported from Europe. This paper built on earlier work by Sloss et al. (1949) and was paralleled independently by Wheeler (1958, 1959a,b, 1963). Later work by Sloss is reviewed in Chapters 2 and 6.

During this period prior to the appearance of seismic stratigraphy (1960s and early to mid-1970s)

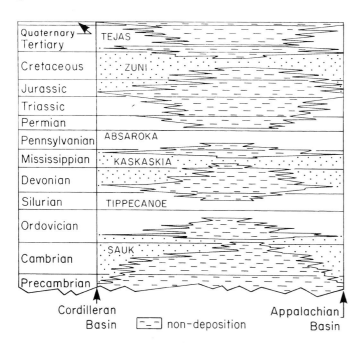

Fig. 1.1. The six sequences of Sloss (1963)

Cordilleran Basin   [`-`] non-deposition   Appalachian Basin

a few other workers were interested in the subject of global stratigraphic correlations, the possibility of eustatic sea-level change, and the geometries of stratigraphic units formed under conditions of fluctuating sea level. For example, Hallam (1963) reviewed the evidence for global stratigraphic events and was among the first to discuss the idea that global changes in sea level may have occurred in response to changes in the volumes of oceanic spreading centers. Many of the ideas incorporated into current models of continental-margin sequence architecture were developed by workers analyzing the Cenozoic record of the Gulf and Atlantic coasts of the United States. Curray (1964) was among the first to recognize the relationships between sea level and sediment supply. He noted that fluvial and strandplain aggradation and shoreface retreat predominate under conditions of rising sea level and low sediment supply, whereas river entrenchment and deltaic progradation predominate under conditions of falling sea level and high sediment supply (Morton and Price 1987). Frazier (1974) subdivided the Mississippi deltaic successions into transgressive, progradational, and aggradational phases (Fig. 1.2), and discussed autocyclic (delta switching) and glacioeustatic sedimentary controls. Brown and Fisher (1977) in a paper that actually deals with seismic data and appears in AAPG Memoir 26, summarized the ideas of an important group of stratigraphers at the Bureau of Economic Geology, University of Texas, that later became an integral part of the Exxon sequence-stratigraphy interpre-

tive framework – the use of regional facies concepts to define depositional systems and systems tracts. Soares et al. (1978) attempted to correlate Phanerozoic cycles in Brazil with those of Sloss, and Hallam continued his detailed facies studies of the Jurassic sedimentary record, leading to successive refinements of a sea-level curve for that period (Hallam 1978, 1981).

Sloss (1963) defined stratigraphic sequences as "rock-stratigraphic units of higher rank than group, megagroup, or supergroup, traceable over major areas of a continent and bounded by unconformities of interregional scope." With the advent of seismic stratigraphic research sequences much smaller than group in equivalent rank were recognized. This raised a nomenclature problem, as discussed below.

Wheeler (1958) is credited with the introduction of chronostratigraphic charts, in which stratigraphic cross sections are plotted with a vertical time axis rather than a thickness axis. These diagrams are now referred to as *Wheeler diagrams* (Sloss 1984). They are a useful way of indicating the actual range in age from place to place of stratigraphic units and unconformities (Fig. 1.3).

Sequence stratigraphy remained a subject of relatively minor, academic interest throughout the 1960s and 1970s (Ross 1991), until the publication of a major memoir by the Exxon group in 1977, which revealed the practical utility of the concepts for basin studies and regional, even global, correlation (Payton 1977). The work was led by Peter R.

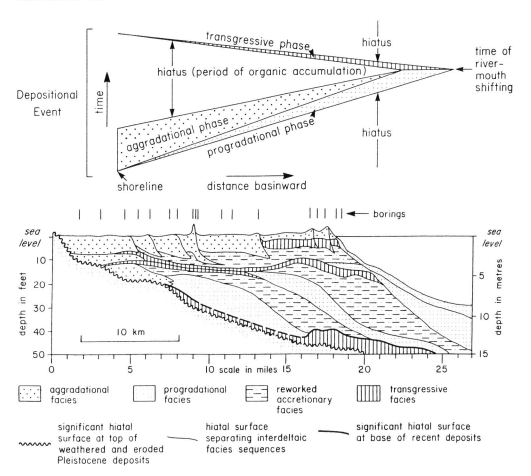

Fig. 1.2. The sequence-stratigraphic concepts of Frazier (1974)

Vail, a former graduate student of Sloss, and the research team also included several other former students of Sloss. Over a period of more than a decade in the 1960s and 1970s Vail and his co-workers studied seismic reflection data, and the methods and results gradually evolved that eventually appeared in AAPG Memoir 26 (Vail 1992). Seismic stratigraphy makes use of the concept that seismic reflections parallel bedding surfaces and are therefore of chronostratigraphic significance, enabling widespread correlation to be readily accomplished, although the reader is referred to Section 2.2.3, where the limitations of this concept are examined. This contrasts with the correlation methods used in conventional outcrop basin analysis, in which lithofacies contacts are known to be typically diachronous. The resulting lithostratigraphic classification may have limited local applicability, and can only, with considerable effort, be integrated into a reliable chronostratigraphic framework (see discussion of mapping methods in Miall 1990; Chap. 5).

Seismic stratigraphy permitted two major practical developments in basin analysis, the ability to define complex basin architectures in considerable detail, and the ability to recognize, map and correlate unconformities over great distances. Architectural work led to the development of a special terminology for defining the shape and character of stratigraphic surfaces. A particular emphasis came to be place on the nature of bedding terminations because of the significance these carry with regard to the processes of progradation, aggradation and erosion (Fig. 1.4). Seismic work by Vail and his team eventually led to the recognition of the sequence-stratigraphic model for the interpretation of seismic records and the building of regional and global stratigraphic syntheses. The basic components of this model are illustrated in Figs. 1.4 and 1.5, and a complete summary of the Exxon models is contained in Chapter 4.

One of the original objectives of Vail's work with Exxon, which presumably reflected the early influence of Sloss, was to attempt to correlate seismic

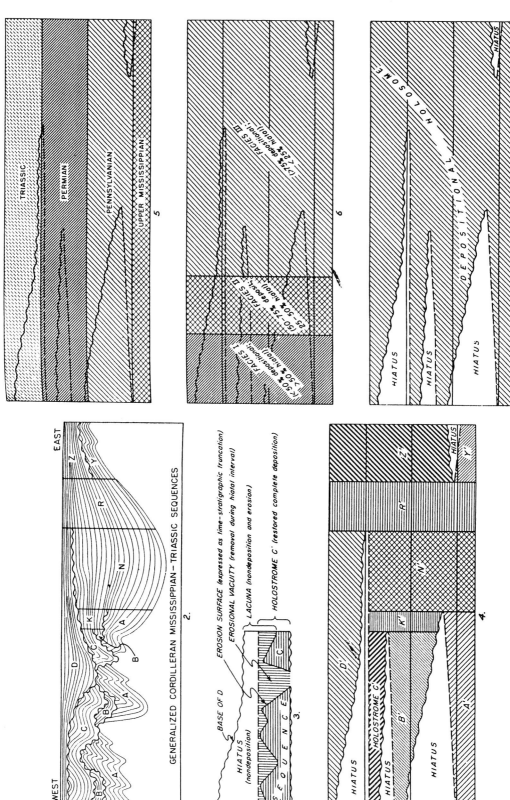

Fig. 1.3. Wheeler's development of chronostratigraphic diagrams, showing a stratigraphic cross section plotted with a vertical time axis to portray accurately the duration of stratigraphic units and unconformities from place to place. In the first diagram a generalized cross section is provided. A time cross section of sequence C is shown below, providing the definition of some of the terms used by Wheeler. Remaining diagrams show conversion of these data into chronostratigraphic charts. (Wheeler 1958, reproduced by permission)

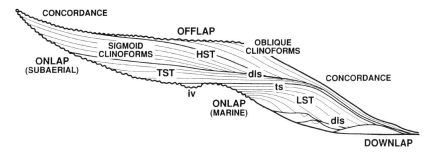

Fig. 1.4. The basic sequence-stratigraphic model, illustrating stratal geometries and terminology. Clinoforms are dipping stratal surfaces that indicate lateral progradation or accretion of the section. *dls*, Downlap surface; *ts*, transgressive surface; *iv* incised valley; *LST* lowstand systems tract; *TST* transgressive systems tract; *HST* highstand systems tract (Christie-Blick 1991, reproduced by kind permission of Elsevier Science-NL, Sara Burgerhartstraat 25, 1055 KV Amsterdam, The Netherlands)

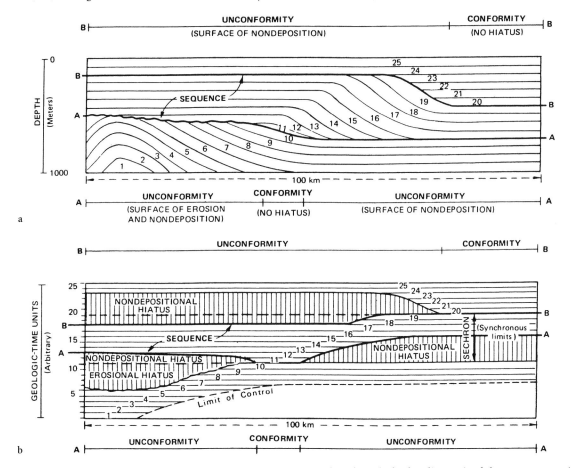

Fig. 1.5. Basic concepts of the depositional sequence. **a** Stratigraphic geometry. Three sequences are shown, separated by unconformities A and B. **b** Chronostratigraphic chart (Wheeler diagram) of the same succession as in **a**, emphasizing the time breaks in the succession (Mitchum et al. 1977b, reproduced by permission)

sequences from basin to basin to test the idea of regional and global cyclicity. As reported by Sloss (1988a), a large data base consisting of seismic sections, well records, biostratigraphic interpretations and related data was assembled within the Exxon group of companies worldwide. From this emerged the famous global cycle chart, which was first published in AAPG Memoir 26 (Fig. 1.6), and has subsequently gone through several revisions and refinements. The chart is introduced in Chapter 5 and discussed at some length in Chapter 14.

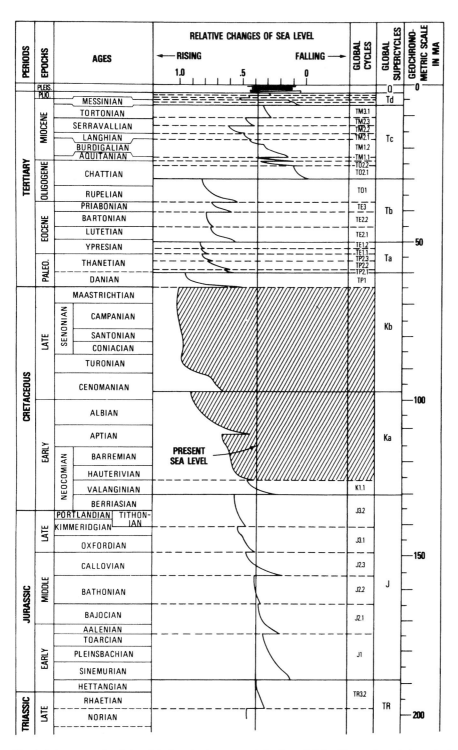

Fig. 1.6. An example of the global cycle chart, as originally published by Vail et al. (1977). (Reproduced by permission)

1.3 Problems and Research Trends: The Current Status

Vail's work, beginning with AAPG Memoir 26 (Payton 1977) has, of course, revolutionized the science of stratigraphy, and Vail himself has been much honored as a result. This is as it should be. However, as with the development of any major new paradigm, many problems, some critical, have developed in the application of sequence concepts, and much research remains to be carried out. Considerable controversy remains regarding the existence of a worldwide sequence framework, and in the interpretation of the origins of several types of sequence. There are several separate but related types of problem:

1. The basic sequence models erected by the Exxon group were intended for a specific type of tectonic setting, that of extensional continental margins, and should be used with caution in other types of setting. The original models were also very simplistic with regard to the nature of the balance between the three major controls of basin architecture, subsidence, sea-level change, and sediment supply, and were developed primarily for siliciclastic sediments. Carbonate and evaporite sequence models require major revisions of the Exxon models (Chaps. 4, 15).
2. The problem of causality is a critical one. Much work has been carried out, and much more remains to be done to investigate the processes that generate unconformities by their effects on sea level or the elevation of the continents. Some mechanisms are regional in scope, others global. Some are rapid in their effects, others slow, even by geological standards of time (Part III). Distinguishing between all these mechanisms involves a consideration of the third problem.
3. The problem of correlation. Are sequences regional or global in scope? Answering this problem is one of the first important tasks in attempts to distinguish between the effects of global processes, such as eustatic changes in sea level, and the results of regional processes, including various types of tectonism. Correlation between sequences and other stratigraphic and tectonic events is critical in answering this question. However, correlation is a problem, because of imprecisions in the geological time scale and difficulties in generating sufficiently accurate dates for any given stratigraphic successions. Serious questions remain about the construction,

meaning, and utility of the Exxon global cycle chart (Chap. 5, Part IV).

Approaches to these various problems have evolved into several categories of research (see also Miall 1995):

1. Theoretical geophysical studies of crust and mantle processes in a search for mechanisms of continent and ocean elevation change (Chaps. 9, 11).
2. Modeling of basins, employing numerical manipulation and graphical computer simulation to integrate the effects of subsidence, sea-level change and sediment supply in various tectonic settings (Chap. 16). Such studies are commonly tested against data from real basins in the process known as forward modeling, using the research described in the next two paragraphs.
3. Detailed stratigraphic studies, employing refined chronostratigraphic methods to assess the significance of unconformities, sequence boundaries etc., and to relate these to regional tectonic events and the global cycle chart (so-called "tests" of the Vail curve). These studies are focusing primarily on the Mesozoic and Cenozoic, for which the stratigraphic record is reasonably complete (Part IV).
4. Stratigraphic studies have become extremely sophisticated for the Cenozoic record, especially the Late Cenozoic, for which the marine stratigraphic record provides excellent undisturbed sections to which many separate techniques can be applied (magnetostratigraphy, strontium and oxygen isotope stratigraphy, refined biostratigraphy), and correlations can be carried out with the record of glacioeustasy (facies and paleoecological changes in the sedimentary record) and the Milankovitch astronomical periodicities (Chap. 10).
5. Investigation of modern and very recent sea-level change around the world, and the effects of glacioeustasy, have been carried out as a means of providing a base-line of well-constrained studies of what is happening now. This very specialized work carries the subject to levels of measurement detail and chronostratigraphic refinement that are beyond the scope of most geological research, and the work is not discussed in this book.

Good overviews of the subject of sequence stratigraphy, from various perspectives, can be obtained from several recent books, special journal issues, and review articles. These include the following.

Books. Miall's (1990) textbook on basin analysis includes a fairly succinct review of sequence stratigraphy and global stratigraphic cycles written at the graduate level. It carries the development of techniques up to about 1988. Hallam (1992) discussed sea-level changes throughout the Phanerozoic, and evaluated the various mechanisms that have been proposed for sea-level change. Walker and James (1992) provided a thorough, updated version of the Geological Association of Canada *Facies Models* book, written at the advanced undergraduate level. It contains extensive discussions of the stratigraphic effects of sea-level change, but there is little discussion of mechanisms in this book beyond a useful introductory chapter. Exxon models and terminology are deliberately avoided in this book.

Research Syntheses (Books and Special Journal Issues). Essential reading is the first major publication on sequence stratigraphy, the AAPG Memoir which established seismic stratigraphy as a major new technique (Payton 1977). A second AAPG collection of articles on seismic stratigraphy appeared in 1984, and constitutes an attempt to examine some of the major premises of the Exxon work, such as the significance of condensed successions in the sedimentary record, and the ability to correlate unconformites using the limited well and seismic data available from specific continental margins (Schlee 1984). Bally (1987) compiled three volumes of seismic-stratigraphic studies in a large, atlas format. These include important introductory papers dealing with Exxon methods, and numerous case studies from many types of basin around the world. Most of these are superbly illustrated with long seismic sections, many in color. Berger et al. (1984) edited a major research compilation dealing with Milankovitch processes, including astronomical and climatic studies and many papers describing the geological evidence.

Two research collections published by the Society of Economic Paleontologists and Mineralogists, one edited by Nummedal et al. (1987), and the other by Wilgus et al. (1988), contain many important papers. The first focuses on studies of Quaternary sea-level change, and provides comparisons of coastal stratigraphy with the ancient record; the second contains a major suite of papers by the Exxon group, dealing with their sequence models. These papers provide the most recent major exposition of the Exxon sequence-stratigraphy research, although much has changed since this book appeared. Practical examples of this work are illustrated by Van Wagoner et al. (1990) in a well-illustrated review of outcrop and subsurface examples of Exxon-type sequences. In my opinion this is the best of the Exxon products because it contains numerous actual examples, and is less "model driven" than the other papers by this group, although there are problems with some concepts and terminology that are addressed later in this book. Another collection of case studies is that edited by James and Leckie (1988), which draws particularly on the wealth of subsurface detail available for the Western Canada Sedimentary Basin. Collinson (1989) edited a compilation of various studies of relevance to the petroleum industry, including techniques of correlation, with examples from the Arctic and North Sea regions. Cross (1990) put together a unique compilation, a collection of research articles describing the various quantitative approaches that are being taken to basin modeling. Many important details of the sequence-stratigraphy story emerged in this book. Ginsburg and Beaudoin (1990) collected research on the Cretaceous system – a synthesis of the first major project of the Global Sedimentary Geology Program. Fischer and Bottjer (1991) provided an introduction to a special issue of *Journal of Sedimentary Petrology* dealing with Milankovitch rhythms.

Macdonald (1991) compiled a suite of case studies of stratigraphic architecture in a variety of convergent and collisional plate settings. One of the major focuses of this book is to examine sequence stratigraphies in a tectonic context. A wealth of stratigraphic detail is contained in this book, but there are no synthesis or overview articles, except that by Carter et al. (1991), who tested the sequence-stratigraphic model and the global-cycle-chart model of Exxon against data from New Zealand. Revelle (1990) edited an important collection of review articles on sea-level change and its causes, including several referenced separately, below. Swift et al. (1991a) published a collection of research articles on shelf sedimentation. This includes a major set of theoretical papers that attempted to establish a quantitative framework for shelf sedimentation within a sequence framework. Cloetingh (1991) edited a special issue of *Journal of Geophysical Research* consisting of a collection of papers that examined the measurement, causes and consequences of long-term sea-level change. Advanced computer modeling and the use of refined stratigraphic data sets are two of the features of this collection. Another collection of papers in this area was edited by Biddle and Schlager (1991). Einsele et al. (1991a) compiled an edited multi-authored compilation of chapters dealing with many aspects of cyclic and event stratification, including several useful overview chapters. An updated version of the Exxon approach to sequence-stratigraphic analysis, including their first realistic appraisal of the importance of tectonism, is included in a chapter by Vail et al. (1991). Franseen et al. (1991) edited a collection of articles on the subject of sedimentary modeling, focusing on high-frequency cycles. Another useful collection of papers on tectonics and seismic sequence stratigraphy is that edited by Williams and Dobb (1993).

Six recent collections of research papers contain much additional data. Two of these, edited by de Boer and Smith (1994a) and House and Gale (1995) focus on orbital forcing and cyclic sequences. Two books edited by Weimer and Posamentier (1993), and Loucks and Sarg (1993) focus on siliciclastic and carbonate sequence stratigraphy, respectively. Eschard and Doligez (1993) edited a suite of papers demonstrating how detailed outcrop studies, including documentation of high-resolution sequence stratigraphy, could be of use in the development of our understanding about petroleum reservoirs. Lastly, the book edited by Posamentier et al. (1993) contains papers dealing with concepts and principles, methods and applications, and a series of case studies. Milankovitch cycles are discussed in a book by Schwarzacher (1993) that provides a detailed treatment of the nature of orbital perturbations and spectral analysis of orbital and cyclic frequencies. Van Wagoner and Bertram (1995) edited a collection of papers dealing with the sequence stratigraphy of foreland basins. The book contains an introductory article by Van Wagoner that provides revised definitions of many of the terms used in sequence stratigraphy.

Review Articles. Useful reviews of Milankovitch-type cycles were given by Fischer (1986) and Weedon (1993). Burton et al. (1987) reviewed the lack of reference frames in attempts to quantify sea-level change, and the fact that it is very difficult to quantify and isolate the effects of the three main depositional controls, subsidence, sea-level change, and sediment supply. Cross and Lessenger (1988) reviewed the methods of seismic stratigraphy, and discussed some of the constraints on the use of seismic data in stratigraphic studies. Two useful companion studies by Christie-Blick et al. (1990) and Christie-Blick (1991) review current work on mechanisms of sea-level change and numerical

modeling techniques. Schlager (1992a) provided a brief but well-illustrated discussion of the sequence-stratigraphy of carbonate depositional systems. Miall (1995) outlined recent developments in research in the field of stratigraphy, including sequence stratigraphy.

This book draws on all these sources, plus numerous individual articles, as referenced throughout the text.

1.4 Stratigraphic Terminology

Sloss (1963) used the term sequence for his packages of strata. The term has had a varied usage since that time, having been employed informally, in a nongenetic sense, as a synonym for "succession" in some literature, and for cyclic or unconformity-bounded units of varying dimensions and time-spans. Sloss's original sequences represent sea-level cycles that were tens of millions of years in duration. They were termed second-order cycles by Vail et al. (1977; see Chap. 3) whereas the term sequence was used for much smaller packages of strata, representing sea-level cycles of a few million years duration (third-order cycles of Vail et al. 1977) in the first seismic work, that of Vail and his co-workers in AAPG Memoir 26 (Payton 1977). Attempts to systematize the terminology have met with mixed success. Chang (1975) proposed the term synthem for unconformity-bounded sequences, the intent being that units would be defined and named using the word synthem in the same way as the term "formation" is used. This proposal was not formally adopted by the North American Commission on Stratigraphic Nomenclature (NACSN 1983), but the International Subcommission on Stratigraphic Classification (ISSC 1987) later approved the suggestion, and added the additional variants supersynthem, subsynthem and miosynthem for groups of synthems, and for two scales of subdivided synthem. Other proposed terms, such as interthem and mesothem were also discussed in this document (ISSC 1987) but were not formally accepted. These terms are given in the new *International Stratigraphic Guide* (Salvador 1994). However, few, if any, geologists, have made use of them, and some (e.g., Murphy 1988) were actively opposed to their adoption.

An alternative approach, contained in the North American code (NACSN 1983), has met with more interest, perhaps only because of the more euphonious nature of the terminology. This is the method of allostratigraphy. "An allostratigraphic unit is a mappable stratiform body of sedimentary rock that is defined and identified on the basis of its bounding discontinuities." (NACSN 1983, p. 865) A hierarchy of units, including the allogroup, alloformation and allomember, was proposed, and rules were established for defining and naming these various types of unit. Allostratigraphic methods enable the erection of a sequence framework that avoids the cumbersome nature of lithostratigraphy, whereby lateral changes in facies within a unit of comparable age, require a change in name. In the past, because of the localized nature of much stratigraphic research, different lithostratigraphic frameworks have commonly been erected for similar successions in separate geographic areas, and this has led to much confusion.

The allostratigraphic method is illustrated in Fig. 1.7, in which it can be seen that using an allostratigraphic approach to the subdivision of a fluvial-lacustrine assemblage the natural subdivision of the succession into four sequences, bounded

by breaks in sedimentation, can readily be formalized in a stratigraphic nomenclature. Several groups of workers are now explicitly employing an allostratigraphic methodology. For example, Autin (1992) subdivided the terraces and associated sediments in a Holocene fluvial floodplain succession into alloformations. R.G. Walker and his coworkers have employed allostratigraphic terminology in their study of the sequence stratgraphy of part of the Alberta Basin, Canada. Their first definition of unconformity-bounded units is described in Plint et al. (1986), where the defining concepts are referred to as event stratigraphy, following the developments of ideas in this area by Einsele and Seilacher (1982). Explicit use of allostratigraphic terms appears in their later papers (e.g., Plint 1990). A recent text on facies analysis that builds extensively on the work of this group recommends the use of allostratigraphic methods and terminology as a general approach to the study of stratigraphic sequences (Walker 1992a). Martinsen et al. (1993) compared lithostratigraphic, allostratigraphic and sequence concepts as applied to a stratigraphic

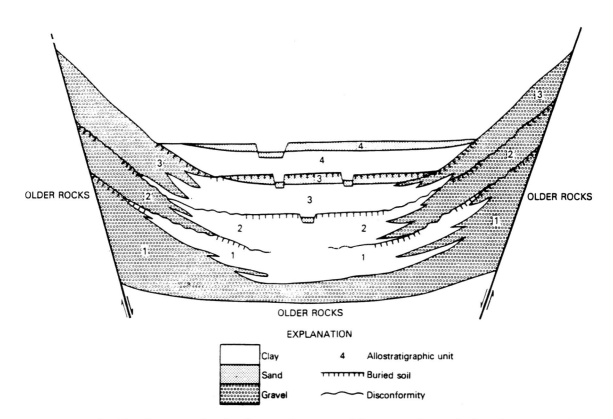

Fig. 1.7. Example of the allostratigraphic classification of a fluvial-lacustrine assemblage in a graben. *Numbers 1–4* correspond to alloformations, which are defined by breaks in sedimentation and cut across facies boundaries.

Using lithostratigraphic methods each gravel, sand and clay unit would typically be given separate names. (NACSN 1983, reproduced by permission)

succession in Wyoming. As they were able to demonstrate, each method has its local advantages and disadvantages.

One of the achievements of seismic stratigraphy has been a development of the ability to trace unconformity-bounded units into areas where the unconformable bounding surfaces are no longer recognizable. Thus Mitchum et al. (1977a, p. 53) defined a depositional sequence as "a stratigraphic unit composed of a relatively conformable succession of genetically related strata and bounded at its top and base by unconformities or their correlative conformities." Recognizing the bounding contacts in a conformable succession might in practice be difficult. The concept of a correlative conformity does not appear in the NACSN code, although some (e.g., Walker 1992a, p. 9) have recommended that it should. As discussed in Parts IV and V of this book, great care needs to be taken in assessing the chronostratigraphic significance of unconformities and correlative conformities. Considerable disagreement exists regarding the correlation of sequence-bounding unconformities from basin margins, where they are commonly of subaerial origin and may be accompanied by major facies shifts, into deeper parts of the basin (Sect. 15.2).

Sequence classifications and allostratigraphic units are based on concepts of sequence scale and duration that are hierarchical in character (first- to sixth-order sequences; synthem and its variants; allo- units). However, it is becoming increasingly clear that in the rock record sequences occur over a wide range of time scales and physical scales (e.g., thicknesses) that show no significant natural breaks such as would justify hierarchical classification (Sect. 3.1). In this book sequences are described with reference to their duration (Chaps. 6–8) but formal hierarchical classifications are largely avoided, except where it is necessary to make reference to earlier literature. At the time of writing this book the IUGS International Subcommission on Stratigraphic Classification had established a Working Group on Sequence Stratigraphy that was wrestling with the problem of how to incorporate sequence concepts into formal systems of stratigraphic nomenclature. My recommendation at present, until considerable further work has been carried out on sequence stratigraphy, is: *don't*. It is to be hoped that once having studied this book readers (including the members of the Working Group) will be aware of the considerable advances in understanding that are required before formal codification should be attempted.

2 Methods for Studying Sequence Stratigraphy

2.1 Introduction

A practical, working geologist faces two successive questions: (a) Is the stratigraphy of the basin fill subdivisible into stratigraphic sequences? and (b) What generated these sequences, regional tectonism, global eustasy, orbital forcing, or some other cause? The first part of this chapter deals with the methods for analyzing the sequence record of a basin fill. These include the following:

1. The mapping of unconformities as a first step in identifying unconformity-bounded sequences.
2. Clarifying the relationship between regional structural geology and the large-scale configuration of sequences.
3. The mapping of onlap, offlap and other stratigraphic terminations to provide information about the internal architectural development of each sequence.
4. The mapping of cyclic vertical facies changes in outcrop or well records to subdivide a stratigraphic succession into its component sequences and depositional-systems tracts, and as an indicator of changes in relative sea level.

The first three steps may be based on seismic-reflection data, well records or outcrops; the last step cannot be accomplished using seismic data alone.

The answer to the second question, regarding generating mechanisms, requires that the sequences be precisely dated and correlated with others in adjacent basins and in tectonically unrelated regions. A detailed chronostratigraphic framework is required to test correlations with regional tectonic events and with known global events, of which eustasy is the most obvious. The question of correlation is dealt with in Chapters 5 and 12–14. In Chapter 5 the Vail method of deriving the so-called global cycle chart from regional records is described. Section 2.2.3 describes the Vail method for constructing local charts of relative changes in sea level, and Section 2.3 reviews other, mostly older

methods that have evolved for assessing regional and global sea-level changes directly from the stratigraphic record. These include the following:

1. The mapping of changes in stratigraphic volumes or areas covered by successions of a specified age range as an indication of transgression and regression and changing rates of subsidence.
2. The use of hypsometric curves to plot broad changes in continental elevation and eustatic sea-level changes.
3. The use of backstripping procedures to quantify basin subsidence and sedimentation histories.
4. The study of paleoshorelines.
5. Graphical and numerical methods for the documentation and analysis of the meter-scale cycles that are very common in some parts of the stratigraphic record (e.g., certain lacustrine and shelf-carbonate successions).

As discussed in several places in this book, one of the principal problems with the assessment of causality is that there are no absolute reference frames for the calibration of sea-level change. The vertical movements of sea level can only be measured by the stratigraphic record left on the earth's crust, but the crust itself is in constant vertical motion, driven by a variety of mantle and lithosphere processes (Burton et al. 1987; Sahagian and Watts 1991). Much of this book deals with the strategies earth scientists have evolved in their attempts to attack and overcome this problem.

2.2 Erecting a Sequence Framework

2.2.1 The Importance of Unconformities

The six sequences of Sloss (1963) were based on the recognition of six craton-wide unconformities within the North American continent, and the mapping of widespread unconformities is obviously an essential prerequisite to the recognition of "unconformity-bounded sequences." Such unconformity surfaces clearly indicate either widespread epeirogenic movements or eustatic sea-level changes. Sloss (1963) emphasized that it is the widespread, interregional extent of the unconformities that is their key characteristic. He noted that "there is no apparent relationship between the prominence of the local evidences and the geographic scope of a given unconformity." Local angular unconformities of tectonic origin may be very prominent in the rock record, complete with evidence of deep erosion, angular discordance, and the presence of basal conglomerates in the overlying strata. However, very few such unconformities can be traced for wide distances along tectonic strike or away from a mobile belt into the adjacent basin or craton. Interregional unconformities may show little or no angular discordance and display little evidence of their importance at the outcrop level. It is only by careful regional mapping and correlation that their importance is recognized.

An important aspect of modern seismic methods, to be discussed later in this chapter (and in more detail in Chap. 13), is the assertion by Vail et al. (1977) regarding the chronostratigraphic significance of seismic reflections, including bedding traces, unconformities, and other reflective surfaces. This idea now receives general support (e.g., Cross and Lessenger 1988), although there are a few exceptions, such as where laterally very rapid and diachronous facies changes can generate what Schlager (1992a) termed pseudo-unconformities

(see Sect. 2.2.3), and the work of Tipper (1993) indicates that this concept must be used with care (Sect. 2.2.3).

Unconformities may be used to define stratigraphic sequences because of two key, interrelated characteristics: (a) The break in sedimentation that they represent has a constant maximum time range, although parts of that time range may be represented by sedimentation within parts of the areal range of the unconformity. (b) The sediments lying above an unconformity are everywhere entirely younger than those lying below the unconformity. These points are readily apparent from Wheeler diagrams, such as Figs. 1.3 and 1.5.

There are a few exceptions to the second rule, that of the age relationships that characterize unconformities. Christie-Blick et al. (1990) described two situations in which diachronous unconformities may develop, such that beds below the unconformity are locally younger than certain beds lying above the unconformity. The first case is that where the unconformity is generated by marine erosion caused by deep ocean currents. These can shift in position across the sea floor as a result of changes in topography brought about by tectonism or sedimentation. Christie-Blick et al. (1990) cited the case of the Western Boundary Undercurrent that flows along the continental slope of the Atlantic Ocean off the United States. This current is erosive where it impinges on the continental slope (Fig. 2.1A), but deposition of entrained fine clastic material takes place at the margins of the main

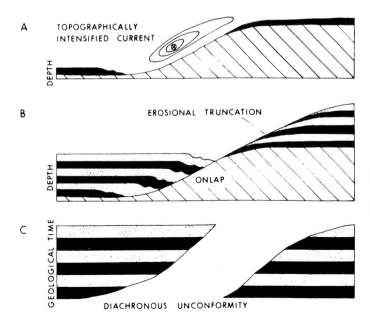

Fig. 2.1. Development of a diachronous unconformity as a result of a shift in the position of an erosive ocean-floor current. **A** View down current, with current strength shown by contours of equal speed. **B** Sedimentation takes place in areas away from the core of the flow, which is gradually displaced up the continental slope. **C** Wheeler diagram of the stratigraphic relationships that result. (Christie-Blick et al. 1990)

current, and the growth of this blanket is causing the current to gradually shift up the slope. The result is onlapping of the deposits onto the slope below the current, and erosional truncation of the upslope deposits (Fig. 2.1B). A chronostratigraphic chart of this relationship indicates the diachronous nature of the unconformity (Fig. 2.1C).

Submarine erosion, and other processes, can generate breaks in sedimentation without any change in sea level. This is particularly the case in carbonate sediments, which are very sensitive to environmental change, and may develop *drowning unconformities* (Schlager 1989, 1992a). Architecturally they may be similar to lowstand unconformities, and care must be taken to interpret them correctly. They may in fact represent an interval of slow sedimentation, with many small hiatuses and interbedded with thin condensed sections, indistinguishable on the seismic record from actual unconformable breaks because of limited seismic resolution. Schlager (1992a) stated:

Drowning requires that the reef or platform be submerged to subphotic depths by a relative rise that exceeds the growth potential of the carbonate system. The race between sea level and platform growth goes over a short distance, the thickness of the photic zone. Holocene systems indicate that their short-term growth potential is an order of magnitude higher than the rates of long-term subsidence or of third-order sea-level cycles.... This implies that drowning events must be caused by unusually rapid pulses of sea-level or by environmental change that reduced the growth potential of platforms. With growth reduced, drowning may occur at normal rates of rise.

Schlager (1992a) pointed to such environmental changes as the shifts in the El Niño current, which bring about sudden rises in water temperature, beyond the tolerance of many corals. Drowning can also occur when sea-level rise invades flat bank tops, creating shallow lagoons with highly variable temperatures and salinities, plus high suspended-sediment loads due to coastal soil erosion. Oceanic anoxic events, particularly in the Cretaceous, are also known to have caused reef drowning. Schlager (1992a) suggested that two Valangian sequence boundaries in the Haq et al. (1987, 1988a) global cycle chart may actually be drowning unconformities that have been misinterpreted as lowstand events (Fig. 14.6). He also noted the erosive effects of submarine currents, and their ability to generate unconformities that may be mapped as sequence boundaries but that have nothing to do with sea-level change. The Cenozoic sequence stratigraphy of the Blake Plateau, off the eastern United States, is dominated by such breaks in sedimenta-

tion that do not correlate with the Exxon global cycle chart (Fig. 2.2), but have been interpreted as the result of erosion by the meandering Gulf Stream.

The second type of diachronous unconformity is that which develops at basin margins as a result of syndepositional tectonism. Continuous deformation during sedimentation may lead to migration of a surface of erosion, and subsequent rapid onlap of the erosion surface by alluvial sediments. Such *intraformational unconformities* were first described in detail by Riba (1976), and subsequent discussions of these structures have been given by Miall (1978) and Anadon et al. (1986). Typically these unconformities are associated with coarse conglomeratic sediments and die out rapidly into the basin. There is therefore little danger of their presence leading to the devlopment of erroneous sequence stratigraphies.

A different kind of problem is that pointed out by Sengör (1992). In areas undergoing active convergence, such as foreland fold-thrust belts and subduction-accretion complexes, the rates of tectonic progradation and aggradation are rapid, and can lead to the development of numerous localized unconformities of short duration. Limits in chronostratigraphic resolution may result in these surfaces being incorrectly conflated into a single surface with an implied regional significance. Unconformities associated with foreland forebulges are typically subtle features within successions of shallow-water sediments, and may be widely traceable along strike. However, as shown by modeling studies of foreland basins (Beaumont 1981; Quinlan and Beaumont 1984) and by stratigraphic studies of the basin fill, the forebulges migrate during active tectonism and during post-tectonic viscous relaxation phases, resulting in diachronous unconformities (Tankard 1986; Knight et al. 1991; Crampton and Allen 1995).

Given the possible complications noted above, it is nevertheless the case that the mapping and correlation of unconformities has become a powerful tool in the recognition of the sequence architecture of basin fills. For example, Plint et al. (1986) recognized the presence of a series of widespread but subtle erosion surfaces in the Cardium Formation (Cretaceous) of the Alberta Basin, Canada, on the basis of the truncation of log markers and by the correlation of gravel horizons overlying the surfaces, as seen in drill core. Commonly these consist of only a thin pebble veneer. Mapping of the erosion surfaces permitted Plint et al. (1986) to erect an

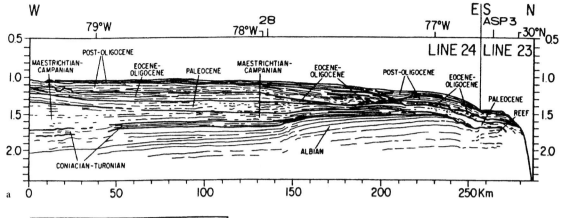

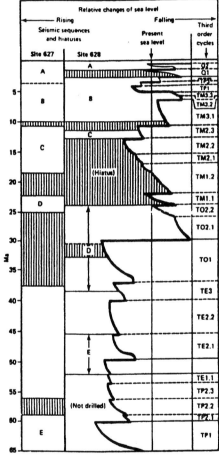

Fig. 2.2. a Cretaceous-Cenozoic sequence stratigraphy of the Blake Plateau, off the eastern United States. **b** Correlation of the sequence boundaries identified in two Deep Sea Drilling Project (DSDP) holes, and a comparison with the global cycle chart of Haq et al. (1987, 1988a). Diagram compiled by Schlager (1992a) from various sources. (After Austin et al. 1988; reproduced with permission of Ocean Drilling Program)

allostratigraphy that has considerably clarified the depositional evolution of the unit (Fig. 8.53), and assisted in the development of its petroleum potential. A second example is given by Van Wagoner et al. (1990) in their reevaluation of the Blackhawk

Formation of the Book Cliffs, Utah. A series of beach sandstone lenses is arranged in a shingled architecture, and was interpreted by Young (1955) as the product of gradual seaward regression, with local scouring and channeling accounting for the

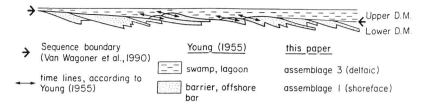

Fig. 2.3. Barrier-bar lagoonal model of the Blackhawk Formation, according to Young (1955) and the sequence re-interpretation by Van Wagoner et al. (1990)

relief on the top of each sandstone unit. Van Wagoner et al. (1990) demonstrated that the scour surfaces can in fact be mapped as a single regional surface and constitute a sequence boundary (Fig. 2.3).

Although unconformities provide the major basis upon which an allostratigraphy should be erected, in places complications may arise. The following are some of the possibilities:

1. Traced seaward a subaerial erosion surface, even one developed during an extreme lowstand of sea level, passes laterally into an area that underwent continuous sedimentation. Vail et al. (1977) therefore suggested that a sequence be defined by its bounding unconformities or their *correlative conformities*. Controversy surrounds the offshore correlation of sequence boundaries into shelf and continental slope successions (Sect. 15.2.2).
2. Depending on the relative rates of subsidence, base-level change and sediment supply, a fall in

sea level may not necessarily result in exposure and erosion, merely a seaward shift in facies belts. In such cases the sequence boundary is not marked by a major erosion surface and may be difficult to detect (the so-called type 2 unconformity: see Sect. 4.4). This problem is particularly likely to be apparent in the nonmarine portion of a sequence, in which channel erosion surfaces may be equally as prominent as regional erosion surfaces in the stratigraphic section.

3. Distinct and prominent erosion surfaces are commonly developed by wave erosion during transgression, and may easily be confused with a sequence-boundary unconformity. Wave-cut erosion surfaces of this type are termed *ravinement surfaces*. Their development is illustrated in Fig. 2.4. They represent the surface of equilibrium on the shelf formed in response to the local wave and current regime. "As sea level rises the shoreface profile translates upward and landward through a process of erosional shoreface retreat.

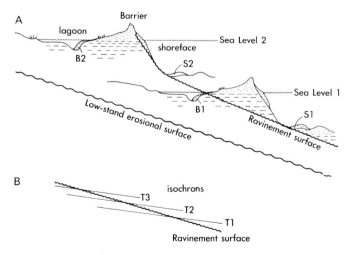

Fig. 2.4. Origin of ravinement surfaces. **A** Storm erosion may occur at time of sea level 1, concurrently with back-barrier (or estuarine or tidal channel) deposition. As sea-level rises to level 2 the storm erosion surface moves landward, stripping away some or all of the underlying coastal deposits formed during earlier times of lower sea

level. Coastal deposits formed behind the barrier at level 2 are contemporaneous with shelf deposits that are formed on the ravinement surface once water level becomes deeper than storm wave-base. **B** The ravinement surface is therefore time-transgressive. (Nummedal and Swift 1987)

The ravinement surface cut by this process will physically rise toward the basin margin. It will also become younger in the same direction" (Nummedal and Swift 1987). The ravinement surface is only one of several types of diastem that may be encountered in transgressive systems tracts. Others include channel scours and marine erosion surfaces. These can make the definition of sequence boundaries difficult. The ravinement surface, in particular, is commonly a major erosion surface that separates markedly different facies, but it forms within the transgressive systems tract and is therefore not a sequence boundary, unless the depth of erosion at the ravinement surface is such that no transgressive deposits are preserved. In the latter case the ravinement surface is coincident with the erosion surface at the top of the underlying regressive deposits.

Demarest and Kraft (1987) illustrated two possible interpretations of a coastal succession that could be made based on different analyses of the major erosion surface that may be present (Fig. 2.5). On the left, the major change in water depth that can be deduced to have occurred at the 8-m level in the section is interpreted as the sequence boundary,

formed at a time of sudden deepening. An erosion surface at 3.5 m is interpreted as a minor channel-scour surface. In the second interpretation, on the right of Fig. 2.5, the erosion surface at 8 m is interpreted as a ravinement surface, and the sequence boundary is placed much lower, at 3.5 m, entirely within a fluvial succession. The erosion surface there formed during the time of maximum lowstand, with the fluvial beds above it having formed much later, when sea level began to rise again. The second interpretation is the one that would be favored by current sequence-stratigraphy models. Note that in this case the sequence boundary is a much less obviously important erosion surface than the ravinement surface.

Recent work has demonstrated that several different types of surface may be generated during falling and rising sea level, such that the placement and correlation of sequence boundaries is not necessarily a simple matter. This is discussed in more detail in Section 15.2.2. Some workers have found that sequences are more easily defined and mapped using the flooding surfaces that develop during sea-level highstands. These concepts are described in Section 4.5, and the sedimentology of flooding surfaces and related systems tracts is described in Section 15.3.4.

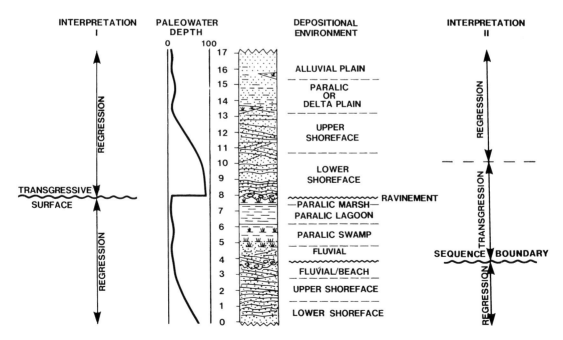

Fig. 2.5. A hypothetical vertical succession with two possible interpretations based on different vertical sequence models. (Demarest and Kraft 1987)

2.2.2 Facies Cycles

Vertical changes in lithofacies and biofacies have long been used to reconstruct temporal changes in depositional environments and, with the aid of Walther's Law, to interpret lateral shifts in these environments. Such an approach comprises the basis of the method of facies analysis, as discussed at length in many textbooks (e.g., Miall 1990; Chap. 4) and review articles (e.g., Wanless 1991). A single example of lithofacies and biofacies analysis will suffice. Other examples are illustrated elsewhere in this report.

Heckel (1986) presented a basic depositional model for the so-called "Kansas cyclothem," and for the environments of deposition occurring on a gently sloping tropical shelf (Fig. 2.6). Beds representing each of these environments are extraordinarily widespread, indicating shifts of environments of hundreds of kilometers. Black, phosphatic shales with conodonts indicate the deepest marine environments. Skeletal and algal wackestones and grainstones were deposited in shallow marine settings, while sandstones and coals indicate non-marine and marginal-marine settings. The succession of these facies indicates transgression and regression. As illustrated in Chapter 10, Heckel

(1986) and Boardman and Heckel (1989) were able to develop a sea-level curve for part of the Pennsylvanian by correlating these cycles and their contained facies changes across the United States Midcontinent from Texas to Iowa.

In the foreland basin of the United States Western Interior Weimer (1986) listed the following criteria for recognizing sea-level changes in the stratigraphic record:

1. Regression of shoreline with incised drainage, followed by overlying marine shale.
2. Valley-fill deposits (of incised drainage system) overlying marine shale:
   A. Root zones at or near base of valley-fill sequence.
   B. Paleosoil on scour surface.
3. Unconformities within the basin:
   A. Missing faunal zone (except where absent for paleoecological reasons).
   B. Missing facies in a normal regressive succession (e.g., shoreface or delta-front sandstone).
   C. Paleokarst with regolith or paleosoil.
   D. Concentration of one or more of the following on a scour surface: phosphate nodules, glauconite, recrystallized shell debris to form thin lenticular limestone layers.

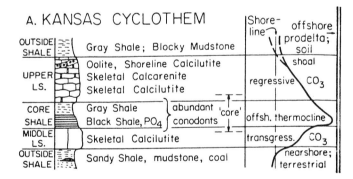

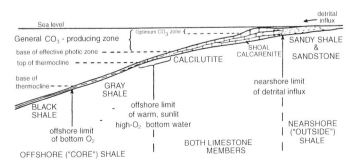

Fig. 2.6. A The typical Kansas cyclothem, showing the interpretation in terms of transgression and regression. **B** Model for deposition on a gently sloping tropical shelf, showing the position of rock types that become superimposed with transgression and regression. (Heckel 1986)

4. Thin, widespread coal layer overlying marine regressive delta front sandstone deposits indicating rising sea level.

A spectrum of short- and longer-term sea-level changes was documented, mainly in Europe and North America, by McKerrow (1979), based on paleobathymetric interpretations of brachiopod communities (Fig. 2.7). In the Ordovician and Silurian, marine shelf faunas consisted mainly of brachiopods, trilobites, corals, stromatoporoids and bryozoans. These comprise what is informally termed the shelly fauna. It contrasts with the graptolitic fauna of deeper water, continental-slope, and abyssal oceanic environments.

Several authors have attempted to subdivide Lower Paleozoic shelf faunas into depth-controlled communities. Ziegler (1965) and Ziegler et al. (1968) recognized five brachiopod-dominated assemblages in the Early Silurian, which he named after typical genera. They are *Lingula*, *Eocoelia*, *Pentamerus*, *Stricklandia*, and *Clorinda*, in order of increasing water depth. These communities map out in bands parallel to the shore in shelf sequences in Wales, the Appalachian Basin, New Brunswick, and Iowa (McKerrow 1979). The communities are not related to distance from shore as the shelf width varies from 5 to more than 100 km, and they are not related to sediment character, as each community occurs in various rock types. However, Cant (1979) has observed that storms can redistribute shallow-water fossil assemblages into deep water, and so some caution must be used in interpreting these data.

Ziegler's faunal differentiation has been established for the Upper Ordovician and the remainder of the Silurian in a few areas. Sea-level changes over the shelf should be accompanied by lateral shifts in these communities, which should be recognizable in vertical sections through the resulting sediments. McKerrow (1979) used this approach, plus supplementary facies data, to construct depth-change curves for the Middle Ordovician to Early Devonian in 13 locations in Europe, Africa, North America, and South America. The results are shown in Fig. 2.7. Some of the depth changes can be correlated between many of the regions examined and may reflect eustatic sea-level changes. Others are more local in scope and were probably caused by regional tectonic events.

McKerrow (1979) distinguished two types of eustatic depth change, slow and fast. Slow changes occurring over a few millions or tens of milions of years include the rise in sea-level during the Llandovery and the fall in the Ludlow and Pridoli. The slow rise and fall during the Silurian took about 40 m.y. to complete. Fast changes include the rapid rise in latest Llandeilo and earliest Caradoc time, a short-lived fall at the end of the Ashgill, and a rise and fall at the beginning of the Late Llandovery. These rapid changes were of 1–2 m.y. duration.

An example of a sequence analysis of a stratigraphic section based on facies studies is given in Fig. 2.8. Sequence boundaries are identified at transgressive surfaces, where deepening took place.

In the subsurface it may be necessary to rely on petrophysical logs to define sequences. The various components of the sequences may have distinctive log characteristics ("log motifs"), which aid in sequence definition, especially if the logs can be calibrated against one or more cores through the succession. For example, condensed sections commonly are revealed by gamma-ray spikes. Cant (1992) and Armentrout et al. (1993) described the methods of analysis. Armentrout et al. (1993) used grids of seismic lines and suites of petrophysical logs tied to the seismic cross sections to correlate Paleogene deposits in the North Sea, and demonstrated that errors in correlation of up to 30 m could occur when markers were traced around correlation loops. In this area wells are up to tens of kilometers apart, and such large errors are not expected in mature areas such as the Alberta Basin or the Gulf Coast.

2.2.3 Stratigraphic Architecture: The Seismic Method

The major breakthrough in the development of sequence-stratigraphic principles was the use of reflection-seismic data, as pioneered by Vail et al. (1977). Seismic cross sections may be analyzed in terms of their geometry and reflection terminations, using the terminology given in Fig. 1.4. Vail et al. (1977) then demonstrated how to use these data in a formal sequence analysis.

The first step is to define the sequence-bounding unconformities. On good seismic records this should be a relatively simple step, except within tectonically very stable areas, such as cratons, where the low angularity of the unconformities may inhibit recognition, and in basin centers, where there may be insufficient acoustic-impedance contrast between beds of similar facies. Prominent seismic markers should then be drawn in, to emphasize the internal architecture of each sequence. These steps

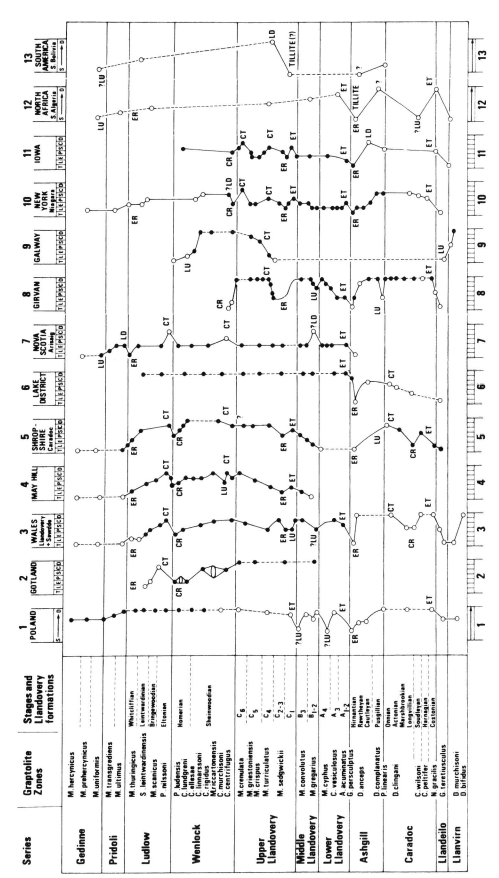

Fig. 2.7. Depth changes on three continents during the Ordovician and Silurian. Symbols at head of column: $S{\rightarrow}D$ shallow→deep; *TLEPSCD* transgression, *Lingula, Eocoelia, Pentamerus, Stricklandia, Clorinda,* regression (see text for explanation). Symbols on graphs: *E* eustatic; *C* continental; *L* local; *T* transgression; *R* regression; *U* uplift; *D* deepening. *Open circles* indicate uncertainty of depth and/or age. *Dashed lines* indicate terrestrial, unfossiliferous, or very deep environments that do not yield good depth control. (McKerrow 1979, reproduced by permission of The Geological Society, London)

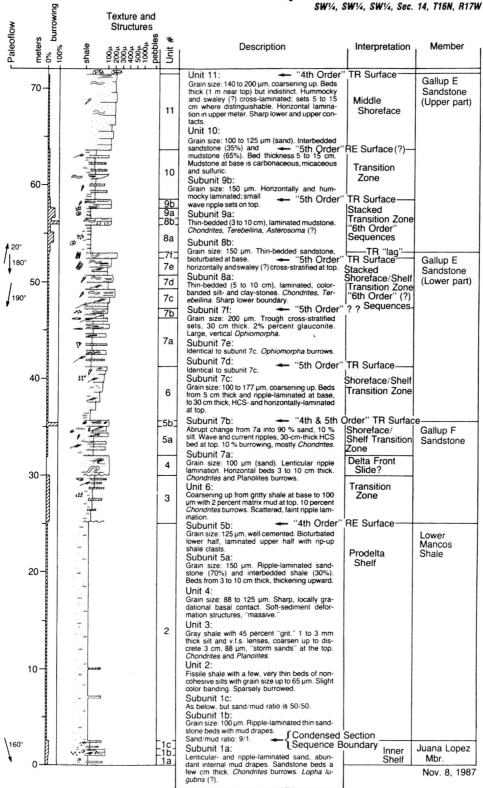

Description	Interpretation	Member
Unit 11: ← "4th Order"	TR Surface	Gallup E Sandstone (Upper part)
Grain size: 140 to 200 μm, coarsening up. Beds thick (1 m near top) but indistinct. Hummocky and swaley (?) cross-laminated; sets 5 to 15 cm where distinguishable. Horizontal lamination in upper meter. Sharp lower and upper contacts.	Middle Shoreface	
Unit 10: Grain size: 100 to 125 μm (sand). Interbedded sandstone (35%) and ← "5th Order" mudstone (65%). Bed thickness 5 to 15 cm. Mudstone at base is carbonaceous, micaceous and sulfuric.	RE Surface (?) — Transition Zone	
Subunit 9b: Grain size: 150 μm. Horizontally and hummocky laminated; small ← "5th Order" wave ripple sets on top.	TR Surface	
Subunit 9a: Thin-bedded (3 to 10 cm), laminated mudstone. *Chondrites, Terebellina, Asterosoma* (?)	Stacked Transition Zone "6th Order" Sequences	Gallup E Sandstone (Lower part)
Subunit 8b: Grain size: 150 μm. Thin-bedded sandstone, bioturbated at base, ← "5th Order" horizontally and swaley (?) cross-stratified at top.	TR "lag" — TR Surface Stacked Shoreface/Shelf Transition Zone	
Subunit 8a: Thin-bedded (5 to 10 cm), laminated, color-banded silt- and clay-stones. *Chondrites, Terebellina.* Sharp lower boundary.	"6th Order" (?)	
Subunit 7f: ← "5th Order" Grain size: 200 μm. Trough cross-stratified sets, 30 cm thick. 2% percent glauconite. Large, vertical *Ophiomorpha.*	? ? Sequences	
Subunit 7e: Identical to subunit 7c. *Ophiomorpha* burrows.		
Subunit 7d: ← "5th Order" Identical to subunit 7c.	TR Surface	
Subunit 7c: Grain size: 100 to 177 μm, coarsening up. Beds from 5 cm thick and ripple-laminated at base, to 30 cm thick, HCS- and horizontally-laminated at top.	Shoreface/Shelf Transition Zone	
Subunit 7b: ← "4th & 5th Order" Abrupt change from 7a into 90 % sand, 10 % silt. Wave and current ripples, 30-cm-thick HCS bed at top. 10 % burrowing, mostly *Chondrites.*	TR Surface Shoreface/Shelf Transition Zone	Gallup F Sandstone
Subunit 7a: Grain size: 100 μm (sand). Lenticular ripple lamination. Horizontal beds 3 to 10 cm thick. *Chondrites* and *Planolites* burrows.	Delta Front Slide?	
Unit 6: Coarsening up from gritty shale at base to 100 μm with 2 percent matrix mud at top. 10 percent *Chondrites* burrows. Scattered, faint ripple lamination.	Transition Zone	
Subunit 5b: ← "4th Order" Grain size: 125 μm, well cemented. Bioturbated lower half, laminated upper half with rip-up shale clasts.	RE Surface	Lower Mancos Shale
Subunit 5a: Grain size: 150 μm. Ripple-laminated sandstone (70%) and interbedded shale (30%). Beds from 3 to 10 cm thick, thickening upward.	Prodelta Shelf	
Unit 4: Grain size: 88 to 125 μm. Sharp, locally gradational basal contact. Soft-sediment deformation structures, "massive."		
Unit 3: Gray shale with 45 percent "grit." 1 to 3 mm thick silt and v.f.s. lenses, coarsen up to discrete 3 cm, 88 μm, "storm sands" at the top. *Chondrites* and *Planolites.*		
Unit 2: Fissile shale with a few, very thin beds of non-cohesive silts with grain size up to 65 μm. Slight color banding. Sparsely burrowed.		
Subunit 1c: As below, but sand/mud ratio is 50/50.		
Subunit 1b: Grain size: 100 μm. Ripple-laminated thin sandstone beds with mud drapes. Sand/mud ratio: 9/1. ← { Condensed Section { Sequence Boundary		Juana Lopez Mbr.
Subunit 1a: Lenticular- and ripple-laminated sand, abundant internal mud drapes. Sandstone beds a few cm thick. *Chondrites* burrows. *Lopha lugubris* (?).	Inner Shelf	

Nov. 8, 1987

Fig. 2.8. Example of facies analysis of a stratigraphic section. Facies successions and the presence of abrupt surfaces of transgression ("TR surface") are used to subdivide the succession into high-frequency sequences spanning 10^4–10^5 years. Gallup Sandstone, New Mexico. (Nummedal et al. 1989)

are illustrated in Fig. 2.9a. The details of onlap, off-lap, truncation, etc., are related to the depositional environment and subsidence history and are considered in detail in Chapter 4.

The sequences must then be dated using the best available biostratigraphic and (if available) radiometric data. Vail et al. (1977) converted biostrati-graphic ages to values in years using a standard time scale. Several age values in millions of years are given for the sequences shown in Fig. 2.9a. These then permit the construction of a chronostratigraphic chart, as shown in Fig. 2.9b. The seismic reflectors within each sequence can be assumed to be time markers and can be straightened out to

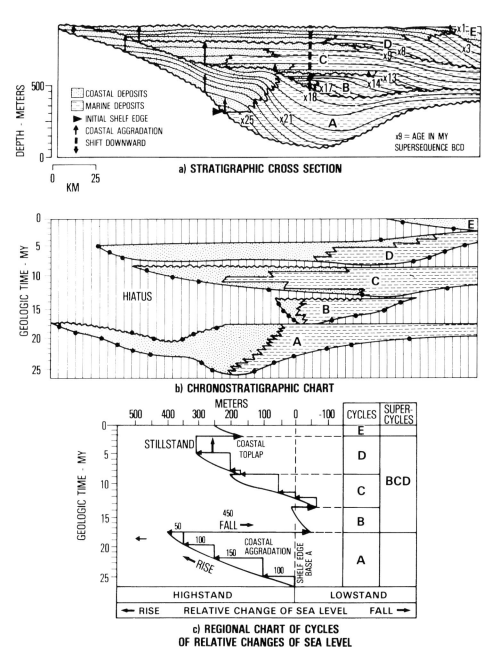

Fig. 2.9. Procedure for analysis of a seismic section. **a** Seismic section redrawn to show major reflectors and sequence boundaries. **b** Chronostratigraphic chart of the same section. **c** Graph of relative changes in coastal onlap (not relative changes in sea level, as in this original version of the diagram). (Vail et al. 1977, reproduced by permission)

permit plotting of the section on a time ordinate. This shows that the boundaries of some of the sequences are strongly diachronous, indicating slow transgression and filling of the basin up to a contemporaneous base level, or local erosion. It also shows that the bounding unconformities are not time planes, but vary in age across the section as a result of differential tilting or slight tectonic disturbance.

Vail and his coworkers have used seismic sections and chronostratigraphic charts, such as those shown in Fig. 2.9a and b, to construct charts showing relative changes in sea level, as given in Fig. 2.9c. Such charts have been drawn by geologists for many years based on conventional stratigraphic and sedimentological analysis (e.g., Fig. 2.6), but Vail et al. (1977) devised a new technique for constructing these charts that purported to generate semiquantitative information about sea-level changes. The method is described briefly below. However, considerable controversy has arisen over the method of construction of the charts and their interpretation, as discussed in Parts III and IV. One of the most important problems is the lack of accuracy and precision in absolute age determination (Chaps. 13, 14).

Figure 2.9c is a graph of relative sea-level change constructed for the section shown in Fig. 2.9a. Vertical aggradation and onlap of marginal-marine sediments is taken to indicate a relative rise in sea level. This could be either a transgression over a stable continent (an actual sea-level rise) or basin subsidence without a change in sea level. From local stratigraphic data alone it is impossible to tell which is the correct interpretation. A relative fall in sea level is indicated by offlapping sequences or by a downward shift in coastal onlap.

Increments of relative sea-level rise and fall can, Vail et al. (1977) suggested, be measured on the seismic section as shown in Fig. 2.9. The starting point of the analysis is the beginning of a major cycle of rising sea level at the base of sequence A. Measurements are made from the assumed position of the shelf edge at the facies contact between coastal and marine deposits. In one million years the initial deposits of sequence A aggrade vertically by 100 m (see vertical arrow at initial shelf edge in Fig. 2.9a). This can be plotted as a corresponding relative sea-level rise, as shown in Fig. 2.9c. Successive increments plotted the same way indicate a total relative sea-level rise of 400 m during the deposition of sequence A. There is then a dramatic fall in sea level before the beginning of sequence B, the total amount is indicated by the heavy dashed arrow in Fig. 2.9a and is 450 m, indicating that sea level dropped to below its starting position at the beginning of sequence A. This is clear from Fig. 2.9c. Remaining sea-level changes are plotted in the same way. The resulting curve is a graphical portrayal of changing relative sea level. It includes a period of stillstand at the end of sequence D time, when because of the unchanging position of sea level, sediments aggraded laterally rather than vertically.

Figure 2.9 is a hypothetical example used by Vail et al. (1977) to illustrate the principles of seismicstratigraphic analysis and construction of a curve for sea-level change. However, it also illustrates a number of problems with the methodology that are not discussed by these authors. They state that measurements of coastal aggradation should be made as closely as possible to the underlying unconformity to minimize the effect of differential basin subsidence. However, very little differential subsidence is indicated in their section. Seismically defined time lines are parallel for much of sequences C and D and nearly so for much of sequences A and B. In practice this is rarely the case. Stratigraphic units almost invariably thicken toward a basin center because of differential subsidence, and so the measurements of relative sea-level change would show quantitative differences depending on where the analysis was carried out within the basin. Second, and related to this problem, units may not show the dramatic onlap relationships illustrated in Fig. 2.9a, but may all taper to a feather edge near a continental hinge line, so that the method described above could not be used. Third, it should be pointed out that coastal deposits may include significant thicknesses of nonmarine sediments deposited above sea level. These must clearly be separated prior to the analysis, but it may be very difficult to do this from seismic data alone. Seismic facies analysis may provide some clues. Miall (1986) summarized some other difficulties with the method. For example, it may not be possible to prove that the so-called "coastal onlap" is coastal at all. Many onlap relationships are developed by deep-marine (e.g., submarine-fan) deposits building across tilted fault blocks (Fig. 11.10) or up the continental slope in a very similar pattern to that shown in Fig. 2.9a. Care must be taken to distinguish these relationships because, of course, deep-marine onlap provides no useful information about sea level. Lastly, as discussed in Chapter 11, similar stratigraphic architectures can be produced by flexural subsidence without any change in sea

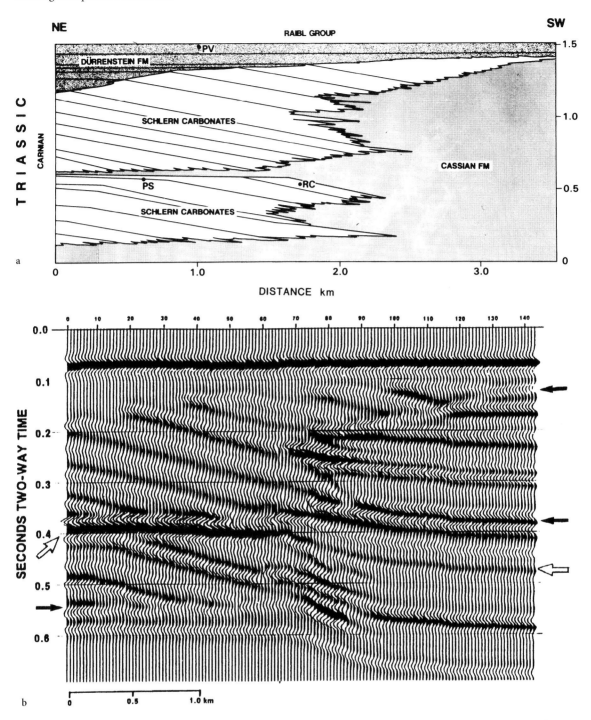

Fig. 2.10. a Example of a platform-to-basin carbonate-clastic facies transition, based on outcrops of Triassic deposits in the southern Alps. **b** Seismic model of this section, showing several apparent unconformable relationships (arrows). The open arrow indicates a true unconformity; the others indicate pseudo-unconformities which are in fact lines of facies change localized along the downdip fringe of carbonate clinforms (Biddle et al. 1992, reproduced by permission)

level. For this reason, later versions of the Vail curves are not labeled as sea-level curves but as curves showing "relative changes in coastal onlap."

As has often been pointed out (e.g., Miall 1990, p. 278; Christie-Blick et al. 1990), there are distinct limits to the ability of the seismic method to resolve thin beds, and resolution decreases with depth, as seismic velocity increases. A general rule is that two reflecting horizons must be a minimum of about one-fourth of a wavelength apart if they are to be resolvable in a typical seismic section. Thus, at shallow depths in loosely consolidated sandstones and mudstones a typical seismic velocity would be on the order of 1800 m/s, and the typical frequency would be around 60 Hz. One quarter wavelength is therefore 7.5 m. Resolution is much poorer in older and deeper formations. Consider, for example, a Paleozoic carbonate section, with a seismic velocity of 4500 m/s and frequency of 15 Hz. The thinnest unit resolvable in this situation would be 75 m. This lower limit is much larger (thicker) than many of the important facies details that can be seen in outcrops and cores. Fulthorpe (1991) demonstrated that under ideal conditions high-frequency cycles less than one million years in duration can be recognized on seismic sections. They require high rates of sedimentation to be distinguishable, and are best seen in prograding clinoform architectures (Fulthorpe 1991). Cartwright et al. (1993) demonstrated that many toplap, downlap, and onlap relationships are not actual truncations, but that the units extend as thin beds that may not be resolvable even on high-quality modern seismic sections. This may cause problems with correlation and dating.

Schlager (1992a) pointed out that because of problems of resolution, and contrary to the important initial assumptions of Vail et al. (1977), seismic reflections *can* develop at diachronous facies contacts, producing what he termed *pseudo-unconformities*. Reflections occur at contrasts in acoustic impedance. Where such contrasts are the result of a vertical facies change, the position of which also climbs rapidly through the section, a diachronous reflection may result (Fig. 2.10). Such pseudo-unconformities are not surfaces of reflection termination, and must not form part of sequence definition. Schlager (1992a) indicated that carbonate platforms are particularly prone to this type of seismic response because of the rapid lateral facies changes that commonly are present.

Tipper (1993, p. 52) demonstrated that this is part of a much more general problem with the assumptions that underlie the analysis of seismic-re-

flection data. He took issue with the statement of Vail et al. (1977, p. 99) that "No physical surface that could generate a reflection parallel with the top of a time-transgressive ... lithostratigraphic unit ... exists in nature." According to Tipper (1993) "That statement may perhaps be true, but it completely misses the point. Reflections parallel to the tops of diachronous units do not have to have been generated at sharp, continuous physical surfaces, because they are generated naturally at the gradational and discontinuous lithofacies boundaries that typically result from unsteady sedimentation." Tipper developed seismic models for reflections from time-transgressive facies contacts similar to those illustrated in Fig. 2.10.

Figure 2.9c and, indeed, all the curves of sea-level change initially published by Vail and his co-workers are asymmetric. Rises in sea level are shown to be relatively slow, whereas falls are shown to be very rapid. In fact, in all the curves, falls are indicated by horizontal lines, suggesting that they took place instantaneously. This is the so-called "sawtooth" pattern, that was the subject of much controversy when it was first published (Kerr 1980). Obviously, a change in sea level takes longer than this, but nevertheless, Vail et al. claimed in their original work that in all the seismic records examined the same pattern is apparent. This has long seemed unlikely to many geologists (Kerr 1980), and independent work by Pitman (1978) indicated that falls in sea level may be just as slow as the rises. A slow fall in sea level should be recognized in the stratigraphic record by offlap, but Vail et al. (1977, p. 72) stated that they had not observed this on seismic data. A partial explanation may be that a fall in sea level results in subaerial exposure and erosion, so that the evidence of offlap may be lost.

A later version of the methods diagram shown in Fig. 2.9 is that given in Fig. 2.11. The sawtooth curve is now labeled a coastal onlap curve, not a "relative change in sea-level" curve. Vail and Todd (1981) observed:

Our studies indicate that coastal onlap for a particular sequence commonly persists until just prior to the accumulation of the overlying onlapping sequence or submarine fan. Thus, the downward shift in coastal onlap appears instantaneous in terms of geological time, as shown in the regional chart [Figs. 2.9c, 2.11c]. In previous papers ... we directly equated relative changes in coastal onlap with relative changes in sea level. Workers such as Pitman (1978) and M.T. Jervey (personal communication) have demonstrated that changes observed in coastal onlap cannot be related directly to relative changes in sea level (particularly the upper part of the cycle), as we have done

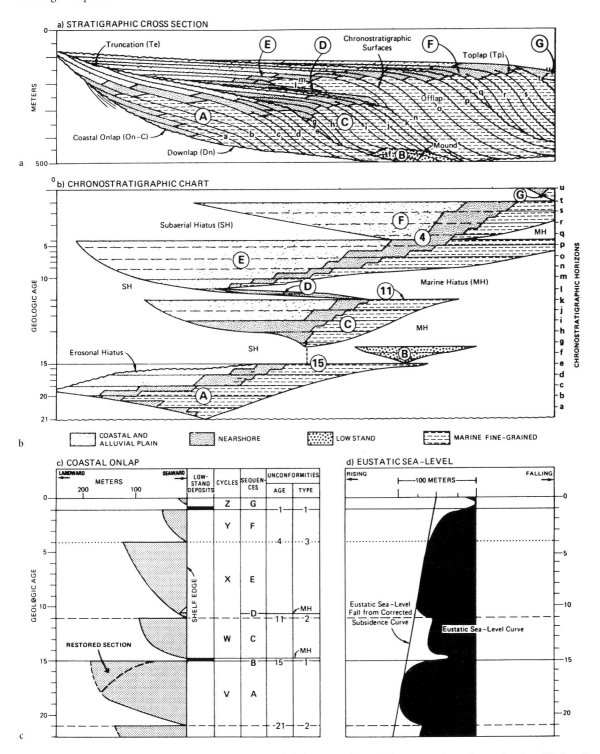

Fig. 2.11a–c. A later version of Fig. 2.9, showing a modified approach to the interpretation of coastal onlap. (Vail and Todd 1981)

in the past. We now refer to our global charts as relative changes in coastal onlap. We cannot relate coastal onlap directly to relative changes in sea level in that our charts of relative changes in coastal onlap extend to the land-

ward edge of the alluvial plain and thus include both the coastal plain and the alluvial plain (with its slightly greater seaward dips). The landward boundary that should be used for plotting relative changes in sea level is the upper limit of the coastal plain. Where an alluvial plain is present, the relative sea-level boundary approximates to the facies change between the coastal plain and the alluvial plain. (M.T. Jervey, personal communication)

As discussed in Chapter 15, this model carries with it assumptions regarding the development of onlap patterns in coastal, nonmarine sequences. Note that in Fig. 2.11a and b facies changes indicate that a seaward progradation begins shortly after onlap commences at the beginning of each sequence. Miall (1991a) argued that this is unlikely to occur.

Christie-Blick (1991) stated that the existence of offlap demonstrates that sequence boundaries develop over a finite interval of time, and not instantaneously, and he suggested that the horizontal segment of the sawtooth curve is an artefact of the methodology, representing the conventional age of the unconformity. He proposed that a modified form of coastal onlap curve be employed, as shown in Fig. 2.12. More recent work has demonstrated the existence in many sections of a distinctive systems tract formed during falling relative sea level. This was not defined in the earlier Exxon work, but can readily be recognized based on regional mapping (Sects. 15.2.2, 15.3.3).

2.3 Methods for Assessing Regional and Global Changes in Sea Level, Other Than Seismic Stratigraphy

Sequence stratigraphy has assumed its present importance primarily because of the advent of seismic methods in the 1970s. In fact, the term *seismic stratigraphy* became widely used following the publication of AAPG Memoir 26, and it is, perhaps,

necessary to point out the misleading nature of this term, by reminding the reader that seismic data are not the only kind of data used in this type of analysis. The so-called "Vail curves," the global cycle charts, have evolved from seismic methods, and are dealt with in Chapters 5, 13, and 14. The remainder of this chapter describes other methods for assessing regional and global sea-level change based on analysis of the stratigraphic record. Other methods of calculating eustatic sea-level change are discussed elsewhere, including calculations of changes in ocean-basin volumes reflecting changes in sea-floor spreading rate (Chap. 9), and the use of oxygen isotope data as a measure of glacioeustatic sea-level changes (Chap. 10).

2.3.1 Areas and Volumes of Stratigraphic Units

In principle the rising and falling of sea level should be recorded by transgressive and regressive deposits, leaving a record of shifting strandlines and of onlap and offlap. A simple measurement technique for tracking these events is to document the changes in the area of the basin or craton underlain by units of successive age. Sloss (1972) employed this procedure to compare the stratigraphic record of the Western Canada Sedimentary Basin and the Russian Platform. As he noted, "the area covered by a given stratigraphic unit is determined by the maximum area of original deposition less the area of post-depositional erosion of sufficient magnitude to remove the unit" (Sloss 1972, p. 25). The record of transgression tends to be better preserved than that of regression, because earlier transgressive deposits are covered by later deposits and thereby preserved, whereas during regression there is the tendency for the deposits to be exposed and eroded. Also, "the

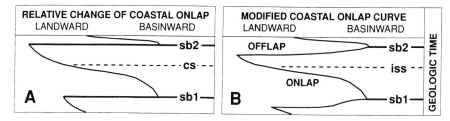

Fig. 2.12. A Diagrammatic coastal onlap curve, generated using the methods of Vail et al. (1977), as explained in the text. B A modified coastal-onlap curve incorporating the existence of offlap beneath each of the sequence boundaries. *sb1* and *sb2* Sequence boundaries; *cs* con-

densed section; *iss* interval of sediment starvation. (Christie-Blick 1991, reproduced by kind permission of Elsevier Science-NL, Sara Burgerhartstraat 25, 1055 KV Amsterdam, The Netherlands)

discovery of an isolated fault block, or diatreme xenolith, or glacial erratic, can shift the purported extent of late-cycle seas by hundreds of kilometers and alter the supposed maximum elevation of sea level by tens to hundreds of meters" (Sloss 1979, p. 462). Given these constraints, nevertheless, Sloss (1972, 1979) found considerable similarities between the stratigraphic sequences of the Western Canada Sedimentary Basin and the Russian Platform (Sect. 3.3, Fig. 3.4).

In order to reduce the error inherent in the analysis of thin feather-edge cratonic remnants Sloss (1979) turned to intracratonic and pericratonic basins, as representing loci of more continuous subsidence. Figure 2.13 shows the volume of sediment perserved per unit time in six Mesozoic-Cenozoic basins in various tectonic settings. The four basins that yielded data on the Triassic-Early Jurassic period indicated a marked acceleration of subsidence in the Late Triassic or Early Jurassic, and there are also peaks in the Middle Jurassic and Middle to Late Cretaceous. These data are combined into a single plot in Fig. 2.14, in which the volume/rate data for each basin have been normalized as a percentage of the median rate for each basin and plotted at successive 10 m.y. increments. The smoothed trend shows a series of cycles about 50 m.y. long, which are presumably the result of interregional or global processes.

A recent application of these techniques was presented by Ronov (1994), based on Russian paleogeographic atlases. This work generated a global "first-order" sea-level curve similar to that of Vail et al. (1977; see Fig. 3.3), but the method is not suitable for application to more detailed studies because of the coarse scale of stratigraphic resolution in the published atlases.

2.3.2 Hypsometric Curves

A hypsometric (or hypsographic) curve is defined as "a cumulative frequency profile representing the statistical distribution of the absolute or relative areas of the Earth's solid surface (land and sea floor) at various elevations above, or depths below, a given datum, usually sea level" (Bates and Jackson 1987). Using a planimeter and an equal-area projection the amount the sea advanced across a continent during any particular time interval can be derived from paleogeographic maps and compared with the curve derived from the present-day continent (Burton et al. 1987). Bond (1976) developed a

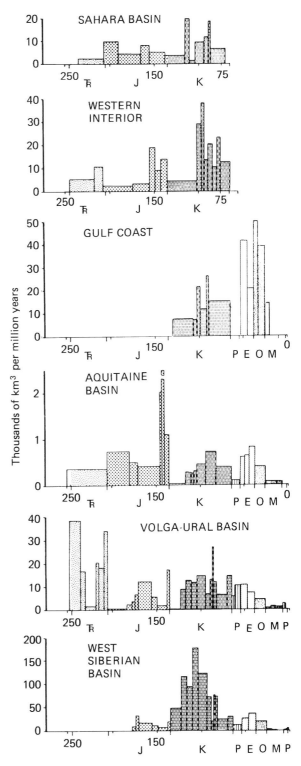

Fig. 2.13. Volume of sediment preserved per unit time in six Mesozoic-Cenozoic basins. *TR* Triassic; *J* Jurassic; *K* Cretaceous; *P* Paleocene; *E* Eocene; *O* Oligocene; *M* Miocene. (Sloss 1979, reproduced by permission)

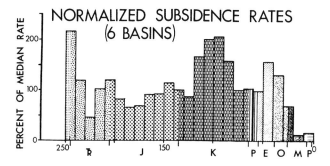

Fig. 2.14. Average Mesozoic-Cenozoic subsidence rates of the six basins shown in Fig. 2.13. (Sloss 1979, reproduced by permission)

method of using hypsometric curves for distinguishing between sea-level changes and vertical motions of large continental surfaces. The principal is illustrated in Fig. 2.15. In practice application of the technique is rendered difficult by the large number of generalizations that must be incorporated. The same difficulties arise as those encountered by Sloss (1972, 1979) in using map distributions, and there is the difficulty in allowing for changes in continental hypsometry over geological time because of plate-tectonic effects (crustal stretching on extensional margins, thickening on convergent and collisional margins).

Bond (1976) was able to demonstrate differences that developed in continental elevations between the major continental blocks during the Middle Cretaceous to Miocene, and also estimated actual (eustatic) sea-level changes during that time. The results are shown in Figs. 2.16 and 2.17. Figure 2.16 indicates the percentages of the continental areas of several continents flooded during several successive time periods. Assuming no change in continental

hypsometry since the time indicated, the rise in sea level required to bring about this percentage of flooding may be read off the graph. These estimates are then replotted in Fig. 2.17A. Clusters of three or four continental points occur at each time period, the close correlation between them suggesting real sea-level changes. However, Africa appears as an anomaly in the Campanian-Maastrichtian, Eocene and Miocene columns, and there is a scatter of points in the Albian column. Bond (1978) suggested that Africa has been uplifted since the Miocene. A 90-m lowering of the Africa point, to the middle of the Miocene cluster (point Af') does not restore the Africa point to the middle of the Eocene cluster, suggesting an Eocene-Miocene uplift in the order of 210 m (point Af"). Other corrections are also indicated in Fig. 2.17B. Bond (1978) concluded that the final clusters indicate overall generalized changes in global sea level, a rise from the Albian to a maximum of about 150 m above present in the Campanian-Maastrichtian, followed by a gradual fall.

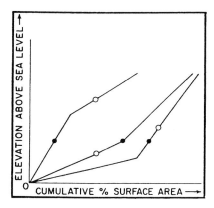

Fig. 2.15. Hypothetical hypsometric curves and various points showing percentages of continental area flooded during a transgression. *Solid points* These all fall at the same elevation, indicating a transgression due to sea-level rise without subsequent change in continental hyposmetries. *Open circles* Transgression followed by substantial change in continental hypsometries. (Bond 1978)

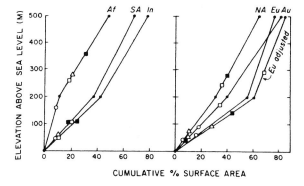

Fig. 2.16. Percentage of flooding plotted on continental hypsometric curves. *Af* Africa; *Au* Australia; *Eu* Europe; *In* India; *NA* North America; *SA* South America. "Europe adjusted" is the curve for Europe with the area south of the Alpine collision zone excluded. Albian: *open squares*; Late Campanian to Early Maastrichtian: *solid squares*; Eocene: *triangles*; Miocene: *open circles*. (Bond 1978)

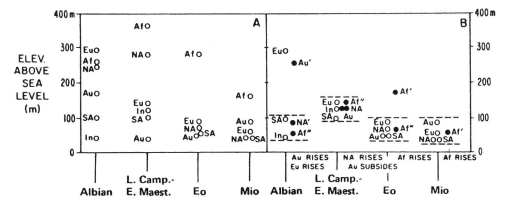

Fig. 2.17. A Sea-level elevations read off the curves in **B** After corrections (*black circles*, corrections are explained in the text), clusters are apparent which suggest eustatic sea-level changes. (Bond 1978)

More recent work on hypsometric curves was summarized by Burton et al. (1987). They concluded that although the technique is useful in providing general estimates of long-term changes in sea level there are too many uncertainties to permit its use for determining detailed, short-term changes. A long-term sea-level curve for the Paleozoic was published by Algeo and Seslavisnsky (1995) based on analysis of recent paleogeographic syntheses. They demonstrated significant differences in the flooding histories of the major continents, and used the results to refine and calibrate the long-term trends indicated by the work of Hallam (1984) and Haq et al. (1987, 1988a). Some causes for the variation in continental elevations over time are discussed in Chapter 9.

2.3.3 Backstripping

Backstripping is a technique for performing a detailed analysis of the subsidence and sedimentation history of a basin. The initial purpose of this type of analysis was to reveal the tectonic driving mechanisms of basin subsidence. The analysis consists of progressively removing the sedimentary load from a basin, correcting for compaction (plus lithification, if necessary), paleobathymetry, and changes in sea level, and calculating the depth to basement. The load may be fitted to an Airy-type or flexural subsidence model, depending on the tectonic setting of the basin, and the residual subsidence that is revealed can then be related to thermal behavior and changes with time in crustal properties. The technique was developed by Sleep (1971) and was first explored in detail by Watts and Ryan (1976). Steckler and Watts (1978) applied the McKenzie

(1978) stretching model to offshore stratigraphic data from the continental margin off New York, and Sclater and Christie (1980) applied the techniques to an analysis of the North Sea Basin, in papers that have become standard works on the subject.

A subsidence curve can be predicted from a knowledge of the tectonic setting of a basin. Departures from the curve can then be interpreted in terms of one or more of the "corrections" that are applied during the analysis, in particular, tectonic events and changes in water depth. Water depth in part represents a balance between subsidence and sedimentation rate, but is also affected by sea-level change. If accurate estimates of water depth during sedimentation can be determined, changes in sea level may then be isolated. For deep-water sediments this is not possible because of the imprecision of paleoecological and other methods of estimating this parameter, However, for shallow-water sediments, such as shelf clastics and platform carbonates, depth corrections are small enough that major changes in sea level may become apparent.

The procedure for backstripping a sedimentary basin starts with the division of the stratigraphic column into increments for which the thickness and age range can be accurately determined (Fig. 2.18). These time slices are then added to the basement one by one, calculating the original decompacted thickness and bulk density and placing its top at a depth below sea level corresponding to the average depth of water in which the unit was deposited. The isostatic subsidence caused by the weight of this sediment can then be calculated, and the depth to the surface on which the sediment was deposited is calculated with only the weight of the water as the basement load (Fig. 2.19). The second unit is then added and adjusted in the same way. The thickness

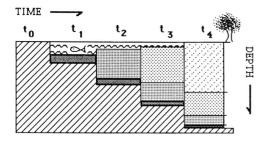

Fig. 2.18. Time slices in the subsidence history of a basin. Note that water depth, sedimentation rate, and compaction all vary with time. Each increment of sediment is compacted beneath the weight of successive increments, and this effect must be removed for each time slice in the backstripping procedure. (Mayer 1987)

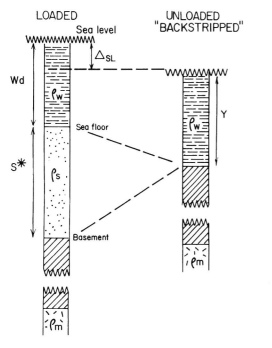

Fig. 2.19. The basic relationships between a loaded sedimentary section and an unloaded, or backstripped, section. Wd Water depth during deposition of given sedimentary unit; S^* total sediment thickness; ρ_m, ρ_s, ρ_w density of mantle, sediment, and water; Y depth of water with no sediment load; Δ_{SL} incremental eustatic change in sea level. (Steckler and Watts 1978, reproduced by kind permission of Elsevier Science-NL, Sara Burgerhartstraat 25, 1055 KV Amsterdam, The Netherlands)

and bulk density of the first unit are adjusted in accordance with the depth of burial beneath the second unit, and so on up the column.

An example of the application of this method to the study of sea-level change was given by Bond and Kominz (1984). These authors were concerned with evaluating the tectonic evolution of the Early Paleozoic Cordilleran miogeocline of western Canada

– the former continental margin. An outcome of their analysis was a model of the geometry of the continental margin based on flexural subsidence, following the methods of Watts (1981; Fig. 2.20). A restored cross section of the margin shows that the actual basin is considerably deeper than predicted, and that the sediments extend much further onto the craton than can be explained by a flexural subsidence model (Fig. 2.20). Bond and Kominz (1984) suggested that thermal effects may account for the greater-than-predicted thickness, and that a regional transgression caused by eustatic sea-level rise during the Middle Cambrian to Early Ordovician must have been responsible for depositing the thin sediment blanket that extends for several thousand kilometers onto the craton.

A major problem with backstripping methodology is deriving an appropriate model for the tectonic driving mechanism. Early work used the pure-shear model of McKenzie (1978), but subsequent studies have revealed numerous complications, including different stretching factors for different crustal levels, the effects of igneous intrusion, with consequent heating and loading at depth, the need for a more sophisticated accounting of heat flow, and the occurrence of more than one style of extensional subsidence, including the simple-shear model of Wernicke (1985), in which crustal plates separate across major detachment surfaces (see reviews in Burton et al. 1987; Miall 1990; Chap. 7). Unless tectonic subsidence can be accurately modeled it is not possible to derive useful information on sea-level change from backstripping curves. The results of Bond and Kominz (1984) described briefly above are obvious enough, but the timing and magnitude of the indicated sea-level rise are only very approximately known. This is an example of a general problem.

Guidish et al. (1984) reported an attempt to correlate backstripping curves from a suite of wells to extract information on sea-level change. Their results are of limited usefulness because of the problems noted above. They used a procedure of weighted averaging to combine the subsidence curves of 158 wells from around the world into a "master subsidence curve," in the hope that "this 'backstripped' basement-motion curve may contain a component or signal that is synchronous on a global scale." Their master subsidence curve is illustrated in Fig. 2.21. Its derivative, the rate of subsidence, is shown in Fig. 2.22, plotted at the same scale as the Vail et al. (1977) curve of global change in sea level. Guidish et al. (1984) claimed

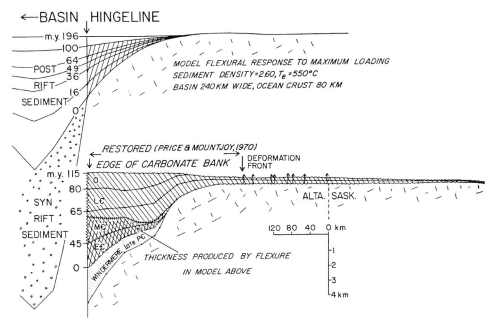

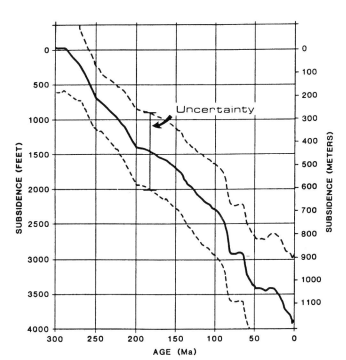

Fig. 2.20. The Cambrian-Ordovician continental margin of western Canada. Comparison of a restored cross section (*below*) with a hypothetical model based on flexural loading of stretched crust (*above*). *Inverted V's* indicate well locations. (Bond and Kominz 1984)

that there are some similarities between the two curves, including a rise in subsidence rates corresponding to a rise in sea level during the Jurassic to Middle Cretaceous and a fall in these two parameters during the Early Tertiary. Subsidence, as derived from stratigraphic thickness information, is a reflection of changes in sedimentary accommodation space, which is affected by changes in sea level as well as by actual changes in basement depth. In detail, however, the two curves are quite dissimilar, and the results of the analysis seem quite ambiguous to this writer.

Fig. 2.21. The global average "master subsidence curve" of Guidish et al. (1984, reproduced by permission)

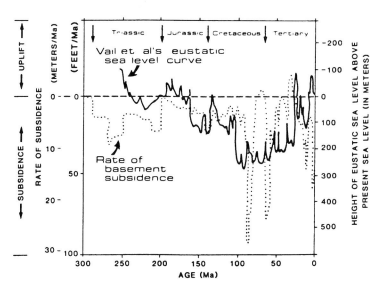

Fig. 2.22. The total-subsidence curve of Fig. 2.21 converted to a rate-of-subsidence curve (*dotted line*), plotted at the same time scale as the sea-level curve of Vail et al. (1977). (Guidish et al. 1984, reproduced by permission)

Bond and Kominz (1991a) and Kominz and Bond (1991) reported a similar, detailed comparison of Early Middle Paleozoic subsidence curves for various continental-margin and intracratonic basins in North America. They suggested that sea-level changes could be isolated by studying the subsidence history of Iowa, a central, highly stable area of the craton which is unlikely to have been affected by any tectonism during this period. This analysis identified three episodes of sea-level rise (Fig. 2.23). Their so-called "Iowa baseline curve" was then subtracted from other curves derived from basinal settings. The plots that resulted could still not be fitted to exponential McKenzie-type subsidence curves, as would have been expected from the tectonic setting of the basins. Additional subsidence is indicated, which the authors related to long-wavelength flexural subsidence induced by intraplate stress, a mechanism discussed in Chapter 11. This result is of considerable importance, because intraplate stress has been proposed as a mechanism that may be capable of generating stratigraphic architectures similar to those arising from eustatic sea-level changes over large continental areas (Cloetingh 1988), and this adds to the questions about the origins and significance of the Exxon global cycle charts (Chap. 11).

As discussed in the next section, one of the best places to study sea-level changes is at paleoshorelines where stratigraphic units thin against stable cratonic areas. This approach guided the choice of Iowa as a "baseline" in the work of Bond and Kominz, discussed above. Sahagian and Holland (1991) followed a similar approach in their use of stratigraphic data from the Russian platform to develop a eustatic curve by backstripping procedures. By avoiding areas of active subsidence the need for complex corrections of doubtful accuracy is eliminated. However, even stable continental interiors are subject to vertical motions. The work of Gurnis (1988, 1990, 1992) and Russell and Gurnis (1994) has demonstrated that the entire surface of the earth, including continents and the floor of the oceans, is subject to broad, gentle epeirogenic movements driven by the thermal effects of deep-seated mantle processes. Gurnis refers to this characteristic of the earth's surface as *dynamic topography*, a topic discussed at greater length in Chapter 9. The important point to note here is that this work demonstrates that no single location can be used as a reference location for assessing absolute sea-level variations.

Attempts have been made to extract additional local and regional detail from one-dimensional vertical profiles using more elaborate backstripping methodologies. The techniques were described by Bond et al. (1989) and Bond and Kominz (1991b), and were also illustrated by Osleger and Read (1993), and consist of a series of reductions of the primary data. The first step, termed R1 analysis, comprises construction of the decompacted, delithified subsidence curve following procedures summarized above. Next, an exponential subsidence curve is fitted to the R1 curve using least-squares methods. This step is of course designed for use where tectonic subsidence is driven by thermal relaxation mechanisms acting over a scale of tens of millions of years, and builds in the assumption that the subsidence follows a simple exponential path.

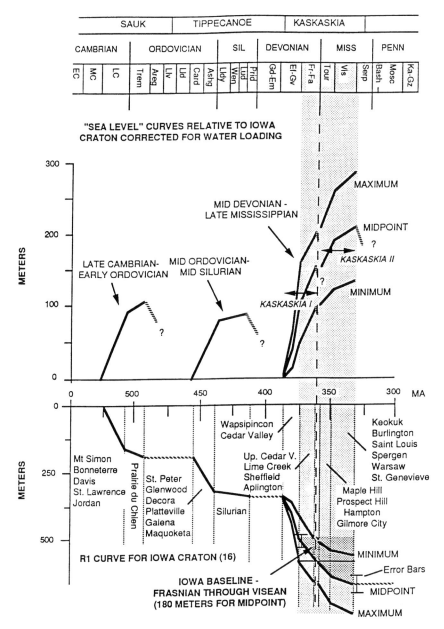

Fig. 2.23. Subsidence curve for Iowa (*lower curve*), from which a sea-level curve (*upper curve*) has been derived by correcting for sediment and water loading. (Bond and Kominz 1991a)

The R2 curve consists of the residuals derived by extracting the exponential curve from the R1 curve. The resulting R2 curve represents the external changes in accommodation superimposed on the thermal subsidence (Fig. 2.24). These changes may be of tectonic and/or eustatic origin, and typically have wavelengths in the million-year range. Comparison of R2 curves from different sections may yield important information on the extent of specific accommodation events. For example, Fig. 2.25 illustrates a set of R2 curves derived for Cambrian sections on the passive western margin of North America. Most of the peaks and troughs can be correlated, suggesting that they are of eustatic origin. A further reduction, termed R3 analysis, fits a polynomial to the R2 curve and plots the residuals. The result is a reflection of the local departures from high-frequency changes in accommodation space, that may be useful in clarifying differences between curves resulting from changes in facies (examples are discussed in Sect. 2.3.5).

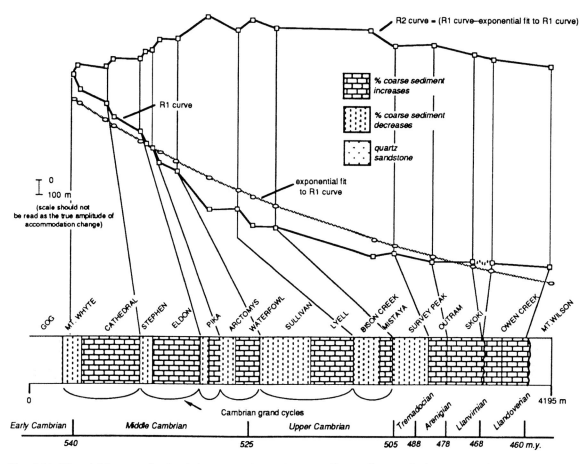

Fig. 2.24. *R1* and *R2* curves for Cambrian succession at Mount Wilson, Alberta, which consists of a series of "grand cycles" (see Sect. 7.7). (Bond and Kominz 1991b)

An evaluation of the many sources of error in these methods, including inaccurate subsidence curves and imprecision in correlation of the sections, indicates that the form of the R2 and R3 curves provides a useful general indication of the changes in accommodation with time, but that they cannot yield accurate measurements of the magnitudes of accommodation events (Bond et al. 1989). It seems likely that these methods would only work well for carbonate sediments because, given conditions suitable for the "carbonate factory" to develop, carbonate facies and thicknesses are sensitive to changes in water depth, whereas in the case of clastic sediments external factors of sediment supply and current dispersion affect resulting thicknesses, and would tend to confuse higher-order data reduction exercises (certainly at the level of R3 analysis).

An alternative approach is to study sediments that are always formed at or near sea level, namely shallow-water carbonates. A special application of

backstripping procedures was reported by Lincoln and Schlanger (1991). They documented the unconformity and solution surfaces that occur in carbonates comprising the platform on which Enewetak and Bikini Atolls are built in the South Pacific Ocean. Gradual subsidence of the atolls, driven by the weight of the sediment and the thermal subsidence of the oceanic crust beneath has preserved a stratigraphic record extending back to the Eocene. Periodic sea-level falls exposed the atoll surfaces, leading to diagenetic changes and the development of karst surfaces. Unless significant erosion takes place during such intervals of exposure, these surfaces are preserved upon subsequent sea-level rise, and provide a record of sea-level history. Figures 2.26 and 2.27 illustrate the principals in the use of atoll stratigraphy to document sea-level change, and Fig. 2.28 illustrates the stratigraphy of the drill holes that were used in their study. Part of the record may be lost to erosion, and the reconstruction of the curve depends on the

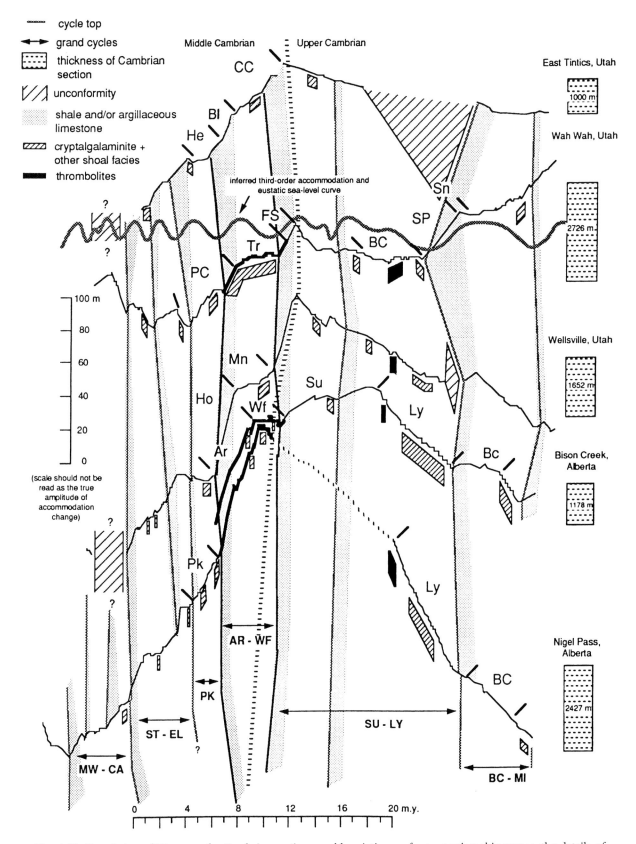

Fig. 2.25. Correlation of R2 curves for Cambrian sections in Utah and Alberta. The fact that most of the peaks and troughs in these curves can be correlated suggests that the accommodation events are of eustatic origin. *Two-letter* abbreviations refer to stratigraphic names, the details of which are not necessary for the purpose of this book. The relevant stratigraphic names in Alberta are given in Fig. 2.24. (Bond and Kominz 1991b)

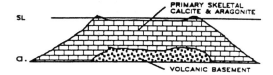

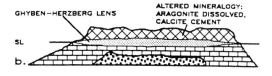

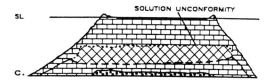

Fig. 2.26. The origin of solution unconformities within atoll stratigraphies. **a** Stage 1, shallow-marine sediments accumulate during a relative rise in sea level. **b** Stage 2, during a fall in sea level primary aragonite and high-magnesium calcite dissolve, while low-magnesium calcite is precipitated in the fresh-water zone (Ghyben-Herzberg lens). Karst surfaces may develop. **c** Stage 3, during a subsequent rise in sea level the karst surface is preserved as a solution unconformity. (Lincoln and Schlanger 1991)

ability to date the sediments accurately and to determine subsidence histories and rates of erosion. Paleodepth corrections must be made for carbonate sediments that are not deposited close to sea level. Lincoln and Schlanger (1991) were able to construct a sea-level curve that tracks the main variations in the Haq et al. (1987, 1988a) global cycle chart reasonably well, but the finer-scale events on the million-year time scale are not detectable (Fig. 2.29).

2.3.4 Sea-Level Estimation from Paleoshorelines and Other Fixed Points

A different approach has been to attempt to locate fixed points, such as shorelines, at specific point in time by detailed study of the local geology, choosing a stable area to eliminate tectonic effects as much as possible, and making every possible allowance for other possible complications by carrying out local corrections. The cratonic interior of North America has been chosen as a reference frame for several important studies of this type, because of its location above the the stable Canadian Shield. For example, Sleep (1976) studied a Cretaceous paleoshoreline in Minnesota, and concluded that sea level was approximately 300 m higher than at present

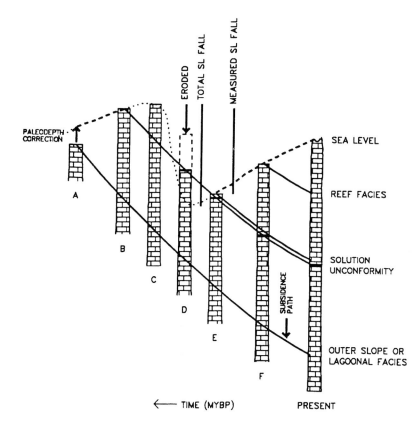

Fig. 2.27. Model for the development of atoll stratigraphy with sea-level changes. *Sloped lines* display subsidence paths of hypothetical atoll from time *A* to *present*. *Dashed-dotted line* is sea-level curve. *Dashed portion* is that part of the curve which can be reconstructed from the atoll stratigraphy; *dotted portion* represents the record lost to erosion. (Lincoln and Schlanger 1991)

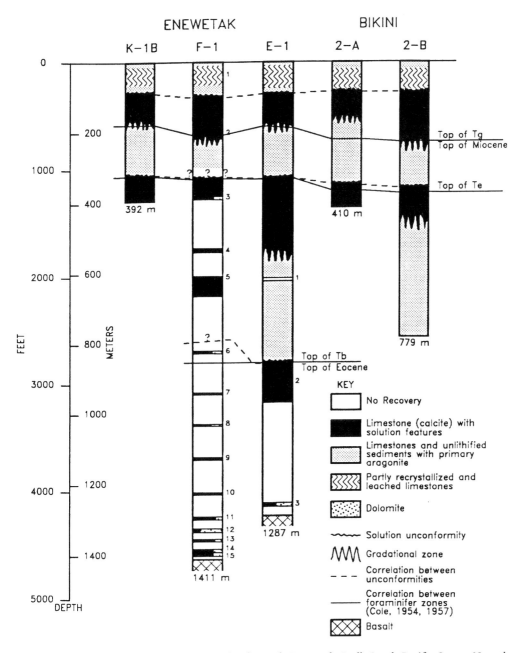

Fig. 2.28. Stratigraphy of carbonate succession beneath Enewetak Atoll, South Pacific Ocean. Note the correlation of solution unconformities, and the alternation of altered and unaltered intervals. (Lincoln and Schlanger 1991)

during the Late Cretaceous. Kominz and Bond (1991) chose Iowa as a "baseline" source of stratigraphic data for their calculations of long-term Paleozoic subsidence and sea-level change (Sect. 9.2). Other research focusing on the study of shorelines on stable cratons has been reported by Sahagian (1987, 1988).

One of the most detailed studies to date of this type is that reported by McDonough and Cross

(1991). The starting point of their analysis is explained in the following statement:

Advantages of this method [the study of paleoshorelines] include its direct measurement of historical sea-level elevation at a point in time; its potential for achieving increased temporal resolution; its requirement of fewer assumptions; its capability for testing and falsifying those assumptions, postulates and results; and its potential as the basis for independent evaluation of assumptions used in other approaches. This method makes a single, sim-

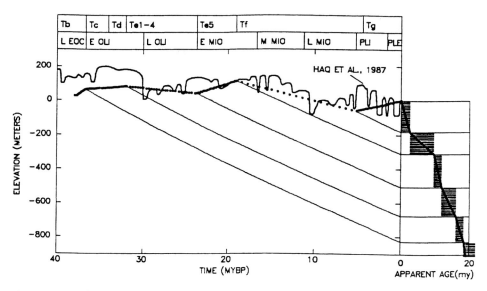

Fig. 2.29. Depth-age profile for Enewetak, interpolated every 6.6 m from biostratigraphic boundaries, is plotted on the *right*. The data are backtracked (*backstripped*) using calculated subsidence and flexure parameters. *Solid circles* indicate the calculated positions of sea level since the Late Eocene. (Lincoln and Schlanger 1991)

plifying assumption: the paleoshoreline has not moved vertically since deposition (or that postdepositional movement can be reconstructed and calculated). Vertical movement of a paleoshoreline may be caused by post-depositional tectonic movement, lithospheric compensation to surface or other loads, and sediment compaction. This assumption is most likely to be satisfied where strata containing a paleoshoreline were deposited high on the margin of a tectonically stable craton of old, cold lithosphere, and where the stratigraphic section is sufficiently thin that sediment compaction is minimized. These two conditions ... are most likely achieved during former high sea levels when continents were flooded to a maximum extent.

These authors returned to Minnesota to collect stratigraphic data along the thin edge of the Cretaceous sedimentary cover, where a single progradational unit could be traced and dated to within approximately 100 ka. This is important because, as Wise (1974) has pointed out, if data from many locations of varying ages are used in paleoshoreline calculations there is a tendency to generalize the results and overestimate the height of eustatic rises in sea level. Earlier work of Sleep (1976) and Sahagian (1987) did not adhere to this guideline, and their results reveal a wide range of estimates.

In Minnesota the single shoreline unit traced by McDonough and Cross (1991) in outcrop and cores varied in present-day elevation from 266 to 286 m above present sea level. Minimal corrections for sediment compaction were required because of the location of these beds above very thin earlier Cretaceous sediments which, in turn, rest on crystalline

basement. Post-Cretaceous movements are thought to have been limited to loading and flexure related to Pleistocene glaciation. Calculations were carried out to eliminate these effects. Possible movements related to dynamic topography effects were not considered. The corrected, average value is 276±24 m above modern sea level. This value was then used to position the Haq et al. (1987, 1988a) chart, as shown in Fig. 2.30. The magnitude of the sea-level rise demonstrated by this work is similar to that calculated by other workers for the Cretaceous (Fig. 2.31), although the wide range of values resulting from this earlier work reflects the broad time range and geographic extent covered by the earlier studies.

A stratigraphic succession may contain many precise records of sea-level elevation, such as shoreline positions. Franseen et al. (1993) and Goldstein and Franseen (1995) showed how these could be used to constrain sea-level curves. The position of sea level at any moment in geological time may be revealed by stratigraphic transitions from marine to nonmarine strata, or by evidence of near-sea-level facies such as tidal-flat or beach deposits, reefs, or surfaces of subaerial exposure overlain by marine deposits. The identification of such stratigraphic records provides a series of *pinning points*, the relative elevations of which yield a series of quantitatively fixed points on a sea-level curve. The method is analogous to the use of onlap seismic terminations in Vail's method (Sect. 2.2.3), but makes use of a wider variety of outcrop in-

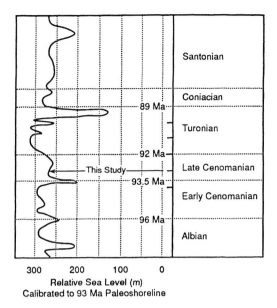

Fig. 2.30. The sea-level curve of Haq et al. (1987, 1988a) calibrated using the Cenomanian paleoshoreline position calculated by McDonough and Cross (1991)

dicators. Corrections must be made for compaction and tectonic tilting, for example, by using geopetal evidence to indicate tilts due to differential compaction.

Figure 2.32 illustrates a stratigraphic cross section in which thirty pinning points have been identified. The sea-level curve constructed from these points is shown in Fig. 2.33. Gaps in this curve indicate time periods for which a reconstruction is not possible because of a loss of the stratigraphic record to erosion. The determination of a few of the individual pinning points is described below to

provide some indication of the types of geological reasoning used in the construction of this curve:

- Pinning point 1: volcanic basement with evidence of subaerial exposure (fissures, spheroidal weathering) overlain by marine deposits.
- Pinning points 7, 8: exposure features on the upper surface of unit DS2 can be traced downslope to their most distal point, interpreted as points on a falling and subsequently rising leg of the sea-level curve.
- Pinning point 10: reef facies encrusted with *Porites*, indicating a reef-crest environment.

2.3.5 Documentation of Meter-Scale Cycles

Regular successions of meter-scale cycles are common in some successions of shallow-marine carbonate and fine-grained clastic rocks. They may be autogenic or allogenic in origin. Their thicknesses commonly vary in a systematic manner, suggesting the influence of a long-term control on thickness variation. Fischer (1964), in a study of Triassic carbonate cycles in the Calcareous Alps, devised a method of plotting cycle thickness as a means of objectively displaying these thickness variations. This plotting technique has come to be termed the *Fischer plot*. There has been considerable renewed interest in these cycles and the use of Fischer plots as a means of documenting cyclicity in the Milankovitch band (see Section 3.1 for an explanation of this term). Recent studies in which the method has been used include Goldhammer et al. (1987), Read and Goldhammer (1988), Osleger and Read

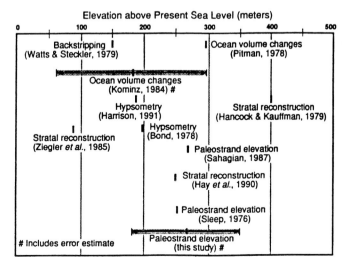

Fig. 2.31. Ranges in estimates of Cretaceous sea level resulting from use of different methods of estimation. (McDonough and Cross 1991)

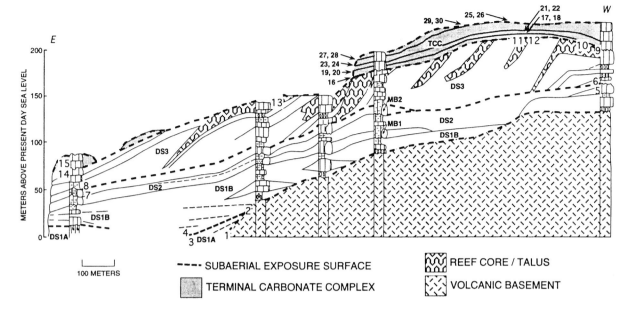

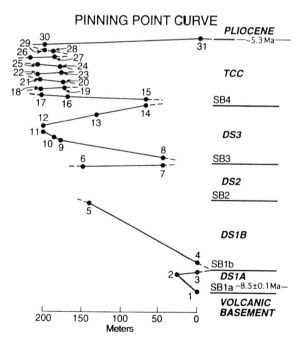

Fig. 2.32. Stratigraphic cross section through a Late Miocene succession in southeastern Spain, showing the location of 30 identified pinning points. (Goldstein and Franseen 1995, reproduced by kind permission of Elsevier Science-NL, Sara Burgerhartstraat 25, 1055 KV Amsterdam, The Netherlands)

(1991, 1993), and Montañez and Read (1992). Excellent discussions of the method, its advantages and pitfalls, have been provided by Sadler et al.

Fig. 2.33. Sea-level curve constructed using the pinning points shown in Fig. 2.32. (Goldstein and Franseen 1995, reproduced by kind permission of Elsevier Science-NL, Sara Burgerhartstraat 25, 1055 KV Amsterdam, The Netherlands)

(1993), Drummond and Wilkinson (1993a), and Boss and Rasmussen (1995).

The basis of the Fischer plot is a zig-zag diagram in which net subsidence is plotted against cycle thickness (Fig. 2.34A). The slope of the subsidence plot is determined from the total thickness of the stratigraphic section divided by the elapsed time for the section. A constant subsidence rate is assumed. The line formed by joining all the cycle tops (heavy line in Fig. 2.34A) rises and falls as an irregular wave train reflecting the changing accommodation space in the depositional environment (Fig. 2.35). A rising slope indicates a succession of thick cycles, suggesting an increase in the rate of generation of accommodation space, such as is brought about by a rise in sea level. A fall in the curve, indicating a succession of thin cycles, suggests a decrease in the rate of generation of accommodation space and a fall in sea level. However, the limitations of the plot need to be borne in mind. A constant rate of subsidence is assumed, and each cycle is assigned the same duration. Neither assumption may be valid. However, considerable information may be derived from the plots if allowance is made for these limitations.

It is important to be clear about what is actually being plotted. As pointed out by Sadler et al. (1993), the vertical axis is in fact a plot of the departure of the individual cycle thickness from mean thickness, and the horizontal axis should be labeled cycle

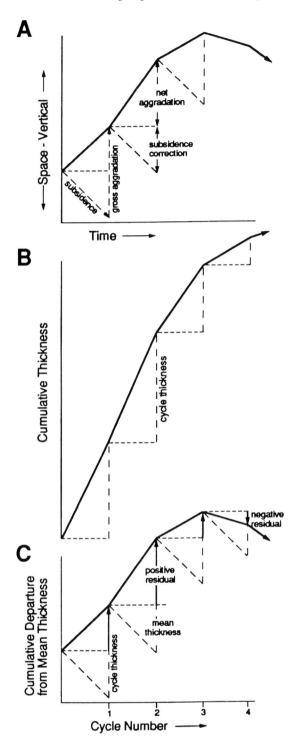

Fig. 2.34. A The basis for the Fischer plot. **B** Replotting the data as a cumulative thickness plot. **C** The same information, labeled in a more objective way. (Sadler et al. 1993)

number, to remove any confusion about its relationship to time (Fig. 2.34C). Because the plots are simply a method of portraying the distribution of total thickness over a given section, the total positive and negative departures from the mean cycle thickness sum to zero, and the graphs always finish at the same vertical position as they start. It is therefore important to base Fischer plots on lengthy sections, or serious distortions may appear, as illustrated in the short segments of the curve replotted in Fig. 2.35B. Sadler et al. (1993) recommended that a plot be based on a minimum of at least fifty cycles, to avoid such difficulties, and to overcome possible statistical distortions of a small sample base. Because of all the qualifications on the shape of the plot, Sadler et al. (1993) prefer to describe the form of the curving trace in terms of "waves" rather than "cycles." Drummond and Wilkinson (1993a) demonstrated that "waves" constructed from short runs of cycle thickness could not be statistically distinguished from random noise.

Practical problems in the operational definition of cycles, reflecting the ambiguity that is common in the stratigraphic record, may lead to difficulties in the definition of cycles in practice. Figure 2.35A illustrates this point. A succession has been subdivided using two different approaches to cycle definition, one that generates relatively thicker cycles, and one that generates thinner cycles. The form of the plot is different, although broad trends of rise and fall are similar in the two plots.

What does the Fischer plot reveal about changes in the rate of generation of accommodation space? Although the rise and fall in the "wave" is a reflection of long-term changes, there are a number of important problems and limitations that need to be considered:

1. The plot cannot account for time that is not represented in the section because of a fall in sea level that exposes the depositional environment (what have been termed "missed beats," as explained in Sect. 8.3). As shown by Drummond and Wilkinson (1993a) where the long-term rate of sea-level fall is greater than the rate of subsidence, long intervals of time (and many cycles) are missing from the section, resulting in severe distortions of the record.

2. Plots cannot be constructed for intervals of the section that are noncyclic.

3. The plots are one-dimensional and cannot differentiate between cycles of local extent that may

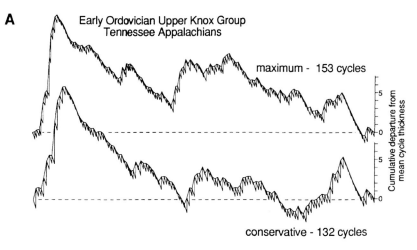

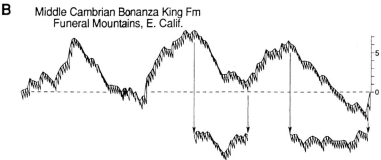

Fig. 2.35. Examples of Fischer plots for peritidal Paleozoic carbonate successions. **A** The same succession measured using two different criteria for cycle definition. **B** Another section, in which two legs of the curve have been replotted as if they constituted the entire data succession. (Sadler et al. 1993)

be of autogenic origin and more regionally extensive cycles of allogenic origin (e.g., Pratt et al. 1992).

4. Distortions may also be introduced by major changes in facies, because sedimentation rates may vary (Pratt et al. 1992).

5. It can be shown that the plots are always asymmetric, because thin cycles are more common than thick cycles, and so falling legs of the wave are longer, and therefore flatter than rising legs (Sadler et al. 1993).

6. As noted by Drummond and Wilkinson (1993a), the plots do not allow for compaction. Differential compaction, reflecting systematic variations in lithology could, theoretically, generate Fischer plots similar to those shown in Fig. 2.35.

Boss and Rasmussen (1995) constructed Fischer plots one cycle in length for the Holocene sedimentary record of the Bahamas carbonate platform, using a seismic transect, and demonstrated that there is no relationship between cycle thickness and accommodation space (depth through the water column and Holocene deposits to the top of the Pleistocene).

Aside from the possibility of differential compaction, regular or episodic changes in accommodation space may be the result of changes in sea level or changes in susbsidence rate, or some combination of both. As discussed in Chapter 11, tectonic mechanisms operate over a wide range of time scales, including the 10- to 100-ka time scale typical of Milankovitch cycles. Therefore, although the Fischer plot provides an objective representation of preserved cycle thicknesses, for all the reasons stated above it offers only a crude representation of changes in accommodation space. Nonetheless, comparisons and correlations between plots for various stratigraphic sections may provide useful insights. Examples were offered by Osleger and Read (1993).

Figure 2.36 illustrates three methods for determining changes in accommodation space. The "paleobathymetry" curve was constructed using qualitative estimates of water depth based on facies interpretations, of the type discussed in Section 2.2.2. A Fischer plot is shown in the middle, and to the right is the "R3" curve derived from subsidence analysis, using the method described in Section 2.3.3. The three curves are quite similar. Exact correlations cannot be expected, because the stratigraphic section and the paleobathymetry curve are plotted against thickness, whereas the Fischer plot

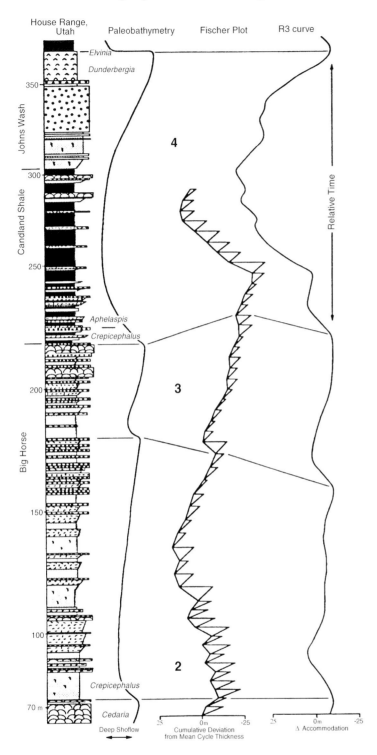

Fig. 2.36. Comparison of three methods for determining variations in accommodation space, as applied to a Cambrian section in Utah. (Osleger and Read 1993)

uses cycle number in the vertical axis, and the R3 curve is plotted against decompacted thickness.

Klein and Kupperman (1992) and Klein (1994) proposed a method for evaluating the relative contributions of tectonic subsidence and sea-level change in the generation of Late Paleozoic cy-clothems. Tectonic subsidence is derived by plotting a backstripping curve and dividing total subsidence (corrected for compaction and loading) by the number of cycles present. Water depth changes are estimated by sedimentological methods and the two results compared.

2.4 Integrated Tectonic-Stratigraphic Analysis

Ideally, data of many types should be integrated to derive an overall basin analysis. Vail et al. (1991) suggested the following sequence of procedures:

1. Determine the physical chronostratigraphic framework by interpreting sequences, systems tracts and parasequences and/or simple sequences on outcrops, well logs and seismic data and age date with high resolution biostratigraphy.
2. Construct geohistory, total subsidence, and tectonic subsidence curves based on sequence boundary ages.
3. Complete a tectono-stratigraphic analysis including:
   A. Relate major transgressive-regressive facies cycles to tectonic events.
   B. Relate changes in rates on tectonic subsidence curves to plate-tectonic events.
   C. Assign a cause to tectonically enhanced unconformities.
   D. Relate magmatism to the tectonic subsidence curve.
   E. Map tectono-stratigraphic units.
   F. Determine style and orientation of structures within tectono-stratigraphic units.
   G. Simulate geological history.
4. Define depositional systems and lithofacies tracts within systems tracts and parasequences or simple sequences.
5. Interpret paleogeography, geological history, and stratigraphic signatures from resulting cross sections, maps and chronostratigraphic charts.
6. Locate potential reservoirs and source rocks for possible sites of exploration.

This book is concerned primarily with steps 1, 3 and 4. Step 2 is dealt with in greater detail by Miall (1990) and Allen and Allen (1990), and steps 4 and 5 are also discussed in detail by Miall (1990). Step 6 is discussed in many books and course notes.

It is interesting that Vail et al. (1991) do not make explicit in this summary the steps involved in relating the products of the analysis to the global cycle chart, although this is inevitably the outcome in any of their own analyses.

Application of Vail's integrated basin-analysis approach is discussed in more detail, with reference to practical petroleum-exploration examples, in Chapter 17. A modified approach is suggested in Section 17.3.3.

3 The Four Basic Types of Stratigraphic Cycle

3.1 Introduction

The duration and episodicity of geological events and stratigraphic cyclicity span at least 16 orders of magnitude, ranging from the repeat time of burst-sweep cycles in turbulent boundary layers (10^{-6} years), to plate-tectonic cycles involving the formation and breakup of supercontinents (10^9 years; Miall 1991b; Einsele et al. 1991b; Figs. 3.1, 3.2). Minor cyclicity may be apparent in the geological record as a result of seasonal changes in weather, fluvial discharge, etc., the so-called *calendar band* of cyclicity (Fischer and Bottjer 1991). Sunspots and other solar processes generate cyclicity on a 10^1–10^2-year scale of cyclicity, the *solar band* of the time scale. These processes and their products are not discussed in this book.

Cyclicity of geological significance begins with *Milankovitch band* cyclicity, over time scales of 10^4–10^5 years. Cycles of Milankovitch-band and longer duration comprise four basic types (types

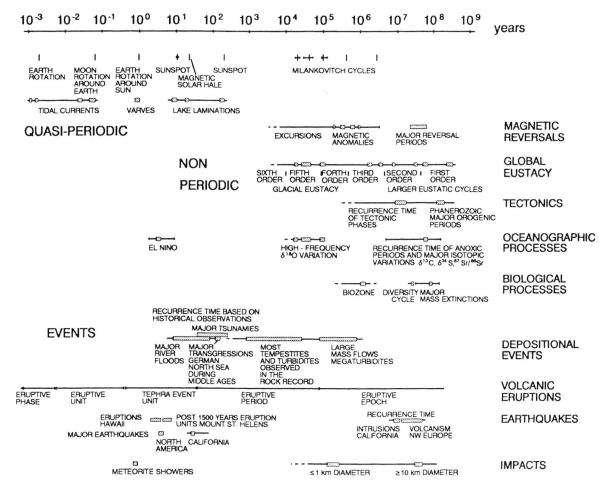

Fig. 3.1. Recurrence time of periodic and episodic processes and events in geology. (Einsele et al. 1991b)

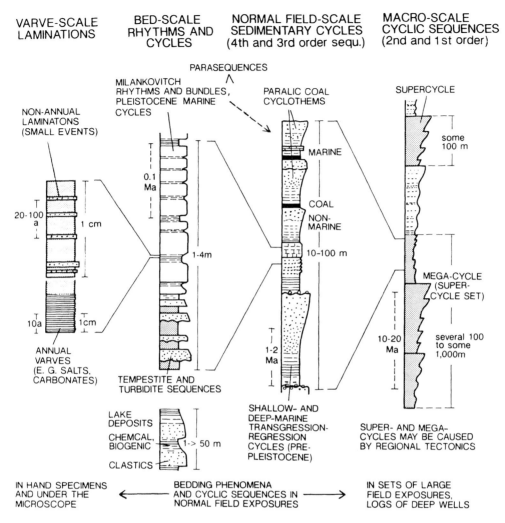

Fig. 3.2. The scales of cyclic sedimentation in the stratigraphic record. This book is concerned with all but the first of these types (varve-scale laminations). (Einsele et al. 1991b)

A–D in Table 3.1). These four types are introduced briefly in this chapter, with detailed discussion and documentation following in Part II. These cycle types reflect the independent operation of at least four type of process, including regional tectonism and various controls on eustasy. Work is still in progress in the analysis of tectonic mechanisms to determine the importance of other processes comparable in scale, duration and areal effect. Important second-order effects triggered or driven by these primary mechanisms include changes in global climate, magmatism, ocean-water circulation patterns, the carbon and oxygen cycles, and biogenesis, all of which have significant, measureable effects on the sediments. Sorting out the various processes and their effects is currently one of the most vigorous areas of basin analysis research.

An empiricial classification of stratigraphic sequences using rank-order designations, based on their duration, was devised by Vail et al. (1977). This classification is widely known but has become increasingly unsatisfactory as more has been learned about sequences and their generating mechanisms. The terminology is shown in Table 3.1 but is no longer recommended.

Carter et al. (1991) pointed out that although a hierarchy of sequences exists in the stratigraphic record, their distinctiveness in terms of duration or recurrence interval is only approximate, the scales of the sequences (for example, their thicknesses) do not define mutually exclusive ranges, and they do not necessarily nest internally in a logical or ordered pattern. They stated (p. 45):

Table 3.1. Stratigraphic cycles and their causes

Sequence type	Duration (m.y.)	Other terminology
A. Global supercontinent cycle	200–500	1st-order cycle (Vail et al. 1977)
B. Cycles generated by continental-scale mantle thermal processes (dynamic topography), and by plate kinematics, including: 1. Eustatic cycles induced by volume changes in global midoceanic spreading centers 2. Regional cycles of basement movement induced by extensional downwarp and crustal loading.	10–100	2nd-order cycle (Vail et al. 1977), supercycle (Vail et al. 1977), sequence (Sloss 1963)
C. Regional to local cycles of basement movement caused by regional plate kinematics, including changes in intraplate-stress regime	0.01–10	3rd- to 5th-order cycles (Vail et al. 1977). 3rd-order cycles also termed: megacyclothem (Heckel 1986), mesothem (Ramsbottom 1979)
D. Global cycles generated by orbital forcing, including glacioeustasy, productivity cycles, etc.	0.01–2	4th- and 5th-order cycles (Vail et al. 1977), Milankovitch cycles, cyclothem (Wanless and Weller 1932), major and minor cycles (Heckel 1986)

It is not self-evident that the SSM [sequence-stratigraphic model] at successive orders can be adequately represented as the sum of a set of topologically similar SSMs that comprise parasequences of the next higher order.... There is also the additional difficulty that in areas of high sediment supply the thickness of sixth or fifth-order sequences may greatly exceed "typical" third-order sequence thickness.... We conclude that it is unlikely that sequences at all orders correspond to a "Russian doll" stacked set, whereby the SSM applies at any level and the sequence at that level is viewed as forming from a large number of finer sequences (parasequences) of the next higher order. None the less, some orders of sequence do indeed embrace lower orders, e.g., the major first-order thermo-tectonic cyle that incorporates the complete sedimentary history of the Canterbury Basin ..., which includes examples of second, third, fourth, and probably fifth-order sequences.

This type of internal stacking is characteristic of basin fills, as described in many examples in this book.

Drummond and Wilkinson (1996) carried out a quantitative study of the duration and thickness of stratigraphic sequences and confirmed the opinions of Carter et al. (1991). Their major conclusion was that "discrimination of stratigraphic hierarchies and their designation as *n*th-order cycles may constitute little more than the arbitrary subdivision of an uninterrupted stratigraphic continuum."

Embry (1995) also found the existing hierarchical classification of sequences arbitrary, and he proposed a more objective approach, based on certain key descriptive characteristics of the sequences themselves, such as the amount of deformation and the degree of facies change at the sequence boundary. He objected that a classification based on measured (or assumed) duration is subjective and

meaningless. His classification is discussed in Section 15.2.

3.2 The Supercontinent Cycle

A very long-term eustatic cycle is generated as a result of the assembly of supercontinents by seafloor spreading and their subsequent rifting and dispersal. The complete cycle takes 200–500 m.y., and the process has been underway for at least 2 Ga. About four complete cycles may have occurred. The assembly of a supercontinent inhibits radiogenic heat loss from the core and mantle, leading to thermal doming, rifting, and breakup of the continent, so that a period of vigorous seafloor spreading ensues. There are significant consequences for global climate and sea level. These mechanisms are discussed in more detail in Chapter 9.

Major cycles of sea-level rise and fall during the Phanerozoic are illustrated in Fig. 3.3. Vail et al. (1977) referred to these as first-order cycles. They include the two extended periods of maximum marine transgression in the Late Cambrian to Mississippian, and the Cretaceous, and a period of maximum regression in the Pennsylvanian to Jurassic. A glance at the geological map of North America confirms the importance of these broad changes. The Canadian Shield is flanked by Cretaceous rocks resting unconformably on a Cambrian or Ordovician to Devonian sequence over wide areas of the craton, from the Great Lakes region across the Prairies into the Beaufort-Mackenzie region and the Arctic Platform. Rocks of Pennsylvanian to Jurassic

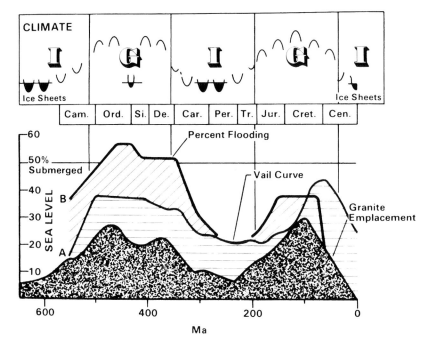

Fig. 3.3. Supercontinent cycles during the Phanerozoic, including sea-level change (from Vail et al. 1977), percentage flooding of the continents (from Fischer 1981), and volume of granite emplacement (from Engel and Engel 1964). The generation of the sea-level curve is discussed in Chapter 5 (Vail et al. 1977). This diagram, which was compiled by Worsley et al. (1984), also shows fluctuations in global climate that accompanied these changes. *I* Icehouse; *G* greenhouse climate (discussed in text). (Reproduced by kind permission of Elsevier Science-NL, Sara Burgerhartstraat 25, 1055 KV Amsterdam, The Netherlands)

age are largely confined to intracratonic basins and the mobile belts flanking the cratonic interior.

Worsley et al. (1984, 1986), Hoffman (1989, 1991), Dalziel (1991), and Rogers (1996) argued that there is fragmentary evidence for several comparable cycles during the Precambrian, back to at least 2 Ga, possibly 3 Ga (earlier work on this subject is summarized by Williams 1981). They reviewed the evidence for four major global episodes of thermotectonic activity comparable in magnitude and extent to the combined Caledonian-Acadian and Hercynian-Appalachian events of the Devonian to Permian. These orogenic episodes may indicate supercontinent assembly. They were followed by the intrusion of major dyke swarms, which probably formed during the initial rifting of the supercontinent. Evolutionary milestones in the biological realm also seem to have postdated the major orogenic episodes, possibly indicating the explosion of biotic diversity in newly flooded shelf seas, caused by increased rates of seafloor spreading 50–100 m.y. after rifting.

Stratigraphic evidence for these Precambrian cycles has yet to be assembled. There are many thick sequences of Precambrian strata around the world, but they have been much disturbed by tectonism, and accurate dating is difficult to achieve. Much work remains to be done in this area.

3.3 Cycles with Episodicities of Tens of Millions of Years

It has now been satisfactorily demonstrated that the original six cycles of Sloss (1963) can be recognized and correlated with comparable cratonic cycles in other continents. In later papers, Sloss (1972, 1979) showed that a similar sequence chronology could be recognized in Europe and Russia. In his 1972 paper, Sloss reported on an analysis of detailed isopach and lithofacies maps of the Western Canada Sedimentary Basin and the Russian Platform. The data source included 29 Canadian maps (McCrossan and Glaister 1964) and 62 Russian maps (Vinogradov and Nalivkin 1960; Vinogradov et al. 1961). Each map was divided into a grid with intersection points spaced about 60 km apart, and the thickness and lithofacies were recorded for each point. From these data, the areal extent and volume of each of the mapped units could be calculated and compared. This approach is subject to possibly serious error because of the high probability of intersequence and even intrasequence erosion. The presence of a single isolated outlier or fault block beyond the edge of the main basin can change the interpreted former area of extent of a map unit by hundreds of square kilometers. Nevertheless, the data from the two areas show remarkable similarities (Fig. 3.4), and the detailed statistical documentation confirms that Sloss's six sequences can be recognized in two widely separated continents that formerly would have been assumed to have undergone a quite different geological history.

These Phanerozoic cycles are clearly global in scope. They are attributed to eustatic changes in sea level in response to volume changes in oceanic spreading centers, an interpretation first suggested by Hallam (1963), in his summary of Late Cretaceous and Cenozoic events. However, eustatic sea-level changes cannot explain all the features of the major sequences. They commonly are separated by angular unconformities, and in many cases, the sequences contain thick continental deposits. A passive rise in sea level would terminate widespread nonmarine deposition, and therefore more must be involved. Geophysical studies have demonstrated that the crust flexes on a regional scale in response to convergent, divergent and transcurrent plate movements. Broad, regional cycles of change in basement elevation accompany this regional flexure in response to thermal changes in the crust and mantle, crustal thickening and thinning, and sediment loading (Chap. 11). A form of regional stratigraphic cyclicity is the result, as documented in Chapter 6.

Detailed backstrippping and other studies, of the type outlined in Chapter 2, show that the crust is constantly undergoing gentle vertical motions and broad continental tilts. This form of epeirogeny has recently been termed *dynamic topography*. It is controlled by thermal processes in the mantle, as discussed in Chapter 9. Intraplate stresses may also be a factor, as discussed in Chapter 11.

3.4 Cycles with Million-Year Episodicities

Detailed stratigraphic and facies studies of well-dated strata of Phanerozoic age have yielded a wealth of information concerning cyclicity over

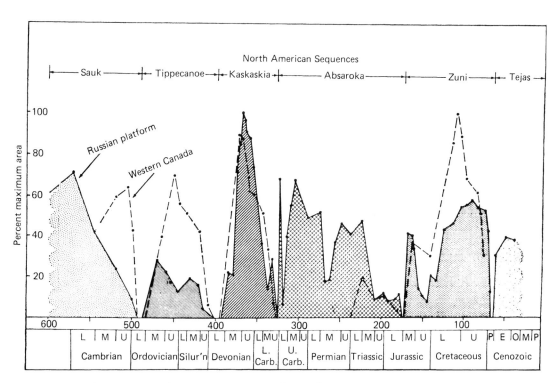

Fig. 3.4. Areas of preservation of units in western Canada and the Russian platform showing relationship to the six sequences of Sloss (1963). *L* Lower; *M* Middle; *U* Upper; *P* Paleocene; *E* Eocene; *O* Oligocene; *M* Miocene; *P* Pliocene. (Sloss 1972)

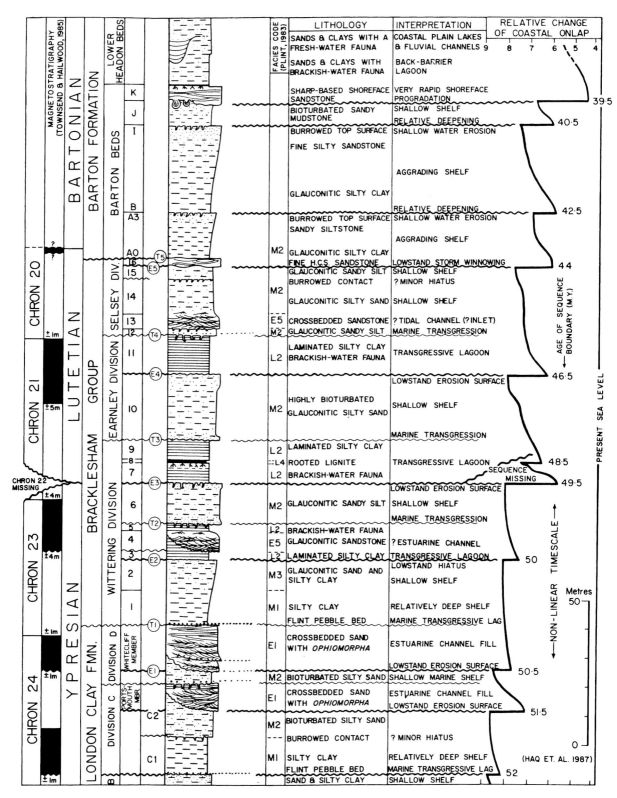

Fig. 3.5. Composite lithological log through the upper part of the London Clay Formation, the Bracklesham Group and the Barton Formation, Hampshire Basin, England, showing magnetostratigraphic correlation, sequence sub-division, and correlation with the chart of relative change in coastal onlap derived by Haq et al. (1987). (Plint 1988b, reproduced by permission of Blackwell Scientific Publications)

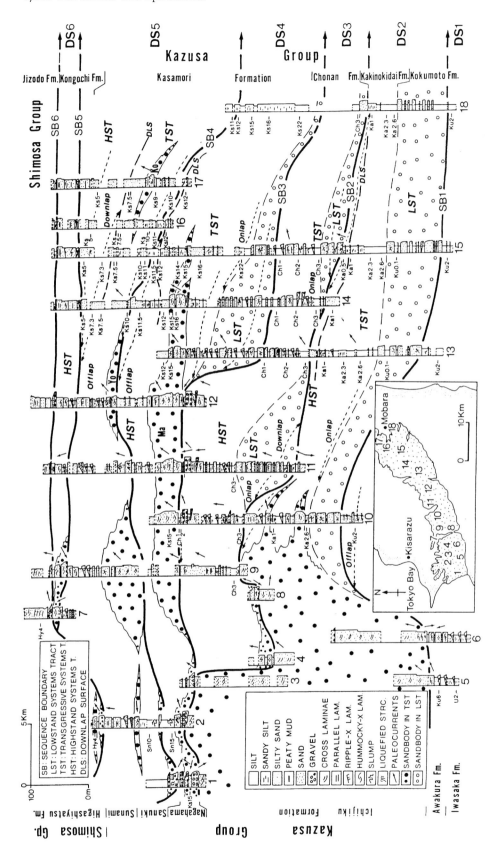

Fig. 3.6. Stratigraphic cross section of the upper part of the Kazusa Group in Boso Peninsula, Japan, illustrating glacioeustatic cycles. *Codes to left of each column* indicate names of volcanic ash layers used for chronostratigraphic correlation. Six stratigraphic sequences are recognized, spanning the period of 0.4–0.8-Ma. (Ito 1992, reproduced by kind permission of Elsevier Science-NL, Sara Burgerhartstraat 25, 1055 KV Amsterdam, The Netherlands)

time scales of 1–10 m.y. Vail et al. (1977) termed these third-order cycles. Sequence stratigraphy models, on which some of this work is based, are discussed in Chapter 4. There is increasing evidence for the occurrence of the same type of cyclicity during the Precambrian (Christie-Blick et al. 1988), but the difficulty of achieving accurate chronostratigraphic correlations in Precambrian sediments has retarded the application of cyclic concepts to Precambrian stratigraphy.

Several different techniques have been used in attempts to reconstruct a chronostratigraphic record of these stratigraphic cycles. The seismic techniques of Vail and his coworkers (Sect. 2.2.3; Chaps. 4, 5) have been applied in analyses of the Mesozoic and Cenozoic sequence record by Vail et al. (1977) and Haq et al. (1987, 1988a). Sloss-type outcrop-area plots have been used by several workers, such as in Hallam's (1975) analysis of the Jurassic record. Detailed stratigraphic reconstructions, emphasizing lithofacies or biofacies data, or both, constitute the third main method of reconstructing sea-level change. Papers in this category are too numerous to list. A few workers have attempted to reconcile the results of these various techniques to come up with average or compromise sea-level curves (e.g., Hallam 1981, 1984, 1992). There are no reliable methods of quantifying the averaging procedures, and so the amplitudes of the curves (the amounts of sea-level change) cannot be taken too literally.

A selective review of research results describing this type of cyclicity in the rock record is given in Chapter 7. The intent is not to provide an exhaustive catalog of either the literature or the results, but to impart an overview of the range of types of stratigraphic cycle that have been described in various tectonic settings, and to provide a flavor of the kinds of research currently being conducted in this area. A single example is illustrated here, that of the Eocene marine section of the Hampshire Basin, England (Fig. 3.5).

Currently there is considerable controversy regarding the origins of these cycles, and whether they are regional or global in scope. Partial success has been achieved in correlating selected sections around the globe with the Vail curves but, as discussed elsewhere (Chaps. 13, 14), such correlations are of questionable value, and it can be argued that it has not yet been proved that any of these cycles are truly global. Several tectonic mechanisms for the development of regional (continent-wide) cycles have been identified, as discussed in Chapter 11.

3.5 Cycles with Episodicities of Less Than One Million Years

There is a complete spectrum of cycles down to those of a few thousand years duration. The classification of these cycles that is commonly used, into those of fourth-order rank (0.2–0.5 m.y. duration) and those of fifth-order rank (0.01–0.2 m.y.) is one of convenience, but is now increasingly seen as arbitrary. As described and illustrated in Chapter 8, there are several distinct types of cycle that can be classified on the basis of their sedimentological character and tectonic and climatic setting, and this range of types suggests that there is more than one, possibly interacting, generative mechanism. Deposits formed in areas distant from detrital sediment sources, such as carbonate shelves and some lakes, commonly show successions of laterally persistent sequences. The useful, informal descriptive term meter-scale cycle is commonly used to refer to such sequences.

Many of the cycles are of climatic origin, caused by changes in the amount of solar radiation received by the earth and its global distribution that are, in turn, brought about by irregularities in the earth's motions about the sun, a process termed *orbital forcing*. Glacioeustasy is the single most important mechanism, as discussed in Chapter 10. The Neogene glaciation generated suites of clastic cycles throughout the world in most tectonic settings. An example is given in Fig. 3.6. The well-known Upper Paleozoic cyclothems of the northern hemisphere are also interpreted as glacioeustatic in origin. Other types of climatic cycle are discussed in Chapter 8, as are other types of cycle of comparable duration that may be tectonic in origin. Cycles generated by orbital forcing are commonly termed *Milankovitch cycles*, after the individual who first developed the detailed theory for their development.

4 The Basic Sequence Model

4.1 Introduction

The purpose of this chapter is to present a succinct summary of sequence concepts, derived mainly from the Exxon work. Vail and his coworkers have developed a complete framework for stratigraphic analysis, including new concepts with their own definitions, and new interpretive methodologies. It is important to be familiar with these, as the terms and concepts are now used by virtually all stratigraphers. However, in recent years there has grown among many stratigraphers a tendency toward the unthinking acceptance of the Exxon models, leading to a lack of focus on important alternative data and ideas. Some of these points are noted below, but the major discussion and critique of the Exxon models is deferred to later parts of this book.

The first presentation of sequence concepts was in AAPG Memoir 26 (Vail et al. 1977), but the most thorough, formal presentations of the Exxon ideas are in the AAPG *Atlas of Seismic Stratigraphy* (Vail 1987; Van Wagoner et al. 1987), and in SEPM Special Publication 42, which is dominated by Exxon contributions (Posamentier et al. 1988; Posamentier and Vail 1988). Another useful source is Vail et al. (1991). Instructive comments on sequence terminology are also given by Walker (1992a).

4.2 Terminology

Sequence stratigraphy is based on the recognition of unconformity-bounded units, which may be formally defined and named using the methods of allostratigraphy, as described in Section 1.3. Some of the basic geometric characteristics of sequences are illustrated in Figs. 4.1 and 4.2, and the terms shown are defined below.

The recognition of *stratigraphic terminations* is a key part of the sequence method. This helps to explain why sequence stratigraphy developed initially from the study of seismic data, because conventional basin analysis based on outcrop and well data provides little direct information on stratigraphic terminations, whereas these are readily, and sometimes spectacularly, displayed on seismic-reflection cross sections.

Among the key geometric characteristics of stratigraphic sequences is the *onlap* that takes place at the base of the succession, and the *offlap* that occurs at or near the top (Fig. 4.1). These architectural, or geometric, characteristics, record the lateral shift in depositional environments in response to base-level change and subsidence (Fig. 4.2). *Downlap surfaces* develop during a transition from onlap to offlap. They typically develop above flooding surfaces, as basin-margin depositional systems begin to prograde seaward. The dipping, prograding units are called *clinoforms*, and they lap out downward onto the downlap surface as lateral progradation takes place. The word *lapout* is used as a general term for all these types of stratigraphic termination.

The broad internal characteristics of stratigraphic units may be determined from their *seismic facies*. This term is defined as an areally restricted group of seismic reflections whose appearance and characteristics are distinguishable from those of adjacent groups (Sangree and Widmier 1977). Various attributes may be used to define facies: reflection configuration, continuity, amplitude and frequence spectra, internal velocity, internal geometrical relations, and external three-dimensional form.

Figure 4.3 illustrates the main styles of seismic facies reflection patterns (Mitchum et al. 1977a). Most of these are best seen in sections parallel to depositional dip. Parallel or subparallel reflections indicate uniform rates of deposition; divergent reflections result from differential subsidence rates, such as in a half-graben or across a shelf-margin hinge zone. Prograding, *clinoform* reflections comprise an important class of seismic facies patterns. They are particularly common on continental margins, where they represent deltaic or con-

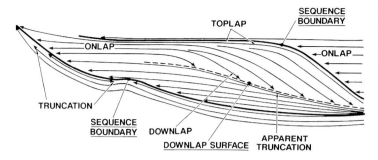

Fig. 4.1. Reflection termination patterns and types of discontinuity in unconformity-bounded sequences. (Vail 1987, reproduced by permission)

tinental-slope outgrowth. Variations in patterns of progradation reflect different combinations of depositional energy, subsidence rates, sediment supply, water depth and sea-level position. *Sigmoid clinoforms* tend to have low depositional dips, typically less than 1°, whereas *oblique clinoforms* may show depositional dips up to 10°. *Parallel-oblique clinoform* patterns show no topsets. This usually implies shallow water depths with wave or current scour and sediment bypass to deeper water, perhaps down a submarine canyon that may be revealed on an adjacent seismic cross section.

Many seismic sequences show a very complex prograding stratigraphy, of which the *complex sigmoid-oblique clinoform* pattern in Fig. 4.3 is a simple example. This diagram illustrates periods of sea-level still-stand, with the development of truncated topsets (termed *toplap*) alternating with periods of sea-level rise (or more rapid basin sub-sidence) allowing the lip of the prograding sequence to build upward as well as outward. Mitchum et al. (1977a) described the *hummocky clinoform* pattern as consisting of "irregular discontinuous subparallel re-

flection segments forming a practically random hummocky pattern marked by nonsystematic reflection terminations and splits. Relief on the hummocks is low, approaching the limits of seismic resolution....The reflection pattern is generally interpreted as strata forming small, interfingering clinoform lobes building into shallow water," such as the upbuilding or offlapping lobes of a delta undergoing distributary switching. Submarine fans may show the same hummocky reflections. *Shingled clinoform* patterns typically reflect offlapping sediment bodies on a continental shelf.

Chaotic reflections may reflect slumped or contorted sediment masses or those with abundant channels or cut-and-fill structures. Disrupted reflections are usually caused by faults. Lenticular patterns are likely to appear most frequently in sections oriented perpendicular to depositional dip. They represent the depositional lobes of deltas or submarine fans.

A *marine flooding surface* is a surface that separates older from younger strata, across which there is evidence of an abrupt increase in water

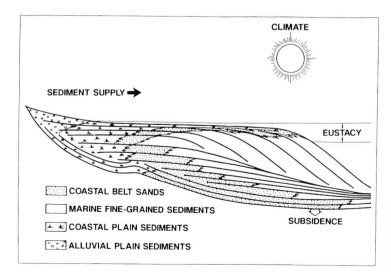

Fig. 4.2. Relationship between seismic architecture and depositional environments, showing the three major controls on stratigraphic architecture. (Vail 1987, reproduced by permission)

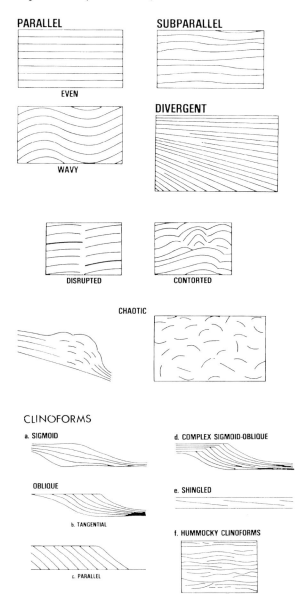

PARALLEL   SUBPARALLEL

EVEN

DIVERGENT

WAVY

DISRUPTED   CONTORTED

CHAOTIC

CLINOFORMS

a. SIGMOID   d. COMPLEX SIGMOID-OBLIQUE

OBLIQUE

e. SHINGLED

b. TANGENTIAL

f. HUMMOCKY CLINOFORMS

c. PARALLEL

Fig. 4.3. Typical seismic reflection patterns, illustrating the concept of seismic facies. (Mitchum et al. 1977a, reproduced by permission)

depth. These surfaces are typically prominent and readily recognizable and mappable in the stratigraphic record. Each of the heavy lines within the lower, retrogradational part of the sequence shown in Fig. 4.2 are marine flooding surfaces. The *maximum flooding surface* records the maximum extent of marine drowning, and separates transgressive units below from regressive units above. It commonly is a surface of considerable regional stratigraphic prominence and significance. It may be marked by a condensed section, indicating slow

sedimentation at a time of sediment starvation on the continental shelf, and may correspond to a downlap surface, as noted above. An example is shown as a dashed line in Fig. 4.1.

Sequences may consist of stacked facies successions, each of which shows a gradual upward change in facies character, indicating a progressive shift in the local depositional environments. The small packages of strata contained between the heavy lines in Fig. 4.2 are examples of these component packages of strata. Van Wagoner et al. (1987) erected the term *parasequence to* encompass "a relatively conformable succession of genetically related beds or bedsets bounded by marine flooding surfaces and their correlative surfaces.... Parasequences are progradational and therefore the beds within parasequences shoal upward." As Walker (1992a) pointed out, "parasequences and facies successions ... are essentially the same thing, except that the concept of facies succession is broader." However, other types of facies succession occur within sequences (e.g., channel-fill fining-upward successions), and the term parasequence is therefore unnecessarily restrictive. Many such successions are generated by autocyclic processes, such as delta-lobe switching and channel migration, that have nothing to do with sequence controls, and to include them in a term that has the word "sequence" within it may be misleading. Walker (1992a) recommended that the term parasequence not be used.

4.3 Depositional Systems and Systems Tracts

The concept of depositional episodes and the depositional-systems basin-analysis methodology were developed largely in the Gulf Coast region as a means of analyzing and interpreting the immense thicknesses of Mesozoic-Cenozoic sediment there that are so rich in oil and gas (Fig. 1.2). The principles have never been formally stated, but have been widely used, particularly by geologists of the Bureau of Economic Geology at the University of Texas (notably W.L. Fisher, L.F. Brown Jr., J.H. McGowen, W.E. Galloway and D.E. Frazier). Useful papers on the topic are those by Fisher and McGowen (1967) and Brown and Fisher (1977). Textbook discussions are given by Miall (1990; Chap. 6) and Walker (1992a).

A *depositional system* is defined as "a three-dimensional assemblage of lithofacies, genetically linked by active (modern) or inferred (ancient) processes and environments" (Posamentier et al. 1988, p. 110). A *systems tract* is defined as "a linkage of contemporaneous depositional systems.... Each is defined objectively by stratal geometries at bounding surfaces, position within the sequence, and internal parasequence stacking patterns. Each is interpreted to be associated with a specific segment of the eustatic curve (i.e., eustatic lowstand – lowstand wedge; eustatic rise – transgressive; rapid eustatic fall – lowstand fan, and so on), although not defined on the basis of this association." (Posamentier et al. 1988). Elsewhere, Van Wagoner et al. (1987) stated that, "when referring to systems tracts, the terms lowstand and highstand are not meant to imply a unique period of time or position on a cycle of eustatic or relative change in sea level. The actual time of initiation of a systems tract is interpreted to be a function of the interaction between eustasy, sediment supply, and tectonics." There is clearly an inherent, or built-in contradiction here, that results from the use in a descriptive sense of terminology that has a genetic connotation (e.g., transgressive systems tract). We return to this problem in Chapter 15.

The Exxon sequence model contains four basic systems tracts (Figs. 4.4, 4.5). Posamentier et al. (1988) and Posamentier and Vail (1988) developed block-diagram models of extensional continental margins, based on the computer simulations of Jervey (1988). Two of these are illustrated in Fig. 4.6. They show the broad character of continental-margin architecture in relationship to the surface depositional systems. Note the timing of each model with respect to the curve showing eustatic sea-level change. The details of this relationship are the subject of discussion and controversy, as described in later chapters of this book.

The *lowstand systems tract* (LST) develops on the continental slope and basin floor at times of low relative sea level. It may contain several components, including: (a) slope fans, (b) basin-floor fans, and (c) a lowstand wedge, consisting of the aggradational fill of incised valleys, and a progradational wedge which may downlap onto the basin-floor fan. The top of the lowstand tract is marked by a transgressive surface (possibly a ravinement surface; see Sect. 2.2.1), above which is the *transgressive systems tract* (TST). This may be a very thin succession of marine shales, a basin-floor gravel lag, or a condensed succession, or it may consist of retrogradational successions of shelf deposits, including marine shale and sandstone, or platform (subtidal-supratidal) carbonates. The top of the transgressive systems tract corresponds to the maximum flooding surface (MFS), and is a downlap surface, above which is the *highstand systems tract* (HST). This third tract forms the top of the stratigraphic sequence, although in some instances it may be considerably reduced in thickness as a result of erosion accompanying the next cycle of fall in base level. The highstand tract is typically aggradational to progradational, and consists of shelf to nonmarine deposits arranged in successive facies successions or parasequences. Clinoform architectures are characteristic.

Shelf-margin systems tracts may be deposited at times of slow fall in relative sea level, when (according to the Exxon models) sea level does not drop below the edge of the continental shelf. This is the main characteristic of a so-called type 2 sequence, as defined in the next section. The shelf-margin tract consists of shelf and slope clastics or carbonates arranged in aggradational or progradational geometries, resting on the sequence boundary and bounded at the top by a transgressive surface.

Recent work has demonstrated that important erosional and depositional events occur during the falling stage of the relative sea-level curve that were not recognized in the original Exxon models. This has led to the recognition of a *falling-stage systems tract* (FSST), and to disagreement regarding the placement of the sequence boundary relative to this systems tract. Discussion and illustration of these additions to the model are provided in Section 15.2.

Carbonate and clastic depositional systems respond very differently to sea-level change (Sarg 1988; James and Kendall 1992; Schlager 1992a). Figure 4.7 illustrates these differences, which are described briefly here. At times of sea-level lowstand terrigenous clastics bypass the continental shelf, leading to exposure, erosion, and the development of incised valleys. Submarine canyons are deepened, and sand-rich turbidites systems develop submarine fan complexes on the slope and the basin floor. Carbonate systems essentially shut down at times of lowstand, because the main "carbonate factory," the continental shelf, is exposed, and comonly undergoes karstification. A narrow shelf-edge belt of reefs or sand shoals may occur, while the deep-water basin is starved of sediment or possibly subjected to hyperconcentration, with the development of evaporite deposits. Evaporites may also develop on the continental shelf during episodes of sea-level fall, when reef barriers serve to block marine circulation over the shelf.

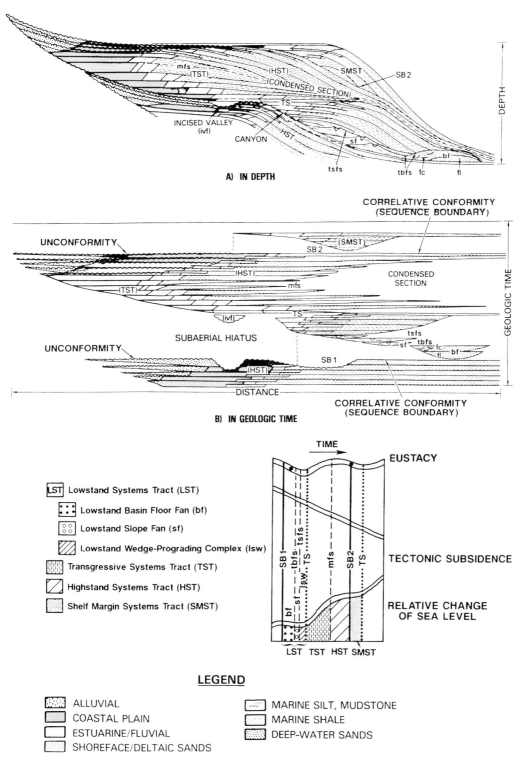

A) IN DEPTH

B) IN GEOLOGIC TIME

LST Lowstand Systems Tract (LST)

Lowstand Basin Floor Fan (bf)

Lowstand Slope Fan (sf)

Lowstand Wedge-Prograding Complex (lsw)

Transgressive Systems Tract (TST)

Highstand Systems Tract (HST)

Shelf Margin Systems Tract (SMST)

LEGEND

ALLUVIAL

COASTAL PLAIN

ESTUARINE/FLUVIAL

SHOREFACE/DELTAIC SANDS

MARINE SILT, MUDSTONE

MARINE SHALE

DEEP-WATER SANDS

Fig. 4.4A,B. Stratigraphic cross section and Wheeler diagram of a siliciclastic sequence succession, showing the subdivision into depositional systems tracts, major depositional environments, and the relationship between eustasy and subsidence. *Abbreviations* are explained in the text (Vail 1987, reproduced by permission)

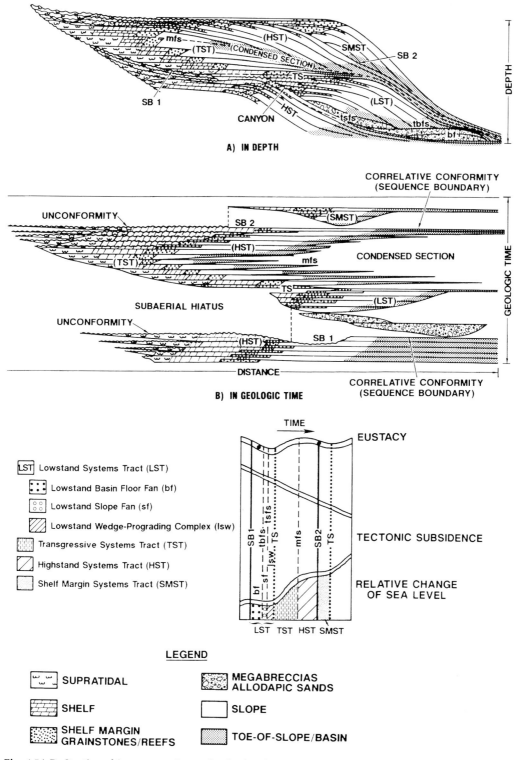

Fig. 4.5A,B. Stratigraphic cross section and Wheeler diagram of a carbonate sequence succession, showing the subdivision into depositional systems tracts, major depositional environments, and the relationship between eustasy and subsidence. *Abbreviations* are explained in the text. (Vail 1987, reproduced by permission)

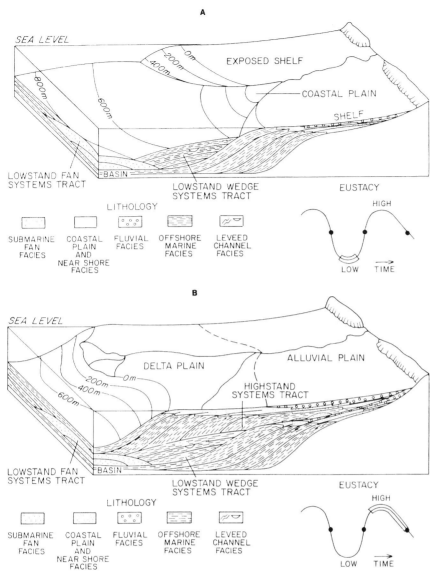

Fig. 4.6. Models of stratigraphic architecture that develop as sea level rises and falls. **A** Lowstand systems tract. **B** Highstand systems tract. Note the timing of the model with respect to the eustatic sea-level curve. (Posamentier et al. 1988)

During transgression of a clastic system incised valleys fill with estuarine deposits and eventually are blanketed with marine shale. There may be a rapid landward translation of facies belts, leaving the continental shelf starved of sediment, so that condensed successions are deposited. By contrast, transgression of a carbonate shelf serves to "turn on" the carbonate factory, with the flooding of the shelf with warm, shallow seas. Thick platform carbonate successions develop, reefs, in particular, being able to grow vertically at extremely rapid rates as sea level rises. At times of maximum transgression the deepest part of the shelf may pass below the photic zone, leading to a cessation of carbonate sedimentation and the development of a condensed section or hardground. The resulting surface is termed a *drowning unconformity*.

Carbonate and clastic shelves are most alike during times of highstand. The rate of addition of sedimentary accommodation space is low, and lateral progradation is therefore encouraged, with the develoment of clinoform slope architectures. Autocyclic shoaling-upward cycles are common in both types of environment (e.g., terrigenous deltaic lobes, tidal carbonate cycles). Schlager (1992a) stated:

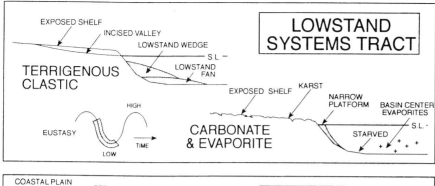

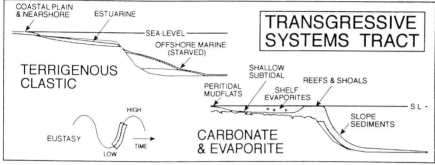

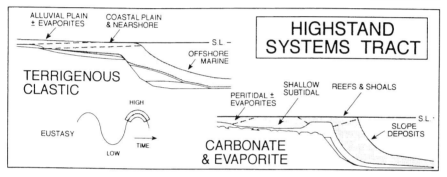

Fig. 4.7. Contrasts in sedimentary response to sea-level change between terrigenous clastic and carbonate-evaporite shelf-slope systems. (James and Kendall 1992)

Prograding [carbonate] margins dominated by offshore sediment transport most closely resemble the classical [siliciclastic] sequence model. They are controlled by loose sediment accumulation and approach the geometry of siliciclastic systems (e.g., leeward margins ...).

Carbonate platforms produce and export most sediment to the continental slope during highstands of sea level, when platforms are flooded and the carbonate factory is at maximum productivity. This process, termed *highstand shedding* is exemplified by the architecture of the Bahamas Platform (Schlager 1992a). It is the converse of the pattern of siliciclastic sedimentation, within which, as already noted, sediment is fed to the continental slope most rapidly at times of low sea level.

It should be noted that the differences between carbonates and clastics discussed above were not incorporated into the original Exxon sequences models for siliciclastic and carbonate depositional systems (Figs. 4.4, 4.5), which appear almost identical. Some alternative models and recent developments are discussed in Chapter 15.

Many ancient shelf deposits are mixed carbonate-clastic successions, containing thin sand banks or deltaic sand sheets and carbonate banks. Galloway and Brown (1973) described an example from the Pennsylvanian of northern central Texas, in which a deltaic system prograded onto a stable carbonate shelf (Fig. 4.8), and other examples are given by Cant (1992), who noted the use of the term *reciprocal sedimentation* for depositional systems in which carbonates and clastics alternate. In the example given in Fig. 4.8, deltaic distributary channels are incised into the underlying shelf carbonate de-

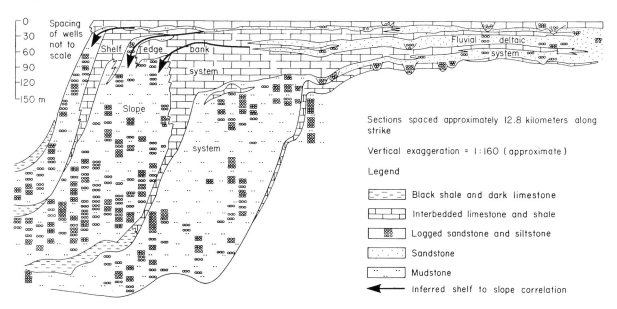

Fig. 4.8. A mixed carbonate-clastic shelf-slope depositional systems tract, formed as a result of alternating periods of high and low sea level. (Galloway and Brown 1973)

posits. Widespread shelf limestones alternate with the clastic sheets and also occur in some interdeltaic embayments. Carbonate banks occur on the outer shelf edge, beyond which the sediments thicken dramatically into a clinoform slope clastics system. This association of carbonates and clastics reflects regular changes in sea level, with the carbonate phase representing high sea level and the clastic phase low sea level. The deltaic and shelf-sand sheets and the slope clinoform deposits represent lowstand systems tracts, while the carbonate deposits are highstand deposits. During episodes of high sea level, clastics are trapped in nearshore deltas, while during lowstands much of the detritus bypasses the shelf and is deposited on the slope (arrows in Fig. 4.8). Additional examples of reciprocal carbonate-clastic sedimentation and sequence development were given by Dolan (1989) and Southgate et al. (1993).

Changes in climate may not have received the attention they deserve as possible controls on sequence style. The nature of the sediment supply, especially the balance between carbonate and clastic sedimentation, is very much dependent on such factors as temperature and rainfall. Ruffell and Rawson (1994) provided examples of this, and documented the effects of climate change over intervals of millions of years in the Jurassic-Cretaceous record of the North Sea basin.

4.4 Sequence Boundaries

In the Exxon models sequence boundaries (commonly abbreviated as SB on diagrams) are drawn at the unconformity surfaces, an approach which readily permits the sequence framework to be incorporated into an allostratigraphic terminology. Vail and Todd (1981) recognized three types of unconformity, but later work (e.g., Van Wagoner et al. 1987) simplified this into two (Fig. 4.9). These are differentiated from each other on the basis of the extent of subaerial erosion and the amount of seaward shift of facies belts.

A *type 1 unconformity* develops where sea-level fall is rapid, more rapid that tectonic subsidence. The coastline may move out to near the shelf edge, and extensive subaerial erosion takes place, with the development of incised fluvial valleys on the shelf, and the deepening of submarine canyons on the continental slope. Clastic detritus is transported down these fluvial and canyon systems to the base of the continental slope, forming extensive lowstand systems tracts. At type 1 unconformities facies belts undergo a substantial basinward shift. Highstand deposits below the unconformity may be deeply eroded. In carbonate systems exposure of the platform may lead to the development of widespread karst systems and internal dissolution, with considerable erosion of the platform margin and downslope movement of carbonate breccias and turbidites (Sarg 1988).

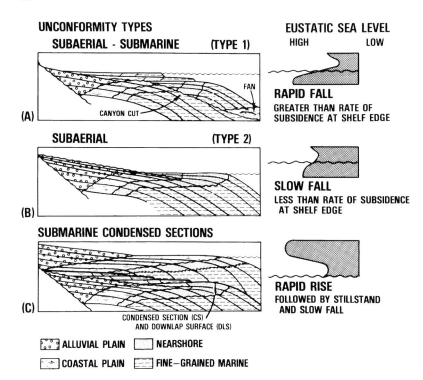

Fig. 4.9A–C. Definition of type 1 and type 2 unconformities. (Vail et al. 1984, reproduced by permission)

A *type 2 unconformity* develops when relative sea level falls slowly, resulting in a gradual seaward shift of facies tracts, but only minor subaerial exposure and erosion. According to Vail et al. (1987, 1991) a shelf-margin systems tract is formed under these conditions. Type 2 unconformities are much more difficult to indentify in seismic and outcrop records because they are not characterized by deep erosion or major facies shifts.

Vail et al. (1991) subdivided sequences into type 1 sequences, which are bounded at the base by a type 1 unconformity and are composed of lowstand, transgressive, and highstand systems tracts, and type 2 sequences, which are bounded at the base by a type 2 unconformity and consist of shelf-margin, transgressive, and highstand systems tracts. This terminology has not received widespread acceptance.

It is important to note that although sequence boundaries develop at the times of lowest *relative* sea level, these may not correspond to the times when *eustatic* sea level is actually at its lowest. This can be seen in Figs. 4.4 and 4.5, in which tectonic subsidence and eustasy are integrated to show an overall curve of relative change in sea level. Sequence boundaries typically represent the time of maximum rate of fall in sea level. Where the eustatic curve approximates a sine curve, as is indicated in most of the Exxon diagrams, the sequence boundary corresponds to the inflection point in the falling curve. After this time, according to the Exxon model, tectonic subsidence outpaces eustasy, and relative sea level rises. This point assumes considerable importance when attempts are made to use sequence boundaries to date eustatic sea-level change and derive global cycle charts. The offset in time between the eustasy inflection point and the time of lowest relative sea level varies, depending on the rate of sea-level change and the rates of tectonic subsidence and sediment supply. These rates both vary widely, depending on the nature of tectonic and eustatic controls, as discussed in later chapters. Generalizations, such as those based on Jervey's (1988) models, are an inadequate basis for developing global cycle charts, because they are not quantitative, and do not take these actual rate variations into account (see Part III).

Care must be taken to map and interpret unconformities correctly. As discussed in Section 2.2.1, not all breaks in sedimentation are caused by changes in sea level, and some erosional breaks are time transgressive. For example, ravinement surfaces are commonly more prominent in a stratigraphic section than subaerial unconformities, yet they form during continuous transgression, and are time-transgressive (Fig. 2.4).

4.5 Other Sequence Concepts

An alternative sequence model, termed the genetic stratigraphic sequence, was defined by Galloway (1989a), building on the work of Frazier (1974; see Fig. 1.2 of this book). Although Galloway stressed supposed philosophical differences between his model and the Exxon model, in practice the difference between them is simply one of where to define the sequence boundaries. The Exxon model places emphasis on unconformities, but Galloway (1989a) pointed out that under some circumstances unconformities may be poorly defined or absent and, in any case, are not always easy to recognize and map. This is particularly likely to be so in the case of type 2 unconformities, and in particular within the marginal-marine depositional systems, where

sedimentation may be virtually continuous across the sequence boundary.

Galloway's (1989a) preference is to draw the sequence boundaries at the maximum flooding surface, which corresponds to the highstand downlap surfaces. He claims that these surfaces are more prominent in the stratigraphic record, and therefore more readily mappable. Differences in sequence definition between the two types of model are illustrated in Fig. 4.10. Galloway's proposal has not met with general acceptance. For example, Walker (1992a) disputed one of Galloway's main contentions, that "because shelf deposits are derived from reworked transgressed or contemporary retrogradational deposits, their distribution commonly reflects the paleogeography of the precursor depositional episode." Galloway (1989a) went on to state that "these deposits are best included in and

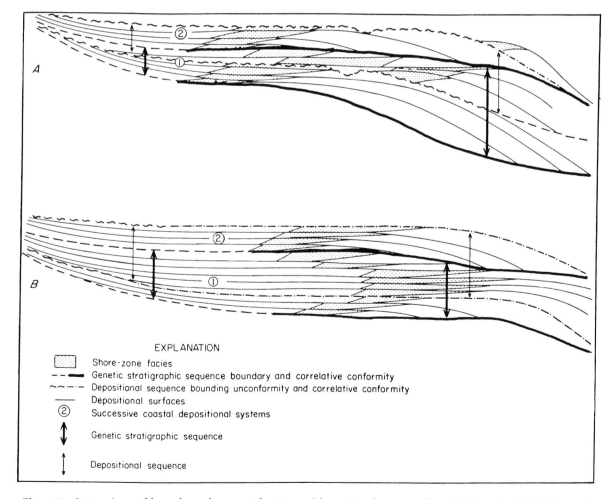

EXPLANATION

▢ Shore-zone facies
– – ━ Genetic stratigraphic sequence boundary and correlative conformity
∼∼–∼– Depositional sequence bounding unconformity and correlative conformity
───── Depositional surfaces
② Successive coastal depositional systems

↕ Genetic stratigraphic sequence

↕ Depositional sequence

Fig. 4.10. Comparison of boundary placement for Exxon "depositional sequence" and Galloway's (1989a) "genetic stratigraphic sequence." **A** Type 1 sequence. **B** Type 2 sequence. (Galloway 1989a, reproduced by permission)

A. PASSIVE MARGIN MODEL

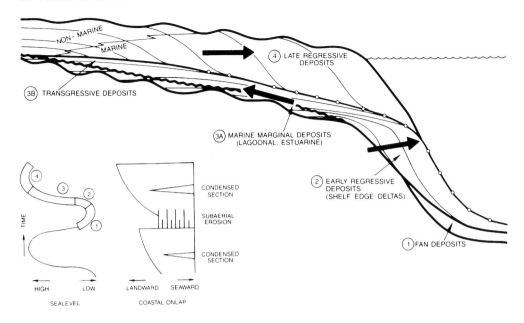

B. FORELAND BASIN MODEL

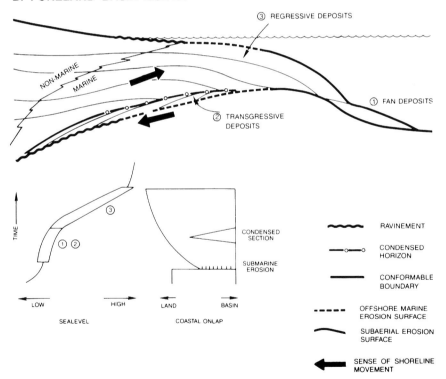

Fig. 4.11. Comparison of sequence architecture on extensional continental margins and in foreland basins. (Swift et al. 1987)

mapped as a facies element of the underlying genetic stratigraphic sequence." However, as Walker (1992a) pointed out, most sedimentological parameters, including depth of water, waves, tides, basin geometry, salinity, rates of sediment suply, and grain size, change when an unconformity or a maximum flooding surface is crossed. From the point of view of genetic linkage therefore the only sedimentologically related packages lie: (a) between an unconformity and a maximum flooding surface, (b) between a maximum flooding surface and the next younger unconformity, or (c) Between a subaerial erosion surface and the overlying unconformity (an incised-valley-fill; Walker 1992a, p. 11).

However, some workers have found Galloway's use of the maximum flooding surface much more convenient for sequence mapping, for a variety of practical reasons. For example, it may yield a prominent gamma-ray spike in wireline logs (Underhill and Partington 1993), or it may correspond to widespread and distinctive goniatite bands (Martinsen 1993), or it may provide a more readily traceable marker, in contrast to the surface at the base of the lowstand systems tract, which may have irregular topography and may be hard to distinguish from other channel-scour surfaces (Gibling and Bird 1994). In nonmarine sections it may be hard to find the paleosol on interfluves that correlates with the sequence-bounding channel-scour surface (Martinesen 1993). In some studies (e.g., Plint et al. 1986; Bhattacharya 1993) it has been found that ravinement erosion during transgression has removed the transgressive systems tract, so that the marine flooding surface coincides with the sequence boundary.

Embry (1993) defined sequences in shelf and slope successions on the basis of their contained transgressive surfaces, and referred to them as T-R or "transgressive-regressive" sequences. This definition of sequences differs from some useages of the Exxon models, as described in Section 15.2.2.

A more important departure from the Exxon model is that of Swift et al. (1987), who pointed out the occurrence of a different type of architectural pattern in foreland basins. Here, the relationship between subsidence and sediment supply is reversed from that prevailing in extensional continental-margin basins. In the latter, sediment enters the basin near the tectonic hinge, and is transported toward the more rapidly subsiding side of the basin. Sequences in foreland basins are *clastic wedges*, as defined by Sloss (1962). "They are sediment prisms poured into the deeper side of a linear half-basin undergoing subsidence along its *landward* margin. Hence it is the proximal (landward) end of the wedge that is most completely preserved." (Swift et al. 1987, p. 449). Accommodation space is generated more slowly in the deeper part of the basin because of slower subsidence there, and so stratigraphic units tend to thin and pinch out toward the center of the foreland basin, whereas many units thicken toward the basin center in the case of extensional continental margins. The major architectural differences between the two types of basin setting are illustrated in Fig. 4.11.

5 The Global Cycle Chart

The major thesis of the Vail et al. (1977) work is that cycles of sea-level change, such as those illustrated in Figs. 2.9 and 2.11, can be correlated around the world, indicating that they are not a response to local tectonic events but the result of global or eustatic sea-level changes. It was stated that much evidence had been amassed to demonstrate this idea, but the only evidence actually offered was a correlation between four Cretaceous to Recent basins in a chart reproduced here as Fig. 5.1. A series of major and minor sea-level changes had been correlated among these basins, producing what is, at first sight, a convincing proof of the eustatic sea-level model. However, this evidence seems suspect. In Fig. 5.1, correlation lines are all based on interpreted sudden sea-level falls. The questionable validity of this interpretation of asymmetric sea-level rises and falls is discussed in Section 2.2.3. Vail and Todd (1981) presented a modified version of the interpretive methodology in which it is assumed that coastal progradation and slow sea-level fall begin during the latter part of each sequence, with an unconformity developing simultaneously around the world at the end of each sequence (Fig. 2.11). It can be argued that this purported relationship between sea-level change and basin architecture is overly simplistic (see Chap. 15), that simultaneous

development of unconformities is in fact unlikely (see Part III), and that, in any case, the data to support the interpretations have never been published, so that independant testing and confirmation of the concepts is not possible (Miall 1986, 1991a, 1992).

Another important point about Fig. 5.1 is that it is rare in geology for the evidence to permit such precise correlation of events that they can be indicated by the straight lines encompassing the globe shown in Fig. 5.1 (Miall 1986). There is still much disagreement over the exact age of magnetic anomalies, and the absolute (numerical) age of biostratigraphic zones. Even in the case of Cenozoic stratigraphy, the determination of chronostratigraphic age is characterized by significant observational and experimental error, and this increases with the age of the sediments. (these problems are discussed in Chap. 13). Yet there is no indication of these possible errors in Fig. 5.1. If such data are not provided, it is difficult to allow for revisions and refinements.

One of the most serious problems that has arisen as a result of the introduction of the Exxon method is the tendency to regard the Vail curves as some kind of approved global chronostratigraphic standard (Miall 1986). Vail and his coworkers have

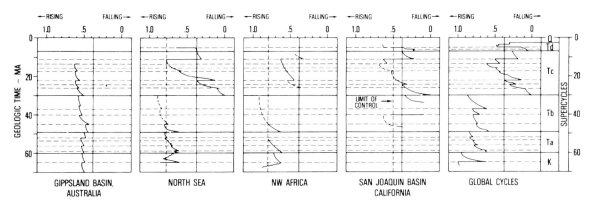

Fig. 5.1. Correlation of cycles of relative change in coastal onlap from four continents, showing how these have been averaged to produce the global cycle chart. (Vail et al. 1977, reproduced by permission)

MAGNETO-CHRONO-STRATIGRAPHY

TIME IN M. YEARS | MAGNETIC ANOMALIES | POLARITY | POLARITY CHRONOZONES

STANDARD CHRONO-STRATIGRAPHY

SYSTEM | SERIES | STAGES

BIOCHRONOSTRATIGRAPHY

① PLANKTONIC FORAM AND CALPIONELLID BIOCHRONO-ZONES
② CALCAREOUS NANNOFOSSIL BIOCHRONO-ZONES
③ MACROFOSSIL BIOCHRONO ZONES (BOREAL) GREAT BRITAIN
④ AMMONOID BIOCHRONO-ZONES (TETHYAN REGION)
⑤ DINO FLAGELLATE BIOHORIZONS

Time	Magnetic anomaly	Polarity chronozone	System	Series	Stage
65	29	C29		PALEOCENE	DANIAN
	30	C30			
70	31	C31			MAASTRICHTIAN
	32	C32			
75	33	C33	CRETACEOUS	UPPER	CAMPANIAN
80					
85	34				SANTONIAN
					CONIACIAN
90		C34			TURONIAN
95					CENOMANIAN
100					ALBIAN
105				LOWER	
110	M0	CM0			APTIAN
115	M1–M3	CM1–CM3			BARREMIAN
	M4–M10	CM4–CM10			HAUTERIVIAN
120	M10N	CM10N			
125	M11–M12	CM11–CM12			VALANGINIAN
130	M13–M16	CM13–CM16			BERRIASIAN / RYAZANIAN
135	M17–M21	CM17–CM21	JURASSIC	UPPER	PORTLANDIAN / TITHONIAN / KIMMERIDGIAN
140					

Numeric boundaries: 66.5, 74, 84, 88, 89, 92, 96, 108, 113, 116.5, 121, 128, 131, 136

SERIES subdivisions: SENONIAN, NEOCOMIAN, VOLGIAN; UPPER, MIDDLE, LOWER

LEGEND

■ NORMAL POLARITY
| REVERSED POLARITY
{ SHORT REVERSED POLARITY INTERVAL

(CRETACEOUS MAGNETIC QUIET ZONE)

(b) HORIZONS
FIRST OCCURRENCE ↳
↱ LAST OCCURRENCE

① AFTER ALLEMAN ET AL (1971); PREMOLI SILVA AND BOLLI (1973); VAN HINTE (1976); PREMOLI-SILVA AND BOERSMA (1977); REMANE (1978); ROBASZYNSKI ET AL (1979,1983); AND CARON (1985)

② AFTER THIERSTEIN (1976); SISSINGH (1977); MANIVIT ET AL (1977); ROTH (1978,1983) AND MONECHI AND THIERSTEIN (1985)

③ AFTER RAWSON ET AL (1978) AND KENNEDY (1984)

④ AFTER VARIOUS AUTHORS IN CAVALIER AND ROGERS (1980); AMEDRO (1980,1981,1984); ROBASZYNSKI ET AL (1983); KENNEDY (1984); AND CLAVEL (1986)

⑤ (COMPILED BY Y Y CHEN N JOANNIDES M MILLIOUD J SHANE L STOVER)

VERSION 3.1A
(January 1987)

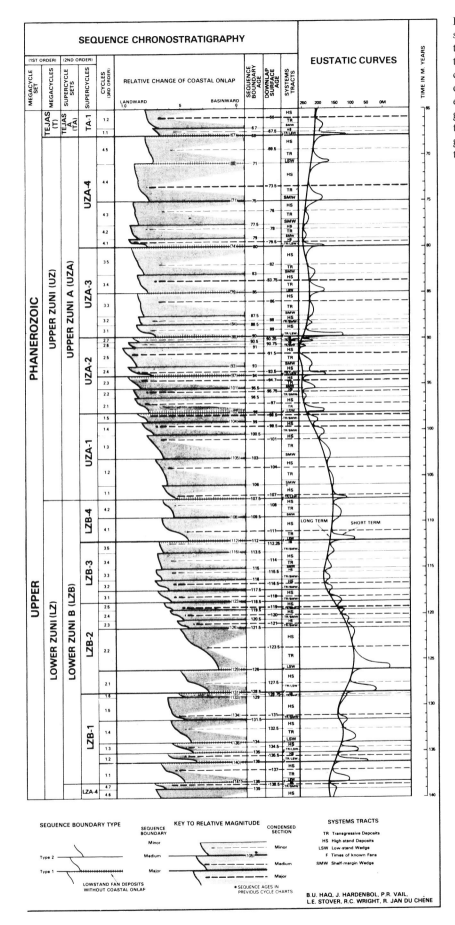

Fig. 5.2. Example of the new sea-level charts developed by the Exxon group. This illustration shows the Cretaceous chart of changes on coastal onlap and the corresponding eustatic sea-level curve, together with some of the details of the chronostratigraphic basis for the correlations. (Haq et al. 1988a)

encouraged this trend and even favor seismic correlation over biostratigraphic correlation when the latter does not appear to fit their models (a series of quotations documenting this philosophy is provided in Sect. 13.2). The global cycle chart has become the primary reference frame against which new data are judged. The danger of circular reasoning in this method are self-evident, and the approach clearly does not lend itself to independent tests of the cycle chart, nor to a search for local or regional anomalies that might have major implications for local tectonic or climatic episodes (Miall 1986). This point is made at greater length in Chapter 13.

It does not help that little of the supporting data have been published. This is a criticism of the Exxon work that has been made several times (e.g., Miall 1986), and Sloss (1988a), among others, has lept to the defense of his former students, pointing out:

[The] common complaint concerning the lack of supporting data and, indeed, a large measure of faith was required of the reader. Complaining parties failed to recognize that the 1977 curves represent two decades of analysis of data, much of it proprietary, including thousands of kilometers of seismic lines, hundreds of subsurface records, and untold man-hours of biostratigraphic work. Further curves showing a higher level of detail in the Cretaceous were not released for publication, and this omission added to the malaise of an ungrateful segment of the public. Many of the deficiencies of the 1977 product have now been corrected in subsequent publications; progress continues with activity now spread over a broad spectrum of industrial, academic, and governmental agencies.

Global correlations such as that given in Fig. 5.1 permitted Vail et al. (1977) to construct charts showing relative sea-level changes that, because of their global correlatability, were interpreted as true, eustatic changes. A detailed chart showing sea-level change relative to magnetic reversal and biostratigraphic events was constructed for the Cenozoic, a less detailed chart for the Late Triassic to Recent, and a generalized chart for the entire Phanerozoic. In these charts the rises and falls for individual regions were averaged to produce a global curve, and this is given in terms of a relative change, because it differs in amount from region to region depending on local tectonic events. The method of global averaging is obscure. These diagrams have become among the most widely reproduced illustrations in the history of geology, because they purport to provide a key as important as that of plate tectonics for understanding worldwide stratigraphic patterns.

The pre-Jurassic cycles were constructed mainly from North American data (Vail et al. 1977, p. 88). For rocks of this age range, correlations with oceanic events are not available (the oldest oceanic crust is probably Middle Jurassic), and so Vail and his coworkers presumably used much the same data as did Sloss, namely, subsurface information from the continental interior. It is therefore hardly surprising that the first four of Sloss's (1963) sequences (Fig. 1.1) are almost identical to the corresponding supercycles of Vail et al. (1977). Younger supercycles differ considerably from those of Sloss because of the present-day availability of a wholly new and more detailed data base, offshore marine seismic and well records.

Sloss (1988a) himself, though supporting the Exxon work and accepting most of its results, disputed the placement of several of the major sequence boundaries in the Exxon global cycle charts. Part of the problem reflects the difficulty in assessing how the relative importance of the sequence boundaries has been determined. The method of averaging the local curves to produce the global chart has not been explained in detail. In some cases an ad hoc approach seems to have been employed. For example, Haq (1991) stated that, "the long-term changes in relative magnitude are estimated using the method described in Hardenbol et al. (1981) with the Turonian high value adopted from Harrison (1986)" (Harrison's paper actually appeared in 1990 following a long delay in publication). Without a rigorous, quantitative method of curve conflation the relative magnitudes of the sea-level excursions cannot be assigned much significance, and therefore it follows that the assignment of certain sequence boundaries in the Exxon curves as "supercycle" boundaries carries little weight.

Updated versions of the Vail curve have been published in several books and papers (Haq et al. 1987, 1988a; Haq 1991). These new curves (e.g., Fig. 5.2) purport to incorporate a large volume of outcrop data, including sections in "western Europe and the trans-Tethys region, the United States Gulf and Atlantic coasts and the Western Interior Seaway, New Zealand and Australia, Pakistan, and Arctic islands of Bjørnoya and Svalbard" (as claimed by Haq et al. 1988a) reflecting work carried out by Exxon workers and their colleagues in academic and government institutions during the 1980s (Haq 1991). Lists of these sections have been published (Haq et al. 1988a), but not the data derived from them, and so the criticism of the lack of published documentation remains. A few location-specific studies have now been made available by the Exxon group (Donovan et al. 1988; Baum and Vail 1988), but these barely begin to tackle the problem of inadequate documentation. Haq et al. (1988a) discussed in some detail their methods for developing a numerical time scale for the global cycle chart, pointing out the possible sources of error in radiometric, magnetostratigraphic and biostratigraphic methods. However, they did not incorporate these sources of error into the chart by providing error bands for the sea-level curve. The implications of this omission are discussed in Part IV.

The new charts include references to various aspects of the sequence-stratigraphic models that have been developed by the Exxon group (Posamentier et al. 1988; Posamentier and Vail 1988), such as the presence of various systems tracts, condensed sections, and types of unconformity. These aspects are defined and discussed briefly in Chapter 4. The presence of these details on the global cycle charts implies that the particular sequence attributes indicated have a global extent. This is unlikely. For example, depositional styles, and rates of subsidence and sedimentation (which determine the type of unconformity) are controlled in part by regional tectonics, and show wide variation from basin to basin. These points are discussed in Part III.

As noted by Miall (1992), the basic premise of the Exxon work is that there exists a globally correlatable suite of eustatic cycles, and that all field stratigraphic data may be interpreted in keeping with this concept. However, the basic premise remains unproven because of the lack of published documentation. All so-called tests of the Vail curve are suspect unless they provide a rigorous treatment of potential error and a discussion of alternative interpretations. In fact, this is almost never done. Some examples of tests of the Vail curve are described in detail in Chapter 13, and comparisons with other sea-level curves are presented in Chapter 14.

II The Stratigraphic Framework

A vast amount of stratigraphic data has been amassed since AAPG Memoir 26 was published in 1977, setting off the current high level of interest in sequence stratigraphy. Some of this consists of regional seismic surveys, but a great deal of work has also been carried out with well records and outcrops. The purpose of this section of the book is to present some of the basic stratigraphic data, with a minimum of interpretation, as a basis for the discussion of mechanisms and controls that constitutes much of the remainder of the book. The wealth of data now available has demonstrated that the original concept of a simple rank ordering of sequences, from first- to fifth- (and higher?) order, is too simplistic. Although sequence generating mechanisms have natural episodicities or periodicities, these overlap, and do not support the use of this simple classification (Table 3.1). A subdivision of sequences according to episodicity is used for discriptive purposes in this part of the book; it forms the basis for the subdivision into three chapters; but no genetic interpretations are implied by this subdivision.

Most of the current interest and controversy regarding sequence stratigraphy focuses on sequences of a few million years duration, and less. For this reason considerably more space is devoted to these cycles in this part of the book (Chaps. 7 and 8).

It has been necessary to make a judicious selection of examples from the wealth of data now available in the published record. The reader is urged to turn to the original publications wherever possible for the details of regional setting, stratigraphy, facies, etc.

- Sequential ordering of beds is found in nearly all stratigraphic successions on various scales. It is thought to be the result from the combination of regional to global causal factors and modifying local environmental processes. (Einsele and Ricken 1991, p. 611)

- The Exxon group's distinction of first-, second- and third-order eustatic cycles is useful. The first-order cycles are the two Phanerozoic supercycles of Fischer (1984), and general agreement exists that the changes in sea level are genuinely global. A fair measure of consensus also exists that the second-order cycles (10–18 Myr duration) have a eustatic origin as well, but the third-order cycles (1–10 Myr) are more controversial. Many stratigraphic experts independent of the Exxon group are, however, persuaded of their reality, on the basis of correlation over extensive regions.... Besides these larger cycles, some would distinguish smaller, fourth- and fifth-order cycles, usually no more than a few meters thick and signifying durations of tens to hundreds of thousands of years. One currently popular idea is that Late Paleozoic and other comparable cyclothems are glacioeustatic phenomena under the ultimate control of an orbital forcing mechanism that affects global climate. (Hallam 1992, p. 204)

- [T]he discrimination of stratigraphic hierarchies and their designation as nth-order cycles may constitute little more than the arbitrary subdivision of an uninterrupted stratigraphic continuum. (Drummond and Wilkinson 1996, p. 1)

6 Cycles with Episodicities of Tens to Hundreds of Millions of Years

6.1 Climate, Sedimentation, and Biogenesis

It is becoming increasingly clear that global climate, oceanic circulation, sedimentation patterns, biotic diversity, and evolutionary trends are all linked to variations in sea level (Fischer 1984; Worsley and Nance 1989; Worsley et al. 1984, 1986, 1991; Veevers 1990). These linkages can be traced through the long-term cycle of supercontinent assembly and dispersal (Fig. 6.1), and many of the same changes can also be observed on shorter time scales. For example, certain trends in sedimentation patterns are related to the position and direction of change in sea level relative to continental margins, and there is some evidence (touched on below) for such variations occurring with episodicities of millions to tens of millions of years.

During supercontinent assembly and dispersal individual continental plates may undergo latitudinal drift, which carries them through various climate belts. This leads to long-term changes in climate and consequent changes in sedimentary styles, of the type described by Cecil (1990) and Perlmutter and Matthews (1990). This is discussed further in Chapter 10.

Flooding of the continents during the fragmentation phase of the supercontinent cycle causes widespread dispersion of marine clastic sediments, followed by chemical sedimentation as the supply of clastics from thermally subsiding continental fragments decreases. Evaporites may be abundant in the newly formed rifted basins. Clastic sedimentation on the continents decreases to a minimum during the maximum-dispersal phase. At times of continental assembly large volumes of terrigenous detritus are produced by erosion of newly emergent orogens, and are deposited as clastic wedges in inland basins (e.g., in foredeeps) and along continental margins. During the stasis phase the presence of large, mountain-rimmed continental-interior basins may lead to climatic extremes in the continental interiors, and continental sedimentation may be dominated by eolian or glacial facies.

The depositional systems tracts described in Chapter 4 are dependent on sea level and its rise and fall over the long and short term. Firstly, erosional and depositional events in submarine canyons and fans are markedly affected by sea level. During highstands of the sea, continental detritus tends to be trapped along the shorelines in deltas and coastal plain complexes (Shanmugam and Moiola 1982). G. deV. Klein (personal communication 1982) suggested that the major tidalite sequences of the Phanerozoic correspond to periods of high sea level, when shelf widths and tidal effects were at a maximum.

Shelf progradation rates are high at times of high sea level, but the deep ocean tends to be starved. Clinoform shelf-slope architecture and offlapping sequences result. Conversely, during periods of low sea level, canyon erosion is active and much detritus is fed directly to the submarine canyons at their mouths. The result is deep-marine onlap of thick fan sequences. Because of the increased sediment supply to the deep oceans and increased thermohaline circulation (see below) during periods of low sea level, contourites are also likely to be more common at these times. Shanmugam and Moiola (1982) attempted to document the major occurrences of thick submarine fan and contourite deposits in the stratigraphic record, and they claimed that most correspond to periods of minimum sea level on the Vail et al. (1977) curves. However, in view of the controversies surrounding these curves this topic needs to be reexamined.

Several recent studies have suggested that many physical, chemical, and biological events in the continents and oceans are correlated with each other and that they change cyclically over periods of 10^6–10^7 a. Funnell (1981) termed this autocorrelation. The subject was explored in depth by Fischer and Arthur (1977), who studied the Mesozoic and Cenozoic record, and later by Leggett et al. (1981),

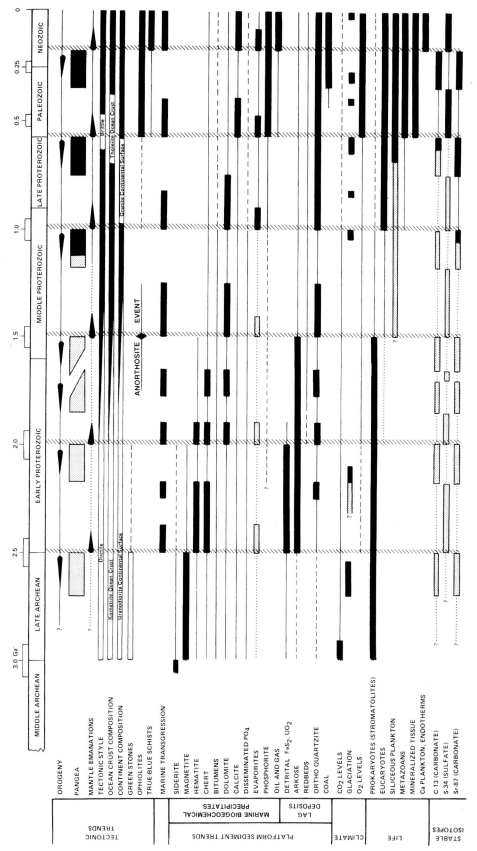

Fig. 6.1. Summary of Late Late Archean to Recent trends in tectonism, platform sedimentation, climate, life, and marine platform stable isotopes. Trends are shown as follows: abundant, intense or heavy: *solid bars*; common or moderate: *dashed or solid lines* (*dotted* where speculative). Proposed supercontinent fragmentation events are indicated by vertical *hatched bars*. (Worsley et al. 1986)

who carried out a similar analysis for the Early Paleozoic. Worsley et al. (1986, 1991) and Worsley and Nance (1989) compiled more recent data and ideas. It was suggested that the main control on autocorrelation is sea level. During periods of high sea level, the world's oceans tend to be warm, with much-reduced latitudinal and vertical temperature gradients. Elevated CO_2 levels are attributed to high rates of volcanic outgassing. Although the area of exposed continent is relatively low, the rate of release of calcium and phosphorus by weathering remains high because of elevated surface temperatures, so that the rate of burial of CO_2 as carbonate remains constant, and the CO_2 content of the atmosphere rises, increasing the climatic greenhouse effect. Oceanic circulation is relatively sluggish, leading to increased stratification and severe oxygen depletion at depth. Deposition of organic-rich sediments in the deep ocean becomes widespread, the carbonate compensation depth rises and faunal diversity increases. These are the times of global anoxic events when widespread black shales are deposited around the world. Increased nutrient supply caused by high rates of midocean volcanism encourages planktonic growth, leading to high general rates of biotic activity, and this may also, in part, account for the development of an unusual number of prolific oil source beds during the Cretaceous (Larson 1991).

During periods of low sea level, global climates are more variable, and the oceans are, in general, cooler and better oxygenated because of better circulation. The CO_2 content of the atmosphere is reduced by greater combination with calcium and phosphorus and burial as carbonates and phosphates, reflecting increased supply of these elements from continental weathering. The climatic greenhouse effect is therefore reduced. Many faunal niches in shelf regions are destroyed by subaerial exposure. Submarine erosion may be intensified. Initiation of these episodes of lower sea level may be a cause of biotic crises, in which faunal diversity is sharply reduced by major extinctions (Newell 1967).

Fischer and Arthur (1977) encapsulated these faunal variations by applying the terms *polytaxic* and *oligotaxic* to periods of, respectively, high and low faunal diversity. The corresponding climatic milieus are referred to as *greenhouse* type when sea levels are high and climates are globally uniform and relatively warm, and *icehouse* type when variable climates accompany times of low sea level. Fischer and Arthur (1977) and Fischer (1981) demonstrated that since the Triassic the oceans

have fluctuated between these two modes during an approximately 32-m.y.-long cycle. These cycles correlate reasonably well with the Sloss-type cycles discussed in Section 3.3, although there does not seem to be much support for the idea of a precise 32-m.y. regularity to the cycles. The climate changes are crudely cumulative, and can be correlated in general terms with the supercontinent cycles of sea-level change, as shown in Fig. 3.3. Major glaciations occurred during the periods of icehouse climate in the Late Precambrian, Late Devonian-Permian, and Late Cenozoic, the first two, at least, corresponding to times of supercontinent assembly when markedly continental climates were to be expected within the very large landmasses. A shorter glacial episode that occurred in the Late Ordovician to Early Silurian does not fit this pattern, but corresponds to a time when North Africa lay over the South Pole.

Carbonate sedimentation trends reflect the long-term oscillations from icehouse to greenhouse climates. Carbonate production is increased during times of high global sea level and greenhouse climate because of the greater extent of shallow shelf seas. A compilation by James (1983) showed that reefs tend to be more abundant at times of high global sea level (those plotted in Fig. 3.3). Lumsden (1985) found that the distribution of dolomite in deep-sea sediments follows the same trend, again, because of the greater production of carbonate sediment when areas of shelf seas are high. Sandberg (1983) suggested that low-Mg calcite is the dominant calcium carbonate mineral precipitated during times of high sea level and greenhouse climates, such as during the Ordovician-Devonian and Jurassic-Cretaceous, whereas high-magnesium calcite and aragonite are dominant at times of low sea level and icehouse climates. Mackenzie and Pigott (1981) and Worsley et al. (1986) documented parallel changes in carbon and oxygen distribution and isotopic composition (Fig. 6.2).

6.2 The Supercontinent Cycle

6.2.1 The Tectonic-Stratigraphic Model

Evidence for two supercontinent cycles in the Phanerozoic, and several more in the Precambrian was assembled by Worsley et al. (1984, 1986), who synthesized a vast and diverse data base. The major elements of their model are illustrated in Fig. 6.2, and are discussed below.

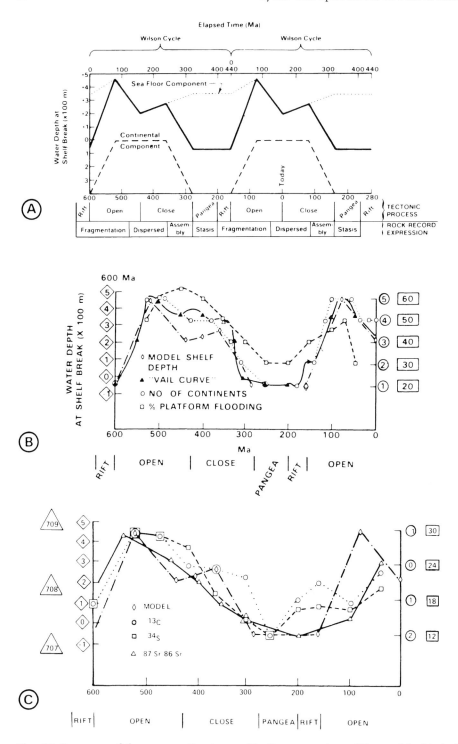

Fig. 6.2. Summary of the supercontinent assembly–fragmentation cycle, and its effects on the Phanerozoic record. **A** Tectonic components of the model, showing two complete assembly–fragmentation cycles. **B** Comparison of the model with the long-term eustatic curve of Vail et al. (1977), platform flooding, and number of continents. **C** Trends in stable isotopes. (Worsley et al. 1986)

The cyclic model has four tectonic phases: the *fragmentation phase* is the interval during which the supercontinent undergoes rifting and dispersion, over a period of approximately 160 m.y. This is a period of active generation of oceanic crust. The average age of the crust therefore decreases. As demonstrated in Chapter 9, this is a cause of rising global sea level, because of the buoyancy of young oceanic crust and the presence of long, active, thermally expanded, sea-floor spreading centers. Orogeny and the emplacement of felsic magma are at a minimum during this period, whereas continental mafic magmatism is at a peak, with the emplacement of mafic dyke swarms and mafic volcanism accompanying the formation of rift systems.

During the *maximum dispersal phase* mature, extensional continental margins face wide, old oceans, as in the case of the Atlantic Ocean and its bordering continents at the present day. The world oceanic crust reaches its maximum average age about 200 m.y. after initial rifting. Atlantic-type oceans may begin to subduct, at which time the presence of old, cold oceanic crust entering the subduction zone leads to subduction-hinge rollback, and the development of backarc spreading. Continental magmatism is at a minimum during this period, and global heat flow is also at a minimum. Sea levels fall during this phase.

The *assembly phase* closes old Atlantic-type oceans, so that the average age of the oceanic crust increases and global heat flow increases to intermediate values. Convergent tectonism and its accompanying felsic magmatism increase globally, resulting in increased continental relief, decreased continental area, and a corresponding increase in ocean-basin volume, with resulting low global sea levels. Terrane-collision events are abundant within the subducting oceans, while backarc spreading occurs along the trailing edges of the converging continents.

The *supercontinent stasis phase* is characterized by epeirogenic uplift of the new supercontinent, as heat builds up beneath it. Oceanic crust is at an intermediate age, and the combination of elevated continental crust and intermediate oceanic depths leads to the phase of maximum eustatic sea-level fall. Collisional orogeny is at a minimum, whereas subduction continues around the margins of the supercontinent.

6.2.2 The Phanerozoic Record

Many events in the Phanerozoic record have for long been suspected to be of global importance. Some of the more important of these are summarized here (Figs. 6.1, 6.2). Worsley et al. (1984, 1986) postulated two supercontinent cycles during the Phanerozoic. Veevers (1990) divided the most recent of these into five stages, as noted below and in Fig. 6.3, column VIII.

Late Proterozoic glaciation, the evidence for which is widespread in Greenland and Scandinavia, may be related to the formation of regional ice caps on rift margins elevated by thermal doming prior to continental separation and the breakup of a Late Precambrian supercontinent (Eyles 1993). The Cambrian transgression may reflect the subsequent increase in the rate of sea-floor spreading as continental dispersal accelerated (Matthews and Cowie 1979; Donovan and Jones 1979). It has been suggested that the Late Precambrian Sparagmite sequence of Norway was formed in rifts representing the incipient Iapetus (proto-Atlantic) Ocean (Bjørlykke et al. 1976). The exceptionally high sea-level stand during the Ordovician might then relate to the rapid widening of the Iapetus Ocean (and probably other world oceans).

Sea-level lowering occurred during the Caledonian-Acadian and Hercynian-Appalachian suturing of Pangea between the Devonian and Permian (Schopf 1974). The drifting of this supercontinent over the South Pole is thought to have been the cause of increasingly continental-type climates, leading to the Late Devonian-Permian glaciations of Gondwana (Crowell 1978; Caputo and Crowell 1985; Eyles 1993; Fig. 10.17) and the widespread Pennsylvanian to Jurassic eolian facies of the United States and Europe (Kocurek 1988a).

Kominz and Bond (1991) and Bond and Kominz (1991a) documented anomalously large subsidence events in North American cratonic basins and on the Cordilleran and Appalachian continental margins during the Late Devonian and Early Mississippian. They suggested that basin deepening and arch uplift may have been caused by intraplate stresses associated with plate convergence toward a region of mantle downwelling during Pangea assembly. This mechanism is discussed in greater detail in Section 9.2.

Worldwide transgressions during the Jurassic and Cretaceous probably reflect the progressive splitting of Pangea (Figs. 9.11, 9.12). North America

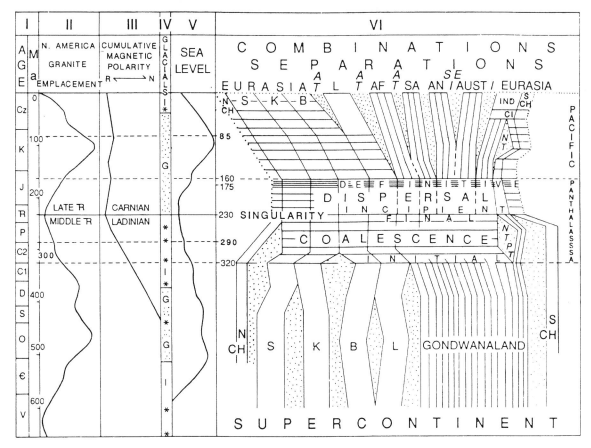

Fig. 6.3. Phanerozoic tectonism, schematic continental assembly and rifting events, and summary of long-term stratigraphic cyclicity in the major continental areas. *Column VI* shows schematically the assembly and rifting of Pangea. Continental fragments, *from left to right*: *N CH* northern China; *S* Siberia; *K* Kazakhstania; *B* Baltica; *L* Laurentia; *Af* Africa; *SA* South America; *AN* Antarctica; *Aust* Australia; *IND* India; *S CH* southern China; *CI* Cimmeria. Oceans: *AT* Atlantic (north, central, south); *SEI* southeastern Indian; *NT* Neotethys; *PT* Paleotethys. Numbered stages in column VIII are discussed in text. (Veevers 1990, reproduced by kind permission of Elsevier Science-NL, Sara Burgerhartstraat 25, 1055 KV Amsterdam, The Netherlands)

and Africa rifted apart in the Middle Jurassic, South America and Africa in the Early Cretaceous; North America and Britain split in the Middle Cretaceous, as did Africa and Antarctica; India and Madagascar rifted apart in the Late Cretaceous, and the North Atlantic split extended northward between Greenland and Scandinavia in the Early Paleocene (summary and data sources in Bally and Snelson 1980; Uchupi and Emery 1991). Larson (1991) stated that "during the Middle Cretaceous, starting in earliest Aptian time (124 Ma), there were eruptions from an extraordinary upwelling of heat and deep-mantle material in the form of one or several very large plumes." Mantle processes are discussed in Chapter 9.

Veevers (1990) subdivided the post-Pangea rift-and-drift cycle into five stages (Fig. 6.3):

- Stage 1 (Late Carboniferous, 320–290 Ma, the platform stage) corresponds to a widespread stratigraphic gap on the continents, caused by the thermal uplift accompanying continental assembly.
- Stage 2 (Permian–Middle Triassic, 290–230 Ma, the sag stage) is represented by the Early Gondwana stratigraphic sequences, and marks the local thinning of the Pangea lithosphere during its initial stretching, with the formation of broad basins or sags.
- Stage 3 (Late Triassic-Late Jurassic, 230–160 Ma, the rifting stage), is represented by ocean-margin rift successions marking the initial break-up of Pangea.
- Stage 4 (Late Jurassic-Late Cretaceous, the spreading stage), marks the time of maximum

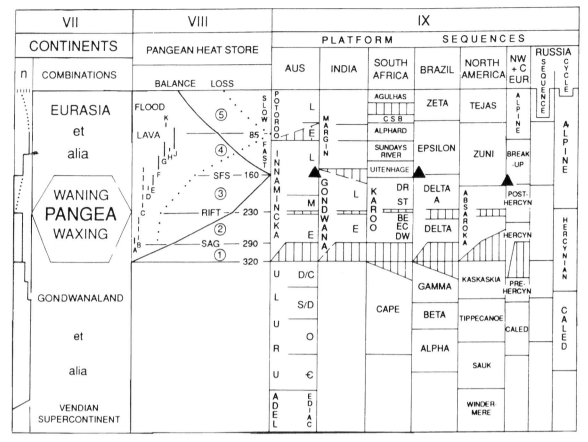

Fig. 6.3 (*Contd.*)

rate of continental dispersal, with high global sea
levels and corresponding widespread shelf sedimentation, and the development of thick exten-
sional-margins successions on the borders of
Atlantic-type oceans.

- Stage 5 (Late Cretaceous-present, 85–0 Ma, a
slow spreading-subduction stage) corresponds to
the end of the dispersion phase of Worsley et al.
(1986), and the beginning of the continental assembly phase, with major collisions occurring
along the Alpine-Himalayan belt.

There is currently considerable discussion regarding the nature of possible earlier (Precambrian)
supercontinent cycles. This is beyond the scope of
the present book. A speculative reconstruction was
offered by Hoffman (1991), who suggested that at
around 700 Ma Laurentia was situated at the center
of a supercontinent that subsequently dispersed and
"turned inside-out" to form Gondwana around the
end of the Precambrian, and subsequently Pangea
in the Middle Paleozoic. Other work on this problem is referenced in Sections 3.2 and 9.2.

6.3 Cycles with Episodicities of Tens of Millions of Years

6.3.1 Intercontinental Correlations

Sloss (1963, 1972) established the six Indian-name
sequences in North America (Fig. 1.1), and demonstrated their correlation with similar sequences
on the Russian platform (Fig. 3.4). Vail et al. (1977)
referred to these as second-order cycles, or supercycles. This section briefly reviews other work on
this type of stratigraphic sequences.

Soares et al. (1978) reported a stratigraphic
analysis of the three major intracratonic basins in
Brazil, namely, the Amazon, Parnaiba, and Parana
basins, all of which contain successions spanning
most of the Phanerozoic. Their interpretation of the
geomorphic behavior of these basins is given in
Fig. 6.4. They recognized seven sequences, which
correlate reasonably closely with those of Sloss, as
shown in Fig. 6.5.

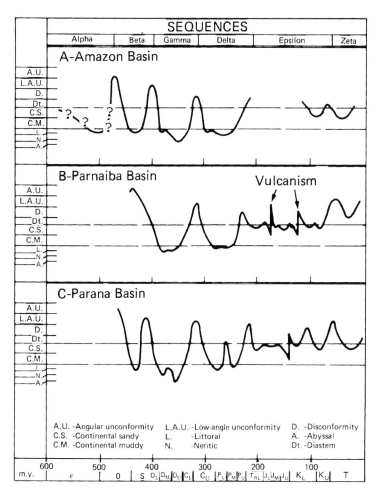

Fig. 6.4. Geomorphic expression of oscillatory movements in three Brazilian basins. (Soares et al. 1978)

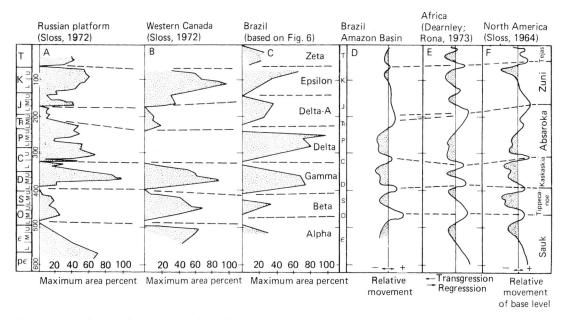

Fig. 6.5. Correlation of sequences of oscillatory movement in North American, Brazilian, European, and African cratons. **A, B,** and **C** are based on preserved sediments. **D, E,** and **F** are based on relative base-level movements. (Soares et al. 1978)

Soares et al. (1978) described an epeirogenic cycle consisting of five phases, which explained stratigraphic events in each of the Brazilian sequences. Note that a tectonic control is indicated for this cycle:

1. Initial rapid basin subsidence with development of nonmarine facies and numerous local unconformities.
2. Basin subsidence slower, with deepening basin centers, marine transgression and differentiation of central marine and marginal nonmarine facies belts.
3. Development of intrabasin uplifts and local downwarps, much local facies variability.
4. Renewed basinwide subsidence, time of maximum transgression, generally fine-grained deposits.
5. Broad cratonic uplift, return to nonmarine sedimentation.

As noted in the previous section, broad sedimentation patterns are controlled by the rise and fall in base level, whether this is controlled by tectonism or eustasy. Shanmugam and Moiola (1982) carried out a preliminary synthesis of stratigraphic data on the distribution of turbidite and contourite deposits, and suggested that many of the major deposits were formed at times of low sea level, as indicated on the Vail curve of first- and second-order cycles (Figs. 6.6, 6.7). This correlation reflects the tendency for greater volumes of terrigenous detritus to be delivered directly to the continental slope during times of low sea level. Oceanic currents capable of transporting and winnowing this detritus to form contourites, are known to be more vigorous at these times, as noted in Section 6.1. In view of the controversies surrounding the Vail curves this synthesis needs to be reevaluated.

Hallam (1984) discussed stratigraphic methods for determining sea-level change, including most of those described in Chapter 2. He also assembled a considerable volume of regional stratigraphic data derived from outcrop and well studies to compare the record of sea-level change with that derived by the seismic method. His resulting curve of sea-level changes in the tens-of-millions-of-years frequency range is shown in Fig. 6.8. Hallam's maps of continental inundation, derived from various sources, suggest that the Vail curve is in error in estimating the all-time highest Phanerozoic sea level to have been in the Late Cretaceous. He suggested that it was in the Late Ordovician, when nearly two-thirds of the continents were inundated. Hallam (1984) pointed out numerous disagreements in detail with

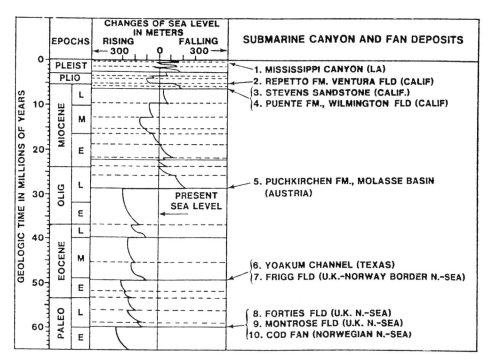

Fig. 6.6. Stratigraphic correlation of some major fan and canyon turbidite deposits with times of low sea level on the Vail curve. (Shanmugam and Moiola 1982)

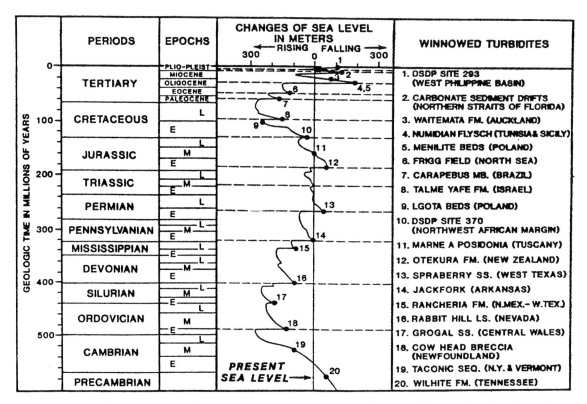

Fig. 6.7. Stratigraphic correlation of some major contourite deposits with times of low sea level on the Vail curve. (Shanmugam and Moiola 1982)

the Vail curve, based on differences in emphasis having been placed on different data sets. According to Hallam (1984), the Vail curve for certain time periods is distorted by the effects of local tectonism, because data from tectonically unstable regions has received too much emphasis in the Vail compilation. As discussed in Chapters 9 and 11, subcrustal thermal effects, crustal stretching, flexural subsidence, in-plane intraplate stresses and simple faulting are all known to have produced regional sea-level changes that have overprinted the eustatic record at different times.

6.3.2 Tectonostratigraphic Sequences

Johnson (1971) emphasized the links between Sloss-type cycles of transgression-regression and regional orogeny (Fig. 6.9). He recognized that

[O]f the four major orogenies that occurred in North America during the Paleozoic and Mesozoic, three began, reached climactic stages, and went through waning stages during the time epicontinental seas were transgressing to their maximum extent and then regressing to form the

great onlap-offlap cycles called sequences.... The correspondence of orogenic events with onlap of the craton is so consistent in general and even in detail, that it must reflect a fundamental relation.

In a series of papers Sloss (1972, 1979, 1982, 1984, 1988b; Sloss and Speed 1974) elaborated the "Indian-name" sequences that he first established in 1963. He developed ever-more detailed isopach maps and discussed the effects of eustasy and tectonism in the development of the sequences. In one of the earlier of these papers (Sloss and Speed 1974; see also Sloss 1984) he attempted to subdivided the six sequences into two broad types. The first type, termed "submergent," and exemplified by the Sauk, Tippecanoe, and Kaskasia sequences,

... is dominated by flexure of the cratonic and interior margins. These sequences exhibit slow regional transgression and onlap, ... gentle subsidence of interior basins separated by less subsident domes and arches, and a lack of widespread brittle deformation manifested by faulting. The emergent episodes preceding each of these flexure-dominated times of deposition are the occasions for developing of the sequence-bounding unconformities and slow progressive transgression of the craton. (Sloss 1984, p. 5)

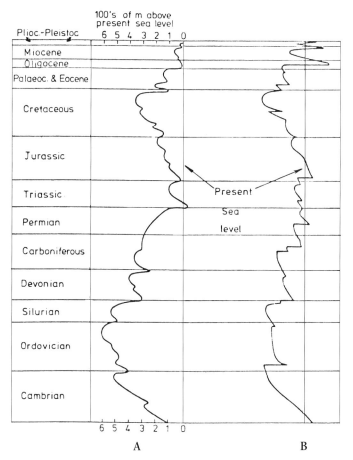

Fig. 6.8. Eustatic sea-level curves for the Phanerozoic. **A** A curve based on stratigraphic synthesis by Hallam (1984). **B** the Vail curve, from Vail et al. (1977). (Hallam 1984, reproduced by permission, from the Annual Review of Earth and Planetary Sciences, vol 12, copyright Annual Reviews Inc.)

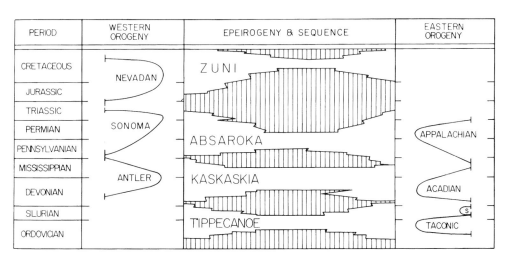

Fig. 6.9. Correlation of orogenic events on the western and eastern margins of the North American continent, with the "Sloss sequences" of the continental interior. (Johnson 1971)

The second type of sequence Sloss termed "oscillatory," and

... is characterized by abrupt termination of a submergent episode through rapid cratonic uplift accompanied by high-angle faulting of basement crystalline rocks and, commonly, by faults that propogate from the basement to fracture and displace the overlying sedimentary cover. The Absaroka Sequence represents a typical oscillatory episode. (Sloss 1984, p. 5)

In his later papers Sloss considered epeirogeny and in-plane intraplate stresses (these are discussed at length in Chaps. 9 and 11) and their effects on sequence architecture. For example, the Late Mississippian sub-Absaroka unconformity is a time of pronounced change on cratonic subsidence and uplift patterns within the North American interior. The Canadian Shield emerged as an important source of sediment, and the southwestern Midcontinent (Texas, Oklahoma, New Mexico, Colorado) underwent pronounced submergence. Sloss (1988b) attributed this major change in continental configuration to intraplate stresses associated with plate convergence along the southern margin of the continent.

In many basins tectonic influences are indicated by the presence of angular unconformities, faults that terminate at sequence boundaries, changes in isopach patterns (indicating changes in sediment transport direction), etc. Such structural features are commonly used as a means to subdivide the stratigraphy into tectonic sequences spanning millions to tens of millions of years. They typically reflect major steps in the plate-tectonic evolution of the area, such as the transition from rifting to thermal subsidence on extensional continental margins. A comparison is offered here between the sequence stratigraphy of some extensional-continental-margin basins, using examples that have been studied by researchers independent of the Exxon group, who did not set out specifically to "test" the Vail curves: Beaufort-Mackenzie Basin, Canada (Fig. 6.10, 6.11; Dixon and Dietrich 1990; McNeil et al. 1990), Beaufort Sea, Alaska (Fig. 6.12; Hubbard 1988), and the Grand Banks, Newfoundland (Fig. 6.12, 6.13; Tankard and Welsink 1987; Welsink and Tankard 1988; Hubbard 1988). In several of these syntheses plate-kinematic events

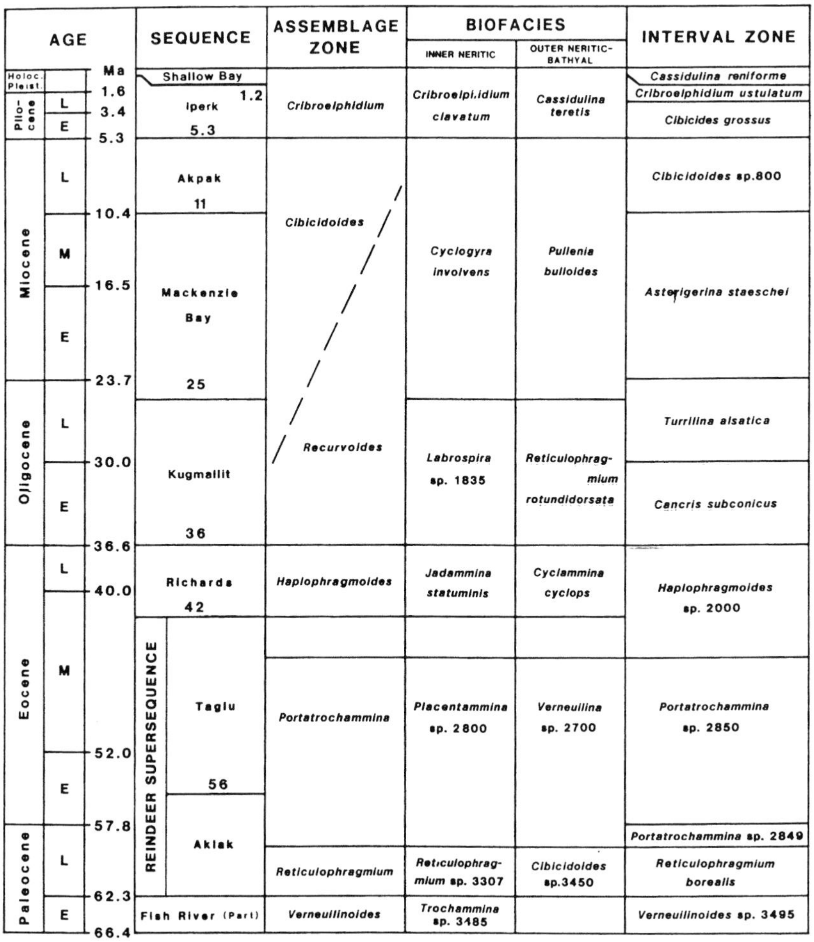

AGE			SEQUENCE		ASSEMBLAGE ZONE	BIOFACIES		INTERVAL ZONE
						INNER NERITIC	OUTER NERITIC-BATHYAL	
Holoc. Pleist.		Ma 1.6	Shallow Bay	1.2	Cribroelphidium	Cribroelphidium clavatum	Cassidulina teretis	Cassidulina reniforme
Plio- cene	L	3.4	Iperk					Cribroelphidium ustulatum
	E	5.3		5.3				Cibicides grossus
Miocene	L		Akpak	11	Cibicidoides	Cyclogyra involvens	Pullenia bulloides	Cibicidoides sp.800
		10.4						
	M	16.5	Mackenzie Bay					Asterigerina staeschei
	E	23.7		25				
Oligocene	L	30.0	Kugmallit		Recurvoides	Labrospira sp. 1835	Reticulophrag- mium rotundidorsata	Turrilina alsatica
	E	36.6		36				Cancris subconicus
Eocene	L	40.0	Richards	42	Haplophragmoides	Jadammina statuminis	Cyclammina cyclops	Haplophragmoides sp. 2000
	M	52.0	Taglu	56	Portatrochammina	Placentammina sp. 2800	Verneuilina sp. 2700	Portatrochammina sp. 2850
	E	57.8						
Paleocene	L	62.3	Aklak		Reticulophragmium	Reticulophrag- mium sp. 3307	Cibicidoides sp.3450	Portatrochammina sp. 2849
								Reticulophragmium borealis
	E	66.4	Fish River (Part)		Verneuilinoides	Trochammina sp. 3485		Verneuilinoides sp. 3495

(Note: In the Eocene–Paleocene rows the left margin is labelled vertically "REINDEER SUPERSEQUENCE".)

Fig. 6.10. Sequences and biostratigraphic scheme for Cenozoic strata in the Beaufort-Mackenzie Basin, Arctic Canada. (McNeil et al. 1990, reproduced by permission of Kluwer Academic Publishers)

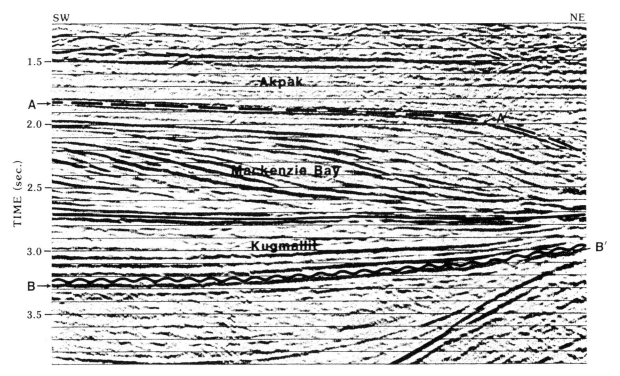

Fig. 6.11. Seismic expression of sequence boundaries in Beaufort-Mackenzie Basin, with the names of the major sequences. *From top to bottom* these boundaries are the 11-, 25-, and 36-Ma events of Fig. 6.10. (Dixon and Dietrich 1988)

and tectonic subsidence phases are indicated. The results from these basins may be compared with the data from the Atlantic margin of the United States and the British Isles, which several workers have used as "tests" of the Vail curves (Fig. 6.14; Poag and Schlee 1984; Poag and Ward 1987; Olsson 1991). A summary of the ages of the major sequence boundaries in these various sources is shown in Fig. 6.15.

The disparity between these various columns is remarkable, especially considering that North Atlantic data were over-emphasized in the construction of the Exxon curves. Very few of the "supercycle" or "supercycle-set" boundaries in the Haq et al. (1987, 1988a) charts are actually represented in the six basin columns shown, unless considerable allowance is made for chronostratigraphic imprecision and error, of the type discussed in Chapter 13. Twenty-three supercycle boundaries appear on the Exxon chart (Haq et al. 1987, 1988a) chart. Of these, only the following appear in at least two of the six basin columns: the 10.5, 30, 39.5, 49.5, and 58.5-Ma events. The 10.5-Ma event is illustrated in a seismic section from offshore New Jersey in Fig. 7.3. It is a particularly prominent erosion sur-

face, indicating a major fall in relative sea level and erosion of the continental slope. The 30-Ma event, in the Middle Oligocene, and the 39.5-Ma event, in the Late Eocene, are both particularly widespread events (if allowance is made for a ±1.5 m.y. margin of error). One or both of these may record rapid build-up of continental ice in Antarctica and is probably a true global eustatic event (or events; Vail et al. 1977; Matthews 1984; Bartek et al. 1991). The comparisons between the Exxon sequence chronology and that of the Atlantic margin of the United States is reasonably good. However, this is hardly surprising, since the Exxon curves were built mainly from this area, plus data from the North Sea and the Gulf of Mexico (Summerhayes 1986). A "test" of the Vail curve from these areas therefore amounts to little more than circular reasoning. Thus, the charts of Olsson (1991) for New Jersey indicate almost total agreement with the Exxon charts (Haq et al. 1987, 1988a).

It is noteworthy that very few of the sequence boundaries on the northern slope of North America (Beaufort Sea, Beaufort-Mackenzie Basin) correlate with the Exxon events. A boundary at about 11 Ma in the Beaufort-Mackenzie Basin, and the 30-, 38-,

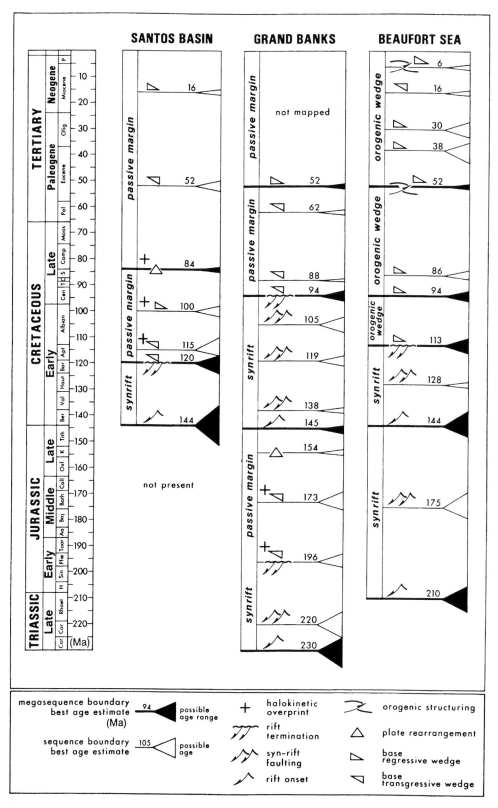

Fig. 6.12. Estimated ages of sequence boundaries in three extensional-margin basins. The Santos basin is in South America, and is not discussed in this book. (Hubbard 1988, reproduced by permission)

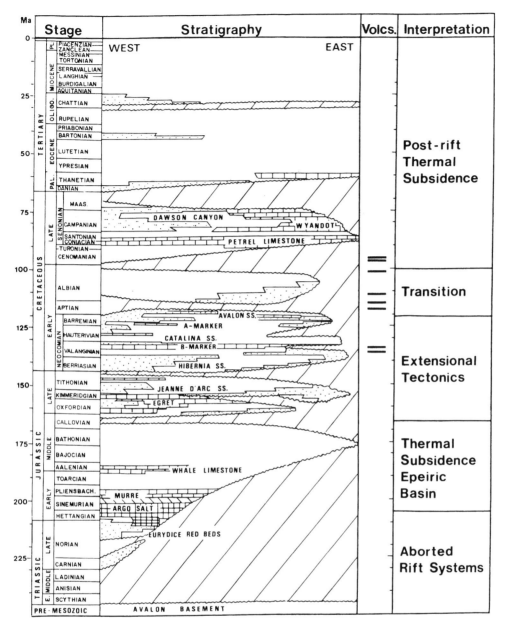

Fig. 6.13. Tectonostratigraphic column for the Jeanne d'Arc Basin, Grand Banks, Newfoundland, showing major episodes in basin evolution, and prominent unconformities. (Welsink and Tankard 1988, reproduced by permission)

113-, 128-, and 210-Ma events in the Beaufort Sea, are the only apparent matches. Whether these correlations indicate eustatic signals or contemporaneous regional tectonism remains to be demonstrated. The question is discussed in Chapter 11. The seismic character of some of these boundaries is illustrated in Fig. 6.11. These are all conformable boundaries, and are categorized as type 2 boundaries by Dixon and Dietrich (1988).

Examples of unconformable sequence boundaries that are clearly related to regional tectonism are illustrated in Fig. 6.16. In this continental-margin section off northern Brazil Petrobras Exploration Department (1988) identified three sequences. The first, of Aptian age (Sequence I), comprises sediment deposited during the rift phase of Atlantic opening. Component seismic unit 1 consists of fluvial and lacustrine sediments, and unit 2 consists

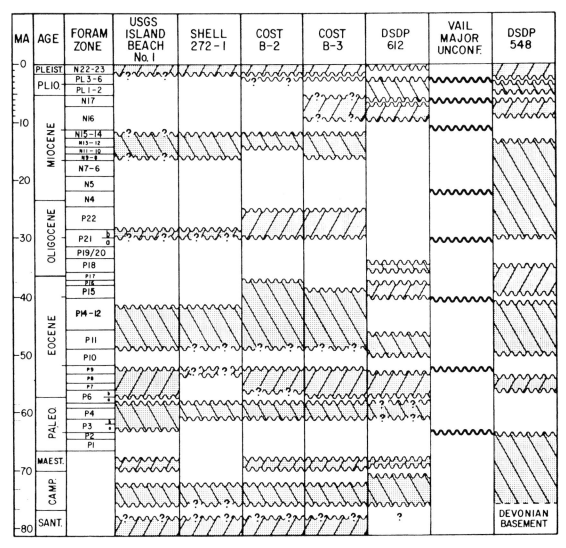

Fig. 6.14. Summary of stratigraphic framework of New Jersey continental margin (five data columns on *left*), offshore Ireland (Deep Sea Drilling Project, *DSDP, hole* *548*), and the major unconformities in the Vail curve. (Poag and Ward 1987)

of sediment deposited during the first marine transgression. Sequence II was deposited during the beginning of the thermal subsidence phase, with the unconformity at the sequence I/II boundary corresponding to the breakup unconformity. Unit 3, of Albian age, consists of onlapping restricted and shallow-marine sediments. Unit 4, of Turonian-Santonian age, is a thick slope clastic succession showing deep-marine onlap. Sequence III was also deposited during the thermal subsidence phase. It is subdivided into six units spanning the Campanian to present. Units 5, 6 and 7 contain prograding platform-edge carbonate deposits, and the remaining units are platform carbonates.

If the sequence boundaries in Fig. 6.15 are compared with the high-frequency cycle boundaries in the Exxon charts (the so-called third-order sequences of Haq et al. 1987, 1988a) allowing for, say, a ±1 m.y. margin of error, almost perfect correlation matches can be made for each of the six basin columns. Dixon (1993) demonstrated that each of his sequence boundaries in the Beaufort-Mackenzie Basin could realistically be correlated with at least two, and in some cases, three of the "events" in the Exxon charts (Fig. 13.13). As Miall (1992) has pointed out, because of the density of events in the Exxon charts almost any succession of events can be found to correlate with them, and so this proves

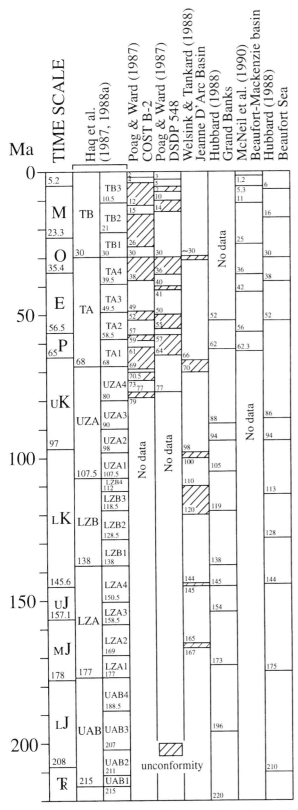

Fig. 6.15. Summary of ages of sequence boundaries in various extensional-margin basins. Data derived from information shown in Figs. 6.10–6.14

nothing. It also begs the question of what differentiates a "cycle" from a "supercycle," in the Exxon terminology, other than their duration. Some of the boundaries in the syntheses by Hubbard (1988), Tankard and Welsink (1987) and Welsink and Tankard (1988) are labeled "megacycle" or "megasequence" boundaries because of their demonstrated relationship to major changes in tectonic regimes. The relationship between regional tectonism and eustasy is a complex one, and is discussed further in Chapter 11.

The comments of the Exxon workers on Hubbard's analysis are instructive. Van Wagoner et al. (1990, p. 50) stated:

Members of the Exxon group have worked in all three basins that Hubbard described and have recognized those sequence boundaries he described. In addition, we described other boundaries that are less prominently developed, but that are important nevertheless in controlling sediment distribution and lithologies within the basin. These occur at the higher frequency expected from the Exxon cycle chart. We certainly agree that Hubbard's megasequence' boundaries, occurring during onset of stages of basin evolution or other structural events, are *tectonically enhanced*, and become the most prominent and important surfaces in structural analysis of a basin.... However, the higher-frequency sequences, when dated as accurately as possible using biostratigraphy, appear to be synchronous between the basins. The presence of these sequences strongly suggests that the higher-frequency eustatic overprint is superposed on the lower-frequency or noncyclic tectonic and sediment-supply controls.

The italics are those of Van Wagoner et al. (1990), and serve to emphasize a claim made repeatedly by the Exxon workers, that unconformities cannot form without a eustatic fall in sea level. The reader is referred to Section 18.6 of this book for additional discussion of the Exxon philosophy, and to Part IV of the book for a discussion of biostratigraphic precision and the ability of the Exxon workers to find global correlations.

In convergent plate margins it is to be expected that tectonism is the dominant control on the long-term development of basin architecture. For example, Seyfried et al. (1991) mapped a series of regional unconformities in Cretaceous-Cenozoic sections of Costa Rica that are clearly related to arc tectonism and volcanism (Fig. 6.17). The spacing of these unconformities indicates the occurrence of tectonic episodes tens of millions of years apart. Seyfried et al. (1991) stated that a cycle of compression, uplift, erosion, unconformity, subsidence (tilting), and basin filling has occurred in Costa Rica and Nicaragua three times since the Late Eocene,

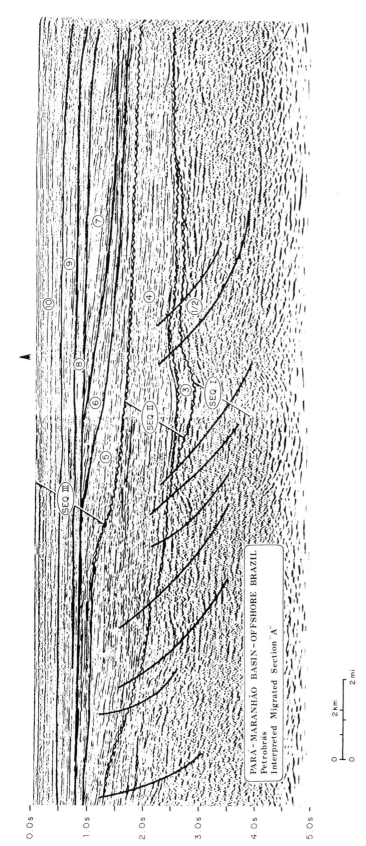

Fig. 6.16. Interpreted seismic section from offshore northern Brazil, showing three sequences and their component seismic-facies units *1–10*. (Petrobras 1988, reproduced by permission)

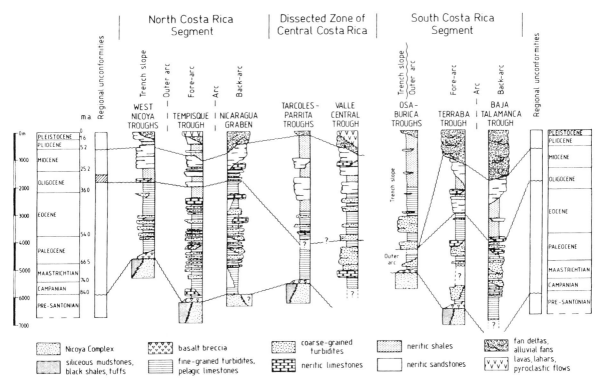

Fig. 6.17. Stratigraphic correlation of Cretaceous-Cenozoic sections in Cost Rica. Note the subdivision of the section into "sequences" by the presence of regional unconformities. However, these are angular unconformities developed in response to regional convergent tectonism. (Seyfried et al. 1991)

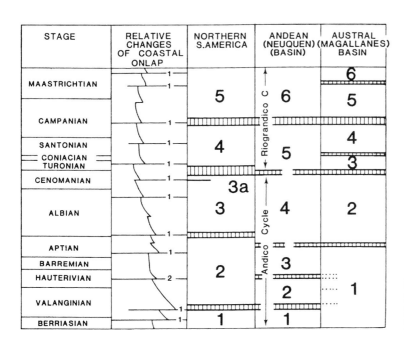

Fig. 6.18. Comparison of long-term Cretaceous cycles in three areas of South America adjacent to the Andes. (Hallam 1991)

and suggested a control by long-term intraplate stress. Similarly, in the Andean backarc basin of Argentina, changes in tectonic regime over intervals of 15–65 m.y. since the Triassic are considered to have been responsible for the main variations in stratigraphic architecture there (Fig. 6.18; Hallam 1991; Legarreta and Uliana 1991). A comparison of Figs. 6.12 and 6.17 shows a similar level of approximate correlation of major stratigraphic events in separate basins within the same large-scale plate regime.

6.4 Main Conclusions

1. Many of the broad characteristics of the global stratigraphic record can be related to the changes in continental scale, climate, latitudinal position and eustatic sea level that result from the assembly and dispersal of supercontinents over a 200–500 m.y. time period.
2. Cratonic sequences of about 10–100 m.y. duration can be traced and correlated between several of the earth's major continental interiors, including the interior of the United States and Canada, Russia and Brazil.
3. Other sequences of comparable time duration are the result of regional tectonism, including the effects of long-term changes in plate-kinematic patterns. These tectonostratigraphic sequences may be correlatable with each other within areas as large as two or three adjacent tectonic plates, the tectonics of which were dominated by the same major events, such as a large-scale plate rifting or collision event. Beyond the effects of these tectonic events corrrelatable sequences are not formed.
4. Almost any sequence boundary can be found to correlate with an event in the Exxon global cycle chart.

7 Cycles with Million-Year Episodicities

A wealth of stratigraphic data has accumulated for cycles having durations and episodicities of a few million years. They have been recorded in a wide variety of Phanerozoic basins in many different tectonic settings. They constitute the main basis of the Exxon global cycle charts, where they are termed "third-order cycles" (Haq et al. 1987, 1988a). A small selection of these is described in this chapter to illustrate stratigraphic patterns and their reflection of tectonic setting. Sequence concepts have also been applied to the study of the Precambrian record (e.g., Christie-Blick et al. 1988), but this work is not discussed here because at this time the record is fragmentary and regional correlations are very limited.

7.1 Extensional and Rifted Clastic Continental Margins

The Atlantic and Gulf Coast continental margins of the United States are classic examples of extensional margins (Figs. 7.1, 7.2). Their stratigraphy and structure were influential in the development of plate-tectonic basin models, and much research has been carried out there to investigate the tectonic history and petroleum potential of the major basins. Techniques for backstripping and for modeling of the geophysical controls of flexural and thermal subsidence were first developed using Atlantic-margin data (see Miall 1990, Chap. 7 for summary).

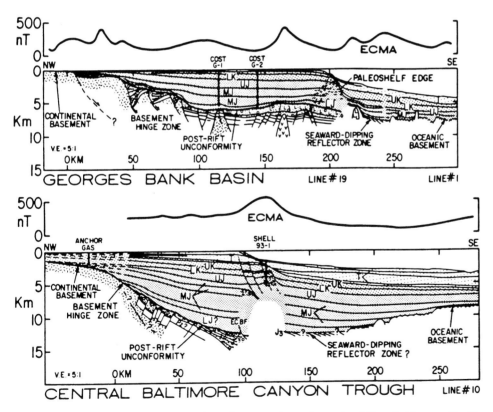

Fig. 7.1. Typical cross sections through the Atlantic continental margin of Canada (Georges Bank) and the United States (Baltimore Canyon Trough). (Sheridan 1989)

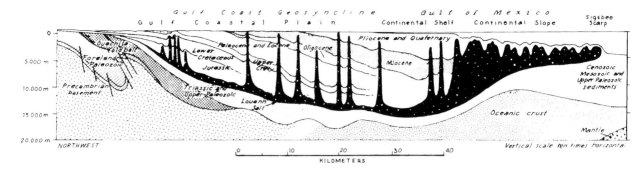

Fig. 7.2. Typical cross section through the Mesozoic-Cenozoic sediments of the Gulf Coast, off central Texas. (King 1977, reproduced by permission of Princeton University Press)

Regional seismic sections showing interpreted sequence stratigraphy have been published for several areas around the Gulf and Atlantic margins (e.g., Greenlee and Moore 1988). Figures 7.3 and 7.4 illustrate lines run offshore New Jersey and Alabama, respectively. Most of the sequences developed by seaward progradation. Pronounced landward and seaward shifts in coastal onlap occur at some levels, and some of these have been correlated with supercycle ("second-order") sequence boundaries. Fluvial and coastal-plain strata show landward thinning and tapering, and in many of the sequences onlap relationships are not observable. Sequence boundaries have been dated with the aid of biostratigraphic data from exploratory wells located on or close to these lines, and ages of these boundaries, correlated to the Exxon chart, are indicated. In a few instances additional sequences were identified, that do not correspond to events on the Exxon chart, including those bounded at the top by the 4.7- and 5.1-Ma events in Fig. 7.4. Figure 7.5 illustrates an enlarged part of Fig. 7.4, showing the extensive slope erosion and onlap of Miocene sequences above the 10.5-Ma sequence boundary. This boundary is interpreted as a supercycle boundary on the Exxon chart. As noted in Section 6.2.2 it has not been identified by workers independent of the Exxon group in some other extensional margins, including the Grand Banks, Newfoundland, or the Beaufort Sea, which calls into question its origin as a eustatic event.

The preserved landward margins of these sequences are exposed in the coastal plains of the eastern and southern United States. Examples of Eocene sequences in Alabama and South and North Carolina were studied by Baum and Vail (1988), and Miocene sequences were extensively studied by Kidwell (1984, 1988, 1989). Figure 7.6 shows the generalized model of shelf sedimentation related to

the sequence concepts of Posamentier et al. (1988) and Posamentier and Vail (1988), and Fig. 7.7 illustrates correlation of outcrop sections in Alabama and the Carolinas with the Haq et al. (1987, 1988a) coastal onlap chart. The Alabama section is predominantly clastic, whereas the sections in the Carolinas are carbonates.

The sequence boundaries in the clastic sections are typically prominent erosion surfaces, and may be cut by incised valleys filled with lowstand valley-fill fluvial or estuarine beds (Fig. 7.6). Subsequent transgressive beds may include shelf, estuarine, tidal or lagoonal deposits, depending on coastal architecture. The thickness of the transgressive beds depends on the balance between subsidence, sea-level change and sediment supply. Nummedal and Swift (1987) provided a set of six stratigraphic columns to aid in the interpretation of all types of transgressive sequence architecture (Fig. 7.8). Note that ravinement surfaces may be prominent within the transgressive beds (the "transgressive surface" of Fig. 7.6; see also Figs. 2.4, 2.5).

Kidwell (1984, 1988, 1989) studied the Miocene part of the succession in coastal outcrops in Maryland, on the eastern coast of the United States. She demonstrated the existence of a series of prominent disconformities that onlap the coastal plain at a low angle (Figs. 7.9, 7.10), and examined the facies of the transgressive units in some detail. They include numerous prolific shell beds formed by winnowing of sand and mud in a shallow-marine setting.

The Sverdrup Basin in the Canadian Arctic Islands is an extensional basin formed near the edge of the North American plate as a result of crustal stretching and aborted rifting. It contains as much as 9 km of Mesozoic strata, which Embry (1988, 1991, 1993) has subdivided into thirty sequences. Johannessen and Embry (1989) and Mork et al. (1989) correlated these sequences with those in other Arctic basins (Fig. 7.11). Figure 7.12 illustrates a cross section developed from well data showing the Norian to Pliensbachian sequences in the western Sverdrup Basin. Note the thinning and truncation of sequences toward the southwestern basin margin. Figures 7.13 and 7.14 illustrate the stratigraphic architecture of these clastic sequences in areas of varying subsidence and sedimentation rate. In areas of high sedimentation rate deposition is continuous across sequence boundaries, and the boundaries cannot easily be recognized within successions of continuous coastal-plain fluvial and deltaic deposits. Embry (1993) compared the ages of transgressive events in the Sverdrup Basin with those given by Hallam (1988) and Haq et al. (1987, 1988a), as discussed in Section 14.3.2. Embry's data also formed the basis for a useful discussion regarding the placement of sequence boundaries (Sect. 15.2.2).

Mork et al. (1989) correlated Triassic sections from Sverdrup basin into Svalbard, northern Greenland, northern Norway and various Russian basins (Fig. 7.15), by comparing the ages of the transgressions that formed the base of each sequence. The precision of the corelation, based largely on ammonite zones, was estimated to be within about ±1 m.y. Most of these transgressions can be correlated with the Exxon chart. Most, but not all, can be recognized in widely separated parts of the Arctic (Fig. 7.16). However, Mork et al. (1989) indicated that several of the transgressions can only be documented in one basin.

Other examples of rifted or extensional basins that include well-described successions of sequences with 10^6-year episodicities are the Viking Graben of the North Sea (Abbotts 1991) and Gippsland Basin, Australia (Rahmanian et al. 1990). The Gippsland Basin is discussed further in Section 17.3.2.

7.2 Foreland Basin of the North American Western Interior

During the Cretaceous, the Western Interior of the United States and Canada formed a vast epicontinental seaway along a foreland basin extending from the Arctic Ocean to the Gulf Coast (Fig. 7.17). The basin was asymmetric, with more rapid subsidence and sedimentation occurring along the

western flank of the basin, adjacent to the fold-thrust belt of the Sevier-Laramide orogen (Fig. 7.18). Up to 5 km of sediments accumulated during the Cretaceous. They constitute a classic "clastic wedge" (Figs. 7.18, 7.19), as this term was defined by Sloss (1962). Weimer (1960) was the first to recognize that the Upper Cretaceous section constitutes a succession of large-scale transgressive-regressive cycles with 10^6-year episodicities. Figure 7.20 illustrates Weimer's (1986) most recent synthesis of the stratigraphy and age of Cretaceous cycles in the Western Interior Seaway, including some of the key stratigraphic names from the Rocky Mountain and other basins. Major interregional unconformities and their ages in Ma are indicated on this diagram. They do not correlate in any particularly obvious way with the sequence boundaries in the Exxon charts, unless allowance is made for errors of up to 1 or 2 m.y., in which case they all correlate. The relationship between transgressive-regressive cycles and tectonism in the foreland basin was discussed by Fouch et al. (1983) and Kauffman (1984), and is considered at some length in Section 11.3.2.

Major regressive sandstone wedges within the succession include the Ferron, Emery, Blackhawk, Castlegate and Price River sandstones (Fig. 7.19). Details of the lowermost of these wedges are illustrated in Figs. 7.21 and 7.22, based on the work of Ryer (1984) and Shanley and McCabe (1991). In the latter, the sequences constitute alluvial-coastal plain facies successions ranging from 16 to 180 m in thickness.

Another well-studied cycle in the foreland-basin clastic wedge is that of the Gallup Sandstone and associated beds in San Juan Basin, New Mexico (Fig. 7.23; Molenaar 1983). The Gallup Sandstone is of Coniacian-Turonian age, and represents approximately one million years of sedimentation (Fig. 7.24). It does not appear on the summary diagrams of Figs. 7.19 and 7.20 because it occurs off the line of section to the south. The sequence stratigraphy of these rocks has been described by Nummedal and Swift (1987), Nummedal et al. (1989), and Nummedal (1990). The Gallup Sandstone can be subdivided into component higher-frequency cycles, as illustrated in Fig. 2.8 and as discussed in Chapter 8. It includes component members formed in a variety of shelf, coastal, and nonmarine environments. The interpreted relationships between these units and their paleogeographic evolution are shown in Fig. 7.25.

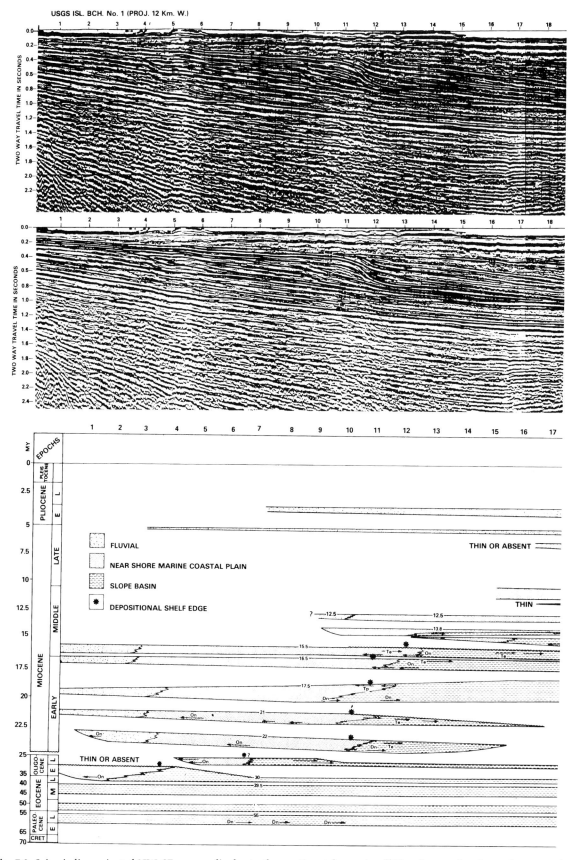

Fig. 7.3. Seismic line oriented NW-SE, perpendicular to the continental margin off New Jersey, showing uninterpreted line, interpreted line, and chronostratigraphic chart. (Greenlee and Moore 1988)

BALTIMORE CANYON TROUGH

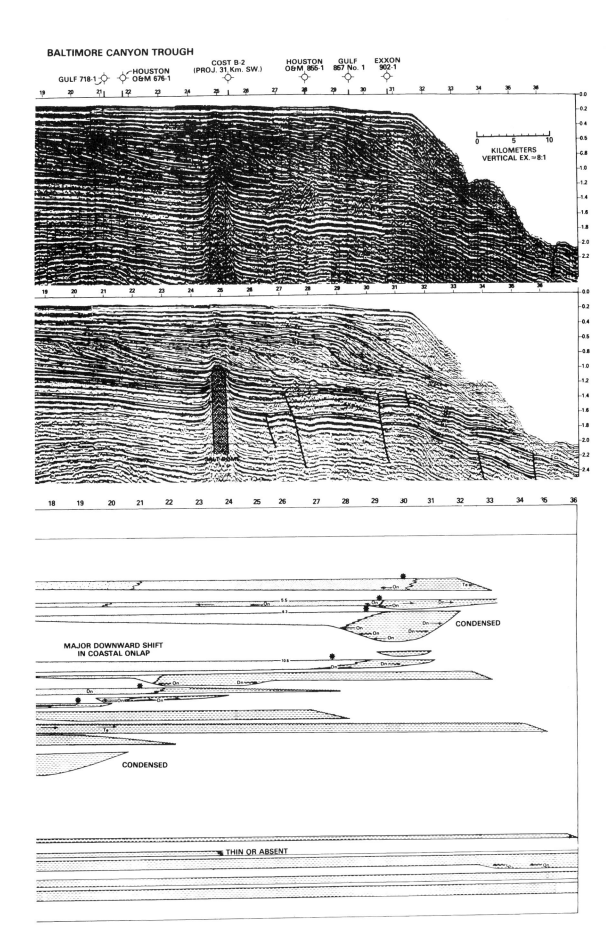

GULF 718-1   HOUSTON O&M 676-1   COST B-2 (PROJ. 31 Km. SW.)   HOUSTON O&M 855-1   GULF 857 No. 1   EXXON 902-1

0   5   10
KILOMETERS
VERTICAL EX. ≈ 8:1

SALT DOME

MAJOR DOWNWARD SHIFT
IN COASTAL ONLAP

CONDENSED

CONDENSED

THIN OR ABSENT

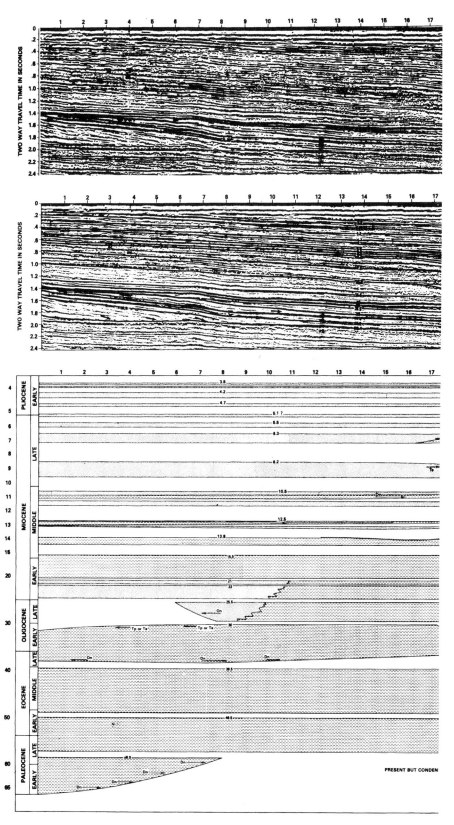

Fig. 7.4. Seismic line oriented NW-SE, perpendicular to the continental margin off Alabama, showing uninterpreted line, interpreted line, and chronostratigraphic chart. (Greenlee and Moore 1988)

OFFSHORE ALABAMA

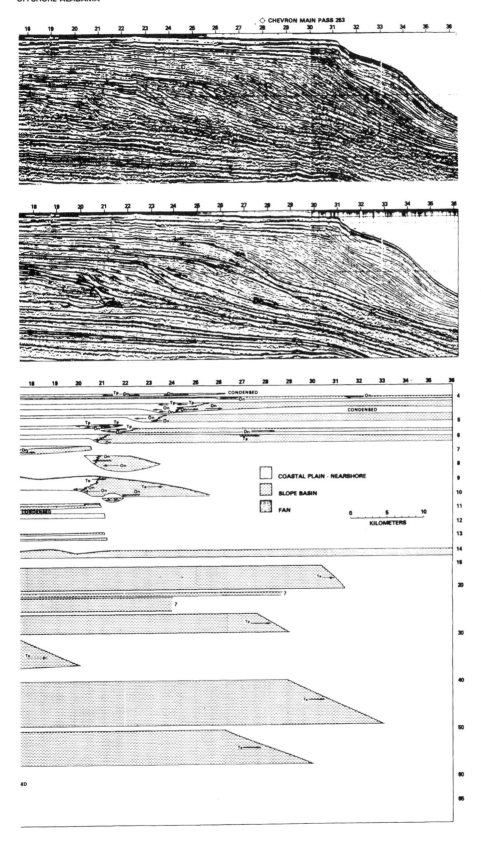

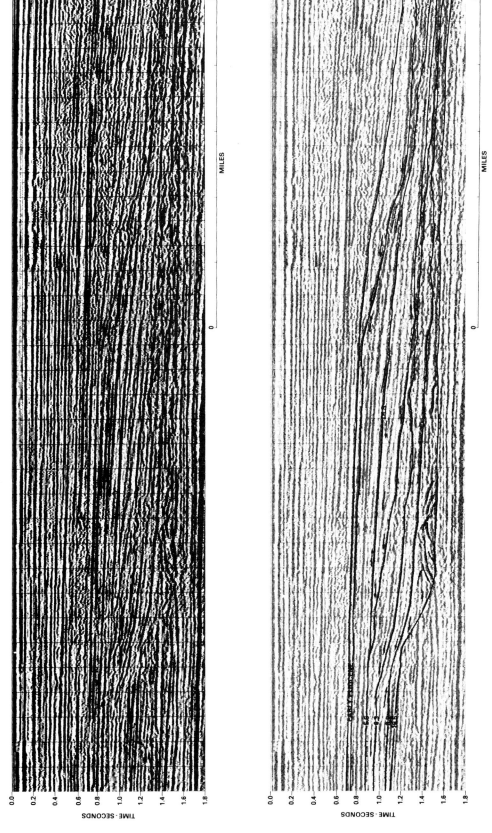

Fig. 7.5. Detail of Fig. 7.4, showing erosion and onlap of the Middle Miocene ("10.5-Ma") sequence boundary. (Greenlee and Moore 1988)

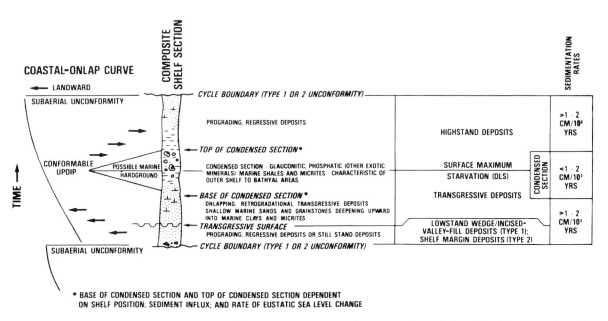

Fig. 7.6. Shelf sedimentary model related to the sequence concepts of Posamentier et al. (1988) and Posamentier and Vail (1988). (Baum and Vail 1988)

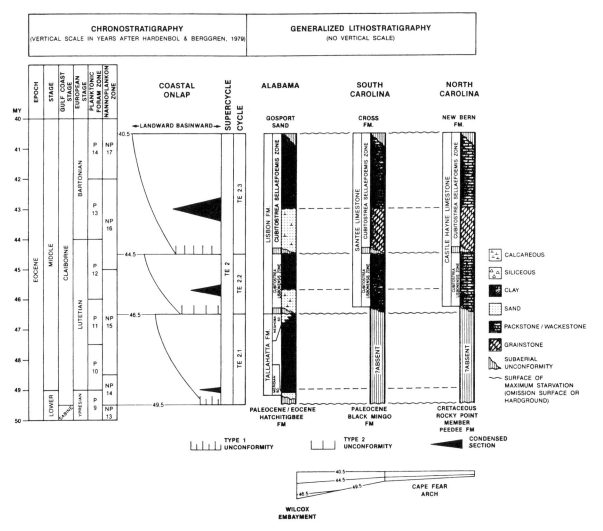

Fig. 7.7. Comparison of chronostratigraphy and lithostratigraphy in outcrop sections in Alabama and the Carolinas, United States. (Baum and Vail 1988)

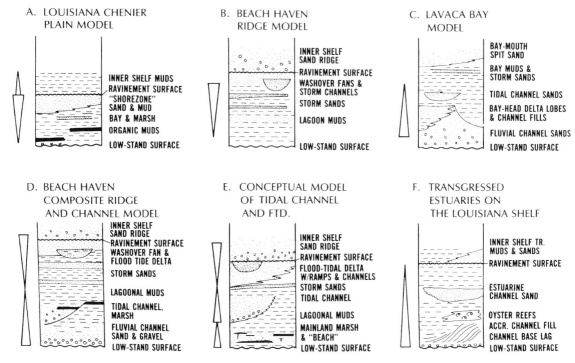

Fig. 7.8A–F. Stratigraphic columns from selected Holocene marginal-marine settings to illustrate transgressive systems tracts. (Nummedal and Swift 1987)

7.3 Other Foreland Basins

Sequence concepts have been employed in the analysis of several other foreland-basin stratigraphic successions. For example, Crumeyrolle et al. (1991) discussed a sequence analysis of the Alpine molasse succession in the Digne Basin, France. This basin (Fig. 7.26) constitutes one of the sub-basins within the Cenozoic Alpine foreland basin of France, Switzerland and Austria. The Oligocene-Miocene succession is more than 2 km thick and

has been subdivided lithostratigraphically, in ascending order, into (Fig. 7.27):

1. Red Molasse (Oligocene): fluvial channel sandstones and channelized fanglomerates.
2. Intermediate Molasse (Aquitanian): thin fluvial and deltaic sandstones and marls.
3. Lower Marine Molasse (Burdigalian): transgressive, wave-dominated deltaic clastics.
4. Upper Marine Molasse (Langhian): wave- and tide-dominated sandstones and marls.
5. Yellow Molasse (Serravallian): prograding succession of channelized congomerates grading

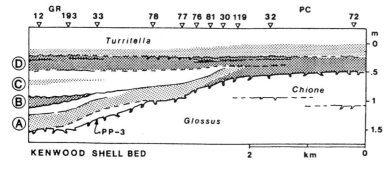

Fig. 7.9. Stratigraphic pinchout and onlapping of shell beds toward the stratigraphic edge of the coastal plain. Density of *stippling* relates to the abundance of shells.

Each shell bed rests on a major erosion surface and sequence boundary. (Kidwell 1989, reproduced by permission of the University of Chicago Press)

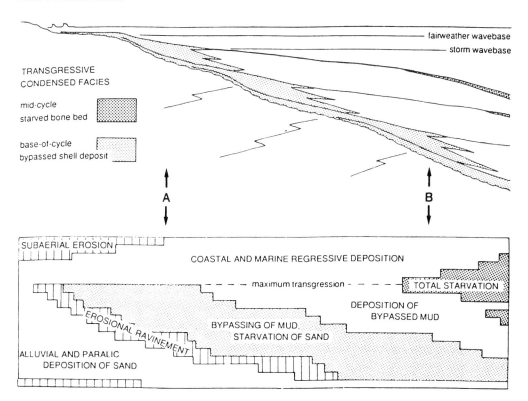

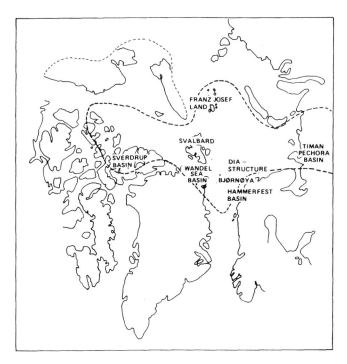

Fig. 7.10. Idealized shelf transect during transgression, showing the landward migration of various depositional processes and facies belts. Shell beds characterize the bypass zone above the ravinement surface. *Top diagram* is an idealized stratigraphic cross section, *bottom section* has time as the vertical axis. Note that the sequence rests on a ravinement surface, which is time-transgressive. (Kidwell 1989)

Fig. 7.11. Location map of Arctic Mesozoic basins. (Mork et al. 1989, reproduced by permission of the University of Chicago Press)

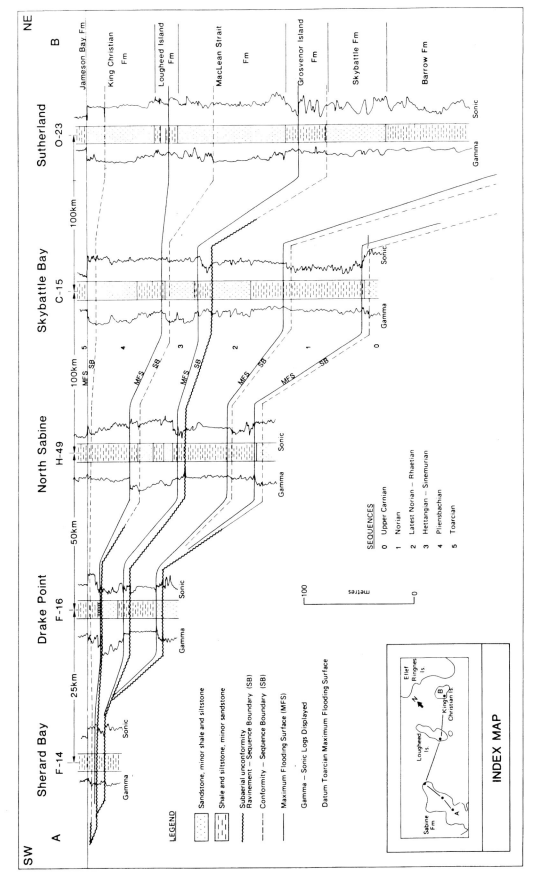

Fig. 7.12. Stratigraphic cross section of Norian to Pliensbachian strata, western Sverdrup basin, Canadian Arctic Islands. (Johannessen and Embry 1989)

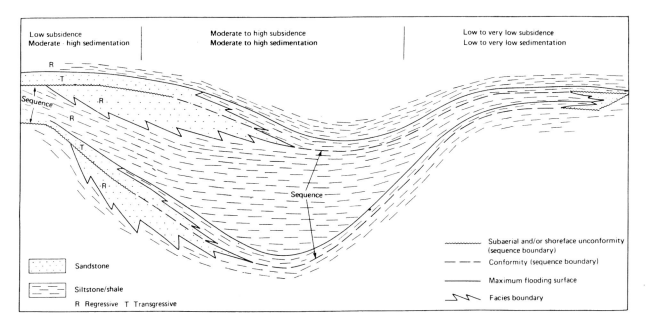

Fig. 7.13. Stratigraphic configuration of a third-order depositional sequence in areas of low to moderate subsidence and sedimentation rate. (Johannessen and Embry 1989)

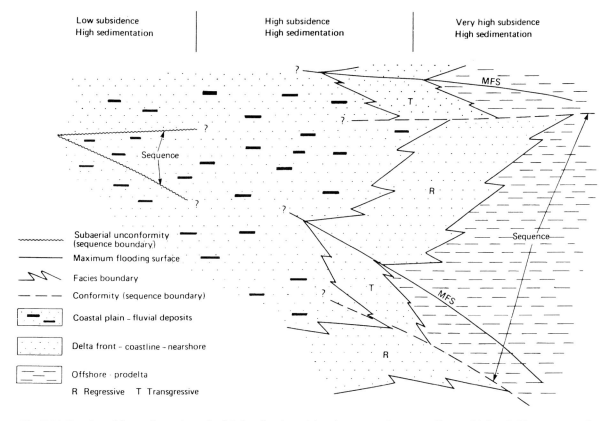

Fig. 7.14. Stratigraphic configuration of a third-order depositional sequence in areas of low to high subsidence rate and high sedimentation rate. (Johannessen and Embry 1989)

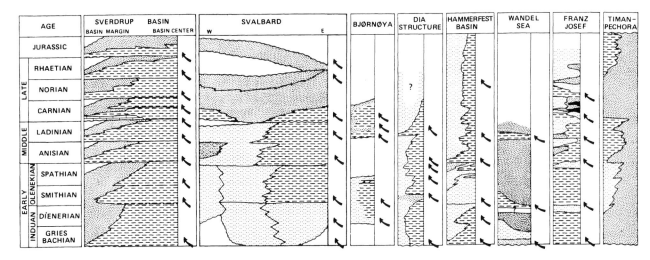

Fig. 7.15. Triassic stratigraphy and correlation across the Arctic from Canada to Russia. *Arrows* indicate transgressions; *stipple* sandstone; *double dots* siltstone; *dashes* shale; *bricks* limestone. (Mork et al. 1989)

Age of transgression	Well dated transgressions (areas)	Fairly well dated transgressions (areas)
earliest Jurassic		Sverdrup Basin , Svalbard
latest Norian		Sverdrup Basin , Svalbard , Hammerfest Basin
earliest Norian	Sverdrup Basin	Franz Josef Land Hammerfest Basin
mid Carnian		Sverdrup Basin
earliest Carnian	Sverdrup Basin , Svalbard	Bjørnøya
early Ladinian		Sverdrup Basin , Svalbard , Bjørnøya , ?Hammerfest Basin
earliest Anisian	Sverdrup Basin , Svalbard	Dia-structure , Hammerfest Basin, Franz Josef L.
late Smithian	Sverdrup Basin , Dia-stucture	
earliest Smithian	Sverdrup Basin , Svalbard , Bjørnøya	Hammerfest Basin Franz Josef Land Wandel Sea Basin
early Dienerian		Svalbard , Bjørnøya , Hammerfest Basin
earliest Griesbachian	Sverdrup Basin , Svalbard , Dia-structure	Hammmerfest Basin , Franz Josef Land

Fig. 7.16. List of Triassic transgressions in the Arctic, and basins where they hav0e been identified. (Mork et al. 1989)

distally into alluvial-plain and shallow-marine deposits.

The Lower Marine Molasse constitutes two third-order stratigraphic sequences (Fig. 7.28). The total section is more than 800 m thick. The lowermost sequence boundary is a ravinement surface. It is followed by a succession of landward-stepping, wave-dominated, deltaic cycles constituting a transgressive systems tract. A condensed interval consists of bored and glauconitized beds, and is followed by progradational deltaic cycles corresponding to the highstand tract. The second se-

quence begins with a bioclastic tidal bar deposit resting on an erosion surface that corresponds to the sequence boundary. The top of the tidal bar is marked by a maximum-flooding surface, and is followed by shelf mudstones. These, in turn are overlain by prograding highstand deltaic deposits. The uppermost of these consist of brackish lagoonal marls containing oysters and pelecypods.

Crumeyrolle et al. (1991) claimed that the Digne Basin deposits can be correlated with the Exxon global cycle chart. They suggested that the overall molasse succession is a long-term trangressive-regressive supercycle induced by foreland-basin tec-

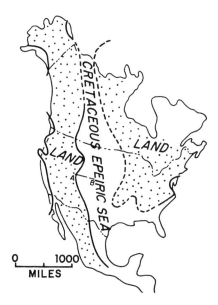

Fig. 7.17. Location of Western Interior Seaway in western North America during the Cretaceous. *A–B* indicates the line of section shown in Fig. 7.18. (Weimer 1970)

tonism. It is punctuated by unconformities and is subdivisible into component sequences, which they suggested may be related to eustatic events.

The South Pyrenean foreland basin of northeastern Spain (Fig. 7.29) contains a mixed carbonate-siliciclastic succession of Paleocene-Eocene age 3 km thick that has been subdivided into nearly twenty sequences having 10^6-year episodicities.

Contrasting analyses of different, but overlapping portions of this succession were provided by Puigdefábregas et al. (1986) and Luterbacher et al. (1991; Figs. 7.30, 7.31). In the first of these papers the authors attributed the development of the sequences to tectonism, the gradual southward overthrusting of the fold-thrust belt leading to a migration and offlapping of depocenters (Fig. 7.30). However, in the second paper it is claimed that correlation of the sequences with the Exxon chart can be carried out, and the authors interpreted eustasy as the main driving mechanism. The contrasts between these kinds of interpretation are discussed in Section 11.3.2.

An idealized cross section of the Eocene-Oligocene portion of the section is shown in Fig. 7.30. The overall structure of the sequences is that of a coarse, nonmarine clastic wedge prograding southward from the thrust front into basin-center marine marls. Shallow-water carbonate deposits are formed on the distal ramp and forebulge of the foreland basin, away from the influence of the fold-thrust-belt clastic source. Carbonates are typically formed in the transgressive stage of each sequence. They may be associated with evaporites formed during the subsequent regressive stage.

Many other studies of the sequence stratigraphy of Pyrenean foreland basins have been published recently. Two contrasting themes commonly are presented in these papers (see in particular Deramond et al. 1993; Millan et al. 1994). On the one

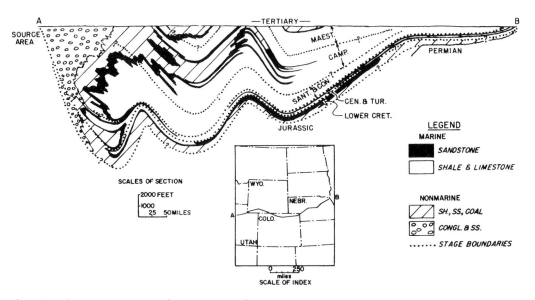

Fig. 7.18. Diagrammatic restored cross section through the Upper Cretaceous rocks of the Western Interior Seaway, flattened on a datum at the base of the Tertiary. Line of section is given by *A–B* in Fig. 7.17. (Weimer 1970)

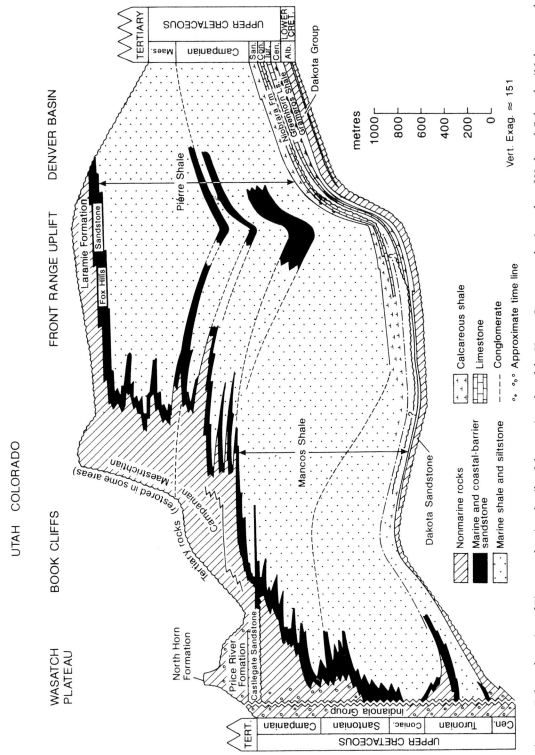

Fig. 7.19. Enlarged portion of Fig. 7.18, showing details of the stratigraphy of the Upper Cretaceous clastic wedge of Utah and Colorado. (Molenaar and Rice 1988)

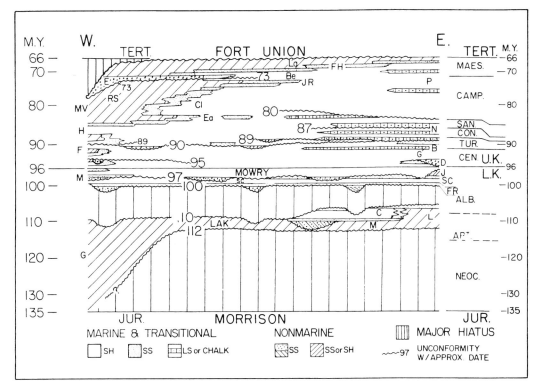

Fig. 7.20. Diagrammatic west-east cross section through the Western Interior Seaway of the Rocky Mountains, from Wyoming-Montana in the west to eastern Colorado-Black Hills-eastern Alberta in the east, showing stratigraphic positions and approximate dates of major transgressive units and interregional unconformities. Formations or groups to the west are: *G* Gannett; *SC* Skull Creek; *M* Mowry; *F* Frontier; *H* Hilliard; *MV* Mesaverde; *RS* Rock Springs; *E* Ericson; *Ea*, Eagle; *Cl* Claggett; *JR* Judith River; *Be* Bearpaw; *FH* Fox Hills; *La* Lance. To the east formations are: *L* Lytle; *LAK* Lakota; *FR* Fall River; *SC* Skull Creek; *J* and *D* sands of Denver basin; *G* Greenhorn; *B* Benton; *N* Niobrara; *P* Pierre; *M* and *C*, McMurray and Clearwater of Canada. (Weimer 1986, reproduced by permission)

hand, considerable structural and stratigraphic evidence is presented to demonstrate tectonic control of sedimentation. On the other hand, close comparison of the sequence-bounding events with those shown in the Exxon chart is claimed, indicating eustatic influences. These ideas are discussed further in Section 13.4.1.

7.4 Forearc Basins

Forearc and backarc basins occur within so-called "active margins," a term which emphasizes the importance of tectonism in controlling stratigraphic architectures. Several recent studies of arc-related basins have examined the basin fills from the perspective of sequence stratigraphy, and in many cases major differences with the stratigraphic styles of extensional and rifted margins, and even with

foreland basins, have become apparent. There is little clear evidence for the existence of cycles caused by 10^6-year eustatic sea-level cycles, unlike higher-frequency cycles, including those of glacioeustatic origin which, presumably because of their rapidity, are locally prominent in arc-related basins and have been mapped and documented in detail, for example, within the Japanese islands (Chap. 8).

Several studies of arc-related basins have been carried out recently in Nicaragua and Costa Rica (Seyfried et al. 1991; Schmidt and Seyfried 1991; Kolb and Schmidt 1991; Winsemann and Seyfried 1991). In general they demonstrate the importance of convergent tectonism as a dominant control in the development of stratigraphic architecture. An example of a sequence-stratigraphic framework in a forearc basin was provided by Kolb and Schmidt (1991, Fig. 7.32). Correlations of these and other sections in Central America with the Exxon global cycle chart seem forced and unconvincing. Bio-

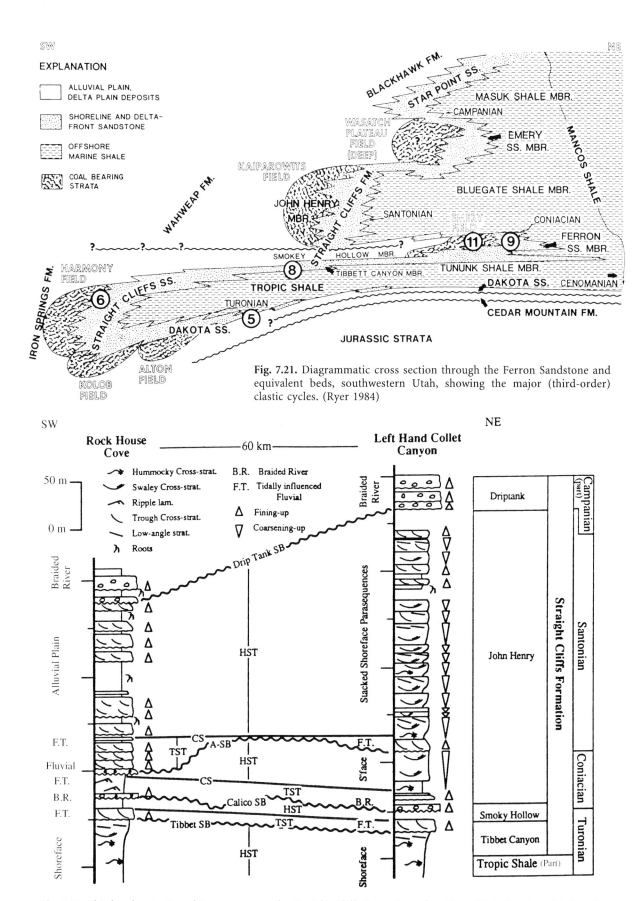

Fig. 7.21. Diagrammatic cross section through the Ferron Sandstone and equivalent beds, southwestern Utah, showing the major (third-order) clastic cycles. (Ryer 1984)

Fig. 7.22. Third-order stratigraphic sequences in the Straight-Cliffs Formation of southern Utah. Stratigraphic location of these sections is shown in Fig. 7.21. (Shanley and McCabe 1991)

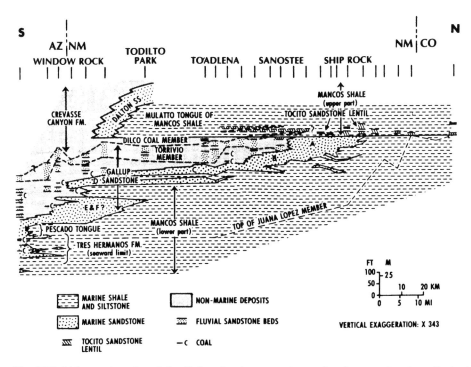

Fig. 7.23. Lithostratigraphy of the Gallup Sandstone and associated strata, San Juan Basin, New Mexico. (Molenaar 1983)

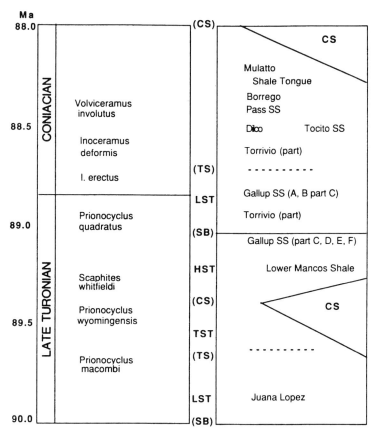

Fig. 7.24. Summary of the sequence stratigraphy of the Gallup Sandstone and associated strata, San Juan Basin, New Mexico. (Nummedal 1990, reproduced by permission of Kluwer Academic Publishers)

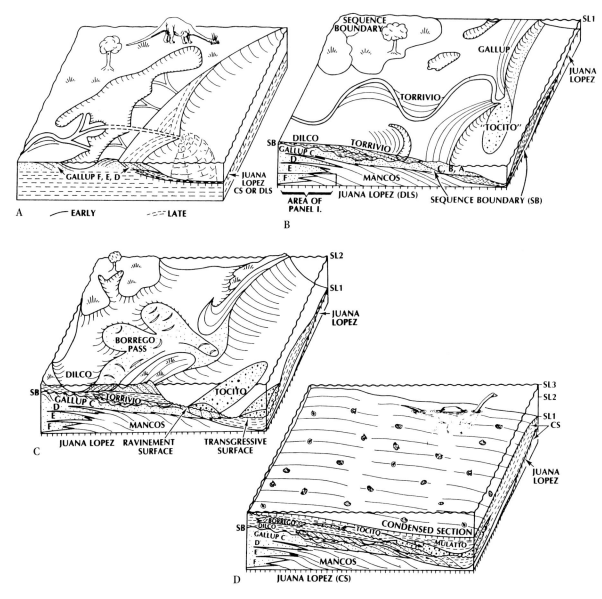

Fig. 7.25. Block diagrams illustrating the development of the Gallup Sandstone and associated strata in the San Juan Basin of New Mexico, during the Coniacian-Turonian. **A** Highstand deposition of the basal Gallup Sandstone tongues (*F, E,* and *D*). **B** Lowstand deposition of tongue *C* and laterally equivalent beds of the fluvial Torrivio Member. **C** Transgressive deposition of the To-cito and Borrego Pass Members. **D** Time of maximum flooding, with the formation of a condensed section corresponding to the Mulatto Tongue of the Mancos Shale. (Nummedal 1990, reproduced by permission of Kluwer Academic Publishers)

stratigraphic evidence for the correlations is extremely limited. In most cases it is evident that folding, faulting, tilting and tectonic uplift and subsidence are the major sedimentary controls (Seyfried et al. 1991; Schmidt and Seyfried 1991). Seyfried et al. (1991) documented the presence of regional angular unconformities that can be mapped for distances of 900 km, as discussed in Section 6.3.2 (Fig. 6.17).

In some cases volcanic control of the sediment supply is the critical factor. Thus Winsemann and Seyfried (1991, pp. 286–287) stated:

The formation of depositional sequences in the deep-water sediments of southern Central America is strongly related to the morphotectonic evolution of the island-arc system. Each depositional sequence reflects the complex interaction between global sea-level fluctuations, sediment supply, and tectonic activity. Sediment supply and

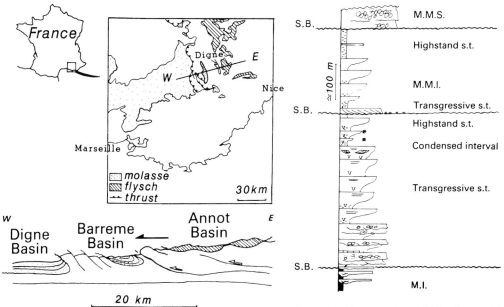

Fig. 7.26. Location map, Digne basin, France, and simplified structural cross section. (Crumeyrolle et al. 1991)

Fig. 7.28. Type section of the Lower Marine Molasse (*M.M.I.*), Digne Basin, showing its subdivision into two stratigraphic sequences. *S.B.* Sequence boundary; *s.t.* systems tract. (Crumeyrolle et al. 1991)

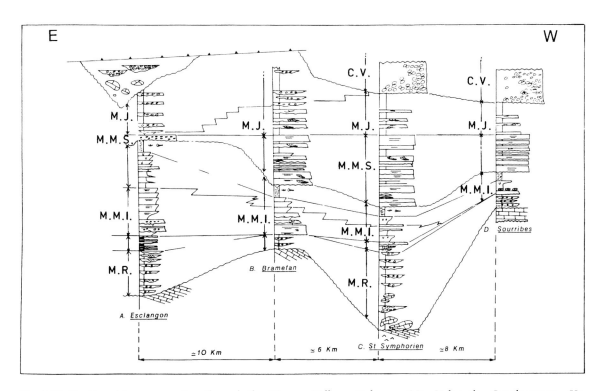

Fig. 7.27. Stratigraphic cross section through the Digne molasse basin. *M.R.* Red Oligocene Molasse; *M.M.I.* Lower Marine Molasse; *M.M.S.* Upper Marine Molasse; *M.J.* Yellow Molasse; *C.V.* Valensole Conglomerate. Unconformities constitute the main sequence boundaries. (Crumeyrolle et al. 1991)

tectonic activity overprinted the eustatic effects and enhanced or lessened them. If large supplies of clastics or uplift overcame the eustatic effects, deep marine sands were also deposited during highstand of sea level, whereas under conditions of low sediment input, thin-bedded turbidites were deposited even during lowstands of sea level.

These sediment-supply considerations are of paramount local importance. They indicate results contrary to the relationship between sea level and deep-water sedimentation that are proposed in the Exxon models (Sect. 4.3) and as have been discussed by Shanmugam and Moiola (1982; Sect. 6.1).

A Jurassic-Cretaceous forearc basin succession in the Antarctic Peninsula is dominated by major facies changes at intervals of several millions of years, suggesting a control by "third-order" tectonism (Butterworth 1991). One major, basin-wide stratigraphic event, an abrupt shallowing, followed by a transgression, gave rise to a unit named the Jupiter Glacier Member (Fig. 7.33), consisting of deepening-upward shelf deposits. Butterworth (1991) suggested a correlation with a major eustatic low near the Berriasian-Valanginian boundary indicated on the Exxon chart (Fig. 7.33). However, this seems somewhat fortuitous. It is the only such correlation that can be proposed for this basin.

A stratigraphic synthesis of the Cenozoic deposits of Cyprus by Robertson et al. (1991) also resulted in tentative (and rather tenuous) correlations of some stratigraphic events with the Exxon global cycle chart (Fig. 7.34). The evidence indicates that subduction and strike-slip deformation had a dominant effect on sedimentation patterns in this area. During the Late Miocene (Messinian) plate-tectonic events led to the isolation and desiccation of the entire Mediterranean basin, an event that is clearly recorded in Cyprus (Fig. 7.34), at a time when global sea levels underwent several fluctuations, according to Haq et al. (1987, 1988a).

7.5 Backarc Basins

The tectonic evolution of backarc basins may compare with that of extensional margins, especially along the interior, cratonic flanks of the basin. Legarreta and Uliana (1991) found that the Neuquén Basin, a backarc basin flanking the Andes in Argentina, underwent an "exponential thermo-mechanical subsidence" pattern, following an Early Mesozoic thermal event. Sediment-supply condi-

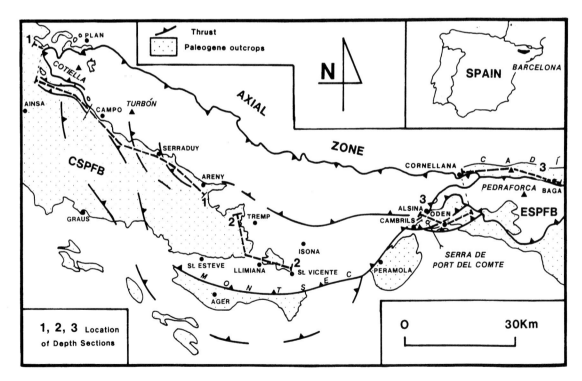

Fig. 7.29. Location of the Central and Eastern South Pyrenean foreland basin (*SPFB*) in northeastern Spain. (Luterbacher et al. 1991)

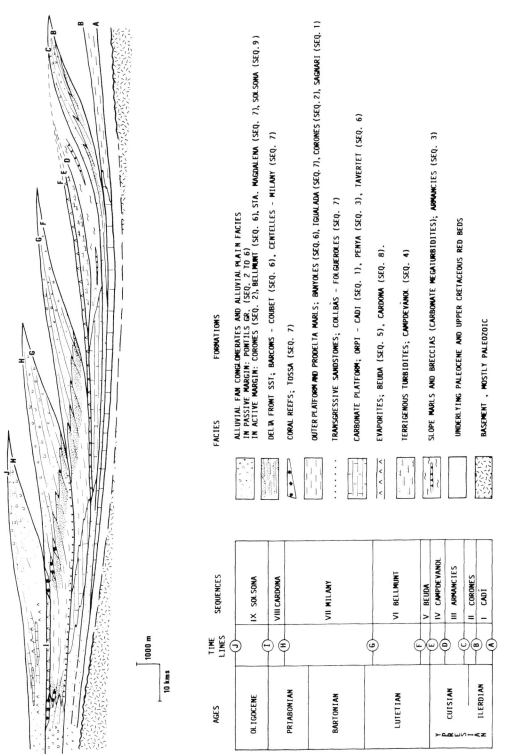

FACIES **FORMATIONS**

ALLUVIAL FAN CONGLOMERATES AND ALLUVIAL PLAIN FACIES
IN PASSIVE MARGIN: PONTILS GR. (SEQ. 2 TO 6)
IN ACTIVE MARGIN: CORONES (SEQ. 2), BELLMUNT (SEQ. 6), STA. MAGDALENA (SEQ. 7), SOLSONA (SEQ. 9)

DELTA FRONT SST; BARCONS - COUBET (SEQ. 6), CENTELLES - MILANY (SEQ. 7)

CORAL REEFS; TOSSA (SEQ. 7)

OUTER PLATFORM AND PRODELTA MARLS; BANYOLES (SEQ. 6), IGUALADA (SEQ. 7), CORONES (SEQ. 2), SAGNARI (SEQ. 1)

TRANSGRESSIVE SANDSTONES; COLLBÀS - FOLGUEROLES (SEQ. 7)

CARBONATE PLATFORM; ORPI - CADI (SEQ. 1), PENYA (SEQ. 3), TAVERTET (SEQ. 6)

EVAPORITES; BEUDA (SEQ. 5), CARDONA (SEQ. 8).

TERRIGENOUS TURBIDITES; CAMPDEVÀNOL (SEQ. 4)

SLOPE MARLS AND BRECCIAS (CARBONATE MEGATURBIDITES); ARMANCIES (SEQ. 3)

UNDERLYING PALEOCENE AND UPPER CRETACEOUS RED BEDS

BASEMENT, MOSTLY PALEOZOIC

AGES	TIME LINES	SEQUENCES
OLIGOCENE	Ⓙ	IX SOLSONA
	Ⓘ	VIII CARDONA
PRIABONIAN	Ⓗ	
BARTONIAN		VII MILANY
	Ⓖ	
LUTETIAN		VI BELLMUNT
	Ⓕ	V BEUDA
	Ⓔ	IV CAMPDEVÀNOL
	Ⓓ	III ARMANCIES
CUISIAN	Ⓒ	II CORONES
ILERDIAN	Ⓑ	I CADÍ
	Ⓐ	

1000 m

10 kms

Fig. 7.30. Idealized cross section of the Eocene-Lower Oligocene sequence stratigraphy of the South Pyrenean basin, located near the eastern margin of the area shown in Fig. 7.29. (Puigdefàbregas et al. 1986)

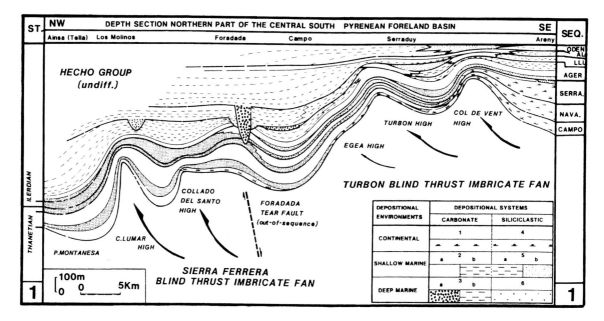

Fig. 7.31. Cross section of the Paleogene strata in the central-southern Pyrenees, showing subdivision into stratigraphic sequences. (Luterbacher et al. 1991)

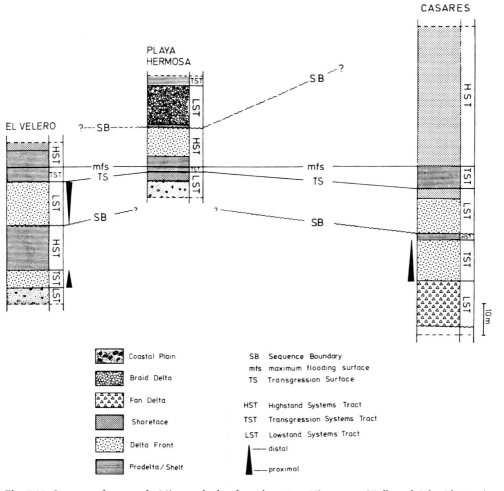

Fig. 7.32. Sequence framework, Miocene beds of southwestern Nicaragua. (Kolb and Schmidt 1991)

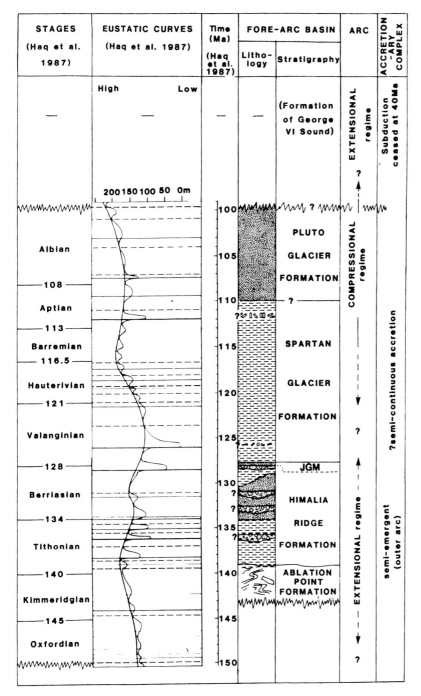

Fig. 7.33. Stratigraphy of the southern Antarctic Peninsula, compared to the Exxon global cycle chart (Haq et al. 1987, 1988a). One stratigraphic event, a shallowing that gave rise to the Jupiter Glacier Member (*JGM*), may correlate with the Berriasian-Valanginian eustatic low on the cycle chart. (Butterworth 1991)

tions along the cratonic flank of a backarc basin, are also likely to be comparable to extensional margins. For this reason it may be expected that these basins may show stratigraphic patterns comparable to those on Atlantic-type margins, including the presence of major carbonate suites (Sect. 7.7), relatively mature clastics, and a sequence architecture containing evidence of cyclicity with 10^6–10^7-year episodicities. Two studies of Andean basins confirm that this is the case. Legarreta and Uliana (1991) described a sequence stratigraphy that they correlated directly with the Exxon chart, while Hallam (1991), summarizing his own work plus that of various South American geologists, developed his own regional sea-level curve that contains transgressive-regressive cycles with 10^6–10^7-year frequencies (Fig. 7.35).

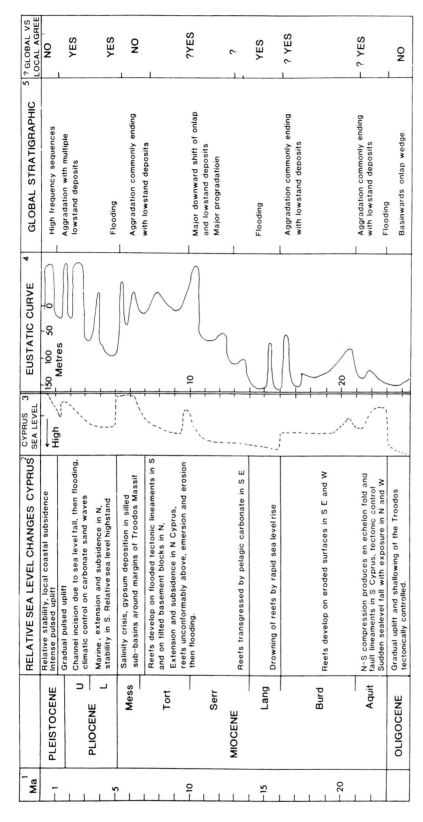

Fig. 7.34. Summary of stratigraphic events in Cyprus, with an interpreted sea-level curve. This is compared to the Exxon global cycle chart (Haq et al. 1987, 1988a). (Robertson et al. 1991)

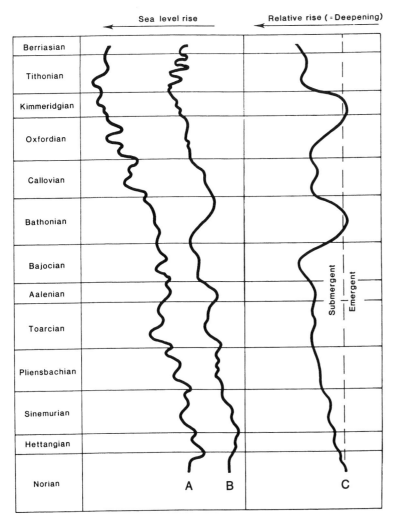

Fig. 7.35. Comparison of latest Triassic to Jurassic eustatic curves from three sources. **A** Hallam's revised Jurassic global curve. **B** The curve of Haq et al. (1987, 1988a). **C** A curve for Andean basins derived from the stratigraphic synthesis by Hallam (1991)

7.6 Cyclothems and Mesothems

A unique type of high-frequency cyclicity characterizes the Carboniferous and Lower Permian strata of much of the North American Midcontinent, northwestern Europe and the Russian platform (Ross and Ross 1988). As discussed in Chapter 10 there is general agreement that these cycles are glacioeustatic in origin. The Carboniferous-Early Permian corresponds to the time when major continental ice caps were forming and retreating throughout the great southern Gondwana supercontinent (Caputo and Crowell 1985; Veevers and Powell 1987), while the areas where cyclothems and mesothems occur lay close to the Late Paleozoic paleoequator.

The term cyclothem was proposed by Wanless and Weller (1932) following their study of these cycles in the Upper Paleozoic rocks of the American

Midcontinent. The term, and the cycles to which it applies, are discussed in Chapter 8. Ramsbottom (1979) noted that unusually extensive transgressive and regressive beds forming the sequence boundaries between some of the cyclothems enable groups of about four or five of them to be combined into larger cycles showing a 10^6-year periodicity, and he proposed the term mesothem for these. Moore (1936) and Wagner (1964) termed groups of cyclothems megacyclothems, but Heckel (1986) showed that these are higher-order cycles than the mesothems discussed here. Holdsworth and Collinson (1988) used the term major cycle for Ramsbottom's mesothems. They compare in duration to the "third-order" cycles of Haq et al. (1987, 1988a). Ramsbottom (1979) identified nearly forty such cycles in the Carboniferous of northwestern Europe, and showed that their average duration ranged from 1.1 m.y. in the Namurian to 3.6 m.y. in the Dinantian. A chronostratigraphic chart of these cycles

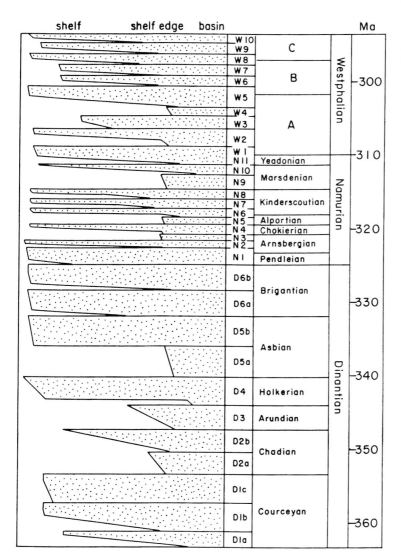

Fig. 7.36. Mesothems in the Carboniferous succession of northern England. (Ramsbottom 1979, reproduced by permission of The Geological Society, London)

as they occur in Britain is shown in Fig. 7.36, and a more detailed Wheeler diagram, showing the main lithostratigraphic components of the Namurian mesothems of part of northern England, is given in Fig. 7.37.

In the Namurian, each mesothem consists of a muddy sequence at the base containing several cyclothems, followed by one or more sandy cyclothems. The lower, muddy parts of the mesothems are broadly transgressive, although the transgressions were slow and pulsed, each cyclothemic transgression reaching further than its predecessor out from the basin on to the shelf (Fig. 7.37). Basal beds in each cycle may contain evidence of high salinities, reflecting the isolation of individual small basins at times of low sea level. Marine beds containing distinctive goniatite faunas are supposedly more extensive at the transgressive

base of mesothems than in the component cyclothems, indicating more pronounced (higher amplitude) eustatic rises corresponding to the mesothemic cyclicity. The regression at the end of each mesothem appears to have occurred rapidly. The sandy phase commonly commenced with turbidites and is followed by thick deltaic sandstones (commonly called Grits in the British Namurian; see Fig. 7.37). The cycles may be capped by coal. On the shelf each mesothem is bounded by a disconformity (Fig. 7.37), but sedimentation probably was continuous in the basins. Deltaic progradation was rapid, approaching the growth rate of the modern Mississippi delta.

Ramsbottom (1979) noted that the Namurian mesothems were of the shortest duration, and many do not extend up onto the shelf. This stage spans the Mississippian-Pennsylvanian boundary, which is

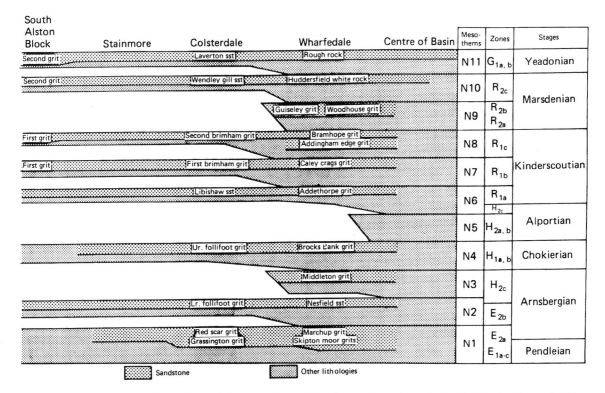

Fig. 7.37. Chronostratigraphic section of the Namurian rocks of part of northern England, showing the main litho-stratigraphic units. (Ramsbottom 1979, reproduced by permission of The Geological Society, London)

designated as the boundary between the Kaskasia and Absaroka sequences in North America, a time at which sea level was at a long-term eustatic lowstand (Sloss 1963; Figs. 1.1, 3.3).

A subsurface analysis of the Middle Pennsylvanian cyclothem record in Kansas reveals an unmistakeable "mesothem" pattern, which Youle et al. (1994) attributed to 10^6-year eustatic cycles. They noted that higher-frequency cyclothems form a transgressive set in which deep-water deposits become more important upwards, and extend successively further onto the craton. The first few cycles overstep each other to onlap Mississippian basement. The highstand sequence set shows a distinctly progradational pattern, recording the long-term gradual drop in average sea level.

The Carboniferous succession in northern England, the type area for the mesothem model, contains evidence for considerable local tectonism, and for lithostratigraphic complexity resulting from local autocyclic controls, such as delta-lobe switching. As a result, mesothemic cyclicity is not everywhere apparent, and Holdsworth and Collinson (1988) provided a critique of the mesothem model based on detailed local studies. In places the evidence does

not suport the simple mesothem model of Ramsbottom (1979; see also Leeder 1988). However, identification of a comparable form of cyclicity in the Late Paleozoic cyclothem succession of Kansas, as noted above, indicates that the mesothem concept should not be discarded. Groups of cycles are more likely to be mappable where tectonic influences are minor, as in Kansas.

Ross and Ross (1988) carried out a worldwide comparison of these Upper Paleozoic cyclic deposits, and proposed detailed correlations between cratonic sections in the United States, northwestern Europe and the Russian platform (Fig. 7.38). They recognized about sixty cycles in the Lower Carboniferous to Lower Permian stratigraphic record. They suggested that similar cyclic successions may occur in contemporaneous continental-margin sections, but that because most of these have been deformed by post-Paleozoic plate-tectonic events the record from these areas is less well known. In view of the scepticism surrounding the mesothem concept, and the difficulties in correlation based on limited chronostratigraphic data, such intercontinental correlation may be regarded as premature.

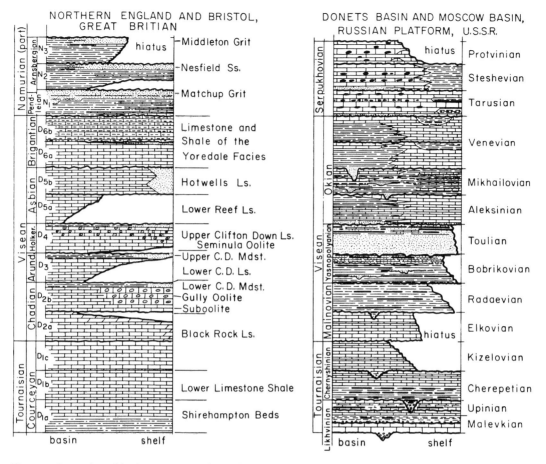

Fig. 7.38. Examples of the intercontinental correlations proposed for Upper Paleozoic mesothems. These diagrams show Lower Carboniferous sections. (Ross and Ross 1988)

7.7 Carbonate Cycles of Platforms and Craton Margins

Carbonate sediments are sensitive indicators of changing sea level for various reasons, including the following three: Firstly, the "carbonate factory" that develops within warm, shallow continental shelves that are free of clastic detritus can produce carbonate sediment at a rate that normally is rapid enough to keep up with the most rapid of sea-level changes. Secondly, within such shallow-water carbonates, depth changes are indicated by a variety of depth-sensitive facies characteristics. Thirdly, within carbonate platforms most carbonate sediments are deposited where they are produced. The complications of sediment redistribution by autocyclic processes that characterize clastic sediments (e.g., the development of delta lobes) therefore are less extreme (on the continental slope it is a different

story; see also Sect. 15.4). For these reasons stratigraphic sequences are commonly well developed in carbonate rocks. Hierarchies of sequences may be present, reflecting the integration of several different generative mechanisms (Sarg 1988; Goldhammer et al. 1990; Schlager 1992a). It is important to note, however, that sequence boundaries can develop as a result of submarine erosion and environmental change, as well as in response to sea-level change (Sect. 2.2). Correlations with the global cycle chart should therefore be viewed with particular caution (Fig. 14.6).

Prograding continental margins provide many examples of 10^6-year carbonate sequence stratigraphy, and these are well displayed in regional reflection-seismic records. Eberli and Ginsburg (1988, 1989) described an excellent example of this, revealing the dramatic lateral growth by progradation of the Bahama Platform. Other examples were given by Sarg (1988), Mitchum and Uliana (1988),

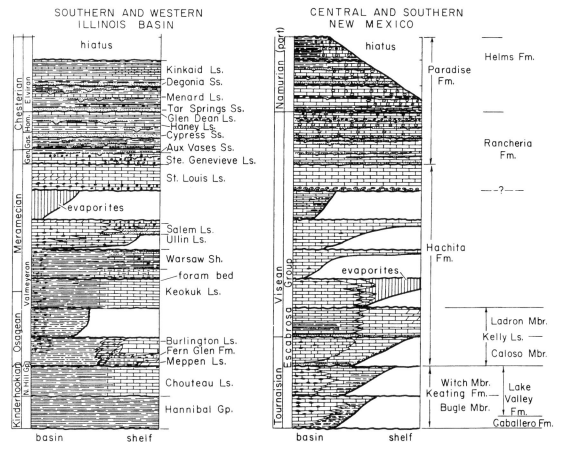

Fig. 7.38 (*contd.*)

and Epting (1989). Bosellini (1984) and Sarg (1988) provided outcrop examples. Many detailed case studies are contained in the book edited by Loucks and Sarg (1993).

Figures 7.39–7.42 illustrate the architecture and sequence stratigraphy of Miocene to Recent carbonate buildups in offshore Sarawak. The buildups are gradually extending seaward, and are being buried by progradation of deltaic clastics. This is a very common pattern in continental-margin carbonate shelves (e.g., Devonian of the Alberta Basin, Canada: Moore 1989). In the Sarawak example, the most

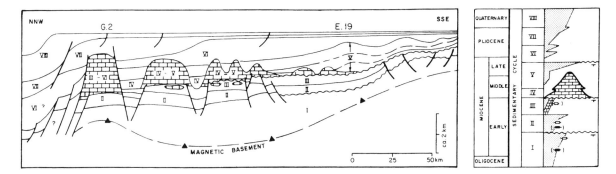

Fig. 7.39. Simplified stratigraphic cross section showing relationship between carbonate buildups and depositional cycles, offshore Sarawak. The carbonate buildups of *cycles III-VI* are buried by deltaic clastics of *cycles V-VIII*.

Seaward stepping of the carbonate buildups and deltaic progradation are continuing to the present day. (Epting 1989, reproduced by permission)

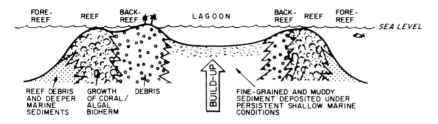

1. RATE OF CARBONATE PRODUCTION IS
   IN BALANCE WITH RISING SEA LEVEL

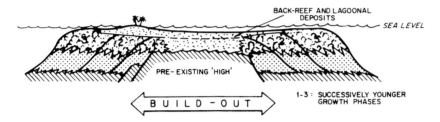

2. RATE OF CARBONATE PRODUCTION
   EXCEEDS RISE IN SEA LEVEL

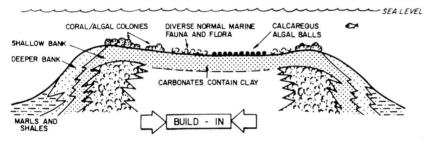

3. RATE OF CARBONATE PRODUCTION CANNOT KEEP PACE WITH RISING SEA LEVEL
   AND/OR IS REDUCED BY TERRIGENOUS INFLUX: SUBMERGED CARBONATE BANK

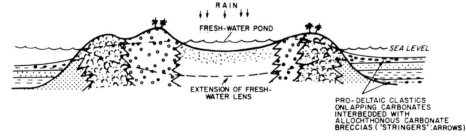

4. CARBONATE PRODUCTION IS TERMINATED BY FALL
   IN SEA LEVEL: SUBAERIAL EXPOSURE AND EARLY DIAGENESIS

Fig. 7.40. The four main stages (*1–4*) of reef growth, Miocene to Recent, offshore Sarawak. (Epting 1989, reproduced by permission)

recent reef buildup, off the section to the north, is not yet drowned by deltaic progradation and is still actively developing.

An example of a standard Exxon-type analysis of a carbonate basin margin in a backarc setting was provided by Mitchum and Uliana (1988; Figs. 7.43, 7.44). It is of interest to note the correlation of this

section with the global cycle chart on the basis of a general positioning of the stratigraphy within the Tithonian-Valanginian interval. This was done by comparison (not detailed correlation) of the subsurface with nearby outcrops, where ammonite zonation has been carried out. No faunal data were available from the wells used to correlate the seis-

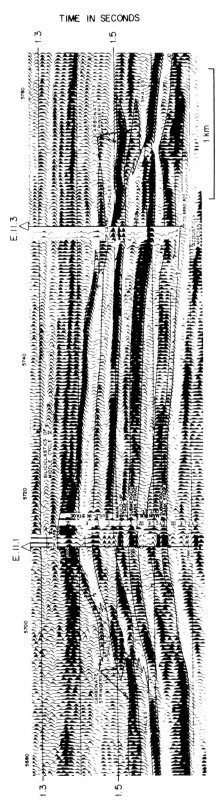

Fig. 7.41. Seismic-reflection line across a carbonate build-up of cycle IV, offshore Sarawak. Internal layering corresponds to alternation of build-up and build-out stage and to the submerged-bank stage. (Epting 1989, reproduced by permission)

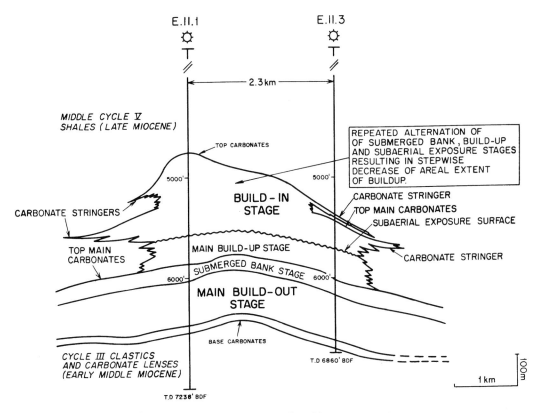

Fig. 7.42. Interpretation of Fig. 7.41. (Epting 1989, reproduced by permission)

mic section! However, the pattern of seismic sequence boundaries is said to match the global pattern for this interval.

Older beds within this same basin are well exposed, and were subjected to a sequence-stratigraphic analysis by Legarreta (1991). Figure 7.45 is a chronostratigraphic chart of these Callovian-Oxfordian beds, and Fig. 7.46 is a simplified cross section, showing the regional architecture and sequence correlation, based on ammonite biostratigraphy. Correlations with the Exxon chart are indicated, but are not discussed here. Each sequence is of 10^6-year type, and preserves the record of minor sea-level fluctuations within a longer-term (10^7-year) transgression and regression. The succession began with areally restricted lowstand clastics and evaporites deposited in eolian, fluvial, and shallow-water hypersaline environments during the Middle Callovian. The first carbonate interval is a thin, transgressive-systems tract overlying a ravinement surface, and denotes slow sedimentation on a gently inclined ramp. As the long-term transgression slowed down carbonate sedimentation was able to catch up with the increase in accomodation space, and the architecture changed from aggradational to progradational. Shelf deposits include

coral-dominated and oolitic grainstones, with bafflestones and framestones developed near the shelf-slope break. Slope to outer shelf deposits consist of oncoidal packstones and grainstones with sponge-algal-coral buildups. Sequence boundaries within this succession are marked by evidence of subaerial exposure and fresh-water diagenesis. Carbonate deposition was abruptly shut off by a fall in sea level in the Late Oxfordian, and the slope was then onlapped by evaporites.

Shelf sediments that were deposited on the margins of the North American craton during the Middle and Upper Cambrian are characterized by prominent large-scale cycles, termed grand cycles by Aitken (1966, 1978), based on his work in the Rocky Mountains of Alberta. Similar cycles have subsequently been documented in the Great Basin, Nevada, the Northwest Territories, and southwestern Newfoundland (Chow and James 1987). The cycles consist of a lower "shaly" half-cycle and an upper carbonate half-cycle. In Newfoundland, the cycles are 100–200 m thick, whereas in the Rocky Mountains, they may exceed 700 m in thickness. Three superimposed cycles in Newfoundland are illustrated in Fig. 7.47. The shaly half-cycles were thought by Chow and James (1987) to represent

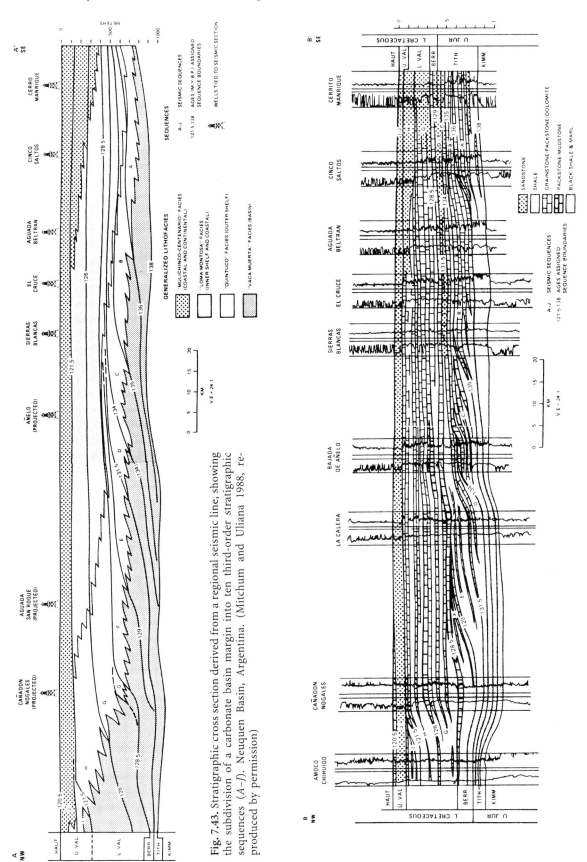

Fig. 7.43. Stratigraphic cross section derived from a regional seismic line, showing the subdivision of a carbonate basin margin into ten third-order stratigraphic sequences (*A–J*). Neuquen Basin, Argentina. (Mitchum and Uliana 1988, reproduced by permission)

Fig. 7.44. The same cross section as in Fig. 7.43, with lithofacies information added from well data, and estimated ages of sequence boundaries, based on comparison with the Haq et al. (1987, 1988a) global cycle chart. (Mitchum and Uliana 1988, reproduced by permission)

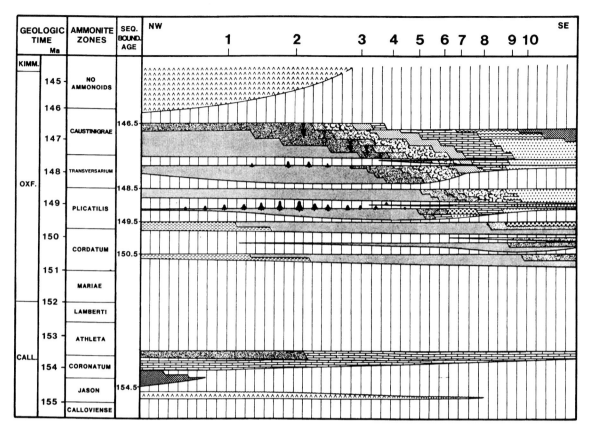

Fig. 7.45. Chronostratigraphic chart of Oxfordian-Callovian third-order sequences in the Neuquén Basin, Argentina. (Legarreta 1991, reproduced by kind permission of Elsevier Science-NL, Sara Burgerhartstraat 25, 1055 KV Amsterdam, The Netherlands)

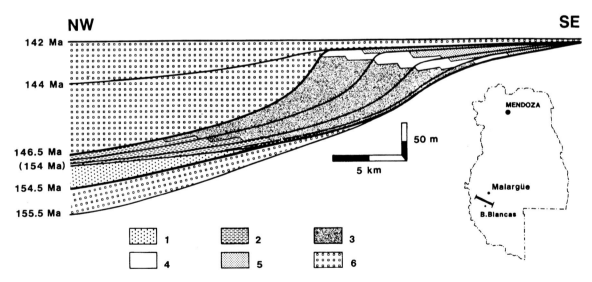

Fig. 7.46. Simplifed cross section of Upper Callovian-Oxfordian carbonate complex, Neuquén Basin, Argentina. *1* Fluvial-eolian clastics; *2* basinal shales and micrites; *3* slope-outer shelf deposits, including oncolitic packstones and grainstones, and sponge-microbial-coral buildups; *4* shelf-margin to inner-shelf coral framestones and bafflestones; *5* inner-shelf to nearshore oolitic grainstones and skeletal rudstones; *6* marine-hypersaline evaporites. (Legarreta 1991, reproduced by kind permission of Elsevier Science-NL, Sara Burgerhartstraat 25, 1055 KV Amsterdam, The Netherlands)

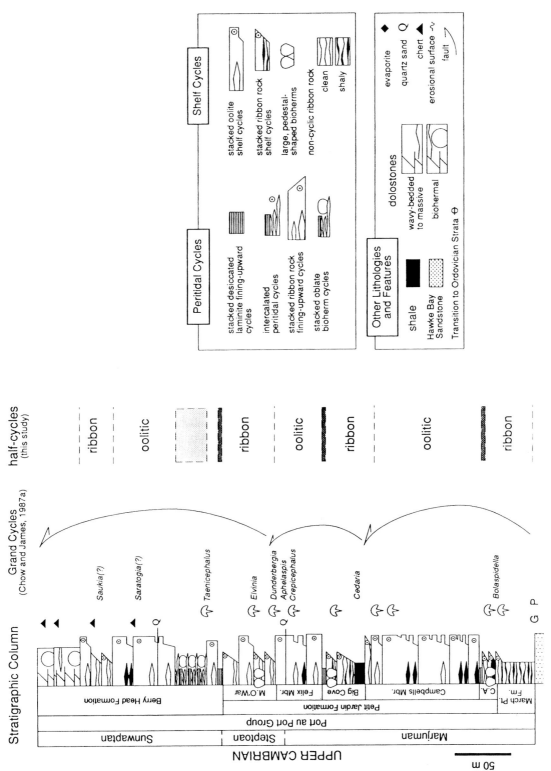

Fig. 7.47. Grand cycles of Middle-Late Cambrian age, southwestern Newfoundland, as originally defined by Chow and James (1987), and showing the new facies subdivision of Cowan and James. (1993)

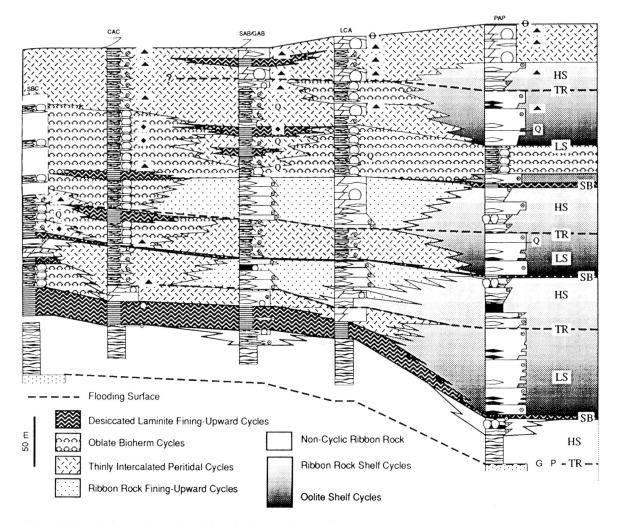

Fig. 7.48. Correlation of Upper Cambrian platform rocks from inboard (*left*) to outboard (*right*) settings. The original grand cycle study of Chow and James (1987) was based on the Port au Port Peninsula section (*PAP*) at right. *SB* Sequence boundary; *LS* lowstand systems tract; *TR* transgression; *HS* highstand systems tract. Compare the sequence classification of the PAP section with the grand cycles of Fig. 7.47. (Cowan and James 1993)

deposition on outer-platform, muddy tidal flats, whereas the upper carbonate half-cycles originated as ooid sand-shoal complexes.

Each of the half-cycles consists of a succession of meter-scale cycles. For example, the lower half-cycle consists of interbedded units of ripple-laminated limestone, flaser- to lenticular-bedded limestone-shale units, and gray shale. In the upper half-cycle, there is a small-scale gradation between ooid calcarenite and carbonate laminite. These small-scale cycles are mostly of autocyclic shoaling-up type.

Chow and James (1987) interpreted the grand cycles as the product of variations in the rate of sea-level rise superimposed on an overall long-term rise. However, their sequence model has recently

undergone a significant revision (Cowan and James 1993), based in part on a more widespread study of the regional stratigraphy that includes correlation of the original grand cycle profile with thinner successions deposited on the inboard area of the platform (Fig. 7.48). In the original Chow and James (1987) model the lower "shaly" half-cycles were interpreted as representing relatively deep-water deposits formed during episodes of rapid rise, whereas the deposits of the upper half-cycles were thought to have formed during periods of slow rise or stillstand, when sedimentation was able to fill the accommodation space. However, Cowan and James (1993) demonstrated that some of the components of the grand cycles are not depth-related facies and cannot therefore be used in the recognition and

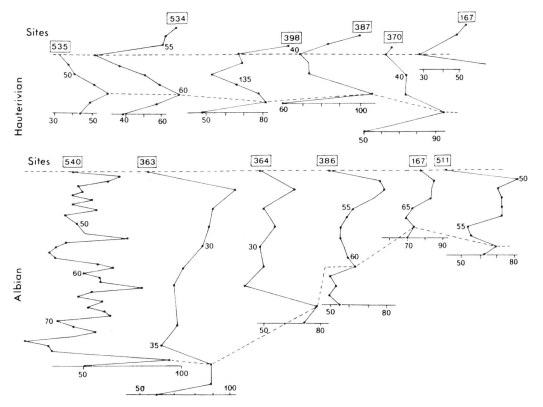

Fig. 7.49. Curves of minor cycle frequency in various DSDP sites in the Atlantic and Pacific oceans. *Dashed lines* indicated chronostratigraphic correlations; *numbers* on graphs are core numbers. (Cotillon 1987, reproduced by kind permission of Elsevier Science-NL, Sara Burgerhartstraat 25, 1055 KV Amsterdam, The Netherlands)

interpretation of grand cycles. The distribution of shale is thought to be related to variations in climate, not to transgression, and sand represents reworking of coastal eolian facies. The new sequence model for the grand cycles assigns quite different interpretations to the various facies components. A facies termed "ribbon rock" has been recognized, consisting of thinly interbedded calcisiltite and either shale or dololutite. This facies is interpreted as forming during transgressive to highstand phases of the sea-level cycle. Much of the shelf was exposed during lowstand, or underwent peritidal sedimentation, commonly under desiccated conditions. Oolitic facies commonly formed on the outboard platform during lowstand.

Variations in thickness and facies between the grand cycles in the different regions reflect variations in sedimentation and subsidence rates and variations in the width of the shelf. Estimates from biostratigraphic correlation of the grand cycles suggest that they each spanned 9–15 m.y. Therefore they may be classified as short second-order cycles or long third-order cycles, using the original Vail et al. (1977) terminology (which demonstrates one of

the inadequacies of a terminology based simply on duration). The cause of the sea-level changes is at present unclear.

7.8 Evidence of Cyclicity in the Deep Oceans

Eustatic sea-level changes would not be expected to affect the deep oceans directly, but some studies of data from the Deep Sea Drilling Project (DSDP) have indicated cyclicity in deep-marine sediments that appears to indicate a 10^6-year eustatic or climatic control. Cotillon (1987) found that the numbers of cycles per unit thickness in DSDP cores of Cretaceous strata from widely separated oceanic sites show predictable fluctuations that permit the cores to be correlated (Fig. 7.49). It was postulated that eustatic controls may be the explanation, changes in sea level "moving the terrestrial sources of suspended, dissolved, mineral and organic products away or towards the basin center, and by modifying the current circulation." Transgressions would trap sediment input on the shelves, whereas regressions would enhance sediment transport to the deep oceans. The fluctuations indicate a cyclicity with a period just under 2 m.y. An alternative mechanism is a long-term climatic forcing process, of the type discussed in Chapter 10. In particular, the work of Matthews and Frohlich (1991) demonstrated the possibility of long-term Milankovitch controls. Their work was concerned primarily with climate change involving glaciation, and the potential for 10^6-year orbital forcing remains largely unexplored.

7.9 Main Conclusions

1. Stratigraphic sequences with 10^6-year episodicities are common in rifted and extensional-margin basins. Their architecture comprises repeated transgressive-regressive packages of siliciclastic or carbonate deposits, of tabular form on continental shelves, or comprising prograding clinoform slope wedges. Mixed carbonate-siliciclastic sequences formed by the process of reciprocal sedimentation are common. These sequences can readily be interpreted using the Exxon systems tract models. In many of these basins close correlation of the sequence framework with the Exxon global cycle chart is claimed.

2. Sequences of comparable duration are also common in foreland basins and can also be interpreted using the systems-tract approach. Siliciclastic deposits are dominant, particularly thick alluvial deposits in proximal settings. Transgressive carbonate successions may occur on the distal ramp and forebulge of foreland basins. Evidence for tectonic control is common, for example, stratigraphic thinning and facies changes related to the position of blind thrusts. However, many authors emphasize correlations with the Exxon global cycle chart.

3. In forearc basins stratigraphic evolution is dominated by tectonic subsidence and faulting events and by volcanism. Few stratigraphic events can be unequivocally related to eustasy, and the correlation of these events with the Exxon global cycle chart is questionable.

4. Backarc basins, especially on their cratonic, hinterland sides, have tectonic and stratigraphic histories comparable to those of extensional continental margins, and commonly contain well-developed records of 10^6-year carbonate or clastic stratigraphic sequences.

5. Some evidence exists of 10^6-year cyclicity in deep-marine sediments, as studied in DSDP cores. Fluctuations in cycle thickness reflect varying sediment supply, which may be controlled by eustasy or long-term climatic rhythms.

6. A distinctive type of cyclicity occurs in Upper Paleozoic rocks of the northern hemisphere. They are of glacioeustatic origin, resulting from the great glaciation of the Gondwana supercontinent. High-frequency cycles, called cyclothems, are the most prominent type of stratigraphic sequence (and are discussed in Chap. 8), but groupings of these into 10^6-year mesothems or major cycles can be recognized, and intercontinental correlation of these has been proposed. Recent detailed stratigraphic studies have thrown doubt on the mesothem concept as applied to the Upper Paleozoic record of northwestern Europe, whereas some evidence for the concept has been obtained from the United States Midcontinent.

8 Cycles with Episodicities of Less Than One Million Years

8.1 Introduction

Cycles of 10^4–10^5 years duration can be grouped into five main types, reflecting their stratigraphic composition, tectonic setting, and age. Many of these cycles are thought to have been generated by global climate changes driven by orbital forcing, the so-called *Milankovitch mechanisms,* as described in Chapter 10. The term derives from the name of the Yugoslavian mathematician who was the first to provide the mathematical basis for the theory of astronomical forcing (Milankovitch 1930). However, there is increasing evidence of the tectonic origin of some of the clastic cycles in foreland basins (Sect. 8.6), as discussed in Chapter 11.

It has been common practice to subdivide high-frequency sequences into those of fourth order, with durations in the 10^4-year range, and those of fifth-order, with episodicities of 10^5 years. This two-fold subdivision has been based largely on interpretations that invoke low- and high- frequency Milankovitch mechanisms. However, this classification should now be abandoned. There is increasing evidence for other sequence-generating mechanisms that do not readily fall into simple temporal classifications, and it has now been demonstrated that sequence thicknesses and durations have log-normal distributions that lack significant modes (Drummond and Wilkinson 1996). As demonstrated by Schwarzacher (1993) orbital forcing includes a periodicity of 2.035 m.y. (related to orbital eccentricity), but few examples of this have been described in the literature. Most examples of orbital periodicities demonstrated from the rock record fall between the 413-ka eccentricity period and the 19-ka precession period.

This book does not deal with cycles in the so-called solar band or calendar band of geological time. The solar band refers to cyclicity in the 10^1- to 10^2-year range, including the sun-spot cycle and its possible geological effects (e.g., El Niño current changes). The calendar band refers to cyclicity relating to earth's seasonal rhythms (freeze-thaw, spring run-off, varves, etc.) and the tidal and other effects driven by the moon (Fischer and Bottjer 1991).

8.2 Neogene Clastic Cycles of Continental Margins

Marine clastic cycles of Neogene age are widespread, occurring particularly in continental shelf and slope settings. They have been identified by detailed outcrop studies, and by analysis of high-quality seismic-reflection data from offshore regions. They have also been documented in deep-sea sediments, using DSDP data. Modern multidisciplinary studies of the stratigraphic record have been able to show correlations of the cycles with the oxygen-isotope record, indicating a direct correlation between sea level and ocean temperatures. These sequences are clearly of glacioeustatic origin.

The Gulf of Mexico is one of the most intensively studied regions of the world, and the Neogene sequence stratigraphy of this area is now well known. Alluvial, deltaic and shallow-marine deposits of the Mississippi valley and delta area were the subject of intensive study by Fisk (1939, 1944), who recognized the relationships between river terraces, erosional valleys and depositional surfaces and attributed them to glacial changes in sea level. Early attempts at sequence-stratigraphic modeling were carried out by Frazier (1974). More recently, Suter et al. (1987) and Boyd et al. (1989) have interpreted the stratigraphy of the Lousiana shelf, including the lobes of the Mississippi delta, in terms of seven Quaternary stratigraphic sequences. Laterally equivalent deposits on the Texas Gulf Coast were described by Morton and Price (1987). The downdip equivalents, comprising the deposits of the Mississippi fan, have been studied by Feeley et al. (1990) and Weimer (1990), who recognized at least

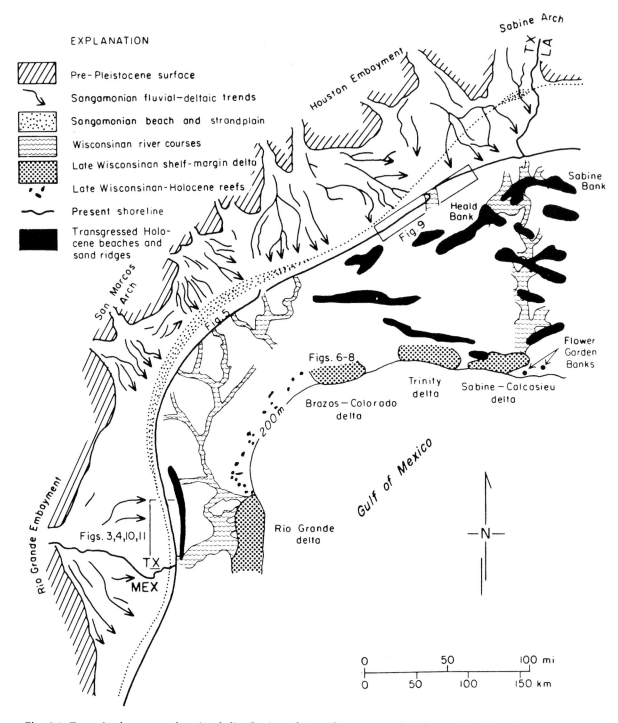

Fig. 8.1. Tectonic elements and regional distribution of Late Pleistocene and Holocene depositional systems, Texas coast. (Morton and Price 1987)

13 sequences representing the last 3.5 m.y. of geological time.

Near the Texas-Louisiana border Pleistocene deposits exceed 3.5 km in thickness, and have extended the continental shelf seaward by nearly 50 km as a result of aggradation and progradation. Fluvial sediment supply from the continental interior was vast, and supplied individual deltaic

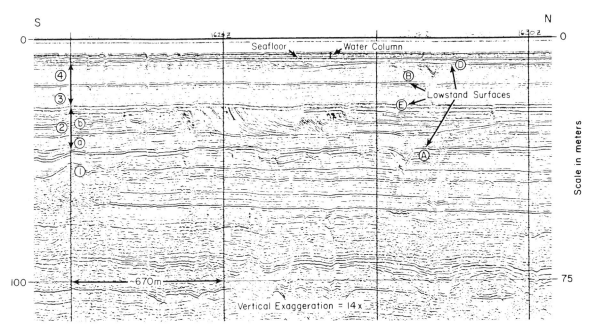

Fig. 8.2. High-resolution seismic profile through Holocene sequences, Lousiana coast. *Uppercase letters in circles* are sequence boundaries; *numbers* are sequences. Sequences *2a* and *2b* represent episodes of transgression and aggradation, respectively. Clinoforms within *2b* represent fluvial levee and overbank deposits. (Suter et al. 1987)

systems up to 70 km across and 180 m thick. Tectonic influences on sediment supply can be recognized in the variations in thickness of individual deltaic complexes along strike around the Gulf Coast. On a local scale the effects of autocyclic delta-lobe switching, and local tectonic disturbances caused by growth faulting and by salt diapirism render sequence recognition and correlation difficult.

Figure 8.1 illustrates the major tectonic elements of the Texas coast, and the distribution of the major Late Pleistocene and Holocene depositional systems. An example of a high-resolution seismic-reflection cross section through these deposits is shown in Fig. 8.2, and a map of the depositional systems that developed during one of the sea-level cycles is shown in Fig. 8.3. Shelf deltas developed during a sea-level stillstand. Clinoform geometry (not visible in Fig. 8.2) indicates that these prograded toward the southwest. A fall in sea level then caused fluvial systems to incise into the shelf, developing the network of channels seen in Fig. 8.3. Near the shelf margin these incised valleys have erosional relief of nearly 60 m and are locally more than 10 km wide.

Six sequences have developed on the Louisiana shelf during the last 700 ka (Fig. 8.4). Dating of the sequences is difficult in the absence of datable samples. Sequence 1 is interpreted to represent the lowstand associated with the Illionian glacial stage. Sequences 5 and 6 correspond to the post-Wisconsinan sea-level rise since 18 ka. The Mississippi delta complex developed during this final stage of sea-level rise (Fig. 8.5). The first three lobes (numbered 1–3 in Figs. 8.5, 8.6) are interpreted as transgressive systems tracts. They backstep onto the shelf because sediment supply to the delta was unable to keep pace with rising sea level. The culmination of the sea-level rise occurred at about 3–4 ka, and resulted in the retreat of the coastline to the mouth of the Mississippi alluvial valley. The St. Bernard lobe represents the first of four highstand systems tracts that have formed since this sea-level highstand. Delta-lobe switching between the various positions has taken place because of autocyclic processes relating to the flattening of the river slope as the lobe progrades seaward.

The Mississippi fan (Fig. 8.7) is the most well-studied fan deposit in the world. It is also one of the largest, consisting of more than 3.5 km of sediment deposited during the last 6.5 m.y. More than 2 km, constituting about 70% of the fan volume, has been deposited since 800 ka. Seismic and sonar surveys and the results of DSDP leg 96 have enabled a very

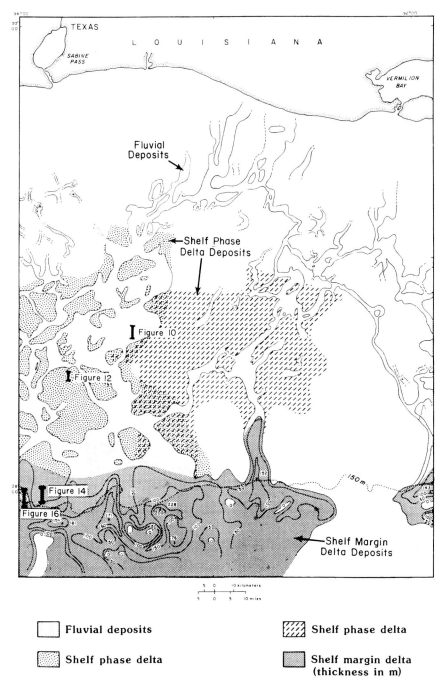

Fig. 8.3. Map of Louisiana shelf (adjacent to NE border of map area shown in Fig. 8.1) showing distribution of fluvial and deltaic systems that developed on the shelf during deposition of sequence 3. (Suter et al. 1987)

detailed sequence analysis to be compiled (Feeley et al. 1990; Weimer 1990). Weimer (1990) documented seventeen seismic sequences within the fan succession, consisting mostly of lowstand channel-levee depositional systems (Figs. 8.8–8.10). Feeley et al. (1990) correlated the sequence stratigraphy with the oxygen isotope curve (Fig. 8.11). However, the ac-

tual chronostratigraphic evidence for the correlations derived from the fan sediments themselves appears to be limited, consisting of a few biostratigraphic picks in several key wells. The correlations shown in Fig. 8.11 are therefore based on the standard Exxon method of pattern matching, constrained by correlation of a few proven fixed points.

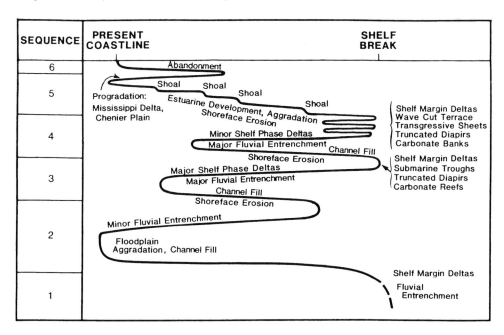

Fig. 8.4. Late Quaternary depositional sequences, sea-level changes since about 0.7 Ma, and associated sedimentary events on the Louisiana shelf. (Suter et al. 1987)

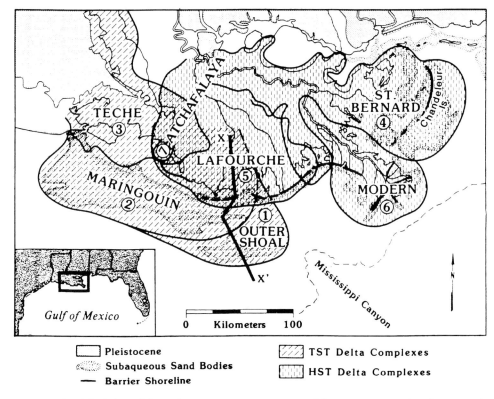

Fig. 8.5. The delta lobes of the Holocene Mississipi River. Lobe 1 is dated at about 9 ka, and represents the first of three trangressive systems tracts formed following the end of the Wisconsinan glacial stage. *Lobes 5–7* are highstand systems tracts. (Boyd et al. 1989)

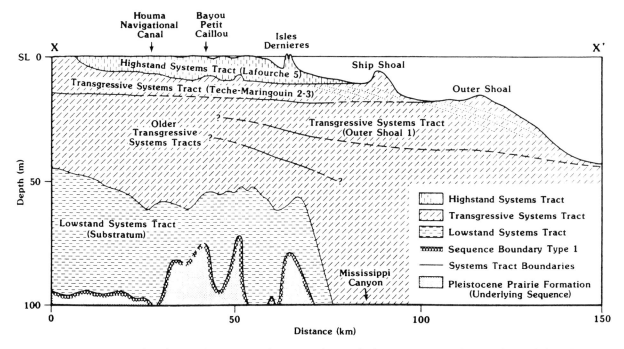

Fig. 8.6. Cross section along line *X-X'* in Fig. 8.5, showing the systems tracts of the Mississippi delta complex. The "type 1" sequence boundary at the base of the complex is dated at about 18 ka, and formed by subaerial erosion during the last Wisconsinan lowstand. It is followed by a lowstand systems tract, formed between 18 and 9 ka. (Boyd et al. 1989)

Other recent submarine fan deposits have also been subdivided into sequences based on seismic-stratigraphic interpretation. They include the Amazon cone (Manley and Flood 1988) and the Indus cone (Ravenne et al. 1988). The chronostratigraphy of these deposits is, however, poorly known because of an absence of well data. Manley and Flood (1988) showed that during times of falling sea level there is a rapid increase in the amount of sediment tipped over the shelf edge. This generates large-scale mass movement and the formation of debris flow and related deposit types on the continental slope, which appear as seismically transparent units, blanketing the underlying topography (at 5.0 and 5.4 s in Fig. 8.12). Continued sea-level fall leads to canyon incision and the development of a point source for sediment dispersion, from which major, bifurcating channel-levee complexes of a large fan develop (at 4.8 and 5.2 s in Fig. 8.12). The sediment becomes finer as sea-level rises, the evolution of the channels slows down, and eventually, as at the present day when sea level is high, sedimentation essentially ceases except for the slow deposition of a pelagic mud blanket. Widespread shale beds in some ancient fan deposits may similarly indicate deposition during times of high

sea level. In the Indus fan Ravenne et al. (1988) documented lateral migration of the main fan channel (Fig. 8.13).

Several studies of the continental shelf and slope of New Zealand have demonstrated a high-frequency sequence stratigraphy in the deposits, and attempts have been made to correlate the successions with the oxygen isotope record (Fulthorpe and Carter 1989; Kamp and Turner 1990; Carter et al. 1991; Fulthorpe 1991). These studies also illustrate several points regarding lower-order cycles and the global cycle chart, as discussed by Carter et al. (1991). During the Neogene, New Zealand underwent active uplift along the transpressive Alpine Fault, and abundant sediment was shed into basins situated in various tectonic settings on the continental shelves surrounding the North and South islands. The Canterbury Basin is located in an extensional, passive-margin setting, while the Wanganui basin is situated in a backarc setting, behind the arc of the North Island Axial Ranges (Fig. 8.14). In the Canterbury Basin the beds above an initial syn-rift fill constitute the transgressive systems-tract of a major long-term thermo-tectonic cycle (Fig. 8.15). Transgression reached a maximum during the Oligocene, with the deposition of a

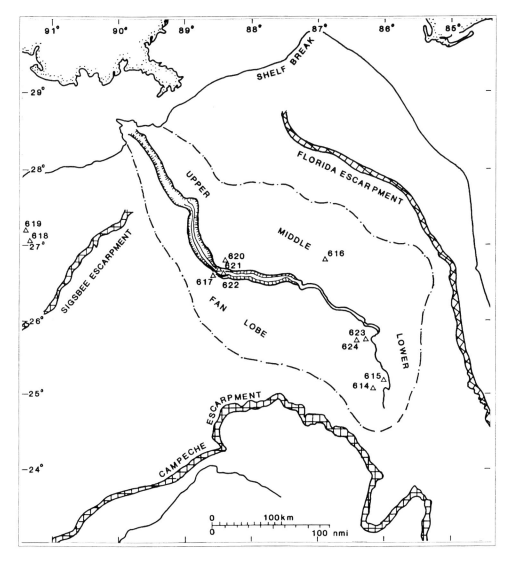

Fig. 8.7. Map of the recent Mississippi fan lobe, outlined by *dot-dash line,* showing position of central erosional channel and location of DSDP leg 96 sites. (Feeley et al. 1990, reproduced by permission)

condensed succession of pelagic to hemipelagic limestone. Transpression along the Alpine fault then led to increased sediment supply and the development of a highstand progradational phase in the Miocene.

Fulthorpe (1991) illustrated a seismic-reflection line from the Canterbury basin that shows the presence of high-frequency sequences in the Miocene section (Fig. 8.16). He pointed out that such cycles (those with episodicities <1 m.y.) cannot normally be identified on seismic-reflection data because of the problems of resolution. These cycles are apparent because of the high sediment supply in the Miocene, which led to the development of a progradational architecture. An example is noted below of the use of shallow high-resolution seismic methods to document near-surface high-frequency cycles.

According to Carter et al. (1991) the Middle Oligocene sea-level high indicated by the Canterbury Basin condensed section is a "spectacular mismatch with the predicted 29-Ma lowstand" of the global cycle chart of Haq et al. (1987, 1988a). However, Loutit et al. (1988, p. 192), discussing the same basin, do not see it this way. They stated that, "the New Zealand Oligocene provides a spectacular example of the effects of subsidence and sea-level movements on stratal geometry." While this is true,

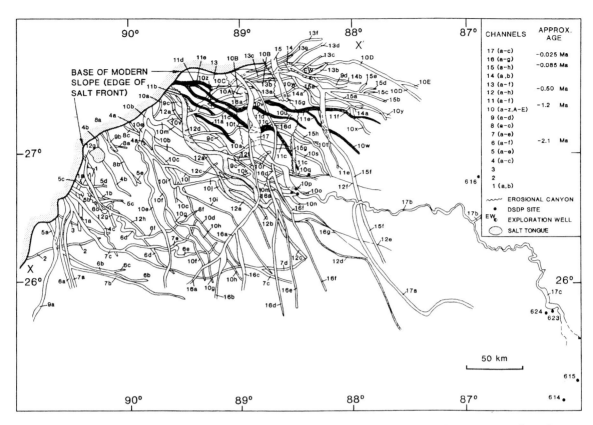

Fig. 8.8. Locations of the channel valleys of the 17 channel-levee systems comprising the Mississippi fan. They are numbered in order from oldest to youngest. (Weimer 1990, reproduced by permission)

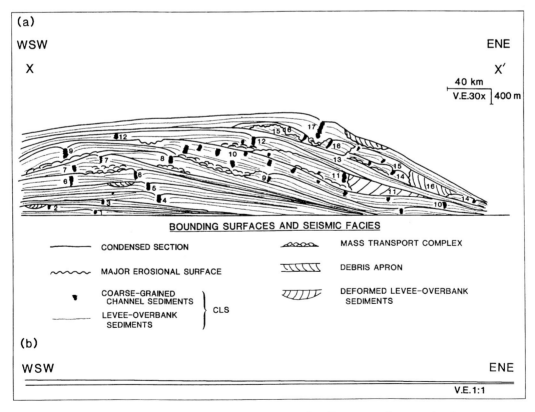

Fig. 8.9. Schematic strike-oriented cross section across the Mississippi fan, showing the mounded pattern of the 17 channel-levee systems. Note the true vertical scale in the diagram at *bottom*. (Weimer 1990, reproduced by permission)

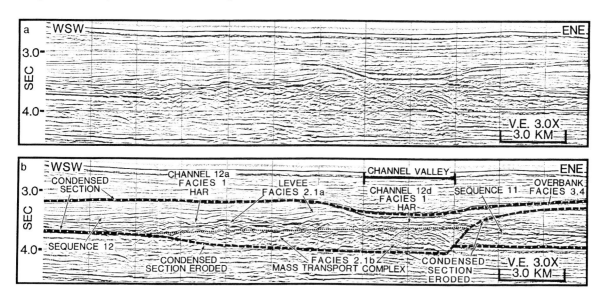

Fig. 8.10. Example of a strike-oriented seismic profile across the Mississippi fan, showing the seismic facies character of some of the channel-levee systems. *Upper section* is uninterpreted; *lower section* shows position of two sequence boundaries, defined by condensed sections (*dashed lines*). (Weimer 1990, reproduced by permission)

it does not explain how they reconcile the relative sea-level changes revealed by the Canterbury Basin stratigraphy with the global cycle chart.

The Wanganui Basin of North Island, New Zealand (location shown in Fig. 8.14), contains a particularly complete Pliocene-Pleistocene sedimentary record. Coastal cliff sections have been interpreted in terms of 10^5-year cyclicity by Kamp and Turner (1990) and by Carter et al. (1991), while correlative cycles also are present in offshore sedi-

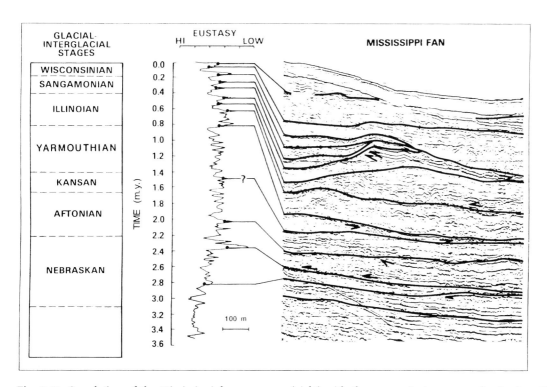

Fig. 8.11. Correlation of the Mississippi fan sequences (*right*) with the oxygen-isotope curve for the Late Pliocene–Pleistocene. (Feeley et al. 1990, reproduced by permission)

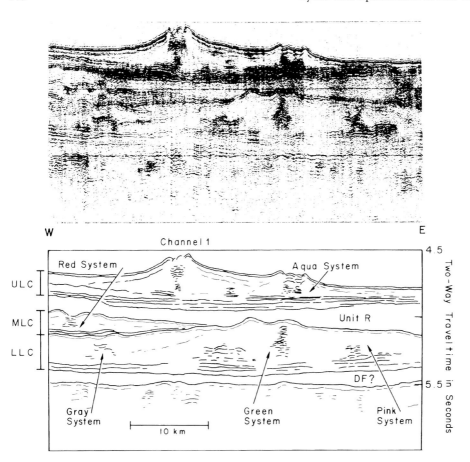

Fig. 8.12. Cyclic sedimentation on the Amazon Cone. (Manley and Flood 1988, reproduced by permission)

ments, as revealed by seismic surveys (Carter et al. 1991). Figure 8.18 summarizes the onshore stratigraphic record. The exposed sediments represent transgressive and highstand deposits, and can be subdivided into sequences 5- to 22-m thick. Each cycle commences with a transgressive sand or gravel above a bored erosion surface. Shell beds occur at the time of maximum flooding, and the highstand deposits consist of marine siltones and silty sandstones. The sequence boundaries represent significant hiatuses. In fact, as revealed by the correlations with the oxygen isotope record (Fig. 8.18), it is of interest to note that in sum the hiatuses represent about as much time as that represented by sedimentation. Even-numbered oxygen-isotope stages correspond to times of sea-level lowstand (glacial stages), and are recorded by the prograding clinoform cycles present at the shelf margin (Fig. 8.19).

Neogene high-frequency cycles have been well documented in forearc and backarc settings on the continental margins of Japan (Ito 1992, 1995; Ito

and O'Hara 1994; Masuda 1994). Examples of 10^4-year cyclicity are illustrated in Fig. 3.6. This succession consists of six sequences spanning the 0.4- to 0.8-Ma period, and estimated to have durations ranging from 45,000 to 50,000 years. Each sequence can be subdivided into systems tracts, comprising mainly progradational slope and shelf deposits. The refined stratigraphic correlations indicated in this outcrop study were made possible by the presence of numerous volcanic ash beds, which can be correlated on the basis of their geochemical signatures and fission-track ages. Ito (1992) showed how seawater temperatures fluctuated during sedimentation, and correlated these data with the oxygen isotope record (Fig. 8.20). He interpreted the cycles to be of glacioeustatic origin (Chap. 10).

A high-resolution seismic survey was used to document high-frequency cyclicity on the continental shelf of part of New Jersey (Ashley et al. 1991). Three persistent reflectors were interpreted as sequence boundaries, corresponding to subaerial erosion surfaces or ravinement surfaces. These de-

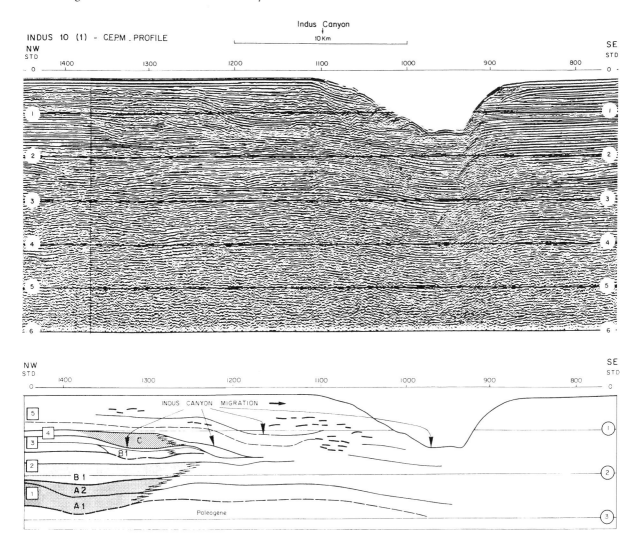

Fig. 8.13. Strike-oriented seismic section across the Indus cone, showing stratigraphic sequences and lateral migration of the main fan channel. *Numbers in circles* are two-way travel times in seconds. *Numbers in boxes* are fine three depositional series; *capital letters* refer to intervals of canyon cut-and-fill. (Ravenne et al. 1988, reproduced by permission)

fine three sequences formed following transgression during interglacial and postglacial sea-level highs. The results provide a more detailed documentation of the Wisconsinian and postglacial sequence record even than the detailed study of the Mississipi fan by Feeley et al. (1990).

8.3 Pre-Neogene Marine Carbonate and Clastic Cycles

A wide variety of 10^4–10^5-year carbonate and clastic cycles has been documented in the pre-Neogene sedimentary record in various tectonic settings (Fischer 1986; Fischer and Bottjer 1991; Dennison and Ettensohn 1994; de Boer and Smith 1994a). These are most spectacularly preserved on carbonate shelves and in deep marine and lacustrine environments, away from clastic influxes, because the autogenic redistribution of detrital sediment tends to obscure regional cyclic patterns. However, high-frequency cyclicity has now also been suggested for many other types of stratigraphic succession, even including some nonmarine strata, as discussed in this section.

The Triassic record of the Dolomites in northern Italy contains a spectacular record of cyclic sedimentation (Goldhammer et al. 1987, 1990; Goldhammer and Harris 1989; Hinnov and Goldhammer

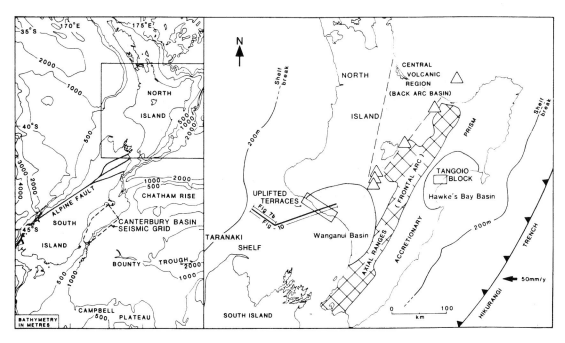

Fig. 8.14. Location of coastal sedimentary basins, New Zealand. (Carter et al. 1991)

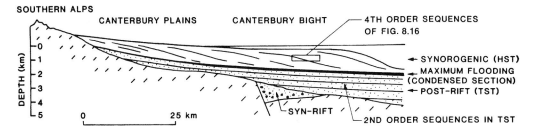

Fig. 8.15. Schematic cross section of Canterbury Basin, New Zealand. *Box* shows position of seismic section given in Fig. 8.16. *TST* transgressive systems tract; *HST* highstand systems tract (Carter et al. 1991)

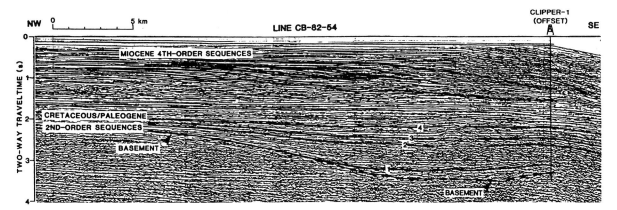

Fig. 8.16. Dip-oriented seismic-reflection profile across Canterbury basin, New Zealand, showing the details of the Cretaceous-Miocene section. (Fulthorpe 1991)

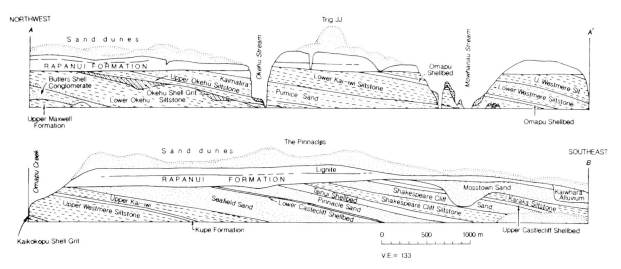

Fig. 8.17. The Castlecliff section, Wanganui Basin, North Island, New Zealand (located south of terraces indicated in Fig. 8.14). (Kamp and Turner 1990, reproduced by kind permission of Elsevier Science-NL, Sara Burgerhartstraat 25, 1055 KV Amsterdam, The Netherlands)

1991; Goldhammer et al. 1993; Figs. 8.21-8.23), including "nested" cycles of 10^4-, 10^5- and 10^6-year episodicity. These cycles are interpreted to be primarily of eustatic origin, with basin subsidence and autogenic marine processes acting to modify the overall succession (Goldhammer et al. 1987, 1990; Jones and Desrochers 1992). The sediments developed on an isolated carbonate platform, named the Latemar Massif. They consist of cyclic, platform deposits passing laterally into prograding slope deposits (Fig. 8.21). The 10^4-year cycles record abrupt shallowing and exposure, with the development of subtidal platform deposits that underwent exposure and vadose diagenesis. The cycles in the Lower and Upper Cycle Facies average 0.7 m in thickness. In the Lower Platform Facies and the Tepee Facies cycle composition and thickness are different, and Goldhammer and Harris (1989) and Goldhammer et al. (1987, 1990) interpreted this as a result of the overprint of slow (10^6-year) eustasy on the higher frequency cycles of sea-level change. During deposition of the Lower Platform Facies the area was undergoing a prolonged long-term rise in sea level. The result was the maintenance of subtidal conditions over the entire platform (approx. 0.5 m.y. in Fig. 8.23):

Subtidal sedimentation is unable to keep up, and thus not every beat of fifth-order high-frequency sea level can "touch down" on the platform top and subaerially expose the top of the sediment column. These "missed beats" of subaerial exposure result in the formation of thick (average approximately 10 m), fourth-order amalgamated megacycles. During this phase of subtidal carbonate ag-

gradation, subaerial exposure occurs infrequently, and lengthy periods of marine submergence promote abundant sydepositional marine diagenesis. (Goldhammer et al. 1990, p. 549).

During development of the Lower and Upper Cycle Facies the rate of long term (10^6-year or "third-order") sea-level change was slow (approx. 0 and 2.3 m.y. in Fig. 8.23). Subtidal sedimentation was able to keep pace with net sea-level change, which was almost entirely due to subsidence. Every high-frequency (10^4-year, "fifth-order") oscillation in sea level is therefore recorded by subaerial exposure. The Tepee Facies records sedimentation during the time of falling sea level on the long-term (10^6-year) cycle (approx. 1.5 m.y. in Fig. 8.23):

Third-order accommodation potential is minimized, but in a realm of continuous subsidence, the net effect is close to but not quite that of a stillstand. Now, sedimentation easily keeps up and very thin cycles form (thinner than those of the lower cyclic facies), which are grouped into fourth-order condensed megacycles with tepee tops representing more lengthy periods of subaerial exposure. The tepees form owing to prolonged subaerial exposure during the falling limb of a fourth-order oscillation. (Goldhammer et al. 1990, p. 549)

Additional details of cycle analysis and the calculation of Milankovitch periodicities in the Latemar Limestone are deferred to Chapter 10. Other carbonate suites comparable to the Alpine Triassic (in episodicity but not necessarily in facies) include a Mississippian example in Wyoming and Montana (Elrick and Read 1991) and various Cambrian examples around the margins of the North American

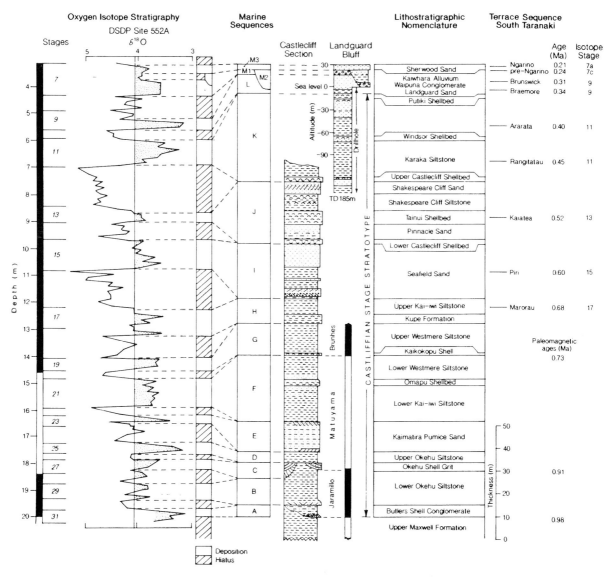

Fig. 8.18. Correlation of the Castlecliff and Landguard Bluff sections, North Island, New Zealand, with oxygen-isotope stratigraphy derived from DSDP data, and with nearby coastal terraces. (Kamp and Turner 1990, re-produced by kind permission of Elsevier Science-NL, Sara Burgerhartstraat 25, 1055 KV Amsterdam, The Netherlands)

continent (Osleger and Read 1991). Weedon (1985) described centimeter-decimeter scale limestone-shale couplets In the Lower Jurassic Lias of southern Britain, and discussed their origin in terms of Milankovitch mechanisms. In the Cretaceous Niobrara Formation of Colorado limestone-shale couplets occur in bundles of 1–12 couplets, each bundle ranging from about 1–5 m in thickness. Laferriere et al. (1987) analyzed these deposits in terms of Milankovitch rhythms.

A mixed carbonate-clastic suite of Milankovitch cycles in the Permian Yates Formation of the De-laware Basin, western Texas, was described by Borer and Harris (1991). The cycles are prominent in a midshelf association, seaward of the inner-shelf evaporite-clastic belt and landward of the shelf-margin Capitan reef belt (Fig. 8.24). The cycles comprise carbonate-clastic couplets. Carbonates consist mainly of algal and peloidal dolomudstones, which yield low gamma-ray readings. Clastics consist of arkosic sandstones and argillaceous siltstones that have high gamma-ray reponses. The alternation between the two main lithofacies types is readily apparent on gamma-ray logs, which reveal the

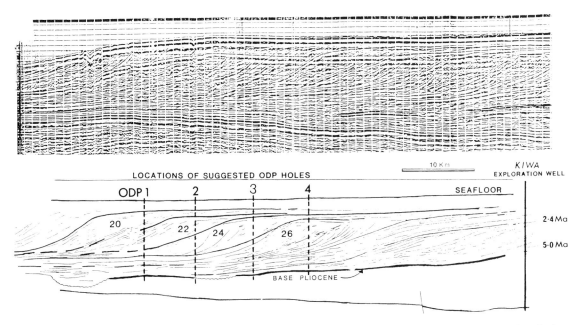

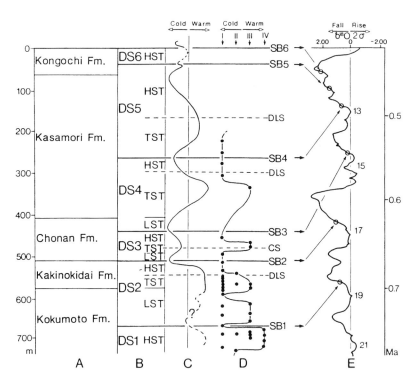

Fig. 8.19. Seismic-reflection line across the shelf-slope break of the Wanganui Basin, New Zealand. Progradational fourth-order cycles have been correlated with the oxygen-isotope stratigraphic record (lowstand isotope stages 20, 22, 24, 26) and can be compared with the on-shore sections summarized in Fig. 8.18. (Carter et al. 1991)

presence of two scales of cyclicity (Fig. 8.25). Clastic intervals were deposited by marine, and possibly eolian processes, during lowstands of sea level, and carbonates were formed during transgressions.

Shallowing-upward cycles have been described in the platform limestones of the Devonian Helderberg Group of New York by Goodwin and Anderson (1985) and Goodwin et al. (1986). They coined the

Fig. 8.20. Correlation of the Kazusa Group, Boso Peninsula, Japan (*columns A, B*), with the oxygen isotope record (*E*). Also shown are temperature records derived from benthic molluscs (*C*) and planktonic molluscs (*D*). Compiled by Ito (1992). (Reproduced by kind permission of Elsevier Science-NL, Sara Burgerhartstraat 25, 1055 KV Amsterdam, The Netherlands)

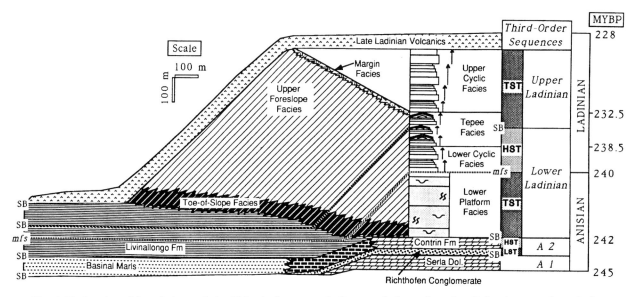

Fig. 8.21. Stratigraphic relations of the Triassic Latemar platform and slope deposits, Dolomites, northern Italy. (Hinnov and Goldhammer 1991)

term *punctuated aggradational cycle* for these deposits. The sequence is interpreted as the product of rapid deepening caused by rapid (geologically instantaneous) sea-level rise, followed by sedimentation and consequent shallowing under conditions of stable or slowly falling sea level (Fig. 8.26). The evidence for sea-level change rather than autogenic causes for the cyclicity is the interpreted lateral persistence for the cycles (tens of kilometers). Goodwin and Anderson (1985) estimated that each cycle represents a period on the order of 10^4 years and suggested that they are glacioeustatic in origin.

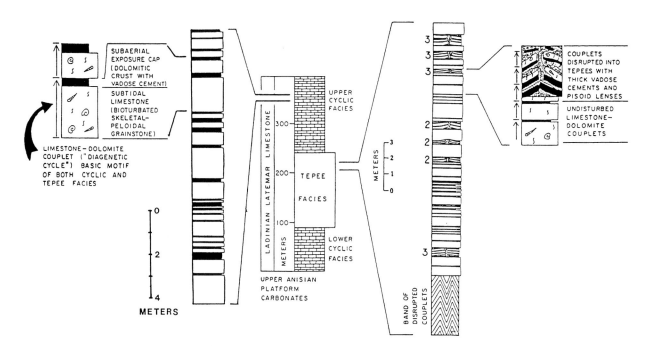

Fig. 8.22. Representative measured sections of the upper cyclic and tepee facies of the Latemar Limestone. Meter-scale fifth-order cycles are arranged in thinning-upward fourth-order megacycles. (Goldhammer and Harris 1989)

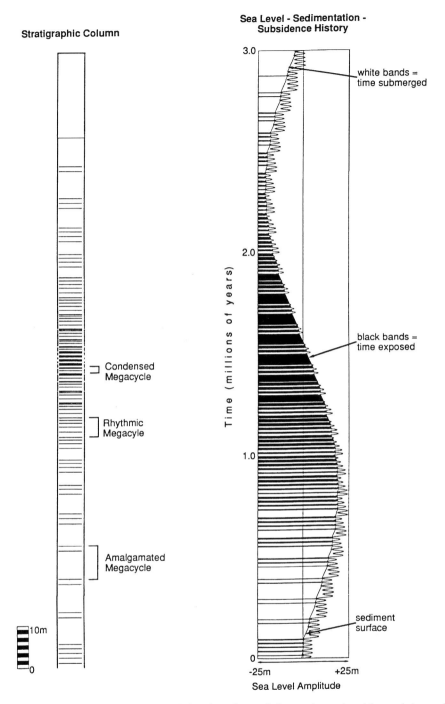

Stratigraphic Column

Sea Level - Sedimentation - Subsidence History

white bands = time submerged

black bands = time exposed

sediment surface

Condensed Megacycle

Rhythmic Megacyle

Amalgamated Megacycle

10m

0

3.0

2.0

1.0

0

Time (millions of years)

-25m   +25m

Sea Level Amplitude

Fig. 8.23. Simulation of third- and fourth-order cyclicity, which explains the variations in cycle thickness and composition in the Latemar Limestone. On the *right* is a simulated sea-level curve, with times of sedimentation shown in *white*, and times of exposure shown in *black*. On the *left* is the resulting stratigraphic section. The details are explained in the text. (Goldhammer et al. 1990)

The proposed asymmetry of the rise and fall in sea level is consistent with what is now known about glacioeustatic sea-level changes in the Cenozoic, as noted in Chapter 10. The periodicity, but not necessarily the asymmetry, is also consistent with the style of rapid subsidence and progradation brought about by thrust-sheet loading in a foreland basin (Sect. 8.6, Chap. 11).

Evaporite deposits are highly sensitive to climatic forcing, because of their dependence for for-

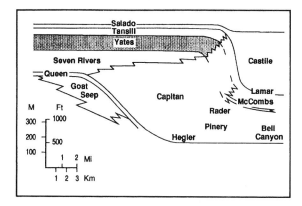

Fig. 8.24. Schematic N-S cross section of the shelf marginal to the Delaware Basin, Texas, showing the stratigraphic relationships of the Yates Formation. (Borer and Harris 1991, reproduced by permission)

mation on the delicate balance between evaporation and the influx of saline and fresh waters. This balance is markedly affected by temperature changes and by changes in base level, of which, of course, eustasy is one of the more important controls. Anderson (1984) reviewed rhythmic and cyclic sedimentation in a number of evaporite basins. Cyclic patterns range from 10^4–10^5-year cycles driven by eustasy, of the type that forms the main subject of this section, to higher-order cycles, including annual varves developed by seasonal insolation changes.

H. Olsen (1990) described a Devonian fluvial section in eastern Greenland containing prominent fining-upward cycles that show systematic variations in channel sandbody thickness (Fig. 8.27). The average thickness of the fluvial fining-upward cycles is 20 m, with a range from 13–30 m. Channel sandstones vary in thickness from 2 to 13 m. Bundles of three to six cycles can be defined, as can thicker trends that trace out megacycles. Megacycles range in thickness from 79 to 135 m, with an average of 101 m. H. Olsen (1990) calculated discharge values for the channels using a suite of geomorphic equations based on channel dimensions. The results are shown in Fig. 8.28, and demonstrate the same cyclic variability. H. Olsen (1990) argued that the simplest explanation for the rhythmic variations is changes in discharge brought about by orbital forcing of climate changes, and he attempted to fit the pattern of thickness and discharge variations to patterns of orbital variation, as discussed in Chapter 10. This work needs to be assessed with caution because of the limited data base upon which the interpretation is built.

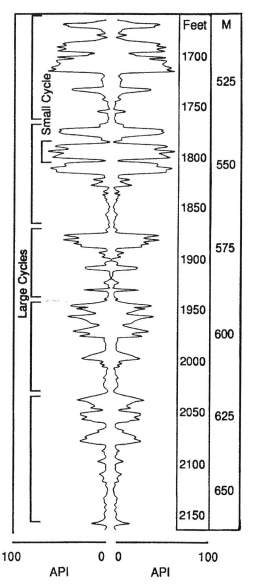

Fig. 8.25. Composite gamma-ray log through the Yates Formation. Log is repeated in mirror image to emphasize cyclicity. (Borer and Harris 1991, reproduced by permission)

Many rhythmically bedded pelagic deposits in deep marine settings can be interpreted in terms of Milankovitch forcing mechanisms. An example is the Pliocene Trubi Marls of southern Italy (Thunell et al. 1991). These beds contain limestone-shale couplets.

Recent studies ... have attributed this sedimentary cyclicity to rapid alternations between wet and dry climatic conditions. These authors associate the darker marls with periods of increased continental runoff, greater terrigenous influx, and higher surface water productivity. In

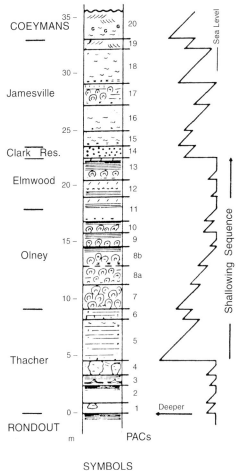

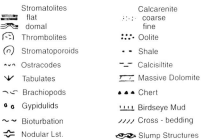

SYMBOLS

Stromatolites		Calcarenite	
	flat		coarse
	domal		fine
	Thrombolites		Oolite
	Stromatoporoids		Shale
	Ostracodes		Calcisiltite
	Tabulates		Massive Dolomite
	Brachiopods		Chert
	Gypidulids		Birdseye Mud
	Bioturbation		Cross - bedding
	Nodular Lst.		Slump Structures

Fig. 8.26. Columnar section of the Lower Devonian Helderberg Group, New York State, with formal stratigraphic nomenclature to *left* of lithological column and numbered punctuated aggradational cycles (PACs) to the *right*. Interpreted water-depth curve is also shown. (Goodwin and Anderson 1985, reproduced by permission of the University of Chicago Press)

contrast, the lighter, carbonate-rich units are attributed to arid climatic conditions and low surface water productivity. (Thunell et al. 1991, p. 1110)

Hallam (1986) warned that rhythmic diagenetic unmixing of argillaceous carbonates may generate a couplet stratigraphy, but this is not thought to be

the case for the Trubi marls. An example of the marl section is shown in Fig. 8.29. Precise chronostratigraphic control of these rocks has permitted a detailed analysis of the cycles in terms of Milankovitch periodicities. Many other examples of this type of rhythmic sedimentation are summarized by Fischer (1986), Fischer and Bottjer (1991), and Einsele et al. (1991a).

8.4 Late Paleozoic Cyclothems

The first cycles to be described in the geological literature were the Mississippian Yoredale cycles of the English Pennines area (Wilson 1975). They represent only part of a lengthy and widespread coal-bearing cyclic succession that spans much of the Carboniferous and Permian and extends throughout northwestern Europe (e.g., Ramsbottom 1979). The classic Pennsylvanian cyclothems of the American Midcontinent were among the first cycles to be described in North America. Wanless and Weller (1932) coined the term *cyclothem* for them. They have also been called *Klüpfel cycles,* after Klüpfel (1917). These cycles in the northern hemisphere are now interpreted primarily as the product of glacioeustasy (Crowell 1978); they developed in response to the major Carboniferous-Permian glaciation of the Gondwana supercontinent (Sect. 10.5). Klein and Willard (1989), Klein and Kupperman (1992), and Klein (1994) have also pointed out the importance of regional tectonism in the development of cyclothems, as discussed below. Groups of cyclothems have been termed mesothems (Ramsbottom 1979; Busch and Rollins 1984; Sect. 7.6).

Klein and Willard (1989) summarized stratigraphic data indicating that in the North American Interior there are essentially three types of cyclothem, an Appalachian type, an Illinois type, and a Kansas type (Fig. 8.30). The types differ mainly in the carbonate:clastic ratio within each cyclothem. Appalachian-type cyclothems were deposited in a foreland basin close to a major clastic source (the Alleghenian orogen) and are clastic dominated. Illinois-type cyclothems were deposited in an intracratonic basin that was partially "yoked" to the Appalachian foreland basin by the flexural effects of foreland-basin thrust loading. This basin was more distant from the sediment supply, and consequently contains a thinner, finer-grained clastic component. Kansas-type cyclothems were deposited within a cratonic area a long distance inboard from clastic

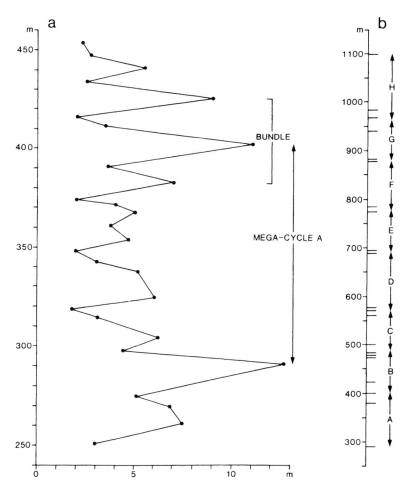

Fig. 8.27. a Variations in sand-body thickness in a Devonian fluvial section in eastern Greenland. **b** Distribution of major sand bodies, consisting of single sandstone beds or bundles of 2–4 beds. (Olsen 1990, reproduced by kind permission of Elsevier Science-NL, Sara Burgerhartstraat 25, 1055 KV Amsterdam, The Netherlands)

sources, and are carbonate-dominated. More than 100 such repetitions have been mapped in Kansas (Moore 1964; Heckel 1986, 1990, 1994). Some individual beds within the cyclothems can be traced for more than 300 km. Crowell (1978) reviewed ideas first proposed in the 1930s interpreting the cyclothems as the product of transgressions and regressions driven by glacioeustasy. He illustrated a typical Illinois-type cyclothem, and suggested the general range of depositional environments that led to the development of the succession during repeated transgressions and regressions (Fig. 8.31). Wilson (1975) developed generalized facies models for the Midcontinent cyclothems and those in Texas, suggesting how the various types might be related by lateral facies changes (Figs. 8.32, 8.33).

Figure 8.34 illustrates how a typical cyclothem in Illinois varies as individual components thicken and thin. Thick sandstones and shales represent deltaic units deposited in regions of local subsidence. Nonmarine sandstones may rest on an erosional, channelized unconformity at the top of the underlying marine units. Coals, underclays, black shales, and marine limestones typically retain considerable uniformity over hundreds of kilometers (Wanless 1964). In cratonic areas away from detrital sources, the nonmarine clastic units may thin to zero, so that the marine limestones rest on each other, as in the Kansas cyclothems.

Heckel (1986) recognized a spectrum of cycle types in the cyclothems of the North American Midcontinent. *Major cycles* record inundations far onto the craton, which deposited widespread conodont-bearing shales. *Minor cycles* lack conodont-rich shales and represent minor transgressions onto the craton margin in the southern part of his project area (Kansas and Oklahoma). Heckel (1986) estimated that the major cycles spanned 235–400 ka,

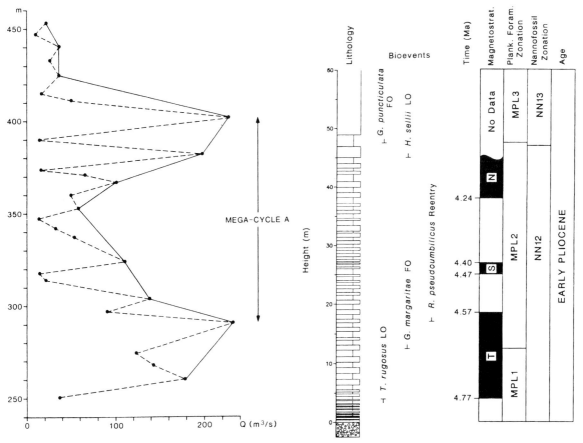

Fig. 8.28. Variations in calculated discharge for sandstone bodies in a Devonian section in eastern Greenland. *Dashed line* connects individual values for each sand body; it defines a form of short-term cyclicity. Maximum values for each cycle are connected by *solid line* and show a longer-term cyclicity. (Olsen 1990, reproduced by kind permission of Elsevier Science-NL, Sara Burgerhartstraat 25, 1055 KV Amsterdam, The Netherlands)

Fig. 8.29. Lithology and chronostratigraphy of a section in the Trubi marls of southern Italy. *Long bars* in the lithology column indicate limestone, and *short bars* represent marl. (Thunell et al. 1991)

and the minor cycles had durations of 40–120 ka. Detailed study of outcrops and cores enabled Heckel (1986) to erect a curve showing regional sea-level changes, as discussed in Section 10.5.

There is by no means universal agreement that cyclothems and mesothems may be correlated over wide areas. For example, George (1978) provided a detailed critique of the work reported by Ramsbottom (1979), based in part on a meticulous examination of the biostratigraphic record. Some geologists reject the concept of the cyclothem entirely. They note the numerous variations in the cyclothem sequence (e.g., Fig. 8.34) and maintain that the concept of an ideal or model cyclothem is a dubious one (e.g., Duff et al. 1967). The development of the facies-model methodology during the last 20 or so years gave geologists an entirely new

way of analyzing cyclic sequences. Application of the process approach, using vertical profiles and Walther's law, has shown that many cyclic relationships are the result of such autogenic processes as point-bar lateral accretion, deltaic progradation, or the shoaling of tidal flats. Ferm (1975) showed that most variations in the cyclothem sequence could be explained by the progradation and abandonment of deltas. He interpreted each cyclothem as a complex clastic wedge, which he termed the *Allegheny duck model,* because of the fancied resemblance of the model cross section to a flying duck. (Ferm also expressed doubts about his own model in the 1975 paper, but it seems a good one to the writer.) That cyclothems can now be explained in modern facies-model terms does not mean that the concept and term cyclothem are no longer valid. The widespread nature of these distinctive sequences, their restriction to rocks of Pennsylvanian and Permian age, and in spite of the

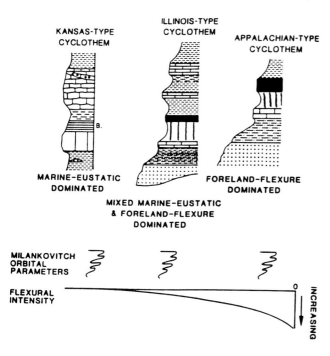

doubts of George and others, the fact that individual cyclothems and mesothems can be correlated for considerable distances call for a special interpretation. As Busch and Rollins (1984) noted, and as illustrated in Fig. 8.34, autogenic controls are certainly important on a local scale. In places extra channel-fill or other units may be present, or minor cycles cut out by channel erosion.

The renewed interest in sequence stratigraphy and eustasy since the late 1970s has revived the credibility of the glacioeustatic model of Wanless and Shepard (1936), and this interpretation is now widely accepted (Crowell 1978; see Chap. 10). However, considerable debate continues regarding the relative importance of tectonism and climate-change in the generation of these cycles (Dennison and Ettensohn 1994). For example, Klein and Willard (1989) and Klein and Kupperman (1992) have argued that tectonic influences were important in the development of the clastic-rich Appalachian-type cyclothems and, to a lesser extent, the Illinois-type cyclothems. They suggested that episodic flexural subsidence driven by thrust-sheet loading within the Appalachian foreland basin was a primary control in generating the clastic detritus and in controlling the transgressive-regressive cyclicity of the depositional environment. In Section 8.6 other classes of foreland-basin cycles are described which, it has been suggested, may also be tectonically driven, and so this is not a unique idea. The tectonic mechanisms are discussed in Section 11.3.

Some models, such as that of Wilson (1975; see Fig. 8.32), indicate that the three cyclothem types are variants within single mappable stratigraphic entitites. However, this raises an interesting question: how can carbonate-dominated, eustatically driven cyclothems within the Kansas craton be correlative with clastic-dominated, tectonically driven cyclothems of the Appalachian foreland basin, unless eustasy and tectonism were exactly in phase on a 10^5-year time scale? Klein and Willard (1989) did not examine this question. According to G. deV. Klein (personal communication 1995) it was not the intention of their work to suggest a synchroneity between tectonism and climate change. The existing correlation framework for cyclothems in the Appalachian foreland basin and the Midcontinent provides only a partial test of such sychroneity (Heckel 1994), and considerable debate continues on this point (Sect. 11.3.2.2).

Recent work in the Western Interior foreland basin has indicated a correlation between some Cretaceous basin-margin clastic sequences and basin-center chemical sequences (Elder et al. 1994), which raises important questions regarding the nature of the controlling processes. We return to this paper in Section 8.6, and the processes are discussed in Chapter 10. The relationship between tectonism and climate change in the Appalachian foreland basin is addressed in more detail in Section 11.3.2.2.

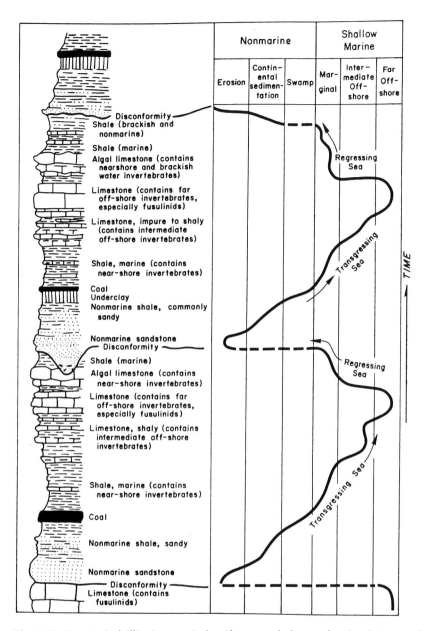

Fig. 8.31. Two typical Illinois-type Carboniferous cyclothems, showing interpretation in terms of transgression and regression. (Crowell 1978; based on Moore 1964)

8.5 Lacustrine Clastic and Chemical Rhythms

Lake sediments are extremely sensitive to changes in temperature, water depth and sediment supply, and are commonly rhythmic, indicating a cyclicity in the external tectonic and climatic controls. The term *meter-scale cycle* is commonly used for this type of sequence. In many cases Milankovitch mechanisms are indicated. Among the best-studied examples are the Triassic-Jurassic deposits of the Newark Supergroup in eastern North America (Van Houten 1964; P.E. Olsen 1984, 1986, 1990), and the Eocene Green River Formation of Wyoming (Bradley 1929; Eugster and Hardie 1975; Fischer and Roberts 1991). The Devonian Orcadie Basin of Scotland also contains a cyclic lacustrine succession (Donovan 1975, 1980). Cycles with 10^4–10^5-year episodicities are discussed in this section. Cyclicity in the solar and calendar bands is common in lacustrine sediments, but is not discussed here.

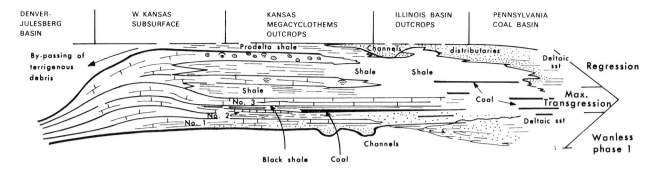

Fig. 8.32. Stratigraphic model of a single Midcontinent cyclothem from the Appalachians to Kansas, showing the relationships of the three main types of cyclothem illustrated in Fig. 8.30. Each cycle is a few tens of meters thick, and may extend laterally for distances of more than 1500 km. (Wilson 1975)

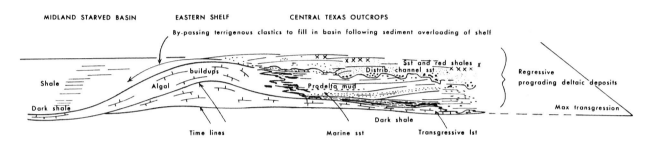

Fig. 8.33. Stratigraphic model of a northern Texas cyclothem. This generalized section extends eastern-west for about 200 km, from the Ouachita orogenic source in the east (*right*) to the Midland starved basin of western Texas (*left*). (Wilson 1975)

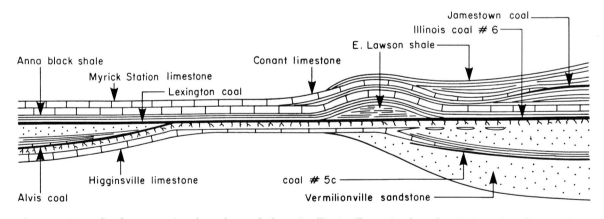

Fig. 8.34. Generalized cross section through a cyclothem, in Illinois, illustrating lateral variations. (Wanless 1964)

The Newark basins are a series of extensional basins developed during the initial, rifting phase of the breakup of Pangea and the opening of the modern Atlantic Ocean. P.E. Olsen (1990) recognized three types of cyclic succession, which he named after the basins where they are well-known (Fig. 8.35). Sedimentological studies of the Newark-type successions indicate variations in depositional environment from deep water to complete exposure, suggesting cyclic variations in lake level by as much as 200 m, as a result of variations in precipitation. The cycles vary in thickness (and therefore in duration) by at least three orders of magnitude, ranging from 1.5 to 35 m in thickness, and from 20 to 400 ka in estimated duration (Fig. 8.36). P.E. Olsen (1990) called these *Van Houten cycles,*

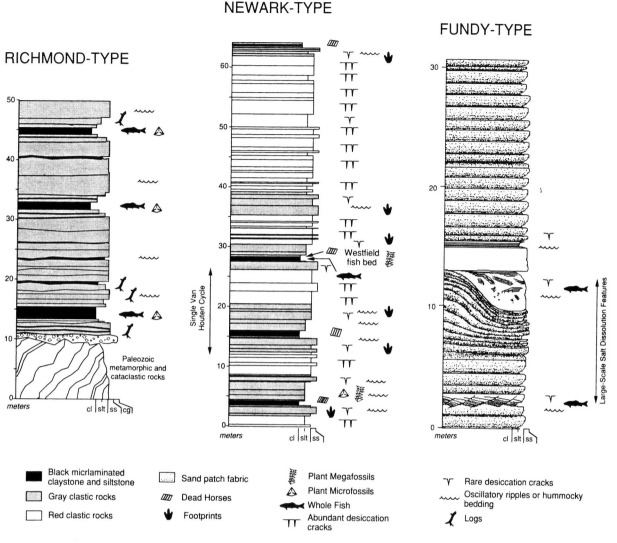

Fig. 8.35. Three types of cyclic lacustrine succession in Triassic Newark-type basins of eastern North America. (P.E. Olsen 1990, reproduced by permission)

after the pioneering work on the deposits by Van Houten (1964). P.E. Olsen (1990) stated that:

Drastic changes in lake level apparently inhibited the buildup of high-relief sedimentary features (other than alluvial fans) within the basin both by wave action during transgression and regression and by the brief time the water was deep. Consequently, lacustrine strata are characterized by extreme lateral continuity and by a tendency for coarse-grained sediment to be absent from deeper water facies and restricted to basin margins.... Large-scale sequence boundaries and large deltas apparently are absent from the main basin fill.

Van Houten (1964) distinguished two main types of cycle (Fig. 8.37), chemical cycles, averaging 3 m in thickness, formed during periods of closed drainage and relatively high evaporation rates, and detrital cycles, averaging 5 m in thickness, that formed during more humid periods, when there was probably a run-off flow-through that maintained low concentrations of dissolved sediment. The detrital cycles are coarsening-upward in type, and are interpreted as regressive sequences formed by basin filling following a rapid rise in lake level.

Richmond-type successions (Fig. 8.35) are characterized by significant coals, and by bioturbated and microlaminated siltstones and sandstones containing no evidence of exposure. The successions are thought to indicate relatively humid environments, in which complete basin desiccation was rare. During some cycles the lake may have remained deep

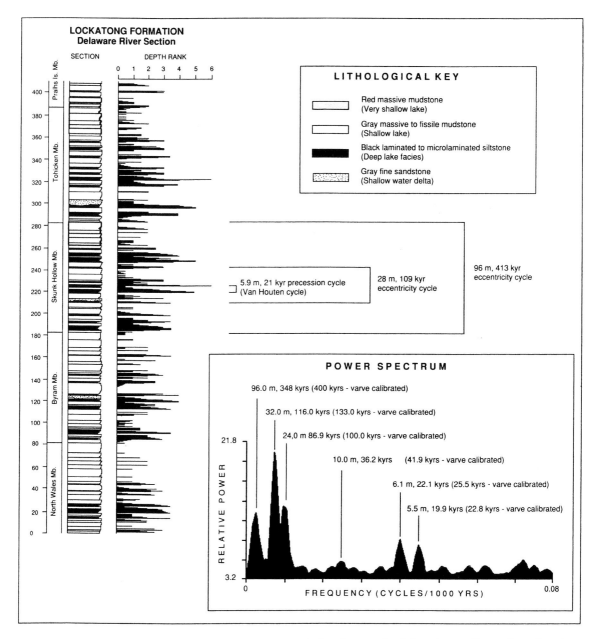

Fig. 8.36. Section of Newark-type lacustrine facies complex in the middle Lockatong Formation, Newark Basin. The power spectrum was derived by Fourier time-series analysis. Depth ranking indicates a ranking of increasing inferred water depth based on sedimentological analysis. Values in kyr are cycle periods in thousands of years. (P.E. Olsen 1990, reproduced by permission)

long enough that high-relief sedimentary features were built, including large, prograding deltas, fan deltas, submarine channels and fans. Large-scale erosional unconformities with deep channeling developed during lake-level lowstands.

Fundy-type successions are "characterized by a cyclicity consisting mostly of what are termed sand-patch cycles.... These represent alternations between shallow perennial lakes and playas with well-devel-

oped efflorescent salt crusts" (P.E. Olsen 1990). Gypsum nodules, salt-collapse structures and eolian dunes are also locally present. Water flow rarely exceeded evaporation in these basins, consequently deep-lake conditions were rarely established, depositional relief remained low, and cycles show considerable lateral continuity. A schematic synthesis of basin stratigraphy and cycle type is shown in Fig. 8.38.

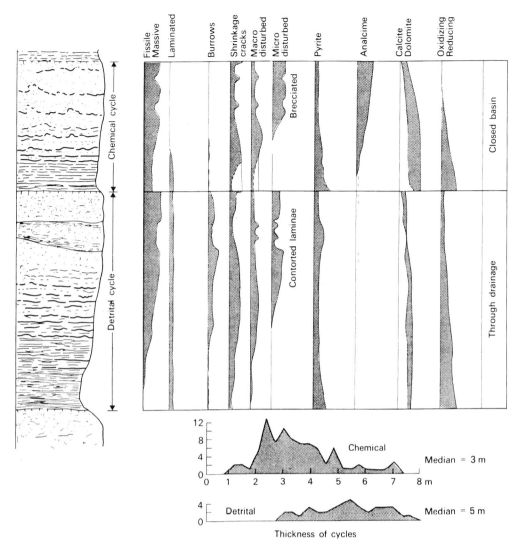

Fig. 8.37. Model of detrital and chemical cycles in the Lockatong Formation (Triassic), Newark Basin, New Jersey. (Van Houten 1964)

The Green River Formation was deposited in several lacustrine basins. Figure 8.39 illustrates the generalized stratigraphy, showing marginal alluvial complexes draining into an arid basin center (Fig. 8.40). Annual varves and solar-band cycles are prominent in these deposits (Fischer and Roberts 1991) but are not discussed here. Milankovitch cycles are well developed (Fig. 8.41), with some cycles traceable for more than 20 km without significant thickness changes. Figure 8.41A shows the basic cycle, consisting of three parts. The oil-shale facies consists of organic-rich dolomitic laminites and breccias, formed by the preservation of algal and fungal matter during relatively wet phases. During drier phases a hypersaline lake developed, forming

the trona facies. Halite is also present in some beds of this facies. The marlstone facies consists of thin-bedded dolomitic mudstones with silt-mud laminae, showing evidence of frequent desiccation in a playa environment. The silt-mud laminae represent occasional sheet floods. The Tipton Member is thought to represent a slightly wetter period than the Wilkins Peak Member. The cyclicity is not visible in core, but is picked up by gamma-ray and sonic-velocity logs, which record subtle variations in uranium and thorium linked to organic matter, potassium in clay, and carbonate content (Fig. 8.41B). Mean cycle thickness is 2.35 m. In the Wilkins Peak Member cycles average 3.7 m in thickness. Bundles of five cycles indicate a longer-

RICHMOND-TYPE LACUSTRINE FACIES COMPLEX

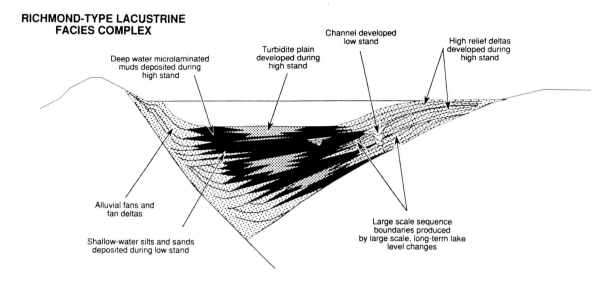

Deep water microlaminated muds deposited during high stand

Turbidite plain developed during high stand

Channel developed low stand

High relief deltas developed during high stand

Alluvial fans and fan deltas

Shallow-water silts and sands deposited during low stand

Large scale sequence boundaries produced by large scale, long-term lake level changes

NEWARK-TYPE LACUSTRINE FACIES COMPLEX

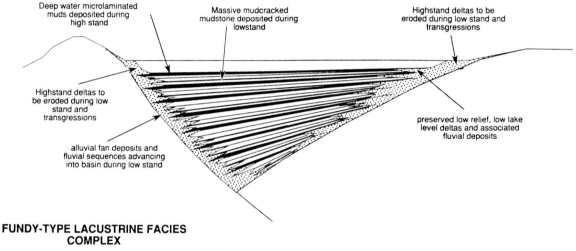

Deep water microlaminated muds deposited during high stand

Massive mudcracked mudstone deposited during lowstand

Highstand deltas to be eroded during low stand and transgressions

Highstand deltas to be eroded during low stand and transgressions

alluvial fan deposits and fluvial sequences advancing into basin during low stand

preserved low relief, low lake level deltas and associated fluvial deposits

FUNDY-TYPE LACUSTRINE FACIES COMPLEX

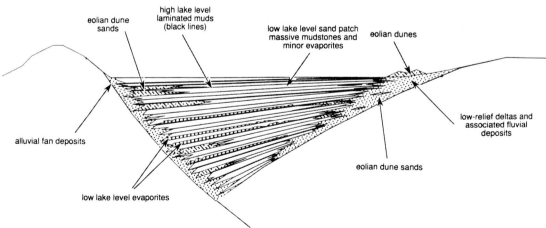

eolian dune sands

high lake level laminated muds (black lines)

low lake level sand patch massive mudstones and minor evaporites

eolian dunes

alluvial fan deposits

low lake level evaporites

eolian dune sands

low-relief deltas and associated fluvial deposits

Fig. 8.38. Idealized lacustrine facies complexes, Newark basins of eastern North America. (P.E. Olsen 1990, reproduced by permission)

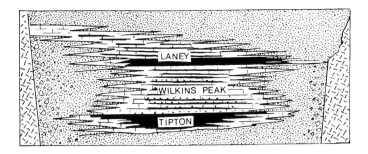

Fig. 8.39. Generalized stratigraphy of the Green River Formation (Eocene), Wyoming. Marginal mountains shed coarse alluvium into an intermontane basin. *Black units* are varved, lacustrine oil shales. *Vertical scale* about 1 km; *horizontal scale* 200–300 km. (Fischer and Roberts 1991)

term cyclicity (Fig. 8.41C), of a type discussed in Section 10.2.5.

The last example discussed briefly here is the Devonian Caithness Flagstone Group of the Orcadie Basin, Scotland (Donovan 1975, 1978, 1980; Allen and Collinson 1986). The basic cycle is shown in Fig. 8.42, but considerable variations occur, which reflect in part the position of the succession relative to the basin margin (Figs. 8.43, 8.44). During low lake levels lake margins were exposed to erosion, and rivers incised their profiles. Cycles are dominantly clastic. During times of rising lake level transgressive beds were deposited, including clastic units deposited by wind-generated currents, desiccation features, and the deposits of inland sabkhas. During times of deep water some degree of lake stratification developed. Carbonates were deposited, and algal remains accumulated in the lake center. Fish faunas may have been exchanged with adjacent basins because of the opening of interbasin connections. The cycles are interpreted as the product of rhythmic climate change, but the

Caithness Flagstones have not been subjected to quantitative cycle analysis, unlike the other lacustrine deposits described in this section, and their Milankovitch periodicites are therefore unknown.

8.6 Clastic Cycles of Foreland Basins

Shallow-marine to nonmarine clastic sequences with 10^4–10^6-year episodicities are common in the Cretaceous strata of the Western Interior of North America. They constitute a distinctive type of cycle, in part because they occur within a foreland basin, with its characteristic pattern of subsidence and sediment supply. It is not known whether the tectonic setting of these cycles is significant in terms of the controlling cyclic mechanisms. Foreland basins are, of course, tectonically highly active, and tectonic influences on sequence development seem to be indicated, as discussed at some length in Section 11.3. However, controversy remains, and

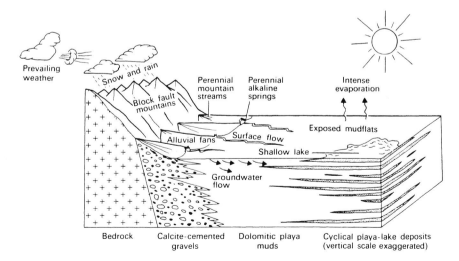

Fig. 8.40. Schematic block diagram showing depositional framework of the Wilkins Peak Member, Green River Formation. (Allen and Collinson 1986, after Eugster and Hardie 1975)

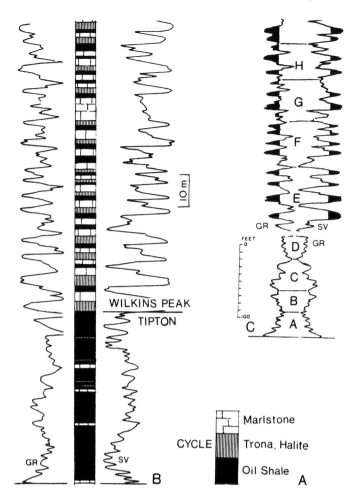

Fig. 8.41. Cyclicity in the Tipton and Wilkins Peak Members of the Green River Formation, Green River Basin, Wyoming. **A** Basic desiccation cycle. **B** Basal 129 m of Green River Formation, showing gamma-ray log (*GR*) and sonic velocity log (*SV*). **C** Log plot illustrating the grouping of cycles into thicker bundles in part of the Tipton Member. *A–D* is a mirror-image GR plot, *E–H* is a GR-SV plot. (Fischer and Roberts 1991)

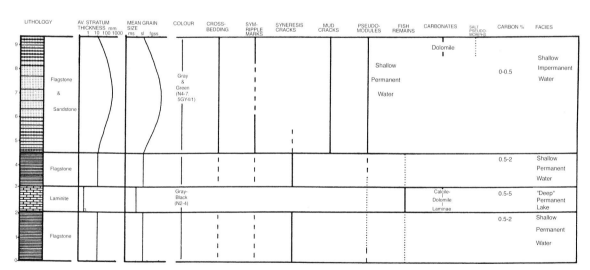

Fig. 8.42. Features of lacustrine cycles in the Devonian Caithness Flagstones, Orcadian Basin, Scotland. These cycles range from 3 to 20 m in thickness with an average of 8–9 m. (Donovan 1975)

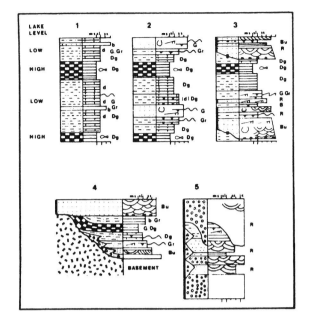

Fig. 8.43. Styles of cyclic sedimentation in the Caithness Flagstones, Scotland. Five types of cycle may be present, depending on the location of the section relative to the lake margin. (Donovan 1978)

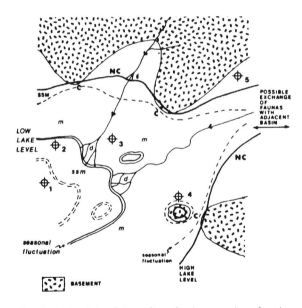

Fig. 8.44. Model of Orcadian basin margin, showing fluctuation in lake levels. *Numbered circles* indicate suggested positions of sedimentological profiles illustrated in Fig. 8.43. (Donovan 1978)

some researchers have refered to possible glacioeustatic causes (e.g., Plint 1991; Elder et al. 1994), despite the very limited evidence for glaciation during the Cretaceous.

Cycles with 10^4–10^5-year episodicities have been described in two main areas of the Western Interior. In Alberta, Canada, many units of Aptian to Santonian age have been subdivided into formal allostratigraphic units based on the mapping of widespread disconformity surfaces. Cant (1989a) provided a summary of the stratigraphy, and several detailed local studies are referred to below. Cretaceous deposits of the Colorado Plateau are also characteristically cyclic. An up-to-date summary of the sequence stratigraphy is not available, although good local descriptions were provided by Nummedal and Wright (1989), and several individual studies are referred to below.

The location of the Alberta Basin is given in Fig. 8.45, and a restored stratigraphic cross section through the foreland-basin fill of Alberta and British Columbia is shown in Fig. 8.46. Units displaying well-developed internal high-frequency cyclicity include the Moosebar, Gates/Falher, Viking, Dunvegan, Cardium, Muskiki and Bad Heart Formations. As shown by Plint et al. (1992) these units, plus several other clastic tongues in this basin, may be compared to the "third-order" cycles of the Exxon global cycle chart. and may be correlated with times of low eustatic sea level on this chart (Fig. 8.47), although the significance of this correlation is questionable, as discussed in Chapter 13.

Cyclicity in the Gates and Moosebar Formations, and their subsurface equivalent, the Falher Formation, of Albian age, was described by Cant (1984, 1995), Leckie (1986), and Carmichael (1988). Seven transgressive-regressive cycles have been mapped within a stratigraphic interval that is interpreted to have lasted for between 3 and 6 m.y. (Fig. 8.48). Leckie (1986) estimated the cycles to have had durations of between 103,000 and 275,000 years. The cycles consist of thin transgressive deposits, including estuarine sandstones and lag gravels, followed by regressive shoreline and deltaic deposits formed during relative sea-level highstands. Evidence of wave and tide activity is present in these deposits, which form coarsening-upward cycles (Fig. 8.49). A depositional model for the cycles is shown in Fig. 8.50.

The Dunvegan Formation represents a major delta complex up to 300 m thick that prograded into the Alberta Basin from the northwest over a period of about 1.5 m.y. (Bhattacharya 1988, 1991; Bhattacharya and Walker 1991). It can be subdivided into seven allomembers, which are separated from each other by widespread flooding surfaces (Fig. 8.51). The allomembers have each been map-

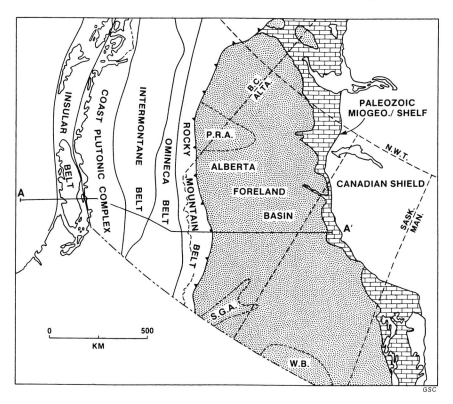

Fig. 8.45. Location map, Alberta foreland basin. *P.R.A.* Peace River Arch; *S.G.A.* Sweetgrass Arch; *W.B.* Williston Basin. (Cant 1989a)

ped over an area in the order of 300,000 km². Each allomember ranges up to 80 m in thickness, and represents a depositional episode about 200 ka in duration. The bounding marine-flooding surfaces are attributed to "allocyclically controlled relative rises in sea level, probably caused by a rate increase in tectonically induced basin subsidence" (Bhattacharya 1991). The allomembers consist of shingled clinoform deposits up to 30 m thick, each shingle comprising a heterolithic deltaic complex representing 10^4 years of sedimentation. An example of a vertical section through a typical shingle is shown in Fig. 8.52. The deltas are of river-dominated type and compare in scale and composition (except for being somewhat more sandy) with the delta lobes of the modern Mississippi delta. Offsetting of the shingles within each allomember probably results from autogenic distributary switching, as in the modern Mississippi.

The Cardium and Viking formations of Alberta each consist primarily of shelf deposits, and both may be subdivided using allostratigraphic methods based on the recognition and mapping of major bounding erosion surfaces. The Cardium Formation was the first unit in the Alberta basin to be sub-

divided in this way (Plint et al. 1986), and recognition of the architectural style of the formation constituted a major breakthrough in foreland-basin geology when the 1986 paper was published. Indeed, the stratigraphic concepts were considered controversial, and some diverging opinions were published (Rine et al. 1987). Subsequent detailed papers on the Cardium Formation include those by Bergman and Walker (1987) and Walker and Eyles (1988). The allostratigraphy of the Viking Formation has been described by Boreen and Walker (1991).

The allostratigraphy of the Cardium Formation, an important hydrocarbon-producing unit in the Alberta Basin, was developed by Plint et al. (1986) as part of a detailed regional surface-subsurface study of the many producing fields in the area. Numerous local studies had, over the years, led to a confusing welter of local informal terminologies for sandstone horizons and marker units within the Cardium Formation and much controversy regarding the depositional environments of the various facies. Routine but meticulous lithostratigraphic correlation of subsurface records led Plint to the recognition that this unit, which is only about 100 m thick, contains at least seven basin-wide erosion

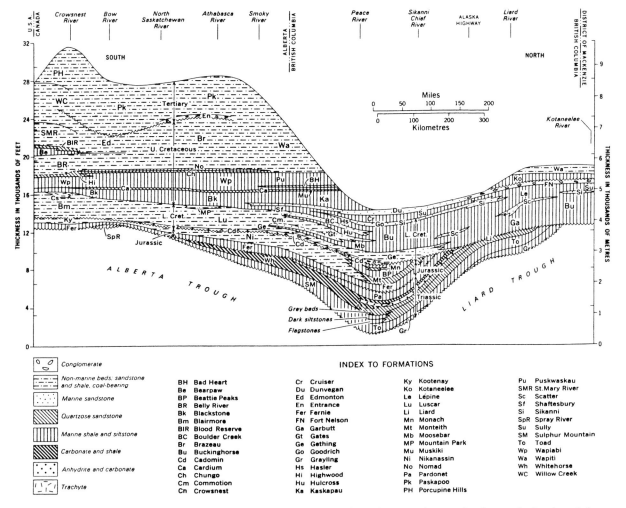

Fig. 8.46. Restored stratigraphic cross section through the Mesozoic rocks of the Alberta Basin. The section runs SW-NE along the axis of the basin, near the fold-thrust belt, from the United States border to the border of the Northwest Territories. (D.F. Stott, in Douglas et al. 1970)

surfaces, indicating the occurrence of this many events of erosion and transgression (Fig. 8.53). Some, at least, of the erosion surfaces can be traced for more than 500 km.

Not only does this new framework provide a rational basis for basin-wide correlation, but it also throws a wholly new light on the depositional history of the formation. The unit consists largely of a series of basin-wide coarsening-upward cycles capped by sandstone units containing hummocky cross-stratification, overlain in some areas by lenticular conglomerate beds. The cycles, up to the level of the sandstones, are readily interpreted as the product of a shoaling shelf environment, with the sediment surface gradually building up to storm wave base. The problem had always been to fit the conglomerate into this interpretation. These coarse deposits are tens to several hundreds of kilometers

from the basin margin, and the problem of transporting coarse detritus out this far in a shelf setting had been discussed in the Cardium literature for many years. The new interpretation by Plint provides a simple resolution to the problem. The conglomerates are not a conformable cap to the coarsening-upward cycles, as had always been thought; each conglomerate lens rests on one of the erosion surfaces. They are therefore, in all probability, beach deposits, which originated as fluvial detritus transported basinward during regression of the shoreline, and concentrated along temporary shorelines during the initial transgression at the beginning of a new cycle of relative sea-level rise. The seven Cardium sequences span about 1 m.y., therefore the average duration of each sea-level cycle is about 140 ka.

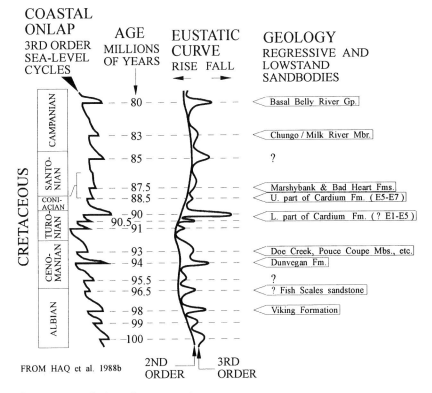

Fig. 8.47. Correlation of major sandstone tongues in the Alberta Basin with the global cycle chart of Haq et al. (1987, 1988a). *Bracket* indicates range of ages assigned to the Coniacian-Santonian boundary in various sources (Table 13.2; see discussion of this figure in Sect. 13.4.1). (Plint et al. 1992)

The Muskiki and Marshybank formations overlie the Cardium Formation in the foothills of Alberta and northeastern British Columbia, and are of Coniacian-Santonian age. The allostratigraphy of the units was established by Plint (1990), and the sequence stratigraphy was discussed by Plint (1991). He demonstrated that the Marshybank Formation includes 12 mappable units bounded by disconformity surfaces, some of which can be traced for more than 380 km along strike from northwest to southeast. Figure 8.54 illustrates the large-scale architecture of the units. The Muskiki Formation and unit A of the Marshybank Formation constitute transgressive systems tracts. Units B-L are progradational highstand systems tracts, forming clinoform units that downlap onto a surface marking the top of unit A. Sedimentation of the Marshybank Formation was terminated by a major fall in relative sea level, resulting in deep erosion. The lightly stippled downlapping units in Fig. 8.54 are coarsening-upward shelf successions capped by hummocky cross-stratified sandstone that range from 5 to 10 m in thickness. Heavy stippling indicates a shoreface facies dominated by "swaley" cross-stratification

(truncated hummocky cross-stratification). This facies formed by progradation over erosion surfaces, as illustrated in Fig. 8.55. Plint (1991) compiled biostratigraphic data for these deposits in an attempt to determine the elapsed time represented by the two formations. He concluded that the small-scale cycles represented by the lettered units in Fig. 8.54 represent, on average, about 100,000 years. These are grouped into poorly defined "midscale" cycles representing several hundred thousand years.

Cretaceous foreland-basin deposits of the Colorado Plateau area of the United States display cyclicity on several scales. The major subdivisions of the clastic wedge (Figs. 7.18–7.24) constitute 10^6-year cycles (Sect. 7.2). Many of these units may be subdivided into smaller-scale high-frequency cycles. There are no useful overviews of the sequence stratigraphy. Molenaar and Rice (1988) outlined the regional stratigraphy, and several individual studies of cyclic sedimentation and sequence stratigraphy are contained in Van Wagoner and Bertram (1995), and the field guides edited by Nummedal and Wright (1989) and Van Wagoner et al. (1991). Other local studies are referred to below.

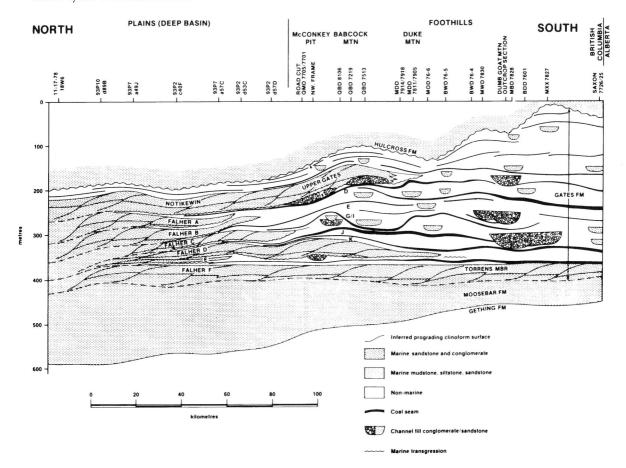

Fig. 8.48. Regional cross section through parts of the Gates and Moosebar Formations, northeastern British Columbia. (Carmichael 1988)

Wright (1986), and Devine (1991) studied cyclicity in tidally influenced shoreline deposits of the Upper Cretaceous Point Lookout Sandstone, in the lower Mesaverde Group of the San Juan Basin, New Mexico. Devine's research dealt with relatively more proximal cycles on the western side of the basin, whereas Wright's paper explored the more distal deposits on the southeastern side of the basin. Precise age control of the cycles is not available, and it is not known if the cycles described in the two papers correlate. They are thought to be of 10^5-year type. Fig. 8.56 illustrates a typical cross section through the deposits and Fig. 8.57 is a depositional model. Note that the lagoonal deposits of time slices 6–8 were deposited during the transgression. The barrier sands that were contemporaneous with them were removed by erosion during that transgression, except for a small remnant at time slices 7 and 8. Shelf sands were deposited at the same time, and are separated from the lagoonal deposits by a prominent ravinement erosion surface.

Swift et al. (1987), Van Wagoner et al. (1990, 1991), Olsen et al. (1995) and Yoshida et al. (1996) studied the Mesaverde Group in the Book Cliffs of Utah, where 10^5-year cyclicity is well developed in shelf, shoreline and alluvial deposits. The focus of these studies was on the Blackhawk and Castlegate Sandstones (Fig. 7.19). Figure 8.58 illustrates a cross section through the fluvial-shelf sequences of the Castlegate Sandstone and equivalent units. In the west the Castlegate Sandstone constitutes a single sequence with a duration of about 5 m.y., consisting of a lowstand braided sandstone sheet, a transgressive systems tract with evidence of tidal activity, and a fluvial highstand systems tract (Olsen et al. 1995). Down depositional dip to the east the Upper Castlegate Sandstone passes into a succession of high-frequency sequences with episodicities in the 10^5-year range, consisting mainly of fluvial and estuarine deposits (Yoshida et al. 1996). These are assigned to the Sego and Neslen formations. Some tectonic control of these sequences is indicated by

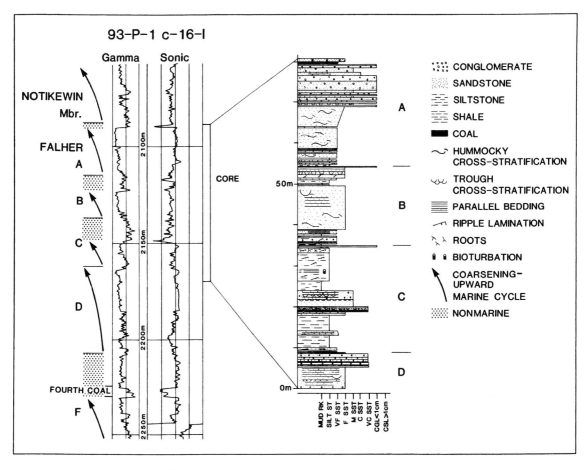

Fig. 8.49. Representative core and wireline log through the Falher cycles of northeastern British Columbia. (Leckie 1986, reproduced by permission)

the angular unconformity that truncates the Buck Tongue northwestward of Green River.

The high-frequency sequences of the Castlegate-Sego-Neslen succession near Green River, Utah, are dominated by "deepening-upward" successions, recording the drowning of broad, possibly estuarine erosional valleys during the gradual rise in base level. Fluvial deposits occur at the base of each sequence, and are overlain sharply by transgressive-marine beds, such as the Buck Tongue, in contrast to the sequence models of Swift et al. (1987) and Posamentier et al. (1988; Fig. 4.6), in which fluvial deposits occur at the top of shallowing-upward, progradational wedges. Only small remnants of

highstand fluvial systems are preserved in the lower part of the succession, with the proportion of nonmarine deposits increasing upward into the Neslen Formation (Yoshida et al. 1996). These deposits therefore constitute a progradational sequence set. Is the lack of fluvial highstand deposits simply a question of preservation, with progradational sequence tops having been lost to erosion during each base-level drop? Or did the coastal plain not advance into the area during each highstand? Most other stratigraphic sequences described from foreland basins are dominated by progradational, highstand deposits (e.g., Plint et al. 1986; Fig. 8.53). The underlying lower Desert Member in

Fig. 8.51. Schematic regional dip-oriented cross section of the Dunvegan Formation, western-central Alberta. *Heavy lines* are regional flooding surfaces that subdivide the formation into seven allostratigraphic units. Within each unit separate offlapping shingles can be mapped, as

shown by the *numbers*. *Root symbol* Nonmarine facies; *light stipple* marine sandstone; *heavy stipple* channel fills; *blank* marine shale. *Arrows* indicate downlap stratigraphic terminations. (Bhattacharya 1991)

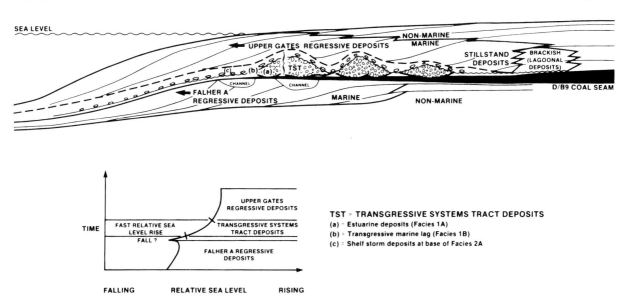

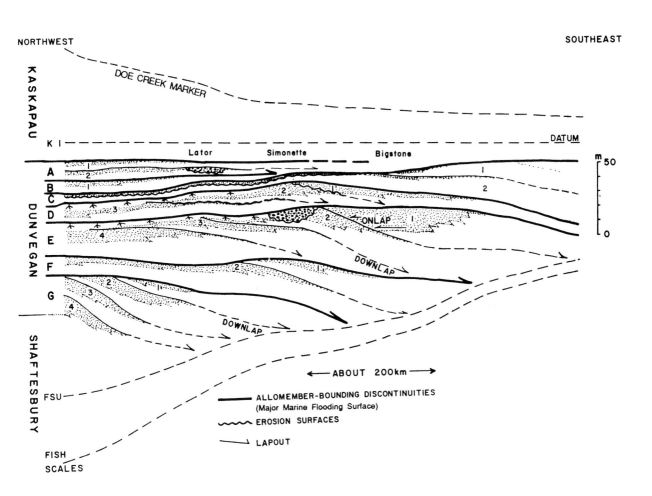

Fig. 8.50. Sequence model for the Upper Gates Formation. The ravinement surface is marked by the transgressive lag (*b*). Estuarine deposits (*a*) are the only beds preserved below the ravinement surface and are commonly thin or absent. (Carmichael 1988)

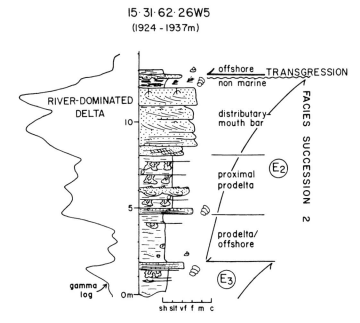

Fig. 8.52. Facies succession through shingle 2 of allomember E of the Dunvegan Formation, Alberta. *sh* Shale; *slt* siltstone; *vf, f, m,* and *c* very fine-, fine-, medium-, and coarse-grained sandstone. (Bhattacharya 1991)

the Green River area also contains thick, progradational, marine highstand deposits (Van Wagoner et al. 1990, 1991). This architecture suggests a rapid rise in base level followed by a long period of highstand or slow fall, whereas the architecture of the Castlegate and Upper Desert Member sequences suggests slow rise followed by a relatively short period of highstand and base-level fall.

The sequence architecture is probably in part a reflection of sediment-supply considerations. The sequence boundaries described by Plint et al. (1986) are blanketed by sparse or patchy conglomerate beds supplied by limited fluvial input and reworked by marine processes during rising base level. In the case of the Castlegate Sandstone the extensive fluvial blankets overlying the sequence boundaries

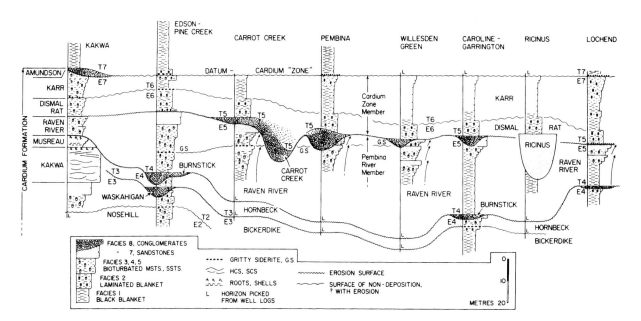

Fig. 8.53. Example of high-frequency cycles in a foreland-basin succession: the Cardium stratigraphy of the Alberta Basin. Surfaces of erosion and transgression are num-bered *E2/T2,* etc. *HCS, SCS* Hummocky and swaly cross-stratification. (Plint et al. 1986)

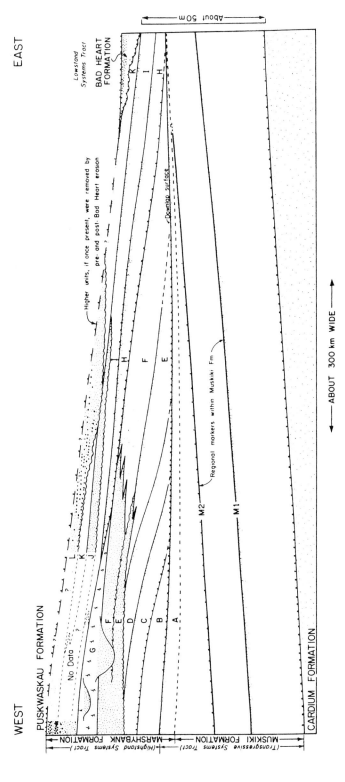

Fig. 8.54. Architecture of the Muskiki and Marshybank formations, northwestern Alberta and northeastern British Columbia. (Plint 1990)

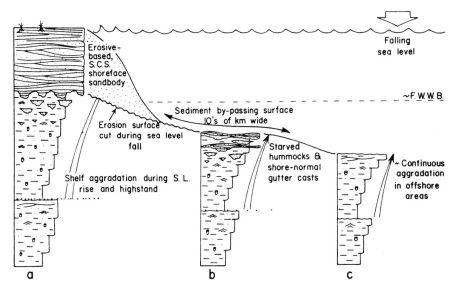

Fig. 8.55. Lateral variation in the facies succession within one of the stratigraphic sequences of the Marshybank Formation. Upward-coarsening successions 5–10 m thick were formed during relative sea-level rise and highstand. **a** shows that the succession was truncated by erosion during the subsequent sea-level fall, after which a shore-face sand prograded across the erosion surface. **b** represents an area of sediment bypass with "starved" hummocks, while continuous aggradation took place in offshore areas. *S.C.S.* Swaly cross-stratification; *F.W.W.B.* fair-weather wave base. (Plint 1991)

suggests that fluvial sediment supply was adequate to keep pace with the increase in accommodation space brought about rising base level, allowing erosional valleys to fill with widespread fluvial deposits. The deltaic channels so well exposed in the Desert Member at Tusher Canyon and Thompson Canyon, and in the Castlegate Member at the latter locality (Miall 1993; Van Wagoner et al. 1990, 1991), represent the distal, marginal-marine fringe of this backfill blanket.

Could sediment-supply controls alone generate the Castlegate and Upper Desert Member sequences? Yoshida et al. (1996) developed a hypothesis that relates the sequence architecture shown in Fig. 8.58 to tectonism acting on two time scales. The influence of eustatic sea-level change cannot be independently demonstrated and may not have been a factor in developing these sequences. This is discussed further in Section 11.3.2.

Ryer (1977, 1983, 1984) and Cross (1988) examined coal seams contained within high-frequency cycles in the foreland basin. Ryer's work dealt with the coals of the Ferron Sandstone in southern Utah. His generalized cross section showing the stratigraphic position of the Ferron Sandstone, as a long-term (10^6-year) regressive tongue, is given in Fig. 7.21. The detailed lithostratigraphy of this

sandstone unit is shown in Fig. 8.59. Ryer (1984) showed that the thickest coal developments occur at times of transgressive maxima and minima on the long-term sea-level cycle, when facies tend to stack vertically. The Ferron examples exemplify a regressive maximum, when 10^5-year deltaic sandstone cycles prograded far into the basin. A possible comparison with Alleghenian cyclothems may be made (Sect. 8.4).

Nummedal et al. (1989) referred to high-frequency cyclicity in the Gallup Sandstone of New Mexico (Fig. 2.8), but they did not attempt a synthesis, their main focus being on long-term (10^5-year) cyclicity (Figs. 7.23–7.25).

A particularly interesting study of 10^5-year sequences in the Western Interior Basin was reported by Elder et al. (1994). They demonstrated that a succession of five Upper Cretaceous clastic strandline parasequences in southern Utah could be correlated with basin-center limestone-marl sequences in Kansas, some 1500 km to the east (Fig. 8.60). Clastic sequences, as discussed in this section, are commonly attributed to tectonic mechanisms or to eustatic sea-level changes, whereas chemical cycles are usually explained in terms of orbital-forcing mechanisms, in which sea-level change is not necessarily a requirement. This correlation therefore

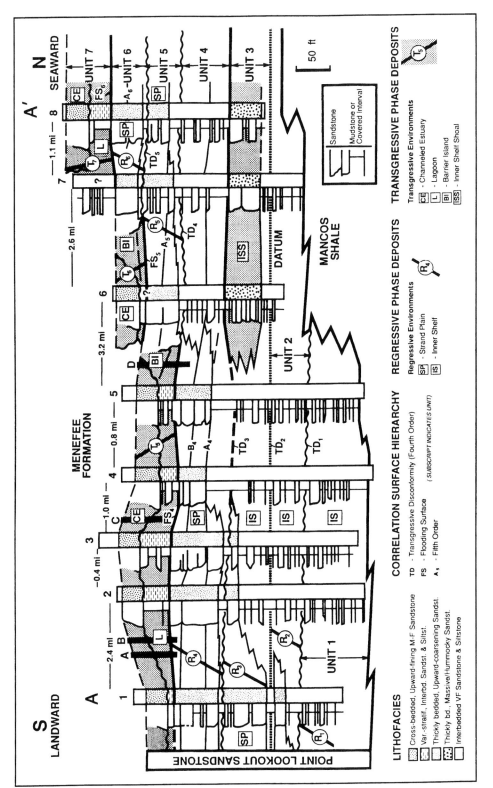

Fig. 8.56. Stratigraphic cross section through the Point Lookout Sandstone (Upper Cretaceous) of northwestern San Juan Basin, New Mexico. (Devine 1991, reproduced by permission)

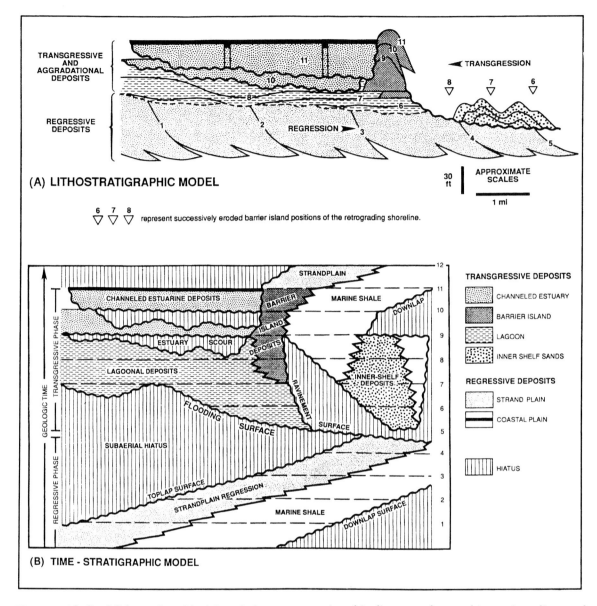

Fig. 8.57. Idealized lithostratigraphic (**A**) and chronostratigraphic (**B**) model of a single regressive-transgressive couplet of the Point Lookout Sandstone, northwestern New Mexico. Numbered events (*1–12*) in the chrono-stratigraphic diagram refer to arbitrary time slices, and correspond to the same numbers in the lithostratigraphic model. (Devine 1991, reproduced by permission)

raises interesting problems of interpretation, that are discussed further in Sections 10.3 and 11.3.2.

8.7 Main Conclusions

1. The Neogene stratigraphic records of continental margins, including the continental shelf, slope, and deep basin, have been intensively studied by reflection-seismic surveying and offshore drilling, including DSDP surveys. Most stratigraphic sections in both carbonate- and clastic-dominated successions are characterized by cycles with 10^4–10^5-year episodicities ("fourth-" and "fifth-order" cycles in the earlier terminology). These can be correlated with the ocean temperature cycles defined by the oxygen-isotope chronostratigraphic record, and are interpreted to be of glacioeustatic origin.

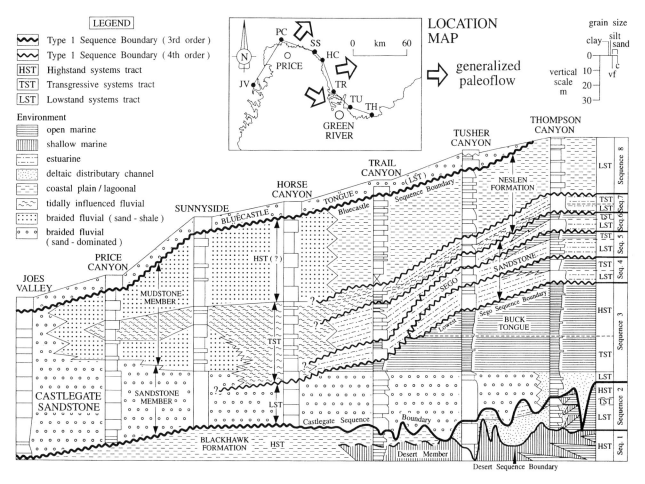

Fig. 8.58. Reconstructed stratigraphic cross section through the Castlegate Sandstone and equivalent units, Book Cliffs, Utah, adapted from Fouch et al. (1983) and Van Wagoner et al. (1990, 1991), with additional mapping data. Note that in the west, the Castlegate Sandstone consists of a single sequence spanning about 5 m.y., but in the eastern it passes into a succession of high-frequency sequences. *Two letter abbreviations* in *inset map* correspond to section location names. *LST* Lowstand systems tract; *TST* transgressive systems tract; *HST* highstand systems tract (Yoshida et al. 1996)

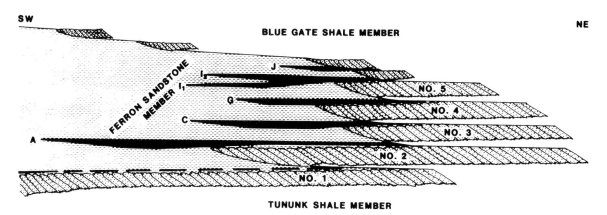

Fig. 8.59. Schematic cross section of the Ferron Sandstone, central Utah. Stratigraphic position of these beds ithin the foreland-basin clastic wedge is shown in Fig. 7.21. (Ryer 1984)

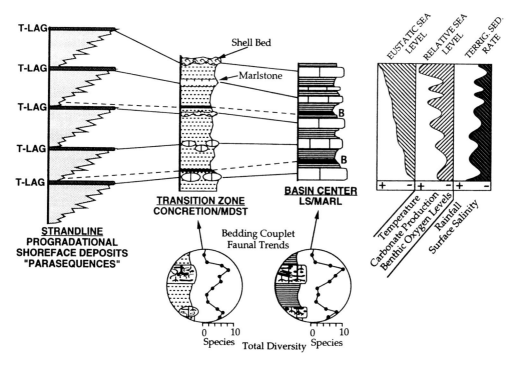

Fig. 8.60. Correlation of clastic sequence in southern Utah (*left*) with limestone-marl sequences in Kansas. (Elder et al. 1994)

2. Many stratigraphic sections in the pre-Neogene record also contain prominent Milankovitch cycles. They are particularly prominent, and have been well-described, in various carbonate-platform successions, notably the Triassic section of the Dolomites, in northern Italy. However, comparable cycles are also present in many other types of succession. Examples of lacustrine, marginal-marine evaporite, pelagic-marine and nonmarine suites are described here. There is controversy regarding the origins of many of these cycles, as there is doubt that glacioeustasy can be appealed to during much of Phanerozoic time, when little or no evidence exists for the presence of large continental ice caps anywhere on earth. However, other types of orbital forcing processes may be responsible, as described elsewhere in this book (Chap. 10).

3. A particularly well-known type of presumed Milankovitch cycle is the Upper Paleozoic record of cyclothems in northwestern Europe and the Midcontinent region of North America. These cycles contain much of the economic coal deposits of the northern hemisphere and have been widely studied since the 1930s. It is now generally agreed that these cycles were driven by the Late Paleozoic glaciation of the Gondwana supercontinent.

4. Many examples of clastic high-frequency cycles have been described from the Cretaceous sedimentary record of the Western Interior of North America, notably in Alberta, Canada, and the Colorado Plateau area of the United States. These cycles were all deposited within a tectonically active foreland-basin setting. Tectonism is known to have been a significant control in the development of the large-scale architecture of the basin fill, and it remains a controversial question whether tectonism also controlled the cyclicity (Sect. 11.3.2). Some chemical cycles of the deep basin may be correlated with clastic cycles of the basin margin, and Milankovitch mechanisms have been suggested as a contributing or controlling factor. A possible comparison may be made between these foreland basin clastic cycles and the Upper Paleozoic cyclothems of the Alleghenian foreland basin of the eastern United States.

III Mechanisms

Research during the last three decades has demonstrated that there are several global tectonic mechanisms and astronomical effects that generate eustatic sea-level changes on several different time scales and of varying amplitude. These are discussed in this part of the book. The processes are summarized in Table 9.1 and Fig. 9.1.

Research has, in addition, demonstrated that there is a range of tectonic processes that generate relative sea-level changes on a local to regional scale. The rates and magnitudes of change are variable, and much research remains to be carried out to quantify these effects. However, it is now clear that many forcing functions of different frequencies and amplitudes are likely to be active at the same time, that they may not be mutually independent, for example, tectonics and climate (uplift affects climate and both affect sediment supply), and that they can produce effects that are very similar to those of eustasy on a regional to continental scale (Table 9.1, Fig. 9.1). This is of critical importance, because it means that sequence stratigraphies recorded in any given basin, however well-documented, cannot be assumed to be global, and therefore representative of a global framework of eustatic cycles. In fact it calls into question the very concept of the global cycle chart based on the tying together of key sections from supposedly representative areas.

The focus of the Exxon models has been on eustatic sea-level change as the primary control of sequence architecture. If eustasy is the primary mechanism, it justifies the use of sequence stratigraphies for the purpose of constructing a global cycle chart, the assumption being that a sequence record anywhere reflects the same global eustatic controls. Tectonism is treated in the Exxon models as a slow background effect that does not substantially modify the stratigraphic response to eustatic sea-level change.

However, sequences of tectonic origin may be expected to vary in age from region to region, because they are generated by plate-tectonic processes and mantle effects that are progressive and diachronous. The only reliable test of a eustatic control is precise correlation. If sequences in widely spaced basins of different tectonic setting are of the same age, they may be assumed to be of eustatic origin. This emphasizes the need for very precise chronostratigraphic control, a subject discussed in Part IV.

Mechanisms, the range of processes that alter sea level, are the subject of this part of the book.

- Allocyclic sequences ... are mainly caused by variations external to the considered sedimentary system (e.g., the basin), such as climatic changes, tectonic movements in the source area, global sea level variations, etc.... Such processes often tend to generate cyclic phenomena of a larger lateral continuity and time period than autocyclic processes. The most characteristic effect of some of the allocyclic processes is that they operate simultaneously in diferent basins in a similar way. Thus it should be possible to correlate part of the allocyclic sequences over long distances and perhaps even from one basin to another. (Einsele et al. 1991b, p. 7)

- "Sea level is an ill-defined concept that incorporates the effects of isostasy, eustasy, ice volume, passive margin subsidence, lithospheric deformation and flexure, erosion and consequent sediment loading, and changes in rate of plate generation and consumption that all serve to alter the elevation of the world seafloor with respect to the elevation of the continental surface. As such, quantitatively defining a universally acceptable definition of sea level to the satisfaction of all has so far proven an intractable problem." (Dockal and Worsley 991, p. 6805)

9 Long-Term Eustasy and Epeirogeny

9.1 Mantle Processes and Dynamic Topography

The major cause of change in the earth's crust is the radiogenic heat engine, which drives mantle convection and generates the geomagnetic field. Convection distributes heat and drives plate tectonics. Oceanic and continental crust are subject to heating close to sea-floor spreading centers where new, hot, oceanic crust is generated, and to cooling overlying subduction zones, where cold oceanic crust descends to the mantle. Widespread cooling occurs over the downwelling zones where continental fragments converge. This differential heat distribution and consequent heating and cooling of different parts of the overlying earth's surface results in broad regional uplifts, downwarps and tilts, because of the effects on crustal densities. These processes maintain what is called *dynamic topography* (Fig. 9.2). The effects of crustal heating are to cause uplift. This occurs along the flanks of new continental rift systems (e.g., parts of present-day Eastern Africa) and above mantle plumes. Thermal doming beneath supercontinents may elevate the crust by as much as 1 km over periods of 100 m.y. Subsidence takes place over cooling areas of the earth's crust, such as areas of aging oceanic crust distant from spreading centers, and over regions of mantle downwelling.

The concepts of dynamic topography are now being explored with the techniques of a new field, called *computational geodynamics*, "in which computer models of mantle convection are used in the interpretation of contemporaneous geophysical observations such as seismic tomography and the geoid as well as of time-integrated observations from isotope geochemistry" (Gurnis 1992). Such models are capable of integrating large volumes of detailed stratigraphic data using the backstripping procedures outlined in Section 2.3.3. These developments are of great significance, and their implications for sequence stratigraphy have yet to be fully realized.

Vertical movement of the crust causes *relative* changes in sea level on a regional or continental scale. This is true *epeirogeny*. However, thermal changes also result in changes in the volume of the ocean basins, which lead, in turn, to *eustatic* changes in sea level. The same broad crust-mantle processes generate both regional and eustatic effects that may or may not be in phase. The result is a highly complex sequence of sea-level changes, and clear global eustatic signals may not always be present in the stratigraphic record.

Only in recent years has this research finally provided a theoretical basis for the process of epeirogeny, a process that has commonly been invoked to explain broad vertical movements of the crust indicated by stratigraphic studies, but which has lacked a basis in the new theories of plate tectonics. Some researchers, including the present writer (Miall 1987), have argued that most regional tectonism is controlled by plate-margin processes and intraplate stresses, which seemed to eliminate the need for epeirogeny. However, continental-scale vertical movements have been amply demonstrated by stratigraphers (e.g., Sloss 1963; Bond 1976, 1978), and remained to be explained until recently.

The processes referred to here are relatively long-term in their effect, and are capable of explaining much of the cyclicity that has been recorded on time scales of tens to hundreds of millions of years. They are the subject of this chapter.

9.2 Supercontinent Cycles

Early work by Wilson, Sutton, Bott, Condie, Windley, and others suggested that the Phanerozoic history of the earth (the broad, long-term stratigraphic patterns outlined in Chap. 6) can be related to the assembly and breakup of the Pangea supercontinent. Most recent workers have adopted this long-term plate-tectonic cycle as the basis for hypotheses of the earth's dynamics (D.L. Anderson

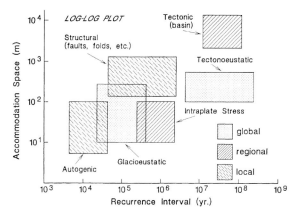

Fig. 9.1. Estimated ranges in accommodation space versus recurrence interval for various processes that generate stratigraphic cyclicity. The regional extent of the processes is indicated by the internal ornamentation of each *box*. (Dickinson et al. 1994)

1982, 1984, 1994; Worsley et al. 1984, 1986; Gurnis 1988; Veevers 1990).

The formation of a supercontinent creates a thermal blanket that inhibits convective release of radiogenic heat from the mantle (Fig. 9.3). Changes in the rotation of the earth's core and in the convective patterns in the mantle may be either the cause or the consequence of these surface events, which also appear to be linked to changes in the earth's magnetic field (D.L. Anderson 1984; Maxwell 1984). The development of the thermal blanket may be the cause of the eventual breakup of the supercontinent, following the establishment of a new pattern of mantle convection (Fig. 9.3). The formation of the thermal blanket beneath a super-

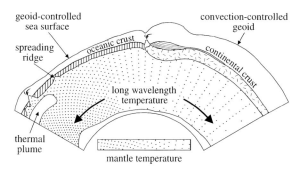

Fig. 9.2. The generation of dynamic topography and controls on sea-level change and regional vertical crustal motion by shallow and deep mantle-convection. Widespread heating beneath supercontinents (*right*) generates continental uplift and an elevated geoid. More localized depression of the geoid is caused by the subduction of cold slabs of oceanic crust. (Gurnis 1992)

continent leads to heating and regional epeirogenic uplift on a continental scale. Gurnis (1988, 1990, 1992) demonstrated that the uplift rate would be 5–10 m/m.y., and could persist for 100 m.y., resulting in an uplift of 0.5–1 km. (This is the first of many processes of sea-level change discussed in this book for which quantitative estimates of rate, duration and magnitude are available. The processes are listed in Table 9.1, which is referred to throughout the remainder of this book).

It has been argued that dynamic mantle uplift is extremely long-lived. It generates a positive geoid anomaly that survives for 10^8 years. Crough and Jurdy (1980) demonstrated the existence of a large positive anomaly beneath Africa (Fig. 9.4). Veevers (1990) showed the position of this anomaly beneath a reconstruction of the Pangea supercontinent (Fig. 9.5). The correspondence is remarkably close, and confirms that Africa was at the center of Pangea. As noted in Section 2.3.2, Bond (1976, 1978) has demonstrated that Africa has undergone anomalous uplift since the Early Tertiary. This is too late to have been caused directly by the heating effect, which would have taken place in the Late Paleozoic or Early Mesozoic following continental assembly. However, it is possible that the uplift relates to intraplate compressive stress generated by the opening of oceans virtually all around the continent. Dispersing continental fragments tend to migrate toward geoid lows, where mantle temperatures are lower, and where relative sea levels rise, leading to exensive platform flooding (Gurnis 1988, 1990, 1992).

The total length of rifting continental-margins and of seafloor spreading centers increases during the breakup of a supercontinent and is accompanied by increased rates of oceanic crust generation, and active subduction, plutonism, and arc volcanism on the outer, convergent plate margins of the dispersing fragments. Spreading rates are episodic, reflecting the structure and behavior of the mantle convection cells that drive them (Gurnis 1988). Major eustatic transgressions occur because of the displacement of ocean waters by thermally elevated young oceanic crust and active spreading centers in the new Atlantic-type oceans. These factors were considered by Pitman (1978), Kominz (1984) and Harrison (1990) in the development of a model for cycles of eustatic sea-level changes over time periods of tens to hundreds of millions of years (see next section).

Heller and Angevine (1985) argued that during the first 50–100 m.y. after the intitiation of the

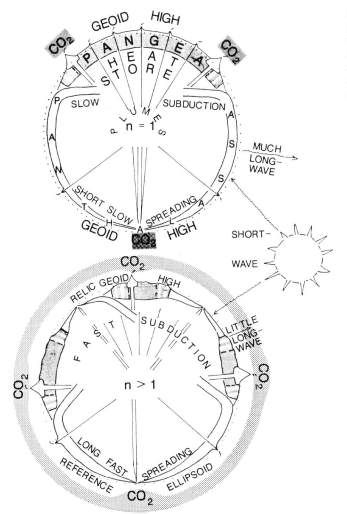

Fig. 9.3. Model of an earth with a supercontinent and a single large ocean (Panthalassa; *above*) versus an earth with dispersed continents and several smaller oceans (*below*). (Veevers 1990, reproduced by kind permission of Elsevier Science-NL, Sara Burgerhartstraat 25, 1055 KV Amsterdam, The Netherlands)

breakup of a supercontinent the global average age of oceanic crust decreases because of the active development of Atlantic-type oceans. This leads to a rise in sea level without any change in global average spreading rate. Dockal and Worsley (1991) developed a simple isostatic model to quantify the effects of changing age of the earth's oceanic crust, simplifying the earth to a two-ocean system, with an opening Atlantic-type ocean replacing a Pacific-type ocean undergoing consumption. They demonstrated that this effect alone can account for a change in sea level of approx. 100 m over about 120 m.y. (Table 9.1).

It has been suggested that the average rates of spreading slow at times of continental assembly, indicating ridge reordering following major continental collision and suturing events. Collision results in crustal shortening, which has the effect of increasing the ocean-basin volume. Therefore at the end of a supercontinent assembly cycle, large areas of old, and therefore cool, and subsided oceanic crust underlie the world's oceans (Worsley et al. 1984, 1986). All these effects lead to enlargement of the world's ocean basins. Times of low sea level might therefore be expected to correlate with, or follow, major suturing episodes (Valentine and Moores 1970, 1972; Larson and Pitman 1972; Vail et al. 1977; Schwan 1980; Heller and Angevine 1985).

The effects of sea-level change on climates, sedimentation and biogenesis are briefly described in Section 6.1.

Although eustatic sea level is predicted to fall during continental assembly, Kominz and Bond (1991) documented a synchronous rise in relative sea level in intracratonic basins and continental margins throughout North America during the Middle Paleozoic (Late Devonian–Middle Mississippian; Fig. 9.6), at the time it is postulated that

Table 9.1. Rates and magnitudes of processes affecting sea-level

Process	Region affected	Type of result	Rate (m/ka)	Duration (m.y.)	Total possible change (m)
Eustatic process					
Age distribution of earth's oceanic crust	Global	Eustasy	0.001	100+	100
Sea-floor ridge volume changes	Global	Eustasy	0.002–0.01	50–100	300
Density changes associated with intraplate stress	Global	Eustasy	1	0.05	50
Continental ice formation	Global	Eustatic fall	1.5	0.1	150
Continental ice melting	Global	Eustatic rise	4–10	0.02–0.04	80–400
Marine ice-sheet decoupling	Global	Eustatic rise	30–50	0.002	6–10
Processes leading to uplift of the continental crust					
Heating beneath super-continent	Hemisphere	Uplift of crust	0.005–0.01	100	500–1000
Thermal doming accompanying rifting	Rift flanks	Thermal bulge	0.012	16	250
Convergent tectonism	Collision zone	Uplift of fault blocks, nappes	0.5^a 10^b	2 0.2	1000 2000
Intraplate stress	Entire plates	Modification of flexural deflections	0.01–0.1	1–10	100
Unsteadiness in mantle convection	Areas of 10^4–10^6 km^2	Regional warping	1–10	0.01–0.1	100
Processes leading to subsidence of the continental crust					
Post-rift thermal subsidence of cont. margin	Continental margin	Hinged subsidence	0.03–0.07^c 0.005–0.03^d	20 200	600–1400 2000–4000
Flexural loading	Foreland basin	Basin subsidence	0.08–1.0	2–15	1000–4000
Intraplate stress	Entire plates	Modification of flexural deflections	0.01–0.1	10	100
Unsteadiness in mantle convection	areas of 10^4–10^6 km^2	Regional warping	1–10	0.01–0.1	100

This table was compiled from numerous sources, as explained in the relevant sections of Chaps. 9–11.
[a] Long-term rate.
[b] Short-term rate.
[c] Initial thermal subsidence.
[d] Decay of thermal anomaly.

the Late Proterozoic supercontinent was dispersing, and Pangea was assembling (Worsley et al. 1984, 1986). Kominz and Bond (1991) attributed the regional rise in sea level to synchronous enhanced subsidence, and argued that this could not have been caused directly by plate-margin processes. Many of the basins are beyond the flexural reach of the continental-margin orogenies that were underway at the time, and some of the data were derived from areas undergoing continental extension where no thermal event has been documented that could explain the timing or rate of subsidence. The synchronous nature of the subsidence (Fig. 9.6) calls

for a continental-scale process, and Kominz and Bond (1991) suggested that the cause was intraplate compressive stress resulting from the movement of the North American plate over a region of mantle downwelling during supercontinent dispersion. Kominz and Bond (1991, p. 59) stated:

[T]he converging limbs of the convection system would increase the in-plane compressive stresses at the base of the lithosphere. For a critical level of stress, all preexisting positive and negative lithospheric deflections (i.e., arches and basins) are enhanced; arches tend to move upward and basins tend to subside. The convection modeling predicts that the maximum compressive stress in a

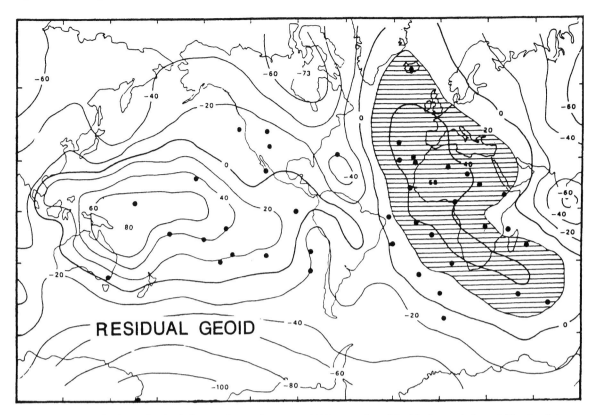

Fig. 9.4. The residual geoid. *Shaded area* shows the positive anomaly located over Africa. A negative anomaly underlies the western Pacific Ocean. (Crough and Jurdy 1980, reproduced by kind permission of Elsevier Science-NL, Sara Burgerhartstraat 25, 1055 KV Amsterdam, The Netherlands)

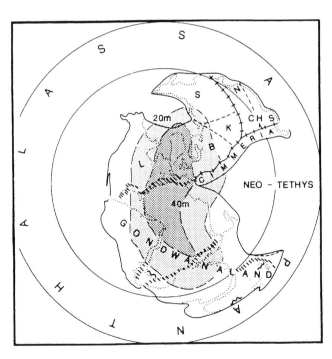

Fig. 9.5. The positive geoid anomaly (*shaded*) shown in relation to a reconstruction of the Pangea supercontinent at about 200 Ma. (Veevers 1990)

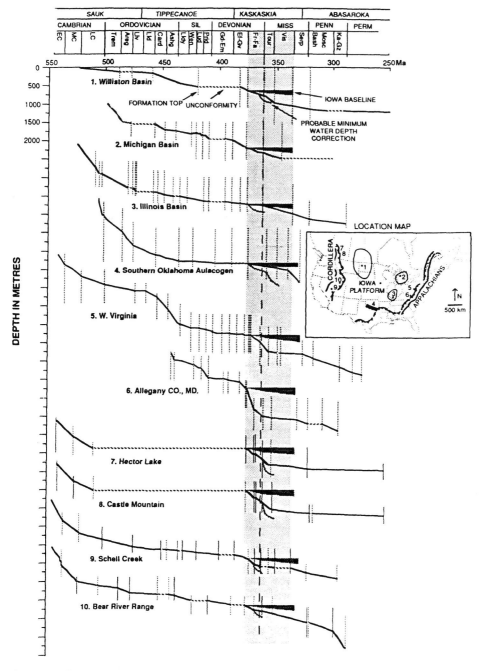

Fig. 9.6. Subsidence curves from various locations in North America, showing period of accelerated subsidence in the Late Devonian to Middle Mississippian. (Kominz and Bond 1991, reproduced by kind permission of Elsevier Science-NL, Sara Burgerhartstraat 25, 1055 KV Amsterdam, The Netherlands)

downwelling region is 60–70 MPa (Gurnis 1988), a range that probably is sufficient to reactivate preexisting deflections, assuming a viscoelastic lithospheric rheology.

The hypothesis of intraplate stress has been developed by Cloetingh (1986, 1988). It may be responsible for regional changes in relative sea level over time scales of tens of thousands to tens of millions of years, and is discussed at greater length in Section 11.4.

In the case of the North American sea-level rise in the Middle Paleozoic, subsidence would have been enhanced by the increase in crustal density overlying a cool downwelling current (the dynamic topography effect). It is not yet clear how much this effect contributed to the overall relative sea-level rise.

As pointed out by Gurnis (1992) there is an ambiguity in attributing causes to long-term sea-level changes. Mantle convection leads to the generation of dynamic topographies, which are reflected in the stratigraphic record by their continental-scale effects on *relative* sea levels. However, the same processes lead to changes in the global average rate of sea-floor spreading, which affect the volume of the ocean basins, and thereby generate *eustatic* sea-level changes. During times of supercontinent splitting, in areas of mantle downwelling, these two processes are in phase, and therefore additive, which makes it difficult to separate and quantify their effects.

9.3 Cycles with Episodicities of Tens of Millions of Years

9.3.1 Eustasy

Hallam (1963) suggested that eustatic sea-level oscillations could be caused by variations in oceanic ridge volumes. Later workers (e.g., Russell 1968; Valentine and Moores 1970, 1972; Rona 1973; Hays and Pitman 1973; Pitman 1978) applied the increasing knowledge of plate-tectonic processes to suggest that variations in seafloor spreading rates, variations in total ridge length, or both are the cause of the volume changes. The average age of the oceanic crust also changes, especially during the assembly and dispersal of supercontinents, as noted in the previous section, and this also affects the volume of the ocean basins.

The oceanic lithosphere formed at a spreading center is initially hot and cools as it moves away from the axis. Cooling is accompanied by thermal contraction and subsidence (Sclater et al. 1971). The age-versus-depth relationship is constant, regardless of spreading history and follows a time-dependent exponential cooling curve (McKenzie and Sclater 1971), as does the overlying continental crust.

Lowstands of sea level would occur during episodes of slow spreading, during which relatively small volumes of hot oceanic lithosphere are being generated. Conversely, episodes of fast spreading would raise sea levels by increasing the ridge volumes. Using the data of Sclater et al. (1971), Pitman (1978) modeled volume changes in a hypothetical

ridge, as shown in Fig. 9.7. The elevation of any part of a ridge can be calculated by converting age to depth, using an appropriate spreading rate. Pitman (1978) showed, for example (Fig. 9.7), that a ridge spreading at 60 m/ka has three times the volume of one spreading at 20 m/ka provided these rates last for 70 m.y. This is the time taken for the oldest (outermost) part of the ridge to subside to average oceanic abyssal depths of 5.5 km, by which time the ridge has achieved an equilibrium profile. The total length of the world midoceanic ridge system is about 45,000 km (Hays and Pitman 1973; Pitman 1978), and Pitman (1978) argued that, allowing for the shape of the continental margins, measured spreading rates can account for eustatic sea-level changes up to a maximum rate of 0.01 m/ka. Later compilations (e.g., Pitman and Golovchenko 1991) demonstrated average rates of 0.002–0.003 m/ka for the Jurassic to Middle Tertiary. This is fast enough to generate Sloss-type cycles, those with episodicities of tens of millions of years.

A rise or fall in sea level is not necessarily the same thing as a transgression or regression. As the rifted margins of a continent move away from a spreading center, they subside as a result of thermal contraction, crustal attenuation, and possibly, phase changes (Sleep 1971; Watts and Ryan 1976). Sediments deposited on the subsiding margin cause further isostatic subsidence. Immediately after rifting and the appearance of oceanic crust, the margins may subside at rates in the order of 0.2 m/ka, decreasing after a few million years to about 0.03–0.07 m/ka and to less than 0.03 m/ka after about 20 m.y. (Watts and Ryan 1976; McKenzie 1978; Pitman 1978). Average rates of tectonic subsidence on extensional margins calculated by backstripping procedures are little more than 0.01 m/ka (Pitman and Golovchenko 1983). Shelf-edge subsidence appears to be always slightly faster than the rate of long-term eustatic sea-level change caused by changes in ridge volumes, and therefore transgressions can actually occur during periods of falling sea level, if the rate of lowering is sufficiently slow. Conversely, regressions can occur locally during periods of rising sea level if there is an adequate sediment supply. Growth of large deltas (such as that of the Mississippi) following the Holocene postglacial sea-level rise is adequate demonstration of this. However, such local regressions are not relevant to the consideration of global stratigraphic cycles. Except on continental margins underlain by unusually old, rigid crust, which subside slowly, it is unlikely that sea-level changes caused by ridge-volume changes could lower sea level to beyond the shelf-break (Pitman and Golovchenko 1983). These tectonic processes that control relative sea level on a regional scale are discussed further in Chapter 11.

Sea-level changes caused by glaciation are, of course, much more rapid, and are not relevant to this discussion of long-term cyclicity, except for the major sea-level fall and cycle boundary in the Middle Oligocene (Fig. 1.6). This was probably caused by the rapid buildup of glacial ice in Antarctica (Pitman and Golovchenko 1983), although the exact timing of this event remains unclear (Matthews 1984; Eyles 1993). More recent work (see Sect. 10.3) suggests two intervals of sea-level fall resulting from Antarctic glaciation, one at about the beginning of the Oligocene (approx. 36 Ma) and one in the Middle Miocene (approx. 15 Ma; Fig. 10.14).

The spreading histories of the world oceans are now reasonably well understood (with the exception of the Arctic Ocean), based on deep-sea-drilling and magnetic-reversal data. Knowledge of worldwide spreading rates enabled Hays and Pitman (1973) to

calculate ridge-volume changes and a sea-level curve for the last 110 Ma. A revised version of this curve was calculated for the period 85–15 Ma by Pitman (1978), based on refinements in oceanic data. It shows a gradual drop in sea level of 350 m at an average rate of 0.005 m/ka (Fig. 9.8). Sea levels rose to a maximum during a period of fast spreading between 110 and 85 Ma (Larson and Pitman 1972), and the subsequent drop reflects slower spreading rates. By calculating the relationship between spreading rates, subsidence rates, and falling sea level Pitman (1978) was able to model a major global transgression during the Eocene (dashed line in Fig. 9.8), as actually documented from stratigraphic evidence by Hallam (1963), and a second transgression during the Miocene. The Middle Oligocene regression is probably related to Antarctic glaciation, as noted above. Vail et al. (1977) used Pitman's curve to calibrate their chart of relative changes in sea level, in the belief that in this way they were adjusting the curve to show true eustatic sea-level change (Fig. 9.9). They suggested that positive departures from Pitmans' curve (where Vail's curve shows a higher sea level than Pitman)

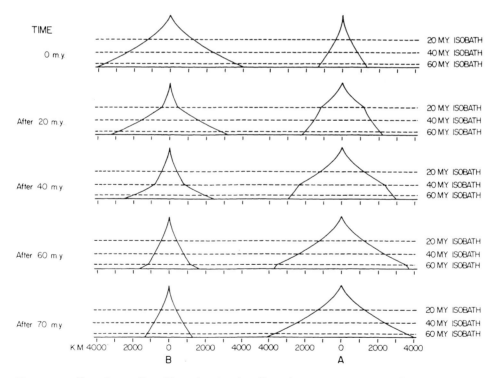

Fig. 9.7. Profiles of spreading ridges showing the effect of different spreading rates on volume at 20, 40, 60, and 70 m.y. after initial condition. **A** Profile of a ridge that has been spreading at 20 m/ka for 70 m.y. and changes to 60 m/ka at time zero. After 70 m.y., the ridge has three times its starting volume. **B** Ridge that has been spreading at 60 m/ka for 70 m.y. and changes to 20 m/ka at time zero. After 70 m.y. its volume has been reduced to one third. (Pitman 1978, 1979)

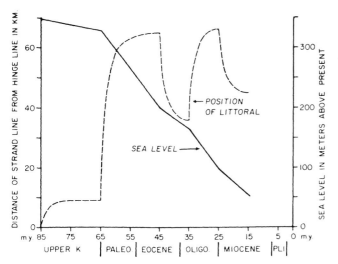

Fig. 9.8. The change in sea level since the Late Cretaceous calculated from a synthesis of sea-floor spreading rates and consequent spreading-center volume changes. Also shown is the movement of the shoreline on an Atlantic-type margin, reflecting the balance between subsidence and sea-level change. This type of curve is discussed in Chapter 11. (Pitman 1978, 1979)

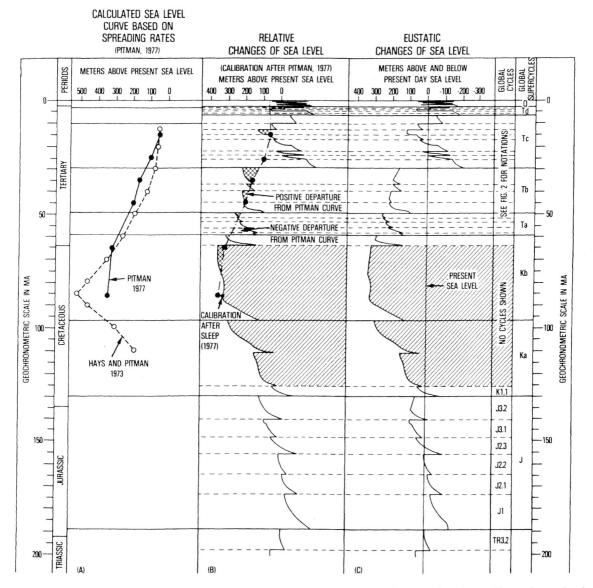

Fig. 9.9. The use of the sea-level curve derived by Hays and Pitman (1973) and Pitman (1978) to calibrate the sea-level curve derived from seismic-stratigraphic onlap relationships. (Vail et al. 1977, reproduced by permission)

are the result of temporary increases in subsidence rates because of sediment loading. Negative departures were attributed to rapid sea-level falls driven by glacioeustasy, a process not factored into Pitman's curve (Vail et al. 1977, p. 92). It is of interest that Vail et al. (1977) mentioned tectonic subsidence as a factor in generating onlap and relative rises in sea level, because the general, indeed overriding importance of this process (discussed in Chap. 11) has been almost completely ignored in subsequent work by the Exxon group, until very recently.

Kominz (1984) reexamined the data on which Pitman's (1978, 1979) curve was based, carrying out her own calculations and incorporating new data, and showed that, because of an incomplete data base, a considerable error must still be allowed for in the development and use of a long-term sea-level curve. She used arbitary, estimated spreading rates for the Tethyan Ocean because, of course, this ocean has now been completely subducted, whereas during the Mesozoic and Early Tertiary it was one of the world's major oceans and its sea-floor spreading history would have had a considerable effect on the eustatic curve. Other errors include inaccurate dating of the sea floor, and missing data; for example, the spreading history of the Arctic Ocean was unknown at the time of her synthesis (and is still incompletely understood). Kominz (1984)

concluded that the range of error is 120 m at 80 Ma, decreasing to about 10 m at present. This has important implications for the accuracy of backstripping calculations used to reconstruct basin subsidence histories. Her curve and its error band are shown in Fig. 9.10, labeled LGP for the Larson et al. (1982) time scale used to calibrate the curve. Other estimate are also shown in this diagram. The discrepancy with Pitman's (1978) curve, which was compiled using the same type of data, results mainly from different assumptions regarding the evolution of ridges in the Pacific Ocean. Harrison (1990) examined the data on ridge-volume changes yet again, and concluded that Kominz's (1984) synthesis was currently the best available.

Dockal and Worsley (1991) examined the effects on the age distribution of the earth's oceanic crust during the formation and breakup of supercontinents. As noted in the previous section, a two-ocean model, in which a Pacific-type ocean (Panthalassa) is replaced by an Atlantic-type ocean, can account for tens of meters of sea-level change over a time scale of hundreds of millions of years. Preliminary modeling of more complex oceanic systems, such as a two-phase opening of the Atlantic Ocean, indicated that second-order effects would also occur with an amplitude up to about 10 m. The opening of other, smaller oceans (e.g., Labrador Sea, Red Sea) would have had similar, if smaller effects.

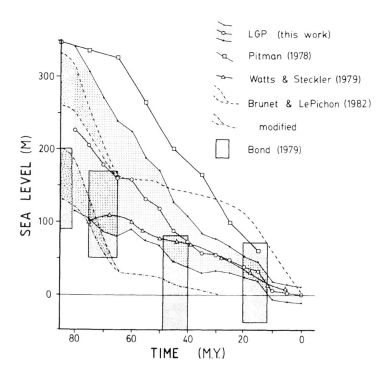

Fig. 9.10. Sea-level curves for the Late Cretaceous to present. The *LGP* curve is that of Kominz (1984), compiled using the time scale of Larson et al. (1982). The curves of Watts and Steckler (1979) and Brunet and LePichon (1982) are derived from subsidence data in the Atlantic margin and Paris Basin, respectively. Bond's (1979) estimates were derived from a combination of stratigraphic data and hypsometry. (Kominz 1984)

A recent compilation of spreading centers associated with the breakup of Pangea is shown in Figs. 9.11 and 9.12. The initiation of each ridge would have had an effect on the global average spreading rates, and on the age distribution of the oceanic crust, with consequences for eustatic fluctuations over time periods of tens of millions of years. However, as discussed in Chapter 11, the same events are now thought to have had a profound effect on intraplate stress regimes, with consequences for regional warping and tilting, and the generation of relative sea-level changes over large continental areas. Estimates of the rate and magnitude of eustatic sea-level changes that can be attributed to volume changes in sea-floor spreading centers were made by Pitman and Golovchenko (1991), based on their earlier work and that of Kominz (1984), and are given in Table 9.1.

Other processes that could possibly affect eustatic sea levels on a time scale of tens to hundreds of millions of years were reviewed by Harrison (1990), who calculated the effects on sea level by relating them to changes in the total volume of the world ridge system. Continental collision increases the thickness of the crust and decreases its area, thus increasing the area of the ocean basins. Major collision and shortening events, such as that between India and Asia, therefore result in a lowering of sea level. The generation and subsequent cooling and subsidence of large oceanic volcanic extrusive masses can be shown to have a modest effect on sea level. Sediment deposited in oceans has an isostatic loading effect which amplifies the subsidence due to crustal aging, but also displaces water. Changes in global average ocean temperature change the volume of the water through thermal expansion and

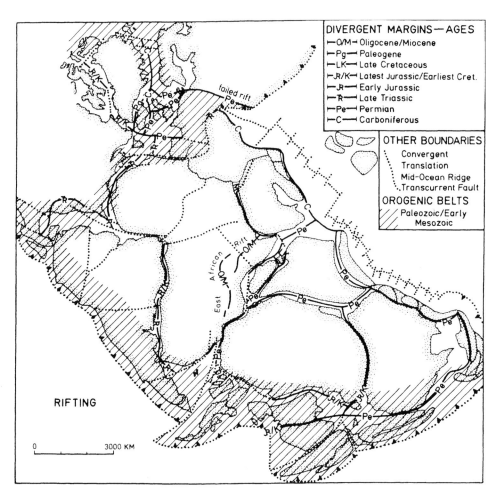

Fig. 9.11. Paleogeography of Pangea during the earliest Triassic, showing the distribution and ages of initiation of rifting on what became divergent continental margins during the Mesozoic and Cenozoic. (Uchupi and Emery 1991, reproduced by kind permission of Elsevier Science-NL, Sara Burgerhartstraat 25, 1055 KV Amsterdam, The Netherlands)

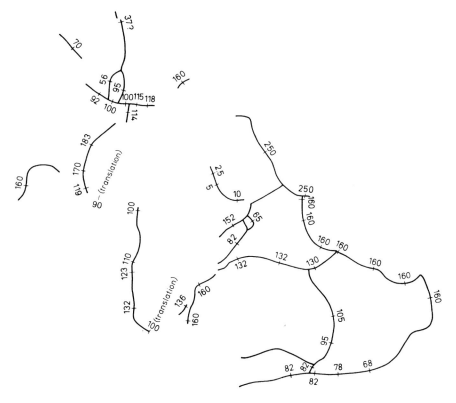

Fig. 9.12. Divergent continental margins that developed within and around Pangea, showing the age of initiation of seafloor spreading. (Uchupi and Emery 1991, repro duced by kind permission of Elsevier Science-NL, Sara Burgerhartstraat 25, 1055 KV Amsterdam, The Netherlands)

contraction, without changing its load. Significant temperature changes have occurred as the earth has cycled between icehouse and greenhouse states. Fairbridge (1961) estimated that a 1° rise in ocean temperatures would lead to a 2-m rise in sea level. All the processes summarized by Harrison (1990) are illustrated in Fig. 9.13. A linear relationship between volume and freeboard can be calculated, based on the use of a standard continental hypsometry.

Sheridan (1987) noted a correlation between calculated seafloor spreading rates, Sloss-type cycles of sea-level change, and the reversal frequency in the earth's magnetic field in the Atlantic Ocean (Fig. 9.14). So-called *quiet zones*, such as that in the Middle Cretaceous, correlate with times of fast

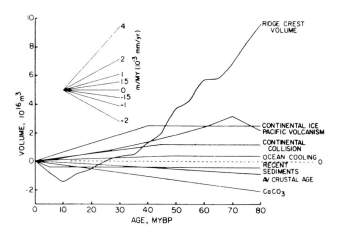

Fig. 9.13. Volume estimates for six phenomena since the Late Cretaceous. A decrease in volume with time produces a sea-level fall (increase in freeboard). Rates of freeboard change are indicated by the slope of the line, as shown in the *top left corner*. (Harrison 1990, reproduced with permission from "Sea-Level Change, " copyright by the National Academy of Sciences; courtesy of the National Academy Press, Washington, D.C.)

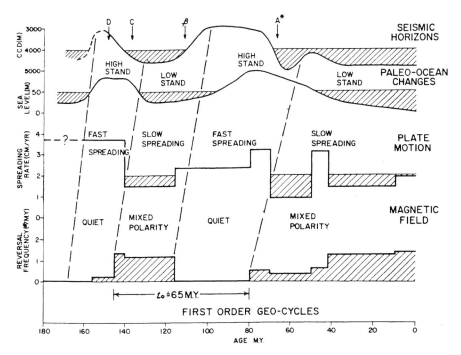

Fig. 9.14. Correlation of magnetic-anomaly reversal frequency, plate spreading rate for the North Atlantic Ocean, sea level, and the position of the calcite-compensation depth (*CCD*). *Arrowed letters at top* refer to ages of major oceanic seismic horizons in the North Atlantic Ocean.

Note the phase lag between changes in the magnetic field and surface processes. (Sheridan 1987, reproduced by kind permission of Elsevier Science-NL, Sara Burgerhartstraat 25, 1055 KV Amsterdam, The Netherlands)

seafloor spreading and high sea level, whereas *mixed polarity zones* occur during times of slow spreading and low sea level. There is in fact a time lag in the correlations, as shown in Fig. 9.14. Information on sea-floor spreading rates is not available for the Paleozoic, but Sheridan (1987) suggested that a correlation between sea level and

changes in the magnetic field could also be made for this time period (Fig. 9.15). He suggested that the breakup of Pangea was driven by a cyclic or pulsing process involving the episodic eruption of plumes from the core-mantle boundary (Fig. 9.16). Plume eruptions are accompanied by relatively smooth, laminar convective flow in the core, and a stable

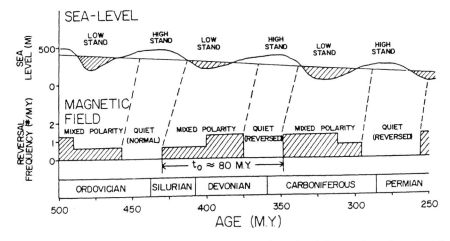

Fig. 9.15. Correlation of magnetic-field changes and sea level during the Paleozoic. (Sheridan 1987, repro-duced by kind permission of Elsevier Science-NL, Sara Burgerhartstraat 25, 1055 KV Amsterdam, The Netherlands)

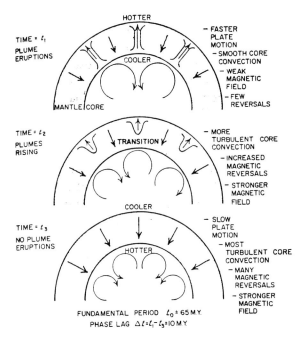

Fig. 9.16. The model of pulsation tectonics, showing the cyclic eruption of plumes from the core-mantle boundary. (Sheridan 1987, reproduced by kind permission of Elsevier Science-NL, Sara Burgerhartstraat 25, 1055 KV Amsterdam, The Netherlands)

magnetic field. The presence of hot convective currents beneath the crust leads to uplift, fast continental breakup and seafloor spreading, and elevated sea levels. Following a phase of plume eruptions there is a time of transition, when convection becomes more turbulent, the magnetic field becomes weaker and reverses more frequently. The phase lag of about 10 m.y. between changes in the magnetic field and seafloor-spreading rates indicates the time required for the plumes to ascend through the mantle and affect the crust.

A similar model of plume eruption was described by Larson (1991), who showed how fast seafloor spreading rates relate to peaks in ocean temperatures and oil generation during the Cretaceous. Anderson (1994) disagrees that Phanerozoic geology can be explained by a cycle of plume activity, but this discussion is beyond the scope of this book.

9.3.2 Dynamic Topography and Epeirogeny

Stratigraphers specializing in the study of continental interiors (e.g., Sloss and Speed 1974) have for a long time appealed to a process that was termed *epeirogeny* by Gilbert (1890). The modern definition of epeirogeny (Bates and Jackson 1987)

defines it as "a form of diastrophism that has produced the larger features of the continents and oceans, for example, plateaux and basins, in contrast to the more localized proces of orogeny, which has produced mountain chains." The definition goes on to emphasize vertical motions of the earth's crust. Many workers have doubted the reality of epeirogeny (e.g., Miall 1987), suggesting that all vertical motions of the crust could be related to contemporaneous plate-margin processes, or to the reactivation of buried crustal elements formed by plate-margin effects during some earlier stage of plate-tectonic history. However, it is now clear that this is incorrect. Modern studies of the thermal evolution of the mantle, supported by numerical modeling experiments, have provided a mechanism that explains the long term uplift, subsidence and tilting of continental areas, especially large cratonic interiors beyond the reach of the flexural effects of plate-margin extension or loading (Gurnis 1988, 1990, 1992; Burgess and Gurnis 1995). These studies have shown that the earth's surface is maintained in the condition known as dynamic topography, reflecting the expansion and contraction of the crust resulting from thermal changes in the underlying lithosphere and mantle (Fig. 9.2). Much work remains to be done to test and apply these ideas by developing detailed numerical models of specific basinal stratigraphic histories, although it is already clear that dynamic topography is affected by both upwelling and downwelling currents on several scales. The following paragraphs describe a range of recent studies.

Bond (1978) provided some of the first important insights into epeirogenic processes by demonstrating that the earth's continents have had different histories of uplift and subsidence since Cretaceous time (Fig. 2.17). It is now possible to explain these differences using the concepts of dynamic topography. One of the most striking anomalies revealed by the hypsometric work is the elevation of Australia. "The interior of Australia became flooded by nearly 50% between 125 and 115 Ma and then became progressively exposed between 100 and 70 Ma at a time when nearly all other continents reached their maximum Cretaceous flooding" (Gurnis 1992). Applying backstripping techniques to detailed isopach maps Russell and Gurnis (1994) estimated that although global sea level was about 180 m above the present level near the end of the Cretaceous, a smaller fraction of the Australian continent was flooded than at the present day. Raising the continent an average of 235 m accounts for the end-

Cretaceous paleogeography superimposed on a 180-m-high sea level. This result is related to cessation of subduction on the northeastern margin of Australia at about 95 Ma. Subduction had generated a dynamic load by the presence of a cold crustal slab at depth, and the cessation of subduction allowed uplift. Then the northward migration of the continent as it split from Antarctica moved Australia off a dynamic topographic high and geoid low, toward a lower dynamic topography, resulting on continental lowering. Russell and Gurnis (1994) also discussed the broad tilting and basin development of the Australian interior. Burgess and Gurnis (1995) developed preliminary models for Sloss-type cratonic sequences invoking combinations of eustatic sea-level change and dynamic topography.

The concepts of dynamic topography have also been invoked to explain cratonic basin formation and subsidence. As noted by Hartley and Allen (1994), there have been at least two major periods in earth history when suites of interior basins formed within large cratons. Both periods are associated with the breakup of supercontinents. The first of these was the Early Paleozoic, when the Williston, Hudson Bay, Illinois, Michigan and other basins formed in the cratonic interior of North America. The second period was the Meszoic breakup of Pangea, when a series of similar basins formed within the continent of Africa. Some of the African basins are undoubtedly related to plate-margin processes, and there has been much debate regarding the importance of reactivation of inherited crustal weaknesses as a cause of the North American basins (Quinlan 1987). However, Hartley and Allen (1994) suggested that small-scale convective downwelling, decoupled from the large-scale motion, may be a significant factor in basin formation. They found strong evidence for this process in the formation of the Congo Basin. The stratigraphic histories of these suites of cratonic basins are similar but not identical (Quinlan 1987), and as with the other examples discussed in this section, the processes that maintain dynamic topography are not thought to generate globally simultaneous (eustatic) changes in sea level.

The thermal consequences of secondary mantle convection above a subducting slab have been invoked as a cause of enhanced subsidence in retroarc foreland basins. Mitrovica et al. (1989) developed this idea as an explanation for the anomalously broad extent of the Western Canada Sedimentary Basin, and a similar idea has been proposed for the Mesozoic basins of eastern Australia (Gallagher et al. 1994). Although this is clearly a plate-margin effect the same concepts of dynamic topography apply as in the case of the broader epeirogenic processes discussed above.

9.4 Main Conclusions

1. A long-term cycle of sea-level change, termed the supercontinent cycle, results from the assembly and breakup of supercontinents on the earth's surface. Eustatic sea-level changes are driven by global changes in sea-floor spreading rate, variations in the average age of the oceanic crust, and variations in continental volumes caused by plate extension and collision.

2. Eustatic sea-level changes on a time scale of tens of millions of years are caused by variations in ocean-basin volume generated by episodic spreading, and by the variations in total length and age of the sea-floor spreading centers as supercontinents assemble, disassemble and disperse.

3. Many other processes have smaller effects on global sea levels through their effects on the volume of the ocean basins. These include oceanic volcanism, sedimentation, ocean temperature changes, and the desiccation of small ocean basins (this last effect is discussed in Chap. 11).

4. Epeirogenic effects are "dynamic topography" resulting from thermal effects of mantle convection associated with the supercontinent cycle. This cycle has long-term consequences, including the generation of persistent geoid anomalies. Vertical continental movements are related to the thermal properties of large- and small-scale convection cells, and can involve continent-wide uplift, subsidence, and tilts, and cratonic basin formation. These movements do not correlate in sign or magnitude from continent to continent.

5. Mantle processes during continental breakup may be "pulsed" on a time scale of tens of millions of years, as suggested by correlations with changes in the character of the earth's magnetic field.

10 Milankovitch Processes

10.1 Introduction

By the end of the nineteenth century it was recognized that during the Pleistocene the earth had undergone at least four major ice ages, and a search was underway for a mechanism that would alter climate so dramatically (useful historical summaries are given by Berger 1988; Imbrie 1985; Dott 1992; de Boer and Smith 1994b). Continued stratigraphic work during the twentieth century produced evidence for many more cycles of ice formation, advance and melting, and it is now known that there were more than 20 such cycles, indicating successive major fluctuations in global climate.

It was an amateur geologist and newspaper publisher, Charles MacLaren, who in 1842 first realized the implications of continental glaciation for major changes in sea level (Dott 1992). The rhythmicity in glaciation is now attributed to *astronomical forcing*. The Scotsman James Croll (1864) and the American G.K. Gilbert (1895) were the first to realize that variations in the earth's orbital behavior may affect the distribution of solar radiation received at the surface, by latitude and by season, and could be the cause of major climatic variations, but the ideas were not taken seriously for many years after this because of a lack of a quantifiable theory and supporting data. Gilbert invoked the idea to explain oscillations in carbonate content in some Cretaceous hemipelagic beds in Colorado (Fischer 1986). Later, Bradley (1929) "recognized precessional cycles in oil shale-dolomite sequences of the Green River Formation in Colorado, Wyoming and Utah, using varves as an unusually precise measure of sedimentation rates" (de Boer and Smith 1994b). Theoretical work on the distribution of insolation was carried out by the Serbian mathematician Milankovitch (1930, 1941), who showed how orbital oscillations could affect the distribution of solar radiation over the earth's surface. However, it was not for some years that the necessary data from the sedimentary record was obtained to support his model. Emiliani (1955) was the first to discover periodicities in the Pleistocene marine isotopic record, and the work by Hays et al. (1976) is regarded by many (e.g., de Boer and Smith 1994b) as the definitive study that marked the beginning of a more widespread acceptance of so-called *Milankovitch processes* as the cause of stratigraphic cyclicity on a 10^4–10^5-year frequency – what is now termed the Milankovitch band (Sect. 3.1). The model is now firmly established, particularly since accurate chronostratigraphic dating of marine sediments has led to the documentation of the record of faunal variations and temperature changes in numerous Upper Cenozoic sections (Sects. 8.2, 10.3). These show remarkably close agreement with the predictions made from astronomical observations.

There is an obvious link between climate and sea level. There is no doubting the efficacy of glacial advance and retreat as a mechanism for changing sea level; we have the evidence of the Pleistocene glaciation at hand throughout the Northern Hemisphere. It has been calculated that during the period of maximum Pleistocene ice advance the sea was lowered by about 100 m (Donovan and Jones 1979). Melting of the remaining ice caps would raise the present sea level by about 40–50 m (Donovan and Jones 1979). Recent history thus demonstrates a mechanism of changing sea level by at least 150 m at a rate of about 0.01 m/year. This is fast enough to account for eustatic cycles with frequencies in the range of tens of thousands of years. Milankovitch processes are now widely accepted as the source of much cyclicity with cyclicities of 10^4–10^5 years in the Late Paleozoic and Late Cenozoic sedimentary record – times when continental glaciation is known to have been widespread (Sects. 10.3, 10.4). Other, more subtle (nonglacial) climatic variations may explain other forms of cyclicity of similar periodicity at times when there is no evidence for widespread continental ice sheets (Sect. 10.2.8). However, there is also increasing evidence for cyclicity of a similar time periodicity generated by tectonic mechanisms, especially in foreland basins

(Sect. 11.3), and it is not safe to assume that sequences of 10^4–10^5-year duration are necessarily generated by orbital forcing. This problem is examined further in Section 10.2.5.

There has been a considerable increase in interest in Milankovitch processes and in cyclostratigraphy in recent years, and a number of books have appeared that treat the subject from various points of view. The astronomical processes that underpin earth's climatic cyclicity are dealt with by Schwarzacher (1993) and by A.L. Berger, in a number of books, including Berger et al. (1984) and Berger (1988). Collections of case studies, with papers explaining orbital mechanics or climatic theory, have been edited by Fischer and Bottjer (1991), Berger et al. (1984), de Boer and Smith (1994a), and House and Gale (1995). Much additional information is contained in the books edited by Franseen et al. (1991) and Dennison and Ettensohn (1994).

10.2 The Nature of Milankovitch Processes

10.2.1 Components of Orbital Forcing

There are several separate components of orbital variation (Fig. 10.1). The present orbital behavior of the earth includes the following cyclic changes (Schwarzacher 1993):

1. Variations in orbital eccentricity (the shape of the earth's orbit around the sun). Several "wobbles," which have periods of 2035.4, 412.8, 128.2,

99.5, 94.9, and 54 ka. The major periods are those at around 413 and 100 ka.
2. Changes of up to 3° in the obliquity of the ecliptic, with a major period of 41 ka, and minor periods of 53.6 and 39.7 ka.
3. Precession of the equinoxes. The earth's orbit rotates as a spinning top, with a major period of 23.7 ka. This affects the timing of the perihelion (the position of closest approach of the earth to the sun on an elliptical orbit), which changes with a period of 19 ka.

Each of these components is capable of causing significant climatic fluctuations given an adequate degree of global sensitivity to climate forcing. For example, when obliquity is low (rotation axis nearly normal to the ecliptic), more energy is delivered to the equator and less to the poles, giving rise to a steeper latitudinal temperature gradient and lower seasonality. Variations in precession alter the structure of the seasonal cycle, by moving the perihelion point along the orbit. This changes the earth-sun distance at every season, thus changing the intensity of insolation at each season. "For a given latitude and season typical departures from modern values are on the order of 5%" (Imbrie 1985). Because the forcing effects have different periods they go in and out of phase. One of the major contributions of Milankovitch was to demonstrate these phase relationships on the basis of laborious time-series calculations. These can now, of course, be readily carried out by computer (Fig. 10.2). The success of modern stratigraphic work has been to demonstrate the existence of curves of

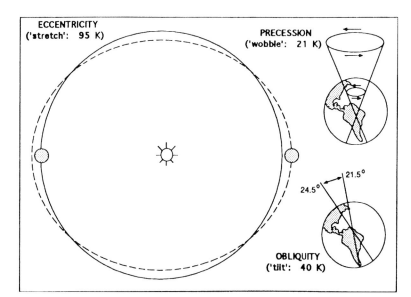

Fig. 10.1. Perturbations in the orbital behavior of the earth, showing the causes of Milankovitch cyclicity. (Plint et al. 1992, after Imbrie and Imbrie 1979)

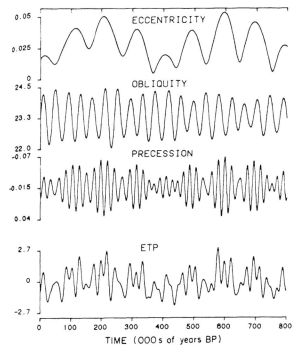

Fig. 10.2. Three major orbital-forcing parameters, showing their combined effect in the eccentricity-tilt-precession (*ETP*) curve at *bottom*. Absolute eccentricity values are shown. Obliquity is measured in degrees. Precession is shown by a precession index. The ETP scale is in standard deviation units. (Imbrie 1985, reproduced by permission of The Geological Society, London)

temperature change and other variables in the Cenozoic record that can be correlated directly with the curves of Fig. 10.2. For this purpose sophisticated time-series spectral analysis is performed on various measured parameters, such as oxygen-isotope content (e.g., Fig. 10.3) or cycle thickness. This approach has led to the development of a special type of quantitative analysis termed *cyclostratigraphy* (House 1985).

10.2.2 Basic Climatology

The major control of global climate is the coupled ocean-atmospheric circulation, which, in turn, controls humidity, rainfall and temperature. Figure 10.4 presents simple models of circulation for three climatic conditions. The earth is encircled by three cells, the Hadley, Ferrel, and Polar cells. Consideration of this broad atmospheric structure and its effects on land and sea leads to a series of useful generalizations regarding regional climate. Where the circulation established by the three cells passes across the earth's surface it sets up persistent wind

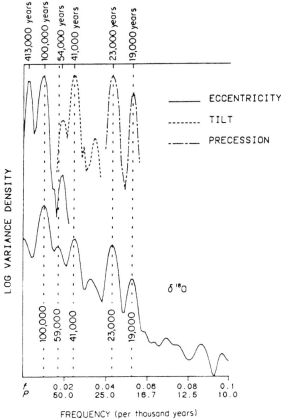

Fig. 10.3. Variance spectra over the past 800 ka of orbital variations (*top*; from data shown in Fig. 10.2) and $\delta^{18}O$ content, measured in foraminiferal tests (*bottom*). Note the close correlation between the two curves. (Imbrie 1985, reproduced by permission of The Geological Society, London)

patterns such as the trade winds. Atmospheric upwelling regions are characterized by greater humidity than downwelling regions, because of adiabatic effects which control air saturation. Land and sea have different heat capacities, and therefore heat and cool at different rates, and this can set up regional atmospheric circulation effects. The cells change position with the seasons, leading to seasonal changes in circulation, such as the monsoonal changes in prevailing wind directions. Onshore winds carry moisture, which is released as rainfall when forced to rise over high relief. Windward slopes therefore tend to be wetter than leeward slopes, and the climate on the eastern and western sides of continents may be quite different.

During climatic minimums the downwelling region between the Hadley and Ferrel cells is located at about 15°–35° latitude, and is an area of warming and high evaporation rates. Deserts are therefore characteristic. Monsoon zones show little latitudinal

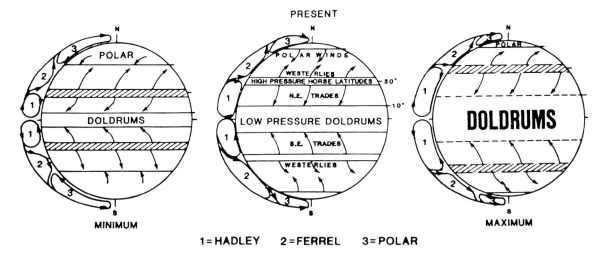

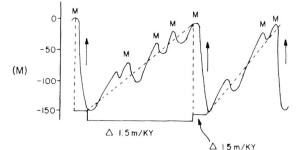

Fig. 10.4. Atmospheric circulation patterns for three climatic conditions. The present climate (*center*) is intermediate between true glacial and interglacial conditions. (Perlmutter and Matthews 1990)

shifting with the seasons, which also limits the amount of moisture transported to this zone. As conditions change to the climatic maximum the Hadley-Ferrel downwelling zone shifts poleward, and the belt of deserts moves to 35°–40°. Hadley and monsoonal circulation increase in efficiency, and the 10°–35° zone becomes more humid. The upwelling arm of the Ferrel-Polar cell moves to about 70° latitude. The importance of these climatic shifts in the generation of nonglacial cyclic processes is discussed in Section 10.2.8.

de Boer and Smith (1994b, p. 5) summarized the orbital effects on climate as follows:

At low latitudes, close to the equator, the influence of the cycle of precession, modulated by the varying eccentricity of the Earth's orbit, is dominant and causes latitudinal shifts of the caloric equator.... In turn, this causes significant shifts of the boundaries between adjacent climate zones. At mid-latitudes (20°–40°) the orbital variations affect the relative length of the seasons and the contrast between summer and winter, and hence of monsoon intensity.... Toward higher latitudes (>40°) the effect of the varying obliquity becomes more prominent.

In detail the response of the climate to astronomical forcing at any point on the earth's surface is extremely complex, reflecting various sensitivity factors, such as the relative positions and sizes of sea and land masses. Their latitude affects air and water circulation patterns and their relative position affects humidity (and hence rainfall), monsoon effects, and so on. These complexities are exemplified by Barron's (1983) discussion of the difficulties inherent in attempting to reconstruct the warm, equable climates of the Cretaceous. In general there

is a lag effect between the astronomical force and the system's response, although on the kiloyear scale under consideration in this chapter the lag effect is probably relatively unimportant. An exception is the build-up and melting of major ice sheets. In general, glacioeustatic transgressions (caused by ice melting) are thought to be much more rapid than regressions (which are caused by continental ice formation; Fig. 10.5). For Pleistocene 100-ka cycles, Hays et al. (1976) found that the sea-level fall occupied 85–90% of the total cycle period.

The rate of sea-level change brought about by the formation and melting of major ice sheets is conventionally estimated at 10 m/ka (e.g., Donovan and Jones 1979). However, it has long been known from

Fig. 10.5. Schematic representation of the asymmetry of glacioeustatic rises and falls. Time moves *from right to left*. Two types of cycle are shown, with 100- and 40-ka (ky) periodicities. Melting events are indicated by *M*. (Williams 1988)

stratigraphic records that the Holocene has been characterized by short intervals of much more rapid sea-level rise. This has been attributed to the breakup of floating ice sheets (Anderson and Thomas 1991). Melting of continental ice sheets results in a eustatic rise that eventually lifts off ice sheets grounded on the continental shelf. Once decoupled, these massive ice sheets are unstable and quickly break up, leading to pulses of very rapid sea-level rise, up to 30–50 m/ka for intervals of up to a few hundred years.

The correlation between Milankovitch periodicities and oxygen isotope content is shown in Fig. 10.3. The linkage between these parameters is as follows. ^{16}O is the lighter of the two main oxygen isotopes, and is therefore preferentially evaporated from seawater. During times of ice-free global climate this light oxygen is recycled to the oceans and no change in isotopic ratios occurs. However, continental ice buildups are preferentially enriched in ^{16}O, while the oceans are depleted in this isotope, with the result that the ^{18}O content of the oceans is increased. Numerous studies since that of Emiliani (1955) have shown that the $^{16}O/^{18}O$ ratio is a sensitive indicator of global ocean temperatures and can therefore be used as an analog recorder of ice volumes (e.g., Matthews 1984, 1988). The measurements are made on the carbonate in foraminiferal tests. Matthews (1984) suggested a calibration value of $\delta^{18}O$ variation of about 0.011‰ per meter of sea-level change. Precise calibrations are impossible because of uncertainties regarding the volume and content of floating pack ice, and diagenesis of foraminiferal tests. It is also important to take into account the lag between the maximum eustatic high and the highest surface ocean temperatures, which occur some 10^3 years later.

10.2.3 Variations with Time in Orbital Periodicities

It is not known how the orbital periodicities determined for the Holocene might have varied in the geological past. Some studies have suggested that small perturbations in the motions of the planets during geological time may have had significant effects on the orbital behavior of the earth (Laskar 1989). Plint et al. (1992) stated that, "over longer periods of time, it is postulated that nonlinear amplifications of even very small perturbations in planetary orbits, as well as changes in the diameter of the core, and of the earth as a whole, leads to chaotic, unpredictable planetary motions." Others

(e.g., Berger et al. 1989, 1992; Berger and Loutre 1994) have disputed this, suggesting that orbital periodicities should have remained fairly stable throughout the Phanerozoic. Some studies argue that as the "eccentricity periods are the product of interplanetary gravitational forces, they have probably remained stable through at least the last 600 m.y." (Algeo and Wilkinson 1988; after Walker and Zahnle 1986), whereas the precession and obliquity periods have changed due to continued evolution of the earth-moon system (Lambeck 1980; Berger and Loutre 1994).

Transfer of angular momentum to the moon has resulted in a decrease in the earth's rotational velocity and an increase in the moon's orbital velocity. The consequent recession of the moon has resulted in attenuation of the periods of the earth's precession and obliquity cycles. Approximation of orbital paleoperiods (Walker and Zahnle 1986) indicate that the period of precesion was about 17,000 years and that of obliquity about 28,000 years at the beginning of the Phanerozoic. (Algeo and Wilkinson 1988)

Calculations by Berger and Loutre (1994) suggested that the precession periods 19 and 23 ka, would have been about 11.3 and 12.7 ka at 2.5 Ga, with little change in amplitude. By contrast, they suggested that the obliquity period would have had considerably greater amplitude at 2.5 Ga, but they predicted no change in period.

Bond et al. (1993) discussed the occurrence of Milankovitch cycles in the distant geological past and faced the problem that we have little hard data regarding Milankovitch periodicities in earlier geological time. In their opinion, astronomical calculations are too beset by assumptions and difficulties, and the best results may come from geological data. The gamma method (described in Sect. 10.2.5) may be the ideal method for examining this problem, and results from Cambrian and Cretaceous successions reported in their paper indicated periodicities consistent with those predicted by Berger et al. (1989, 1992).

There is clearly considerable disagreement and uncertainty in this area, and scepticism should be exercised in attempts to relate cycle periodicities in pre-Pleistocene rocks to specific orbital parameters, as has been done in some studies (e.g., Triassic of the Dolomites: Goldhammer et al. 1990; Pennsylvanian cyclothems of the North American Midcontinent: Heckel 1986. See Chap. 8 for stratigraphic documentation of these examples). House (1985) and House and Gale (1995) have even suggested using calculated periodicities to calibrate time

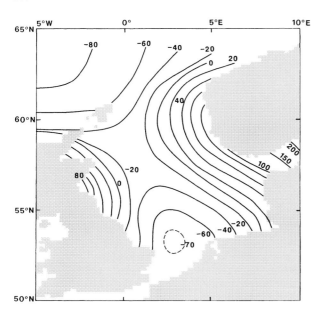

Fig. 10.6. Sea levels (m) relative to present-day sea levels in the North Sea, 20 ka ago, prior to the melting of the British and Scandinavian ice caps. (Lambeck et al. 1987)

scales, and discussed the use of interpreted Milankovitch cycles in the rock record to refine the geological time scale determined by biostratigraphic and radiometric means. At present this seems premature, because of the uncertainties regarding orbital changes in the past. In addition, as discussed in Chapter 13, there remains considerable uncertainty regarding the precision of the geological time scales within which Milankovitch refinements are to be calculated.

10.2.4 Isostasy and Geoid Changes

An additional complication concerning glacioeustasy is that isostatic and geoidal effects must be taken into consideration before changes in ice volume can be translated into changes in sea level. The geoid is defined as a sea-level surface encompassing the globe as if it extended continuously through the continents. Fjeldskaar (1989) stated that, "under the continents the geoid can be thought of as the surface defined by the water level in narrow canals cut to sea level through the land masses." The geoid reflects gravity anomalies, and is therefore not spherical. Major continental ice sheets are several kilometers in thickness, and have a significant isostatic loading effect on the crust. Upon melting, the crust rebounds much more slowly than the rise in sea level brought about by melting. Locations close to the ice cap (within a few hundred kilometers) therefore undergo an initial rapid rise in sea level, followed by a slow fall, over periods of tens of thousands of years, a fall which is not recorded in areas beyond the isostatic reach of the ice cap. The ice mass also has a gravitational attraction that affects the shape of the geoid. Lambeck et al. (1987) modeled these two forces and produced the map shown in Fig. 10.6. The geoidal effect alone is shown in Fig. 10.7. This map shows the depression of the geoid (sea level) that would occur if the 20-ka ice cap underwent instant melting. The conclusion to be drawn from the studies of geoidal and isostatic effects is that the sea level signature in the stratigraphic record close to major ice caps cannot be subjected to simple, straightforward interpretations of eustatic sea-level change. However, at distances of a few thousand kilometers from the edge of the ice cap these effects are small enough to be negligible.

10.2.5 The Nature of the Cyclostratigraphic Data Base

Milankovitch cycles have been identified in the rock record primarily on the basis of observed rhythmicity in the rock record, particularly the occurrence of meter-scale or thinner (varved) cycles, typically in rocks of carbonate shelves, and fine-grained pelagic and lacustrine sediments. Some clastic or mixed carbonate-clastic successions, such as the Upper Paleozoic cyclothems, are also attributed to Milankovitch processes. However, the presence of rhythmicity and an estimation of cycle durations of the appropriate magnitude does not prove orbital forcing as the cycle-generating mechanism. As stated by Algeo and Wilkinson (1988):

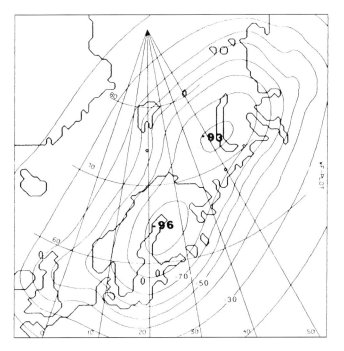

Fig. 10.7. Theoretical deflection of the geoid over northwestern Europe caused by instantaneous deglaciation of the 20-ka ice sheet. Contour interval is 10m. (Fjeldskaar 1989)

Despite an often-claimed correspondence between cycle and Milankovitch orbital periods, factors independent of orbital modulation that affect cycle thickness and sedimentation rate may be responsible for such coincidence. For example, nearly all common processes of sediment transport and dispersal give rise to ordered depositional lithofacies sequences that span a relatively narrow range of thicknesses.... Further, long-term sediment accumulation rates are generally limited by long-term subsidence rates, which converge to a narrow range of values for very different sedimentary and tectonic environments (Sadler 1981). In essence, the spectra of real-world cycle thicknesses and subsidence rates are relatively limited, and this in turn constrains the range of commonly determined cycle periods. For many cyclic sequences, calculation of a Milankovitch-range period may be a virtual certainty, regardless of the actual generic mechanism of cycle formation.

Algeo and Wilkinson (1988) studied cyclicity in more than 200 stratigraphic units, determining periodicities from cycle thickness, sedimentation rate, depositional environment and age range. They found that calculated periodicities are randomly distributed relative to all the Milankovitch periods except the 413-ka eccentricity cycle. They concluded that, "if, in fact, a calculated average period within the broad range of Milankovitch periodicities is not a sufficient test of orbital modulation of sedimentary cycles, demonstration of such control becomes significantly more difficult than hitherto appreciated." Peper and Cloetingh (1995) demonstrated by numerical modeling the significance of the reverse problem, that certain non-Milankovitch pro-

cesses, such as slope failure and intraplate stress changes, can generate perturbations that distort or obscure the stratigraphic response to Milankovitch processes.

Demonstration of a hierarchy of periodicities may be a useful indicator of orbital control. For example, where there are two orders of cycles with recurrence ratios of 5:1 a precession (20 ka) and eccentricity (100 ka) combination may be indicated. Fischer (1986) discussed examples of this.

Van Tassell (1994) provided a good example of the kind of problem described by Algeo and Wilkinson (1988). A wide spectrum of cycles is present in the Devonian Catskill Delta of the Appalachian Basin, but proving they are of Milankovitch type (as claimed by the author) is problematic. Their thickness and duration fits that of Milankovitch and autogenic cycles, as Algeo and Wilkinson (1988) predicted would occur in these types of rocks. Demonstration of cycle duration is difficult, because of uncertainties about their age range. No Milankovitch-type bundling can be demonstrated, and the proof of correlation of cycles across major facies boundaries is very limited. A comparable, partly correlative suite of cycles was described by Filer (1994), based mainly on subsurface gamma-ray log correlations. His correlations seem unconvincing to this writer. No clear cyclic signatures can be traced from well to well. These date would benefit from application of the gamma technique, described below.

Drummond and Wilkinson (1993b) modeled carbonate cycles to study the effect of "lag time" (the time taken for the carbonate factory to reach full productivity rates when a platform is flooded). They showed that several discrete cycles could be produced during one smooth leg of sea-level rise, as the factory overcomes the lag, races to deposit sediment, then is shut down as the platform is built to sea level. Therefore a one-for-one correspondence between sedimentary cycles and sea-level cycles cannot be expected, and cycle durations and hierarchies cannot be directly related to Milankovitch frequencies. However, this is just a numerical model, and they did not offer evidence of this from the rock record.

The requirements for the establishment of orbital forcing are two-fold. Firstly it is necessary to demonstrate that the cycles are widespread, by examining their long-distance correlation. As noted by Pratt and James (1986) and Pratt et al. (1992), many shallow-marine cycles are generated by autogenic processes, and are laterally impersistent. Studies based on single vertical sections, however, detailed, are inadequate for the purpose of testing models of orbital forcing. Then, given a demonstration of lateral persistence, it is necessary to demonstrate a persistent regularity of cycle periods, and cycle bundling. This may be done by precise measurement of facies thicknesses or parameters such as carbon or calcium content over lengthy sections (see many examples in de Boer and Smith 1994a). Spectral mapping methods may be particularly useful (Melnyk et al. 1994; Schwarzacher 1993). Cycle hierarchies and cycle stacking ratios have been discussed by Goldhammer et al. (1987, 1990) and Schwarzacher (1993). The use of Fischer plots for the documentation of meter-scale cycles is discussed in Section 2.3.5.

A method for independently testing the periodicity of meter-scale cycles, termed the *gamma method*, was devised by Kominz and Bond (1990). Cycles are subdivided into lithofacies according to a predetermined classification, and the thickness of each facies unit is measured in each cycle. It is assumed that each facies has an approximately constant accumulation rate, reflecting constant depositional conditions. A value for gamma for each facies is given by gamma=elapsed time/thickness. If it can be assumed that each cycle has the same duration (the absolute value of which does not need to be known) a set of equations can be written that determines the proportion of elapsed time represented by each facies in each cycle. If these va-

lues are reasonably consistent, then periodicity has been demonstrated. If a method of determining absolute age is available, for example, by radiometric dating of bentonites that bracket the section under study, then the absolute duration of the cycle period can be determined, and if appropriate, this can be related to orbital frequencies. Given that orbital frequencies may have changed over time (Sect. 10.2.3), and that other sedimentary processes may generate cycles of comparable duration, these quantitative methods of cycle documentation are of considerable value.

10.2.6 The Sensitivity of the Earth to Glaciation

The modern literature on paleoclimatology (e.g., Barron 1983; Barron and Thompson 1990) indicates a considerable complexity and uncertainty in reconstructions of past climates. Although the control of Milankovitch mechanisms on glacial cyclicity is now widely accepted, it is still not entirely clear why major glaciations start in the first place. Factors involved in sensitizing the earth to climatic change include the current climate, the plate configuration of the earth, atmospheric content (volcanic dust, O_2, CO_2), vegetation cover, and the nature of oceanic and atmospheric circulation (Barron 1983; Barron and Thompson 1990). Low levels of carbon dioxide and correspondingly high oxygen levels are thought to be characteristic of times of low sea level during supercontinent assembly phases, and tend to favor icehouse climates because of the reduced atmospheric greenhouse effect (Worsley et al. 1986, 1991; Worsley and Nance 1989; see Sect. 6.1). Glaciation is favored under such conditions, especially where large continental landmasses are located over polar regions, as in the case of the Late Paleozoic glaciation (Sect. 10.4). Eyles (1993) made a strong case for the importance of tectonic control in the triggering of widespread continental glaciation. Uplift accompanying continental collision, and the widespread uplift of rift flanks during supercontinent dispersal, both lead to the development of broad, high-altitude areas where major Alpine ice-caps can easily form. Once formed, feedback effects, such as increased albedo, can lead to continued cooling. The geological record is replete with examples that can be interpreted in this way. For example, Eyles (1993) attributed the widespread Late Proterozoic glaciation to tectonic uplift of rift flanks during the dispersal of the supercontinent of which Laurentia was the center (see Hoffman 1991). The assembly of

Pangea was accompanied by the uplift of many orogenic highlands, and was followed by the long-lasting Gondwana glaciation (Sect. 10.4). Orogenic highlands formed in the Miocene, such as Tibet, may also have been instrumental in triggering of the Late Cenzoic glaciation, although this is controversial, as discussed in Section 10.3. Another major contributing factor in the case of Cenozoic glaciation was undoubtedly the plate-tectonic separation of Australia from Antarctica, which opened the Drake Passage and led to the development of circumpolar currents, which essentially isolated Antarctic weather systems from the rest of the southern hemisphere (Kennett 1977).

Given these broad, long-term (10^6–10^7-year) regional to continental controls, Milankovitch processes then govern the amount of climate change and its periodicity within the Milankovitch time band. Considerable variation can be detected at this level. Imbrie (1985) carried out a spectral analysis of the $\delta^{18}O$ variations in deep sea cores for the last 782 ka. It shows a change in sensitivity at about 400 ka ago. The oxygen isotope record for the 782–400-ka record contains a major peak corresponding to a 100-ka period, and a lesser peak for a 23-ka period. The younger part of the record shows somewhat more variance, with stronger peaks in the longer-term periods, including a strongly represented 100-ka period, and an additional peak corresponding to a 41-ka period. Moore et al. (1982) studied the variation in carbonate concentrations in Upper Miocene and Quaternary deep-sea sediments. The variations are attributed to the degree of preservation of calcite microfossils, which, in turn, is related to carbonate solubility and the corrosiveness of bottom waters. Both factors are climate dependent. Moore et al. (1982) found that periods of 41, 100, and 400 ka were dominant in their spectral analyses. However, the 400-ka period is far more strongly recorded in the Miocene data, with the 100-ka period the dominant one in the Quaternary record. Imbrie (1985) interpreted the strong 100-ka peak as a combination of the two distinct eccentricity peaks at 95 and 123 ka.

Matthews and Frohlich (1991) extended Imbrie's (1985) analysis back to 2.5 Ma, and forward to a hypothetical 2.5 m.y. in the future. They demonstrated that the 100-ka eccentricity and 19-ka precession components combined to produce a 2-m.y. modulation, although the amplitudes of the various peaks varied according to input conditions. This result suggests that glacioeustasy may be capable of explaining longer-term cyclicity with million-year periodicities (equivalent to Vail's "third-order" cycles). To this analysis Matthews and Frohlich (1991) then added the effects of the presence of warm, saline waters in midlatitude positions, such as would have occurred in the Tethyan Ocean before this water body was eliminated by subduction of its sea floor.

Warm water upwelling adjacent to a cold continent should create a very efficient mechanism for producing ice on the continent, thereby lowering sea level. Air descending over the cold continent flows out over warm water and picks up water vapor. Return flow aloft delivers this moisture to the interior of the continent as snow. (Matthews and Frohlich 1991).

It is postulated that thermohaline currents would have delivered this water to Antarctica, and could have accounted for large build-ups of snow there, with a calculated episodicity of 1.6–2.4 m.y. Qualitatively, therefore, a glacioeustatic mechanism exists to explain million-year ("third-order" or "mesothemic") cyclicity during times when the earth was sensitive to glaciation. However, the lack of evidence for major continental ice caps throughout much of geological time means that this cannot be regarded as a universal explanation.

Variations with time in the stratigraphic signal amplitude of the various Milankovitch periods may relate to the varying character of glacial regimes. In the Miocene, only the Antarctic ice cap existed. This is mainly a land-based ice cap and is entirely surrounded by polar waters. In the Quaternary, the Northern Hemisphere ice cap was also present and consisted of several smaller centers, some bordered to the north or south (or both) by water masses. It seems likely that the sensitivity of the ice caps to climate forcing would have been quite different in these different settings, with the large, stable Antarctic ice cap tending to respond much more slowly than the Northern Hemisphere ice centers.

Another example of the varying styles of glaciation affecting the styles of cyclicity was discussed by Wright (1992). He classified platform-carbonate cycles into two broad types. Type 1 consist of meter-scale "keep-up" cycles dominated by peritidal carbonates. They are common in the Cambro-Ordovician, Early Devonian and Middle Triassic to Cretaceous sedimentary record. Type 2 cycles are 5–10 m thick. They consist mainly of shelf and deeper-water carbonates, with common paleokarstic and paleosol horizons (vadose caps). They are common in the Carboniferous sedimentary record. Wright (1992) suggested that the differences in cycle type may have been caused by differences in the type of glacial cycle, reflecting differences in the size of ice caps and their consequent sensitivity to change. He postulated that type 1 cycles represent small-scale ice caps, such as those formed during "greenhouse" climatic phases (Fig. 3.3).

[W]hen continental configuration or other factors did not facilitate the generation of major ice sheets. During such periods the amplitudes of these climatically induced sea-level changes were small, only slower third-order changes influenced the types of small cycles developed. In the absence of ice-buildups no rapid decameter-scale falls or rises in sea level occurred. The falls in sea level triggered by changes in insolation were small in amplitude and slow enough to be outpaced by subsidence, creating minor accommodation space, readily filled by peritidal cyclothems.... The subsequent rises were slow enough to be outpaced by carbonate production to produce more peritidal cyclothems. (Wright 1992, p. 2).

As noted elsewhere (Sect. 10.2.7) there is, however, doubt regarding the existence of significant ice caps for much of Phanerozoic time, and this mechanism for type 1 cyclothems remains unproven. Wright (1992) suggested that type 2 cyclothems developed in response to the growth of large continental ice caps during "icehouse" climatic phases, such as during the Carboniferous Gondwana glaciation, and the Late Cenozoic glaciation. At such times high-amplitude sea-level falls lead to exposure of platform tops and to extensive meteoric diagenesis. Rapid rises lead to drowning in which only reefal organisms are capable of keeping pace with sea-level rise. Platform interiors are drowned, becoming areas of subtidal sedimentation, in which peritidal and reef sedimentation are restricted to nucleation sites on earlier topographic highs. This is the pattern of so-called "catch-up" sedimentation (see Sect. 15.4.1 for additional discussion of carbonate sequence variation).

10.2.7 Glacioeustasy in the Mesozoic?

One major school of thought holds that all global high-frequency cycles with periodicities of 10^6 years and less, are glacioeustatic in origin. Vail et al. (1991) made this assertion with regard to the "short-term eustatic curve" of Haq et al. (1987, 1988a), despite the lack of convincing evidence for continental glaciation for much of geological time (Frakes 1979, 1986; Hambrey and Harland 1981; Barron 1983; Francis and Frakes 1993). As argued elsewhere in this book there is in fact doubt regarding the very existence of a global framework of cycles in the 10^6-year range that would require such

explanation (Chaps. 11, 13), and nonglacial Milankovitch mechanisms can be invoked to explain many types of high-order cycle (Sect. 10.2.8). Nevertheless, glacioeustasy remains a popular hypothesis for all types of high-order cycle.

Various authors have examined the question of glaciation, or other evidence of cold climates during the Mesozoic, when most available evidence suggests that the earth was in a greenhouse climatic state (Fig. 3.3). As summarized by Barron (1983) and Frakes (1986), the evidence indicates that ocean-water temperatures were higher than at present during the Cretaceous (Fig. 10.8; see also Fig. 10.14), with a smaller equatorial-polar temperature gradient. Frakes and Francis (1988) and Francis and Frakes (1993) provided summaries of the evidence for cold polar climates in the southern hemisphere during the Cretaceous. They noted that there is ample evidence for locally cold temperatures, perhaps of seasonal character, in many Lower Cretaceous high-latitude deposits. Many ice-rafted deposits of this age have also been recorded from such as areas as central Australia, northern Canada and northern Russia, which would have been located in high latitudes at the time. At the very least this indicates the presence of significant pack-ice fields. Frakes and Francis (1988) stated:

Given that the Earth is a sphere which receives solar insolation unequally over its surface and that heat transport in both atmosphere and oceans has its physical limits, it is difficult to explain how the polar zones could ever have been warm enough to melt all traces of ice and snow there. The problem would be further exacerbated if "normal" topographic relief to 1–2 km elevation existed on polar continents, producing much colder temperatures at high elevations. These factors suggest that at least seasonal ice might be expected to form in high latitudes throughout most of the Phanerozoic.

Frakes and Francis (1988) reported that "the geological literature reveals that reports of ice-rafted deposits exist for every period of the Phanerozoic Era except the Triassic." While these data and climatological considerations support the idea of high-frequency orbital forcing at all times during the Phanerozoic, the evidence for significant continental glaciation and consequent glacioeustasy remains scanty for the Mesozoic and much of the Early and Middle Paleozoic. Eyles (1993) pointed to the existence of temperate ice caps in Iceland and Alaska at the present day, and stated that the evidence cannot exclude ice caps in the Mesozoic. However, this seems to be a special pleading, and he concluded that there is no convincing evidence for a

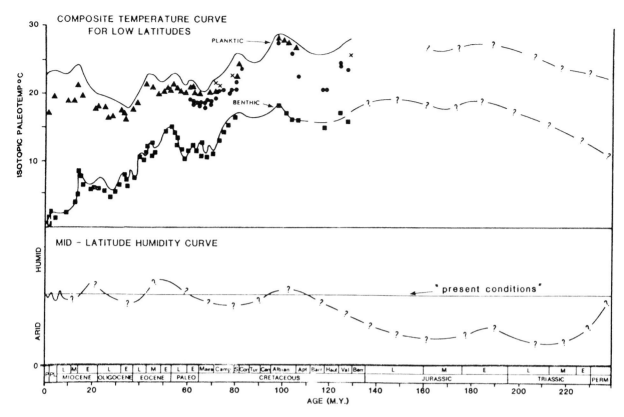

Fig. 10.8. Low-latitude temperature and midlatitude humidity curves for the Mesozoic-Cenozoic. Ocean temperatures are from oxygen isotope determinations on surface (planktic) and bottom-dwelling (benthic) foraminifera. Humidity curve is an estimate. (Frakes 1986)

glacioeustatic cause of the observed high-frequency cyclicity in the Mesozoic. Plint (1991) described some Upper Cretaceous clastic cycles in the Alberta foreland basin, Canada, which appear to indicate a periodicity of base-level change in the Milankovitch band, and speculated about glacioeustatic control. Leithold (1994) offered a similar speculation regarding some Upper Cretaceous high-frequency cycles in Utah. Plint (1991) summarized the evidence for glaciation during the Cretaceous, and concluded that the available evidence is consistent with small glacioeustatic fluctuations, in the order of less than 10 m, that would explain his observed cycle architecture.

Rowley and Markwick (1992) tackled the problem of glacioeustasy during greenhouse climatic phases in a different way. They calculated the changes in oceanic water volume that would be required to yield changes in sea level comparable to those indicated by the "third-order" cycles that comprise the main basis for the Exxon global cycle chart. Assuming glacioeustatic mechanisms as the dominant process, they calculated that it would require an Antarctic-sized ice sheet to evolve and dissipate about every 1–2 m.y. to create the magnitude of sea-level changes contained in this chart, but could find no evidence for this. Changes in oxygen isotope measurements from Jurassic through Middle Cenozoic sediments in DSDP cores show quite different patterns to those predicted from a glacioeustatic mechanism. Rowley and Markwick (1992) were unable to identify generative causes for such changes during nonglacial times, such as during the Mesozoic and Early Cenozoic. An important additional point is that the periodicity of glacial freeze and melt is not in the million-year range, but takes place over periods in the 10^4-year band.

10.2.8 Nonglacial Milankovitch Cyclicity

Orbital variations may be capable of causing climatic change and consequent sedimentary cyclicity, even when the global cooling effects are inadequate to result in the formation of continental ice caps.

For example, because of the interaction between the eccentricity and precession effects, there are phases when one hemisphere experiences short, hot summers and long, cold winters, while the other hemisphere undergoes long, cool summers and short, mild winters. These effects alternate between the hemispheres over periods of 10^4 years, and the intensity of the effect waxes and wanes as the orbital variations go in and out of phase (Fischer 1986). Such changes undoubtedly have dramatic effects on surface air- and water-temperature distributions, oceanic circulation and wind patterns. There is therefore considerable potential for subtle sedimentological effects, without any change in sea level.

Cecil (1990) provided a general model of the response of depositional sytems to climate change. As climate belts shift as a result of orbital forcing, or the continents drift through climate belts as a result of plate motions, climate changes lead to changes in clastic sediment yield and patterns of chemical sedimentation. Arid climates are times of low sediment yield and are accompanied by formation of pedogenic carbonates and evaporites. Sediment yield increases with increasing precipitation, leading to a predominance of clastic sedimentation. Maximum yields occur under temperate, seasonal wet/dry climatic conditions. Very humid climates are characterized by thick vegetation cover, which results in a reduction in sediment yield and an increasing importance of peat/coal formation. In Section 10.4 this model is applied to Late Paleozoic cyclothems.

Variations in organic productivity, in the depth of carbonate compensation (CCD), and in the degree of oxidation of marine waters are the kinds of effects that are controlled by these climatic variables and have direct consequences for sedimentation patterns (Fischer 1986). For example, the rate of skeletal supply relates to organic productivity. Erba et al. (1992) documented rhythmic variations in species abundances in a Cretaceous clay using Fourier analysis, and interpreted these as "fertility cycles" driven by changes in oceanic temperature and circulation in response to variations in seasonality. Variations in the supply of clay due to variations in source area weathering and marine circulation affect the proportion of biogenic components in the resulting sediments. The carbonate fraction varies if the seafloor is close to the CCD. Various examples of laminated pelagic sediments in the geological record were discussed by Fischer (1986) as illustrations of these processes (Fig. 10.9).

Weltje and de Boer (1993) showed that the thicknesses and mineralogy in some rhythmic Pliocene turbidites in Greece can be attributed to precession-induced changes in precipitation and runoff feeding deltaic sediment sources. Redox cycles are another type of cycle resulting from varying sediment fluxes and oxygenation levels (de Boer and Smith 1994b).

Many processes show a significant lag between changes in the forcing function and the sedimentary response. This is a well-known characteristic of the "carbonate factory" (Sect. 15.4), but may be a factor in other environments. For example, de Boer and Smith (1994b) cited examples of Pliocene carbonate-marl cycles and anoxic-aerobic cycles in the Mediterranean region that suggest a lag in the response of monsoonal systems to changes in insolation. Similarly, peat and methane production may not reach a peak for thousands of years after the related insolation peak has passed. Complex interrelationshops and feedback loops were summarized by Mörner (1994).

A particularly interesting study of Milankovitch cycles is that by Elder et al. (1994; see Fig. 8.60), who demonstrated a correlation between 10^5-year clastic cycles on the margin of the Western Interior Seaway in Utah with carbonate cycles in the basin center in Kansas, over a distance of about 1500 km. They developed a model based on small-scale sea-level changes (10–15 m) driven by orbitally forced climate changes. Their model included the formation and melting of small glaciers, but this component of the model does not seem to be necessary to this writer. Elder et al. (1994) suggested that during warm, dry phases thermal expansion of surface waters, glacial melting, or transfer of stored groundwater to oceans, could have contributed to transgression, flooding, and enhancement of carbonate productivity in the basin center. During cool, wet phases sediment supply from the Sevier highlands to the west would have increased, leading to shoreface progradation, and a clay-rich phase in the basin center. There is no evidence of diachroneity of the sequence boundaries, as in some comparable sequences from this basin in Alberta (Sect. 11.3.2), therefore the rolling flexural wave model described for these examples cannot be invoked in this case. However, there is some evidence of thickness changes at faults, and strandlines parallel the faults, so some tectonic influences were active during sedimentation.

Hallam (1986) urged caution in that some cyclicity may relate to diagenetic effects and may not be

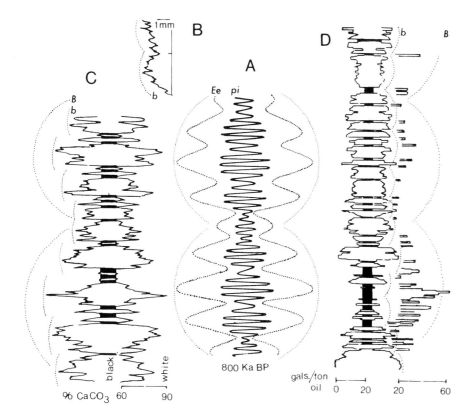

Fig. 10.9. Comparison of Milankovitch cycles plotted at similar time scales. **A** Milankovitch parameters plotted for the last 800ka. *pi* Precession; *Ee* long and short cycles of orbital eccentricity. **B** Variations in thickness of evaporite varves, Castile Formation (Permian), New Mexico. **C** Rythmicity in Fucoid Marl (Cretaceous), Italy. Decimeter-scale limestone-marl couplets are shown by fine detail in carbonate variation and are driven by precession. These are grouped into 50-cm bundles (*b*) by a redox cycle, in which marls are periodically black and sapropelic as a result of bottom stagnation, related to an eccentricity cycle. Superbundles (*B*) may reflect a long-term orbital eccentricity cycle. **D** Variations in oil yield in the Parachute Creek Member, Green River Formation (Eocene), Colorado. Kerogen-rich layers formed during humid lacustrine phases, whereas kerogen-poor intervals reflect arid playa phases. Bundling in response to various orbital controls is shown by dotted lines. (Fischer 1986, reproduced by permission, from the Annual Review of Earth and Planetary Sciences, vol 12, copyright Annual Reviews Inc.)

a reflection of primary sea-level or climate controls. He discussed the case of limestone-shale cycles, such as the decimeter-scale couplets in the Blue Lias of southern England. House (1985), Weedon (1986), and others have studied these rhythms using refined biostratigraphic analysis to ascertain time spans and have performed spectral analysis on the cycle thicknesses in search of rhythmicity. Weedon (1986) claimed success in this endeavor and suggested a correlation with precession and obliquity effects. However, Hallam (1986) showed that in at least some of the cyclic sequences rhythm thickness is constant regardless of the overall sedimentation rate. He suggested that in such cases the rhythms may have been formed by diagenetic unmixing of calcium carbonate during diagenesis.

Diagenetic unmixing was ruled out by De Visser et al. (1989) in their examination of the Pliocene Trubi marls of Sicily (Fig. 10.10). This detailed study of a succession of calcareous rhythmites included quantitative analysis of foraminiferal and palynological contents, stable isotopes, carbonate and carbon content, trace elements and granulometric data. The beds consist of alternations of carbonate-rich and carbonate-poor units. Gray layers are interbedded with white layers within which a beige-colored, less indurated layer is intercalated. The gray layers are the most organically rich, and "were probably formed under relatively humid and also relatively warm climatic conditions with increased local run-off of nutrient-rich, sediment-loaded river water." The beige layers are co-

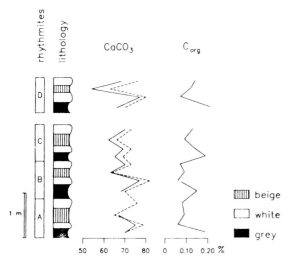

Fig. 10.10. Four rhythmites of the Trubi Formation, Sicily. Variations in calcium carbonate were determined volumetrically (*dashed line*) and by atomic absorption spectrometry (*solid line*). Organic carbon conent is also shown. (De Visser et al. 1989, reproduced by kind permission of Elsevier Science-NL, Sara Burgerhartstraat 25, 1055 KV Amsterdam, The Netherlands)

lored by the presence of fine clastic particles. They are the poorest in carbonate, and "are thought to have been deposited under relatively dry and cooler climatic conditions" when eolian dust was being transported from North Africa. White layers are the most carbonate rich and are thought to represent climatic and other conditions intermediate between the other two. The Trubi Formation therefore is thought to represent a complex interaction of a terrigenous dilution cycle and a carbonate productivity cycle.

Hay and Leslie (1990) considered possible changes in the volume of water stored in groundwater reservoirs as a speculative mechanism for causing eustatic sea-level changes. They estimated that existing sediments contain 25×10^6 km^3 of pore space. Emptying or filling this space with water, which is thereby removed from the runoff-evaporation-precipitation cycle, could change sea level by an estimated 50 m, allowing for isostatic adjustment. Climate changes brought about by orbital forcing could alter the balances in the water cycle, and changes in continental elevation resulting from plate-tectonic processes would alter the volume of potential water storage. However, there is as yet no concrete evidence for changes in the mass balances of the water cycle.

10.3 The Cenozoic Record

Many successions of Late Cenozoic age have now been correlated with the oxygen-isotope record, demonstrating a glacioeustatic control of the sequence stratigraphy. Seismic sections illustrating this type of architecture are available from many continental margins (e.g., Figs. 8.11, 8.19). Bartek et al. (1991) provided a review of this stratigraphic record, with a particular reference to the evidence for the earliest glaciations, in Antarctica. They compiled a composite stratigraphic cross section showing the characteristic features of the stratigraphic architecture of the Cenozoic record in the Ross Sea area of Antarctica (Fig. 10.11).

A widespread sea-level fall in Middle Oligocene age has been suggested in many stratigraphic records. It appears on the Exxon global cycle charts, and also in many stratigraphic records that have been independently compiled for specific basins and continental margins (e.g., Fig. 6.15). However, according to Eyles (1993) "the earliest input of glacial sediment to the Ross Sea margin of Antarctica occurred at about 36 Ma when temperate wet-based glaciers reached the Victoria Land rift basin.... A continental ice sheet was probably in existence at this time ... but a large continental ice mass did not become a permanent fixture of Antarctica until about 14 Ma." Isotopic evidence, referred to below, suggests that the major sea-level falls occurred at these two times.

Bartek et al. (1991) compiled seismic-stratigraphic evidence from Antarctica, which indicates that an extensive erosional event occurred on the Antarctic shelf during the Late Oligocene, as a result of the grounding of an expanded ice sheet on the Ross Sea. By contrast, Pliocene sediments in parts of Antarctica contain marine faunas and floras and fossil wood, indicating fluctuating climates during the younger Cenozoic. The erosional events in the Oligocene, and again in the Middle-Late Miocene, developed broad, troughlike scours 70–80 km wide, which filled with till tongues (Fig. 10.12). This seismic evidence is important, as it appears to have settled a long-standing controversy regarding the extent of continental ice in Antarctica in the Early Cenozoic, with many regarding the continent as essentially ice-free prior to the Middle Miocene (see discussion in Matthews 1984).

In the northern hemisphere alpine glaciation in interior regions such as Alaska began about 10 Ma. Extensive sea-ice cover and coastal glaciation ap-

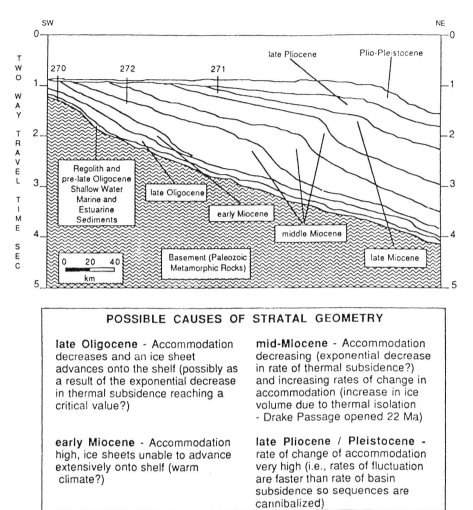

POSSIBLE CAUSES OF STRATAL GEOMETRY

late Oligocene - Accommodation decreases and an ice sheet advances onto the shelf (possibly as a result of the exponential decrease in thermal subsidence reaching a critical value?)

mid-Miocene - Accommodation decreasing (exponential decrease in rate of thermal subsidence?) and increasing rates of change in accommodation (increase in ice volume due to thermal isolation - Drake Passage opened 22 Ma)

early Miocene - Accommodation high, ice sheets unable to advance extensively onto shelf (warm climate?)

late Pliocene / Pleistocene - rate of change of accommodation very high (i.e., rates of fluctuation are faster than rate of basin subsidence so sequences are cannibalized)

Fig. 10.11. Composite stratigraphic cross section compiled from the eastern Ross Sea, Antarctica. (Bartek et al. 1991)

peared at about 5 Ma, and the earliest significant occurrence of ice-rafted debris in the North Atlantic Ocean has been dated at 2.5 Ma (data and summary in Eyles 1993, p. 165).

Various authors since Emiliani (1955) have documented the oxygen isotope record and attempted to calibrate it with reference to changes in ocean-water temperature and sea level (e.g., Savin 1977; Matthews and Poore 1980; Chappell and Shackleton 1986). Useful reviews which compared this record to that of seismic stratigraphy were provided by Matthews (1984) and Williams (1988). The Pleistocene to Recent stratigraphic record has been subdivided into isotope stages, based mainly on detailed analysis of DSDP cores (Fig. 10.13; Williams et al. 1984; Shackleton and Hall 1984). Examples of seismic and outcrop sections tied to this chronology are given in Figs. 8.11, 8.18 and 8.20.

The older Cenozoic record is not yet as well known from the point of view of oxygen isotope stratigraphy. Williams (1988) compiled the record of $\delta^{18}O$ data for the last 70 m.y., time-averaged at 1-m.y. increments to filter out high frequency events (Fig. 10.14; an earlier compilation was carried out by Frakes 1986; see Fig. 10.8). The results were compared to the timing of million-year ("third-order") events on the global cycle chart, indicating some degree of correlation (Fig. 10.14). However, the significance of such correlations is questioned, as discussed in Chapter 13. The evidence in Fig. 10.14 does not support the idea of a single Middle Oligocene fall in sea level, indicating, instead, two major falls, at about 35 Ma and 14 Ma. Miller et al. (1987) suggested additional lowstand events at 31, 25, and 10 Ma, based on this evidence. These five events do not correlate well with the Exxon global cycle chart.

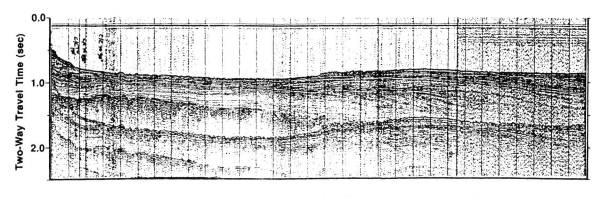

PD - 90 - 49B

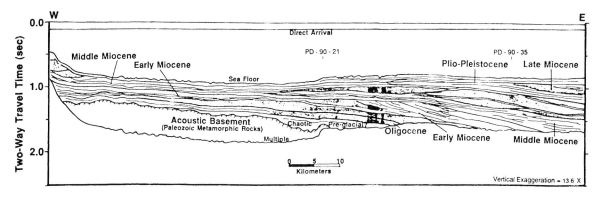

Fig. 10.12. Seismic section in Ross Sea area, Antarctica, and interpretation, showing architecture of till tongues (*stippled layers*), developed by shelf erosion during glacioeustatic lowstands. (Bartek et al. 1991)

Did plate-tectonic events in the Miocene, such as the uplift of the Himalayan ranges and the Tibetan Plateau, lead to regional cooling and diversion of monsoonal air masses that then helped to trigger the northern hemisphere glaciation? Or did global cooling brought about by other causes lead to increased rainfall and decreased upland vegetation coverage, that then led to more rapid erosional unroofing and isostatic uplift of mountains and plateaux? Undoubtedly both effects are important, but the nature of the ultimate controls is a subject of ongoing debate. Various lines of evidence including isotopic data, paleobotany, fission-track data relating to rates of uplift, and the plate kinematics that led to collision and crustal shortening, have all been brought to bear on this topic. Eyles (1993, pp. 165–169) provided a summary of this debate to which the reader is referred.

10.4 Late Paleozoic Cyclothems

It is now widely accepted that the Upper Paleozoic cyclothems are the product of high frequency eu-

static sea-level fluctuations induced by repeated glaciations with 10^4–10^5-year periodicities. This is an old idea (Wanless and Shepard 1936; Wanless 1950, 1972) that was recently been revived by Crowell (1978), Heckel (1986), and Veevers and Powell (1987). Crowell (1978) and Veevers and Powell (1987) showed that the Late Devonian-Permian glaciation of Gondwana had the same time span as the cyclothem-bearing sequences of the Northern Hemisphere. The lower part of each cyclothem was formed during a transgression, which drowned clastic coastal-plain complexes and developed a sediment-starved shelf on which carbonates were deposited (Figs. 8.31–8.33). This phase represents the melting of continental ice caps. With renewed cooling and ice formation, regression began, with the initiation of rapid deltaic progradation. The erosion surface at the base of the nonmarine sandstone (Fig. 8.31) may represent local deltaic channeling or widespread subaerial erosion. Variations in thickness and composition of the cyclic sequences were caused by local tectonic adjustments (epeirogenic warping, movement on fault blocks, etc.) and by proximity to clastic sources

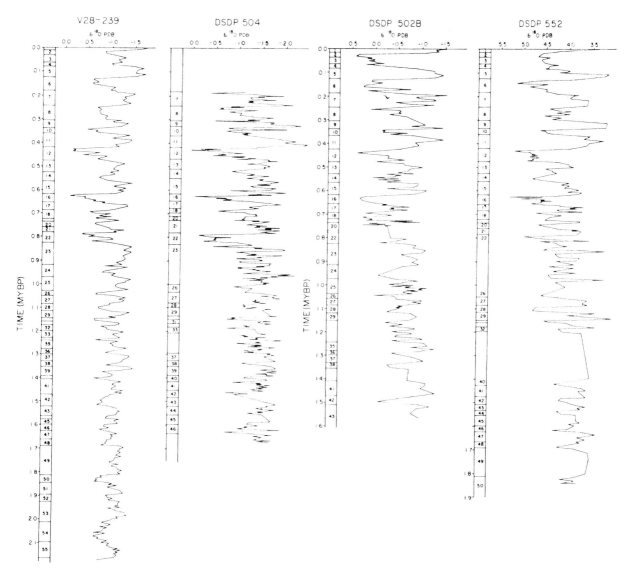

Fig. 10.13. Empirical correlation of oxygen isotope stage zonation in four DSDP cores from the Pacific Ocean, Caribbean Sea and Atlantic Ocean. (Williams 1988)

(Fig. 8.34). A similar model of transgression and regression was developed for the Kansas cyclothems by Heckel (1986; Fig. 2.6). He distinguished major and minor cycles and estimated that the major cycles spanned 235–400 ka, while the minor cycles had durations of 40–120 ka. Detailed study of outcrops and cores enabled Heckel (1986, 1990) to erect a curve showing regional sea-level changes in the United States Midcontinent (Fig. 10.15). However, the duration of these cycles may be questioned because of uncertainty regarding the periodicity of orbital parameters in the distant geological past, as discussed in Section 10.2.3.

Heckel's (1986) sea-level curve was correlated with stratigraphic successions in Texas by Board-man and Heckel (1989), and tentative correlations have been extended into southern Arizona, northern Mexico, and the Paradox Basin, Utah, by Dickinson et al. (1994). As pointed out by Dickinson et al. (1994), that the cycles retain similar character and thickness throughout the entire North American interior, across cratonic areas and basins characterized by very different tectonic styles and subsidence histories, is convincing evidence against a tectonic origin for the cycles, and therefore supports the glacioeustatic model.

Van Veen and Simonsen (1991) attempted to extract a longer-term (million-year periodicity) signature from Heckel's curve, and used this to suggest correlations into Russia based on simple

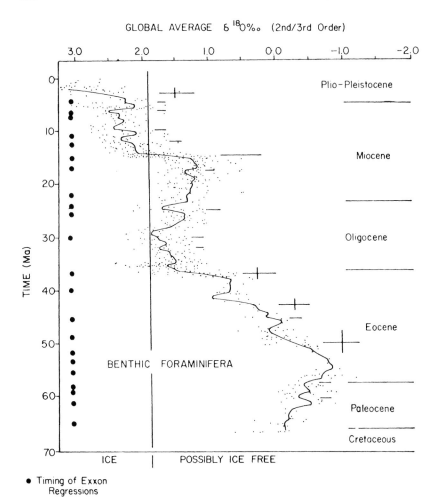

Fig. 10.14. Compilation of $\delta^{18}O$ data for the last 70 m.y. of geological time, averaged at 1-m.y. increments. The graph shows generalized shifts in the isotopic values, and major "events" are indicated by the lines or crosses to the right of the curve. *Black dots at left* are ages of "third-order" events on the global cycle chart of Vail et al. (1977). (Williams 1988)

sequence matching. Boardman and Heckel (1991), in their comment on this proposal, noted the incompleteness of the data base, and expressed scepticism of the ad hoc way the correlations had been done, in the absence of biostratigraphic control. It left out many other factors that could have affected cycle development on longer term time scale, especially tectonism.

Major regressive events during the Devonian to Permian (events 1–5 in Fig. 10.16) may correspond to times of exceptionally widespread Gondwanan glaciation. Veevers and Powell (1987) suggested that, at these times, regional episodes of tectonism may have uplifted broad areas of Gondwana, so that the cooling effects of elevated altitude were added to the climatic effects of the high latitudinal position of the supercontinent. The closure of the paleo-Teth-

yan Ocean as Gondwana collided with Laurentia, and the uplift of western South America and eastern Australia would have been particularly significant triggering events (Eyles 1993, p. 129).

A synthesis of the chronology and extent of glacigenic deposits in Gondwana by Caputo and Crowell (1985) provided an explanation for the widespread glaciations of Late Ordovician and Late Devonian to Permian age. They related the distribution of glacial deposits to the polar wandering paths established by paleomagnetism (Fig. 10.17). Glacial episodes occurred when the South Pole lay over a major continental area. Between the Middle Silurian and the Middle Devonian, the supercontinent drifted away from the pole, which lay over the paleo-Pacific Ocean (Panthalassa). There was therefore no major continental glaciation at this

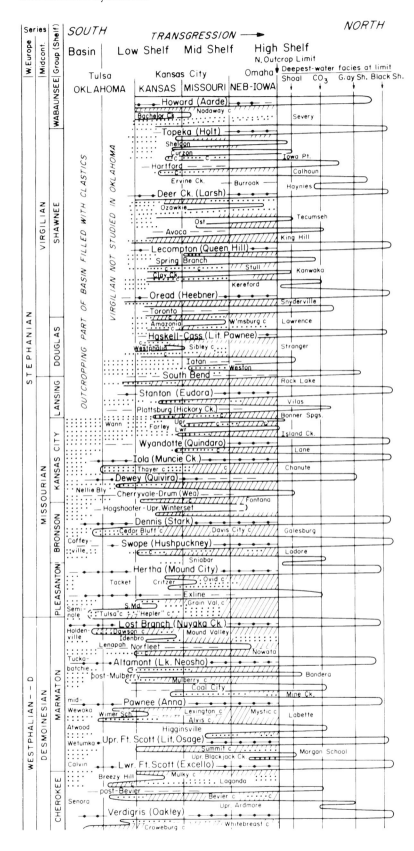

Fig. 10.15. Sea-level curve for part of the Middle-Upper Pennsylvanian sequence of the North American Midcontinent, which ranges from 260 m thick in Iowa to 550 m thick in Kansas. The lateral extent of erosion surfaces at the base of the cycles is shown by *diagonal hatching*. Fluvial deltaic complexes are shown by a series of *dots*, and conodont-bearing shales by *lines with dots*. (Heckel 1986)

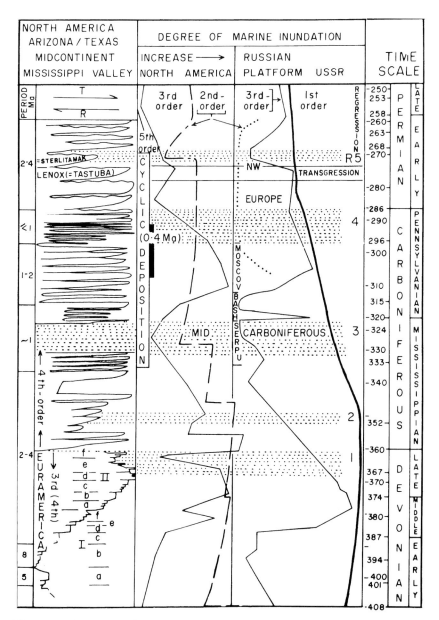

Fig. 10.16. Summary of Devonian-Permian "first-" to "fourth-order" cyclicity in Europe, Russia, and North America. Position of "third-" and "fourth-order" cycles of Heckel (1986; Fig.10.15) in *column at left* is shown by vertical black bar. Devonian cycles (*a–f*) of Johnson et al. (1985) are shown *at bottom of left-hand column. Dashed* "*second-order*" *curve* is that of Vail et al. (1977). Major regressive episodes are shown by *dashed areas* and *numbers 1–5 at right*. These correspond to periods of extensive Gondwana glaciation. (Veevers and Powell 1987)

time. Similarly, the continent drifted away from the pole in the Triassic, and glaciation ended, not to return to the Southern Hemisphere until the major cooling episode of the Middle Tertiary, when the Antarctic ice cap developed.

Studies of the changing sensitivity of the earth to orbital forcing in the Cenozoic (Sect. 10.2.6) suggest useful analogies for the interpretation of the

Gondwana glacial record. The high-frequency cyclothemic periodicity determined by Heckel (1986) and others appears to fit closely the Cenozoic results discussed above. Polar wandering resulted in a progressive shifting of glacial centers across Gondwana (Fig. 10.17), which would have led to continuous variations in the sensitivity of the ice masses to climate forcing. The continuous varia-

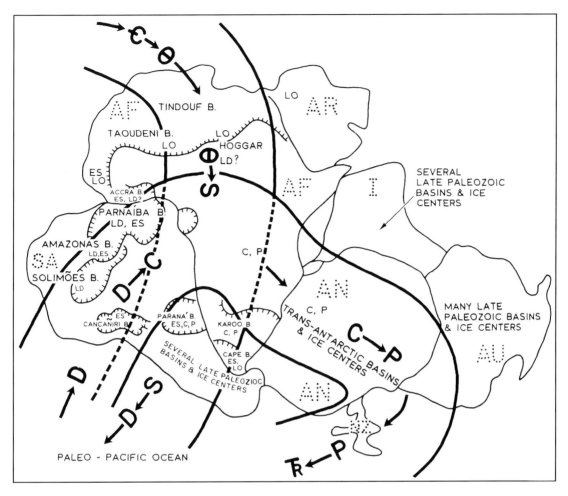

Fig. 10.17. Major glacial centers in Gondwana during the Ordovician to Permian showing the polar-wandering path for the same period. (Caputo and Crowell 1985)

tions in cycle length shown in the compilation by Veevers and Powell (1987; Fig. 10.16) are therefore readily explained.

Crowley and Baum (1991) estimated the magnitude of sea-level changes that might have occurred during the Gondwana glaciation. They developed three scenarios for ice cover, ranging from a minimum to a maximum, based on basin analysis of the glacial deposits and tectonic considerations. Calculations of ice volumes from relationships developed for the Cenozoic glaciation, and conversion of these estimates into volumes of water, indicated possible total glacioeustatic fluctuations of between 45–75 m (minimum ice cover) and 150–190 m (maximum ice cover) allowing for isostatic adjustments. These are consistent with estimates derived from stratigraphic studies of the cyclothems, as shown by Klein (1992).

Klein (1992) also demonstrated that long-term climate change and tectonism also contributed to

the total base-level changes that were responsible for developing the Late Paleozoic cyclothems of the North American Midcontinent. It is a considerable challenge to sort out these various complexities. Cecil (1990) developed a model (discussed in Sect. 10.2.8) to explain variations in the importance of coal through the Late Paleozoic cyclothems of North America. Both long- and short-term variations in climate can be detected. The long-term changes occur over periods of tens of millions of years, and are therefore related to the drift of the continent through climatic belts. Latest Devonian-earliest Mississipian cyclothems tend to contain thick clastics, suggesting a wet climate. An arid interval occurred in the Late Osagean and Early Meramecian, characterized by restricted clastic input and deposition of evaporites. North America then drifted into the tropical rain belt in the Early Pennsylvanian, resulting in an important phase of coal development.

Superimposed on this long-term trend are the short-term climatic cycles that accompanied the glacial-interglacial fluctuations. In Nova Scotia, calcareous paleosols occur at cyclothem sequence boundaries, formed at times of lowstands of sea level, indicating seasonally dry climates during glacial episodes in the southern hemisphere. This contrasts with the humid (interglacial) climates indicated by coal-bearing deposits of the highstand systems tracts (Tandon and Gibling 1994). Another study of paleosols that revealed fluctuations in climate related to the southern hemisphere glacial-interglacial cycles was reported by Miller et al. (1996). Yang and Nio (1993) demonstrated an oscillation between arid conditions during lowstand and more humid conditions during highstand in the eolian-fluvial cycles of the Lower Permian Rotliegend Sandstone of the southern North Sea Basin. These cycles were also driven by Gondwana glacioeustasy. A further discussion of the attempt to separate the effects of tectonism and glacioeustasy in the Appalachian foreland basins is given in Section 11.3.2.2.

10.5 The End-Ordovician Glaciation

A short-lived glaciation occurred at the end of the Ordovician, centerd on a continental ice-cap in the present area of the Sahara desert in North Africa. The field evidence was summarized by Eyles (1993, p. 116). The worldwide bathymetric and isotopic evidence for the resulting eustatic fall in sea level, which probably only lasted a few hundred thousand years, was synthesized by Brenchley et al. (1994). Long (1993) provided a description of the resulting sequence boundary in Anticosti Island, Canada, with references to other work. The short-lived break in sedimentation appears as a regional unconformity in many stratigraphic records, for example, see the brachiopod fauna diagram in Fig. 2.7, which indicates a short fall at the end of the Ashgill Series.

What caused this glacial episode during a long phase of global greenhouse climatic conditions (Fig. 3.3)? Paleomagnetic reconstructions show that North Africa drifted across the pole during the Ordovician (Fig. 10.17). Eyles (1993, p. 118) suggested that glaciation was triggered by broad platform uplift in response to Taconic orogenesis along the West African plate margin. Brenchley et al. (1994) argued that migration of Gondwana toward the pole led to a rapid feedback effect of increasing

snowfall, decrease in albedo, and the development of cold ocean bottom waters. High rates of upwelling and productivity, leading to absorption of CO_2 and a negative greenhouse effect then ensued. The arrival of Gondwana at the pole may have decreased precipitation and paradoxically reduced ice cover.

10.6 Main Conclusions

1. The evidence for Milankovitch control of climate and sedimentation in the Late Cenozoic is now overwhelming. Stratigraphic cyclicities in many Late Cenozoic sections have been correlated with the oxygen-isotope record that tracks ocean-water temperature changes, and it now generally agreed that the $\delta^{18}O$ record can be used as an analog recorder of sea level, bearing in mind the lag between the eustatic high and the attainment of the highest temperatures. This body of knowledge permits chronostratigraphically very precise studies of Late Cenozoic sequence stratigraphy. The methods of cyclostratigraphy, as it is now termed, include sophisticated time-series analysis of cyclic parameters, such as cycle thickness and carbon content.

2. The earth's sensitivity to climate change depends on a wide range of parameters and feedback mechanisms, which make reconstruction of past climate change and of Milankovitch controls very difficult. Of particular importance, as demonstrated by the initiation of the Gondwana glaciation and the Early Cenozoic glaciation of Antarctica, is the plate-tectonic control of large continental masses relative to the poles, and the oceanic and atmospheric circulation around them. Major glacial episodes may be initiated by regional uplift caused by collisional uplift of plateaus and mountain ranges, and by thermal uplift during supercontinent fragmentation.

3. Controversy remains regarding the periodicity of Milankovitch parameters in the geological past. Major changes may have occurred during the Phanerozoic, because of changes in the orbital behavior of the earth. In addition, sedimentological studies indicate that various autogenic mechanisms can generate apparent periodicities that simulate Milankovitch effects, some tectonic processes can also generate sequences with similar periodicities, and there is imprecision in biostratigraphic dating of most sequences. For these reasons, it is considered unwise to attempt

to correlate suites of cycles to specific periodicities or to invoke Milankovitch mechanisms purely on the basis of interpreted 10^4–10^5-year cycle frequencies.

4. Glacioeustasy is markedly affected by isostatic and geoidal changes within hundreds to a few thousands of kilometers of the edge of a major continental ice cap. The timing, direction and magnitude of sea-level changes within this region may be entirely different from the signatures elsewhere.

5. The postulation of Milankovitch controls for any given cyclic sequence should only be attempted following rigorous spectral analysis of one or more cyclic parameters within the succession, because other autogenic and allogenic mechanisms can generate cycles with similar thickness and repetitiveness. Demonstration of the presence of a hierarchy of cycle types with characteristic periodic ratios, such as the 1:5 precession:eccentricity combination, is addi-

tional evidence of Milankovitch controls, although the cautions noted in paragaph 3, above, need to be borne in mind.

6. The earth was entirely ice free for only short periods during the Phanerozoic, but the evidence for major continental glaciation and consequent significant glacioeustasy outside the periods occupied by the Late Paleozoic Gondwana glaciation and the Late Cenozoic glaciation is at present weak.

7. Glacioeustasy is not the only cycle-forming mechanism driven by orbital forcing. Variations in climate and oceanic circulation have significant effects on such parameters as sediment yield and organic productivity, which have generated many successions of meter-scale cycles throughout the geological past. These are especially prominent in the deposits of many carbonate shelves and in fine-grained clastic deposits formed in lakes and deep seas.

11 Tectonic Mechanisms

11.1 Introduction

Sea-level changes within a basin may be caused by movements of the basement rather than changes in global sea level (eustasy). In Chapter 9 we examine long-term mechanisms of basement movement driven by deep-seated thermal processes, including the generation of dynamic topography, and we also discuss eustatic sea-level changes driven by changes in ocean-basin volume. These processes act over time periods of tens to hundreds of millions of years and are continental to global in scope. In this chapter we discuss relative sea-level changes caused by continental tectonism that are regional to local in scope, and act over time periods of tens of millions to tens of thousands of years (possibly less). Tectonism of this type is driven primarily by plate-tectonic processes. Extensional and contractional movements accompanying the relative motions of plates cause crustal thinning and thickening and changes in regional thermal regimes, and this leads to regional uplift and subsidence. These stresses and strains are generated primarily at plate margins but, because of the rigidity of plates, they may be transmitted into plate interiors and affect entire continents.

By their very nature these mechanisms of sea-level change are regional, possibly even continental in extent, but cannot be global, because they are driven by processes occurring within or beneath a single plate or by the interaction between two plates. They affect the elevation of the plate itself, rather than the volume of the ocean basins or the water within them. This category of sea-level change is therefore not eustatic. However, as described in this chapter they can simulate eustatic effects through the full range of geological episodicities over wide areas, and it is now thought by many researchers that tectonic mechanisms were responsible for many of the events used to define the global cycle charts of Haq et al. (1987, 1988a). An important additional point is that because the earth

is finite, regional plate-tectonic events, such as ridge reordering or adjustments in rotation vectors, may result in simultaneous kinematic changes elsewhere. It is possible therefore for tectonic episodes to be hemispheric or even global in scope. However, such episodes would take a different form (uplift, subsidence, extension, contraction, tilting, translation) within different plates and even within different regions of a given plate. It is possible therefore that simultaneous events of relative sea-level change occurring over wide areas but of varying magnitude and the same or opposite direction could be genetically related. Possible examples identified by Embry (1993) and Hiroki (1994) are described in Section 11.4.

The Exxon school holds that tectonism is not responsible for the generation of sequence-bounding unconformities. As recently as 1991 Vail et al. (1991, p. 619) stated:

[E]ven in ... tectonically active areas, the ages of sequence boundaries when dated at the minimum hiatus at their correlative unconformities match the age of the global eustatic falls and not the plate tectonic event causing the tectonism. Therefore tectonism may enhance or subdue sequence and systems tract boundaries, but does not create them.

Here Vail and his coworkers place primary emphasis on the dating of the sequence boundaries and their global correlatability, a methodological approach that has guided all their research. Problems of chronostratigraphic dating which bring this approach into serious question are discussed in Part IV.

Other workers examining the sequence-stratigraphic record have reached different conclusions regarding the driving forces for sea-level change. For example, Embry (1990) proposed a set of general guidelines for identifying tectonism as a major control of relative sea-level change. His proposal was based on his examination of the Mesozoic stratigraphic record of Sverdrup Basin, Arctic Canada, a succession up to 9 km thick, which he subdivided into 30 stratigraphic sequences in the million-year frequency range. He suggested, however, that the points have a general applicability. They are as follows:

(1) [T]he sediment source area often varies greatly from one sequence to the next; (2) the sedimentary regime of the basin commonly changed drastically and abruptly across a sequence boundary; (3) faults terminate at sequence boundaries; (4) significant changes in subsidence and uplift patterns within the basin occurred across sequence boundaries; and (5) there were significant differences in the magnitude and the extent of some of the subaerial unconformities recognized on the slowly subsiding margins of the Sverdrup Basin and time equivalent ones recognized by Vail et al. (1977, 1984) in areas of high subsidence.

To these points could be added that truncation of entire sequences beneath sequence boundaries indicates tectonic influence (e.g., Yoshida et al. 1996), and that a sequence architecture consisting of thick clastic wedges cannot be generated by passive sea-level changes (Galloway 1989b).

One of the key areas in the construction of the original Vail curves was the North Sea Basin, and outcrop sections in areas adjacent to the North Sea were used extensively in the revisions of the curves published by Haq et al. (1987, 1988a). It is therefore of considerable significance that the Mesozoic-Cenozoic geology of the region has now been shown to have been strongly affected by almost continous extensional tectonism throughout this period (Hardman and Brooks, 1990), the effects of which appear not to have been accounted for in the Exxon work. Vail and Todd (1981, p. 216–217) stated:

Tilting of beds as a result of fault-block rotation and differential subsidence occurred almost continuously throughout the Jurassic. Unconformity recognition is enhanced by periodic truncation of these tilted strata by lowstand erosion and/or onlap which developed during the subsequent rise. We find no evidence that the regional unconformities are caused by tectonic events within the northern North Sea basins, only that unconformity recognition is enhanced by the truncation and onlap patterns created by tectonic activity.

Most workers today would not agree with this view. Some of the new data from this area, and its implications for sequence stratigraphy, are discussed in Section 11.2.2.

It is simplicity itself to demonstrate that tectonism may be an important factor in the timing of sequence boundaries. A modification of one of the standard Exxon diagrams can be used to illustrate this point (Fig. 11.1). Figure 11.1A shows how rates of subsidence and eustatic sea-level change may be integrated to develop a curve of relative changes in

sea level. Sequence boundaries correspond to lowstands of relative sea level. Figure 11.1B has been constructed by integrating curves for extensional subsidence with a curve of sea-level change showing eustatic cycles with wavelengths of 10^7 years. It is easy to see from this figure that changing the slope or position of the subsidence curve changes the position of lows and highs on the accommodation curve. Given that tectonism is not necessarily the slow, steady, background effect described in the Exxon work, but varies in rate and duration markedly from location to location, within and between basins, it seems unlikely that relative sea-level lows and highs would ever tend to occur at the same times in adjacent basins, let alone on different continents. In fact, it can be stated that if rates of eustatic sea-level change and tectonic subsidence are comparable (of the same order of magnitude) there can be no synchronous record of eustasy in the preserved stratigraphic record. For example, Fig. 11.1B demonstrates that if an Atlantic-type margin undergoes progressive rifting along strike ("unzipping") 10^7-year eustasy does not generate synchronous sea-level highs and lows. Parkinson and Summerhayes (1985) made a similar point.

The rates and magnitudes of various tectonic processes leading to uplift and subsidence are listed in Table 9.1 and are discussed in this chapter. Long-term and short-term rates, and magnitudes of total change, are in fact comparable to the rates and magnitudes of long-term and short-term eustatic sea-level change, which are also shown in Table 9.1. Therefore tectonism cannot be relegated to a background effect of secondary importance. Eustasy and tectonism may be considered as independant curves of uplift and subsidence that go in and out of phase in a largely unpredictable manner. As discussed by Fortuin and de Smet (1991), this means that a careful tectonic analysis must be performed in an area of interest, including the calculation of subsidence histories by the methods of geohistory analysis or backstripping, before any conclusions can be drawn regarding the importance of eustasy. Vail et al. (1991) also referred to the need for detailed "tectono-stratigraphic analysis" and went to considerable lengths to describe how this should be done. However, they did not take the next step, which is to show how the results should be tested against and integrated with the predictions made from the eustatic model of sequence stratigraphy.

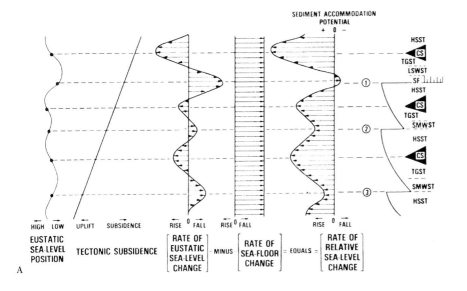

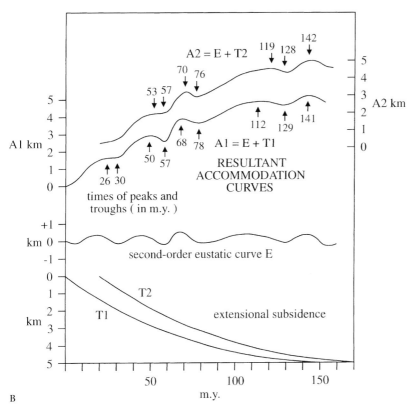

Fig. 11.1. A The standard Exxon diagram illustrating the relationship between eustasy and tectonism. Integrating the two curves produces a curve of relative change in sea level, from which the timing of sequence boundaries can be derived (*events 1–3, right-hand column*). However, changing the shape of the tectonic subsidence curve changes the shape and position of highs and lows in the relative sea-level curve, a point not acknowledged in the Exxon work. This version of the diagram is from Loutit et al. (1988). *CS* Condensed section; *HSST* highstand systems tract; *LSWST* lowstand wedge systems tract; *SF* submarine fan, *SMWST* shelf-margin wedge systems tract; *TGST* transgressive systems tract. **B** Illustration of the shifts in peaks and troughs of accommodation space induced by changes in the rate of subsidence. A simple asymptotic subsidence curve, with rates of subsidence derived from the values given in Table 9.1, is shown in two positions, *T1* and *T2*, shifted 20 m.y. relative to each other. A curve simulating irregular eustasy with a 10^7-year frequency is shown at center, and at *top* the results of integrating this curve with the subsidence curve are shown. Comparison of the two accommodation curves *A1* and *A2* shows that if extensional subsidence is delayed by 20 m.y. but then follows the same pattern, peaks in the accommodation curve occur several million years later, whereas troughs tend to occur somewhat earlier

11.2 Rifting and Thermal Evolution of Divergent Plate Margins

11.2.1 Basic Geophysical Models and Their Implications for Sea-Level Change

The evolution of extensional continental margins is shown diagramatically in Fig. 11.2. There is an initial rapid phase of continental stretching, which may be completed within a few million years, in the simplest case (although some basins, such as the North Sea, undergo protracted or repeated rifting events over tens of million of years). This is the rift phase shown in Fig. 11.2. Deformation typically is brittle at the surface, taking the form of extensional faults. These may consist of repeated graben or half-graben faults, and may be listric, allowing for considerable extension of the crust. Beneath the brittle crust the lithosphere is plastic and extends by stretching. This causes hot asthenospheric material

to rise closer to the surface, resulting in heating and uplifting of the rift margins, as shown in Fig. 11.2. The uplift is equivalent to a relative fall in sea level, the rate and magnitude of which has been modeled by Watts et al. (1982), and is given in Table 9.1. The rift phase is then followed by a much longer cooling and flexural-subsidence phase, which lasts for tens of millions of years. Thermally driven subsidence, accompanied by water and sediment loading, leads to downflexure of the continental margin. The flexural phase involves a much broader area of the continental margin than the rift phase, which accounts for the classic steer's head cross section of extensional-margin basins (Fig. 11.2). Cooling and subsidence is rapid at first, but the rate decreases asymptotically. The result is a subsidence curve that is concave-up (Fig. 11.3).

The basic physical model illustrated in Fig. 11.2 evolved from the work of Walcott, Sleep, Watts, and McKenzie, and is described in more detail in textbooks, such as Miall (1990) and Allen and Allen

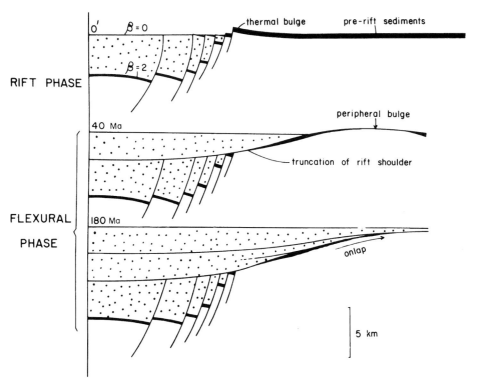

Fig. 11.2. The steer's head or Texas longhorn model of basin geometry. An initial rift phase is generated by rapid crustal stretching, resulting in listric normal faulting and ductile flow of the lower lithosphere, with Airy-type isostatic subsidence. The thermal anomaly generated during this phase causes an uplifted thermal bulge, which typically is eroded subaerially, forming an unconformity.

The thermal anomaly then decays, resulting in relatively slower subsidence. The flexural strength of the lithosphere causes the sediment load to be spread over a wider area than during the first phase of basin subsidence, creating the horns of the model. (Dewey 1982, reproduced by permission of The Geological Society, London)

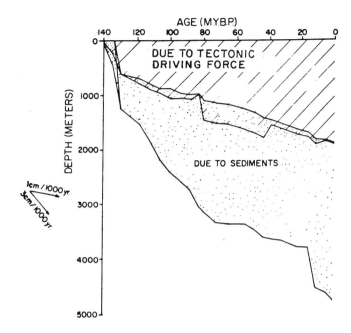

Fig. 11.3. Typical subsidence curve for extensional continental margin. Note the concave-up shape. This curve was developed by backstripping procedures for the Atlantic margin of the United States off New York. Subsidence has been divided into that due to thermal subsidence (the tectonic driving mechanism), and that due to sediment loading. (Steckler and Watts 1978, reproduced by kind permission of Elsevier Science-NL, Sara Burgerhartstraat 25, 1055 KV Amsterdam, The Netherlands)

(1990). The mechanics of the model (heat flow, effects of water and sediment loading, crustal strength and elasticity, flexure) are by now well known and have been simulated using numerical modeling techniques. Stratigraphic and other data are backstripped to reveal tectonic driving mechanisms, and these are fitted to equations of crustal behavior involving subtle modifications of the flexural model to incorporate possible geographical and temporal variations in flexural rigidity, viscous versus elastic behavior, internal heat sources, such as radioactivity, and the thermal-blanketing effects of the sediments, which have low heat conductivity (Watts 1981, 1989; Watts et al. 1982).

In the original models it was assumed that one of the most important attributes of the thermal behavior of the lithosphere is that as it cools its flexural rigidity increases. The incremental depression of the crust that accompanies each additional load of water and sediment therefore becomes progressively wider but shallower (Fig. 11.2). Watts (1981) illustrated two simple models of a continental margin that assume different stratigraphic behavior under similar conditions of gradually increasing flexural rigidity (Fig. 11.4). The upbuilding model is typical of carbonate-dominated environments, whereas the outbuilding model shows the characteristic clinoform progradation of a clastic-dominated environment. In the upbuilding model, the load is added in the same place, and a virtually horizontal shelf develops. Deeper (outboard) sediments become backtilted toward the shoreline. In

the outbuilding model, the load is progressively shifted seaward, and the backtilting effect does not occur. Figure 11.5 is an enlargement of the landward edge of one of these models, showing how the increase in flexural rigidity of the crust with age leads to the spreading of the sediment and water load over progressively wider areas of the crust, causing coastal onlap. Coastal onlap is also a feature of the basic steer's-head model shown in Fig. 11.2.

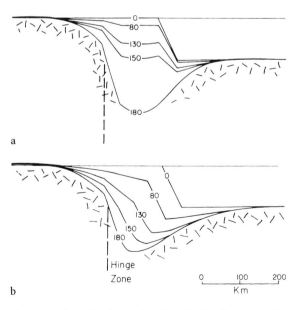

Fig. 11.4. Two simple models for the development of a continental margin, based on **a** upbuilding and **b** outbuilding. (Watts 1981, reproduced by permission)

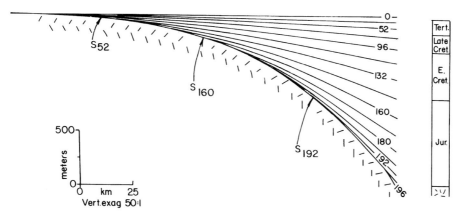

Fig. 11.5. Enlargement of the landward edge of the stratigraphy generated in flexural models, such as those illustrated in Fig. 11.4. Note the progressive onlap of younger sediments onto the coastal plain. (Watts 1981, reproduced by permission)

A comparison of this model with a cross section through an actual continental margin (Fig. 11.6) shows that the model successfully mimics real stratigraphic architecture. This has some very important implications. Coastal onlap is one of the key indicators used in seismic stratigraphic models of sea-level change to postulate a rise in sea level. We can see that there is in fact no need to invoke such a change in sea level. This is a critical point with respect to Exxon's global cycle charts.

More recent work (Watts 1989) has suggested some revisions to the original steer's head model. It is now thought that continental crust does not show a significant increase in rigidity with time, in part because the sediment load inhibits heat loss and therefore slows the process of cooling. Cooling and subsidence of the underlying mantle may be contributary factors in the development of flexural subsidence and coastal onlap, but a more significant effect is the flexural load of the sediment wedge itself. Considerable accommodation space is available for sediment to be deposited by lateral progradation of the continental margin, and Watts (1989) has now shown that this load can account for

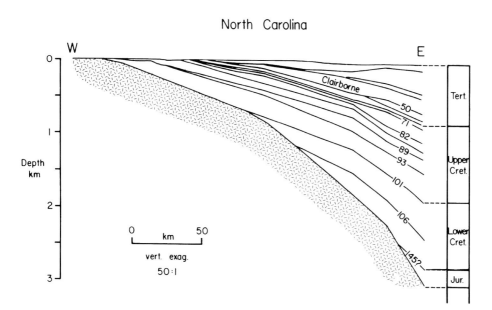

Fig. 11.6. Observed stratigraphy of the Atlantic coastal plain of North Carolina. Note the coastal onlap of Cretaceous strata and compare this with the synthetic model of Fig. 11.5. The significance of this is discussed in the text. (Watts 1981, reproduced by permission)

much of the onlap pattern observed on the Atlantic continental margin and elsewhere.

Bond and Kominz (1984) modeled the shape and subsidence history of the Early Paleozoic divergent margin of southwestern Canada using the techniques of Watts (1981) and compared it with a palinspastically restored cross section through this margin. They found that the flexural model could not account for the great thickness of Lower Paleozoic section actually preserved, nor for the presence of a thin wedge of sediments extending hundreds of kilometers beyond the margin onto the cratonic interior (Fig. 2.20). The thickness problem probably relates to the use of incorrect parameters for crustal thickness and thermal properties and to inadequate information concerning the geometry and subsidence history of the westernmost, now-deformed part of the margin. However, the existence of the extensive cratonic cover beyond the limits of the basin is probably the result of regional or global sea-level rise and consequent transgression. That sea-level change can be deduced from flexural basin models is important. This was one of the first quantitative attempts to document such effects in an ancient continental margin, and was the beginning of the important work by these same researchers (described in Chap. 9) in which regional subsidence curves were used to unravel the various causes of subsidence across cratonic and continental-margin areas of North America during the Paleozoic (see Fig. 9.6).

Summerhayes (1986) pointed out that the Exxon global cycle charts are biased by data from the North Atlantic and from the North Sea and the Gulf of Mexico – both areas bordering the Atlantic Ocean. The question arises whether the results reflect not eustatic sea-level change but regionally synchronous tectonic effects resulting from widespread rifting and flexural subsidence. Summerhayes (1986) noted that many of the Cretaceous-Tertiary cycles on the Vail curve are not present in the Australian stratigraphic record, which appears to confirm this point. As noted above, Watts (1981, 1989) and Watts et al. (1982) demonstrated that coastal onlap could be caused by increases in flexural rigidity or sediment load with time on an extensional continental margin (compare Figs. 11.5 and 11.6), whereas times of low relative sea level may represent times of marginal thermal uplift accompanying rifting events. Watts et al. (1982) constructed a synthetic curve of apparent sea-level change to demonstrate what might be the effects on sea level of a succession of rifting and thermal-

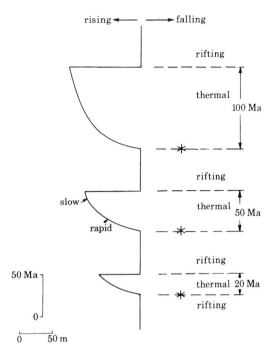

Fig. 11.7. Apparent sea-level changes (coastal onlap and its subsequent downward shift) inferred from the flexural history of an extensional basin, based on the steer's head model of rifting and thermal subsidence. (Watts et al. 1982, reproduced by permission of the Royal Society)

subsidence events bordering a major ocean, such as the Atlantic, and the result is very similar to a Vail curve (Fig. 11.7). They demonstrated that in fact many of the curve segments in Vail et al.'s (1977) original chart can be related to rift-drift events in Atlantic Ocean spreading (Fig. 11.8). Considerable additional stratigraphic information is now available from these areas, and a more complete appraisal of the eustasy-versus-tectonism debate can be attempted, as in Section 11.2.2.

Guidish et al. (1984) pointed out that sea-level changes introduce significant modifications to the shape of subsidence curves compiled by geohistory analysis (backstripping). If curves from several different structural settings could be integrated for successive time slices, eustatic effects should be additive and should therefore show up on the combined, averaged curve. To test this idea, they compiled data from 158 wells around the world and derived a curve showing the global average rate of basement subsidence for the last 300 Ma (Fig. 2.21). One way of interpreting this curve is to argue that sea-level rise produces increased accommodation space for sediment, which should show up as an apparent increase in the basement subsidence rate.

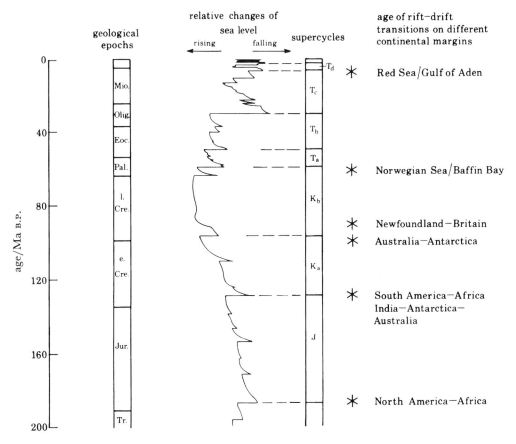

Fig. 11.8. Part of one of the original global cycle charts of Vail et al. (1977), with details of rifting events mainly in the region of the Atlantic Ocean. Note the correspondence of times of rifting with the initiation of cycles of rising relative sea-level. (Watts et al. 1982, reproduced by permission of the Royal Society)

Guidish et al. (1984) plotted a derivative of this curve, the rate-of-subsidence curve, against the Vail curve for the same time period (Fig. 2.22) and claimed an approximate correspondence in the shape of the two curves. They stated:

On Figure [2.22], we see some similarity between Vail et al's eustatic sea-level curve and the master rate of basement subsidence curve. In general, the sea-level low of the Late Tertiary and the sea-level high of the Middle Tertiary are both reflected in the rate of subsidence curve, as are the rises in sea level in the Early to Middle Cretaceous and during the Jurassic. However, in detail the correspondence between the two curves is not as good. In particular, we notice differences between the two during the Late Cretaceous, and correlating individual events on a one-to-one basis is difficult. Unaccountable major differences, on a gross scale, can be seen in the Late Tertiary. However, the major discordance in the Triassic is a reflection of the paucity of data both for ourselves and Vail et al. (1977). Thus, the overall trend of Vail et al's sea-level curve appears to be reproduced in the master rate of subsidence curve, although details are seldom correlatable. (Guidish et al. 1984, p. 171)

I suggest that in fact there are few similarities between these curves. This does not necessarily disprove that eustatic sea-level changes occurred, but might only reflect the crudeness of the techniques or the imprecision of available seismic and well data.

The problem of what ultimately causes synchronous episodes of coastal onlap is a tricky one. Rifting events along the margins of major new oceans are followed by flexural subsidence and consequent local transgression, but are also followed by a period of rapid seafloor spreading, which causes a rise in sea level because of the thermal expansion of spreading centers. The disparity between the Vail and Watts models may therefore be more apparent than real, because both mechanisms for generating coastal onlap tie back to the same original causes, a point made by May and Warme (1987). However, Pitman and Golovchenko (1983) and Parkinson and Summerhayes (1985) examined the dependence of local relative rises and falls in sea level (actual transgressions and regressions) on the balance between eustatic change and local tectonic (thermal) subsidence. They showed that if basins are in different stages of subsidence, having rifted at different times, eustatic effects may be masked to the extent that local sea-level changes may be completely different in each basin. This point was made in the introduction to this chapter, with reference to Fig. 11.1.

The breakup of the supercontinent Pangea, which initiated a long-term cycle of sea-level change, was accompanied by the development of extensional margins bordering the future Atlantic, Indian and Southern Oceans (Fig. 9.11). Segments of extensioal margin were initiated at intervals of tens of millions of year between 250 Ma and 5 Ma (Fig. 9.12), and each underwent a history of flexural subsidence. A complex pattern of tectonically driven (Watts-type) onlap and offlap was therefore superimposed on the long-term eustatic effects caused by this history of sea-floor spreading. It seems unlikely to me that this combination of events could be capable of generating a clear, overriding global signal of sea-level change, although the stratigraphic record of 10^7-year cyclicity (Chap 6) appears to indicate that intercontinental correlation of cratonic sequences has been achieved.

11.2.2 Some Results from the Analysis of Modern Data Sets

A great deal more data have accumulated since the mid-1980s that permit a closer examination of the relationships between regional tectonism, relative sea-level change and eustasy. In particular, the release into the public domain of extensive regional seismic surveys, coupled with an increase in the quantity and precision of biostratigraphic data have permitted much more sophisticated examination of regional histories than was possible before.

A survey of the documentation of the tectonic history of various divergent continental margins based on recent research by individuals not connected (either in an employment sense or philosophically) with the Exxon group is provided in Section 6.3.2. The results indicate litle global, or even regional, correlation between the successions of events with 10^7-year episodicities, and very little correspondence with the Exxon global cycle charts (Fig. 6.15). Most unconformities relate to the regional history of rifting and flexural subsidence. In many cases the regional stratigraphy has been divided into tectonic sequences of 10^7-year duration

and correlated with specific plate-tectonic events in the adjacent ocean. This type of correlation is particularly well illustrated in the case of the Grand Banks of Newfoundland, the development of which reflects a long-continued history of successive rifting and thermal events as the Atlantic Ocean gradually "unzipped" northward, around the promontory now comprising the Grand Banks continental margin (Fig. 11.9; Tankard and Welsink 1987; Welsink and Tankard 1988).

It is worthwhile focusing at this point on the North Atlantic margins and the North Sea basin, as the birthplace of much of the Vail curve and the controversy surrounding it. Extensional continental margins are commonly termed "passive" margins, to contrast them with "active" convergent margins. However, this is misleading. Such margins, including associated rift basins, are typically far from passive. As I wrote in 1986:

They may undergo repeated tectonic disturbance. For example, during the Jurassic, the northern North Sea was affected by extensional faults, wrench faults, diapiric movement of Zechstein salt, volcanic activity, and regional isostatic subsidence (Glennie 1984). To separate the stratigraphic effects of this activity from those due to eustatic sea-level change requires meticulous documentation of local structure and stratigraphic architecture, with an emphasis on the departures from the Vail curves. If Vail and his coworkers have done this, it has not been published. (Miall 1986, p. 136)

Similar comments regarding the importance of tectonism in the North Sea Basin were made by Hallam (1988).

Several attempts at detailed stratigraphic documentation have now been published. Thus Badley et al. (1988) used a large seismic data set to examine the structural evolution of the North Viking Graben, and documented the ongoing nature of extensional tectonism during the Mesozoic. Hiscott et al. (1990) examined the Jurassic-Cretaceous rifting history of the basins of Maritime Canada, Iberia, and the continental margin west of the British Isles. Underhill (1991) reexamined the seismic data base of the Moray Firth, North Sea, on which much of the Jurassic part of the Vail curves was based. Underhill and Partington (1993) demonstrated the occurrence of a significant thermal heating and uplift event in the Middle Jurassic. A book edited by Hardman and Brooks (1990) on the tectonic evolution of the British continental margins provides a wealth of data on this topic. The consensus is that vertical movements accompanying rift faulting and thermal effects were ubiquitous. Badley et al. (1988) stated:

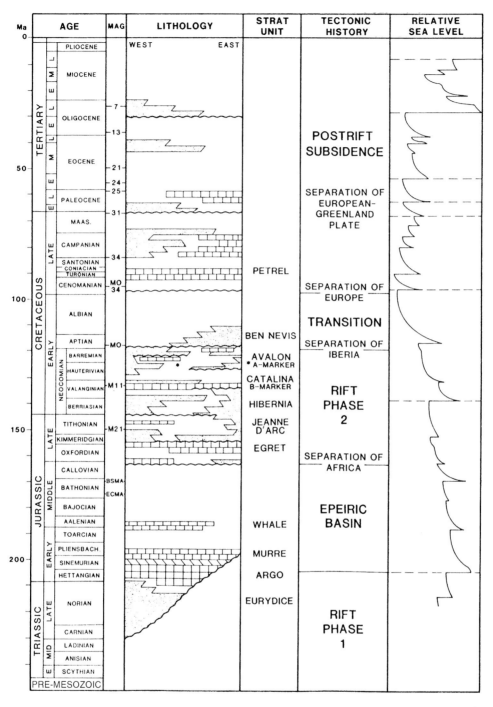

Fig. 11.9. Tectonic sequence analysis of the Jeanne d'Arc Basin, Grand Banks of Newfoundland, showing the five major unconformity-bounded tectono-stratigraphic sequences interpreted on the basis of seismic-stratigraphic analysis, with the Exxon global cycle chart at *right,* for comparison. (Tankard and Welsink 1987, reproduced by permission)

Numerous authors have discussed unconformities and their causes in the northern North Sea (e.g., Vail and Todd 1981; Rawson and Riley 1982; Ziegler 1982; Miall 1986). The structural evolution of the northern Viking Graben presented below shows extension-related tec-

tonics to have been the primary causal mechanism in the development of unconformities in the area. Sea-level changes appear to have played only a secondary role in unconformity development. Two types of tectonically induced unconformity are commonly recognized in extensional basins: regionally developed, angular stratal relationships related to differential block tilting, and local stratigraphic breaks on the crests of fault blocks elevated by footwall uplift, often involving erosion. (Badley et al. 1988, p. 460)

Badley et al. (1988) documented numerous examples of unconformities related to faulting, and wedge-shaped stratigraphic units developed by onlap during postrift thermal subsidence. They recognized rift-thermal subsidence phases during the Triassic-Early Jurassic, Bathonian-Kimmeridgian, Kimmeridgian-Ryazanian, Albian, Turonian-Cenomanian and Late Cretaceous-Eary Tertiary. They pointed out that most of these phases of movement were local, and should not be correlated with regional tectonic pulses, such as the Cimmerian, sub-Hercynian, Austrian or Laramide movements, as has been suggested by some writers (e.g., Ziegler 1982).

The definitive documentation of repeated regional rifting and thermal-subsidence events based on extensive analysis of seismic-reflection data (e.g., Badley et al. 1988), would seem to leave little "accommodation space" for eustasy as a cause for the observed regional unconformities. Indeed, it seems that, fortuitously or not, the Exxon seismic line in the Moray Firth, off northeastern Scotland, one of two used in the major analysis by Vail and Todd (1981) and Vail et al. (1984), did not cross the major faults whose movement, it is now thought, was re-

sponsible for developing the unconformities that they observed. Underhill (1991) showed that the major faults are oriented northeast-southwest in this area, and this is the orientation of the Exxon line! He demonstrated that rotation of fault blocks is a primary cause of unconformable onlap in this area (Fig. 11.10). When seen perpendicular to the fault trend this is revealed by a distinctive fanning of stratigraphic units, caused by block rotation during fault-induced subsidence and sedimentary onlap (Fig. 11.10). However, this is obscure in fault-parallel sections.

Another aspect of stratigraphic architecture that may be of tectonic or eustatic origin or may be unrelated to either is the downlapping pattern developed by the progradation of submarine fans. As pointed out by Miall (1986) onlapping by itself is an unreliable indicator of a relative rise in sea level, unless it can be demonstrated that the facies showing the onlap are coastal in origin, which is difficult to accomplish from seismic-facies data alone. However, most of the strata analyzed by Vail and Todd (1981) are deep-marine fan deposits. Fan progradation may lead to onlap of tilted basement surfaces, and could be interpreted as a product of lowstand fan growth (Sect. 6.1). However, fan rejuvenation may be triggered by fault movement, and fan switching can occur entirely as a result of autogenic channel-switching processes. A quantitative

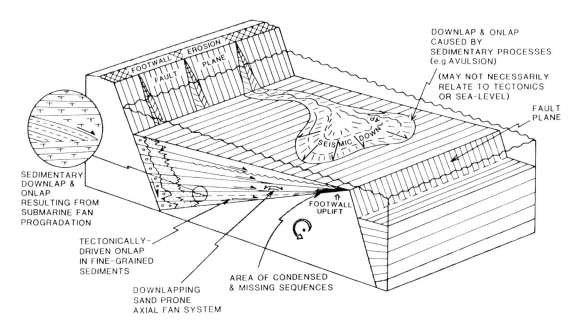

Fig. 11.10. Diagram to illustrate how local sedimentary and tectonic processes can develop stratigraphic architectures that may be interpreted as sequence boundaries. Based on analysis of seismic data from the Inner Moray Firth, off northeastern Scotland. (Underhill 1991, reproduced by permission of Blackwell Scientific Publications)

model of half-graben subsidence that explored the variables of sediment input, block rotation and isostatic response was offered by Roberts et al. (1993), who showed how the model could be applied to specific aspects of the Viking Graben. Prosser (1993) examined the paleogeography of evolving rift basins and the resulting seismic facies and architecture, and proposed a series of models of tectonic systems tracts linked to stages in rift evolution.

Based on a more complete understanding of regional tectonics and stratigraphic architecture than was available to Vail and Todd (1981), Underhill (1991) argued that the Inner Moray Firth sequence data could not be used to verify a global sea-level curve, thus undermining one of the original foundations of the Late Jurassic portion of the Vail et al. (1977) curves. Boldy and Brealy (1990) developed a similar view of the ubiquity of extensional tectonism in the Outer Moray Firth, and, together with the synthesis by Badley et al. (1988) dealing with the nearby Viking Graben, the consensus is clearly that the entire northern North Sea area was dominated by the effects of extensional tectonism. In retrospect it now seems to have been a poor region to have chosen as a basis for the investigation of eustasy.

Underhill and Partington (1993) demonstrated that the intra-Aalenian "Middle Cimmerian" unconformity in the North Sea basin, which corresponds to the "first-order" 177-Ma sequence boundary in the Exxon global cycle chart of Haq et al. (1987, 1988a), is also a regional event. Underhill and Partington (1993) pointed out that this event in the Exxon chart was based exclusively on stratigraphic sections within the North Sea Basin, and must therefore be questioned as a "global" event. Careful regional mapping of marker beds, particularly maximum-flooding surfaces, demonstrated that the unconformity developed as a result of subaerial erosion following thermal doming centerd on a triple-point junction in the central North Sea (Fig. 11.11). Uplift resulted in removal of Jurassic strata and exposure of Triassic or older deposits at the center of the uplift, with erosion to shallower levels extending out over a distance of at least 500 km (Fig. 11.11). Underhill and Partington (1993) concluded that a transient plume developed at the point from which now radiates the Viking and Central Graben and the Moray Firth Basin. The uplift rapidly decayed during the later Aalenian, leading to progressive onlap of the unconformity.

A detailed analysis of the sequence stratigraphy and of the tectonic-eustatic control of the Jurassic-Cretaceous succession in eastern Greenland was reported by Surlyk (1990, 1991). Rift basins there underwent a similar type of history of repeated rifting and thermal-flexural subsidence to that in the North Sea Basin, and Surlyk reached similar conclusions regarding the origin of the sequence of relative sea-level changes there. Regarding the Wollaston Forland Embayment he stated:

A major phase of rotational block-faulting was initiated in the region in the Middle Volgian, lasted throughout the Late Volgian-Ryazanian, and gradually faded out in Valanginian time.... The tectonic overprint is so strong for this period, that it is a major challenge to separate the tectonic, eustatic, and regional sea-level signals. The Middle Volgian-Valanginian time interval is characterized by major fluctuations in the eustatic sea-level curve of Haq et al. (1987). This curve is, however, strongly biased towards the North Sea where the same tectonic phase occurred; the apparent sea-level changes are probably controlled mainly by regional tectonics (cf. Hallam 1988). (Surlyk 1990, p. 80)

Elsewhere Surlyk stated:

Substituting tectonism with sea-level change as the dominating controlling factor on development of the Wollaston Forland Group is too simplistic. The basic aspects of the system in terms of geometry, environment, and processes were undoubtedly controlled by tectonism. The large-scale cyclicity probably is controlled to a large extent by sea level rises. However, I suggest that the sea-level fluctuations were controlled by regional tectonism and that each major episode of faulting is associated with a rapid relative sea-level fall followed by a longer sea-level rise, reflecting increasing subsidence after rifting. This pattern would explain the presence and apparent contemporaneity of the numerous high-amplitude, short-wavelength fluctuations in relative sea level in the North Atlantic region during the tectonically very active interval spanning the Jurassic-Cretaceous boundary. (Surlyk 1991, pp. 1483–1485)

A comparative analysis of the rift basins flanking the North Atlantic region (Maritime Canada, Iberia, western British continental shelf) by Hiscott et al. (1990) demonstrated, again, the importance of local tectonism. The authors attempted to relate their sequence record to the global cycle chart of Haq et al. (1987, 1988a), but the documented record of sequence boundaries and varying subsidence rates showed few inter-basin correlations, and the authors appealed to various mechanisms of rift faulting, thermal doming, postrift thermal subsidence, and transtensional faulting to explain the stratigraphy of each basin. Coupled with this was the suggestion that intraplate stress (Sect. 11.4) would have led to the transmission of many of these stress events to adjacent basins.

Janssen et al. (1995) showed by backstripping analysis of over 200 well sections that subsidence

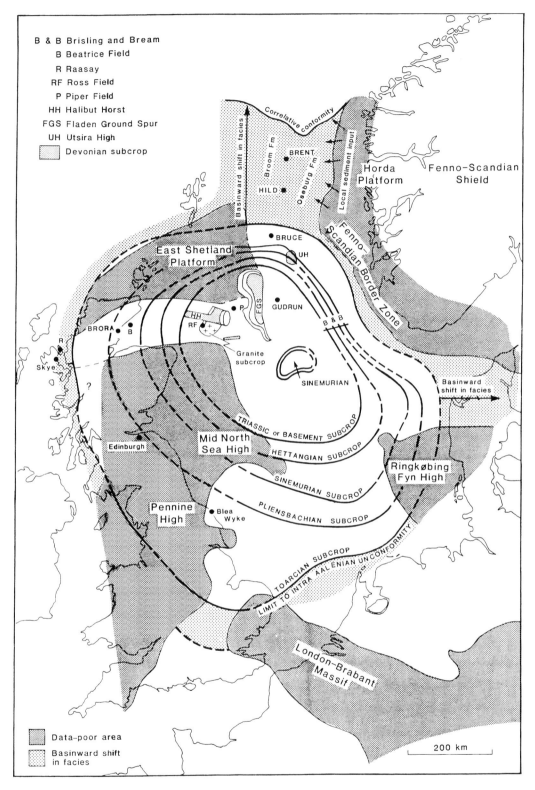

Fig. 11.11. Subcrop map of the strata beneath the Middle Cimmerian unconformity of the North Sea basin. Maximum removal of strata below this Middle Aalenian surface occurs at the junction of the Viking and Central Grabens and the Moray Firth Basin. Uplift radiated outward over a distance of about 500 km. Beyond this, facies underwent a significant basinward shift. (Underhill and Partington 1993, reproduced by permission)

patterns in rifted basins within and on the flanks of Africa can be related to changing intraplate stress patterns as the oceans surrounding Africa opened between Late Jurassic time and the present.

Loup and Wildi (1994) reported on a subsidence analysis of the Paris Basin and a comparison with other northwestern European basins. None show the simple concave-up subsidence-versus-time curve that would be expected from a simple extensional basin (e.g., Fig. 11.3). Most are characterized by short concave segments, with periods of slow or no subsidence, intervals of accelerated subsidence, and intervals of uplift. Few of the events are synchronous over wide areas of northwestern Europe. These results are interpreted as the product of rifting, transtension, and basin inversion resulting from regional plate-tectonic movements, and subcrustal processes, including mantle plumes, together with intraplate stresses resulting from all these processes.

Submarine erosion of continental margins reduces the sediment load and allows for flexural rebound. Stratigraphically the result may resemble a fall in relative sea level. McGinnis et al. (1993) examined the supposed Late Eocene-Early Oligocene global fall in sea level, which has long been attributed to the development of an Antarctic ice sheet. They suggested that one important result of a global climatic cooling (which undoubtedly occurred in the Early Cenozoic; see Fig. 10.14) would have been enhancement of latitudinal thermal gradients, leading to strengthening of bottom currents and increased global submarine erosion. The amount of eustatic fall associated with this cooling event therefore may have been significantly exaggerated.

11.3 Tectonism on Convergent Plate Margins and in Collision Zones

11.3.1 Magmatic Arcs and Subduction

It is a truism to state that areas of convergent tectonism are areas of active tectonics. Several distinct processes lead to local and regional relative changes in sea level, which may be recorded as transgressive and regressive sedimentary cycles in the major basins overlying and flanking the zone of convergence.

In and adjacent to arcs magmatic intrusion results in regional doming over areas a few tens of kilometers across. Eruption is followed by deflation. The result is a lowering of relative sea level during the filling of the magma chamber, followed by a rise accompanying deflation of the chamber. Sloan and Williams (1991) documented a suite of transgressive-regressive cycles in the Devonian of western Ireland which they interpreted as the product of this process. Regressive offshore-barrier sequences a few tens of meters thick developed during the inflation of magma chambers and consequent uplift. The sequences are capped by volcanic intervals 1–190 m thick, recording the eruptions, and the next sedimentary units are deeper water offshore deposits indicating a rise in relative sea level following deflation and collapse of the magma chamber. The age and duration of these sequences is not clear, but a 10^4–10^5-year episodicity seems probable.

A study of an Upper Cretaceous volcaniclastic forearc-basin succession in Baja California led Fulford and Busby (1993) to conclude that tectonic basin subsidence and source-area uplift were by far the most important controls on the architecture of the basin fill. The overall succession indicates retrogradation, suggesting rapid basin subsidence along bounding faults. However, the succession includes a coarsening-upward sequence that contains a significant change in sandstone composition relative to older parts of the succession, and this was interpreted as indicating source-area uplift and unroofing, coupled with an interpreted reduction in the fault-controlled subsidence of the basin.

The long-term subsidence (and sea-level) history of forearc basins depends on the evolution of the subduction zone. Subduction angles change in response to changes in the age of the subducting oceanic crust, and this leads to changes between extensional and contractional arc types (Dewey 1980). For example, Moxon and Graham (1987) showed how the outer margin of the Great Valley forearc basin of California (the side distant from the arc) underwent uplift as the subduction angle decreased in the Late Cretaceous. The arc itself retreated eastward, away from the subduction zone, and the arc flank, underlying the inner edge of the forearc basin, subsided asymptotically as the arc magma cooled. The subsidence pattern that developed in the overlying sedimentary basin is similar to that of extensional basins (concave up, cf. Fig. 11.3). The rate and magnitude of the regional, tectonically driven relative sea-level changes in this basin are comparable to those of eustatic changes of 10^7-year type.

Forearc regions of arcs are also affected by the nature of the material undergoing subduction. The entry of seamounts or aseismic ridges into the subduction zone results in local uplift. Forearc basins may therefore be characterized by cycles of uplift followed by extensional collapse. This process characterizes much of the western coast of South America. Tide-guage records over the last 30 years record rapid neotectonic movements that, if continued for geologically significant periods of time, could readily account for the development of high-frequency sequences over 10^4–10^5-year periods (Fig. 11.12). A network of extensional faults is present along much of the coast, and accommodates extensional movements during the process of subduction of an irregular oceanic plate. Flint et al. (1991, p. 101) stated:

The fore-arc uplift – collapse process could result in a coupled lowering/raising of relative sea level over relatively short periods (c. 100,000 years). Furthermore, as the above studies both show that negligible accretion/obduction occur during seamount chain/ridge subduction, the inherent morphological/structural irregularities in

these structures could result in cycles of minor fore-arc uplift/subsidence at such 100,000 – 200,00 year frequencies. This tectonic mechanism could thus produce relative sea-level changes at a frequency equal to glacio-eustatic processes. These tectonically driven cycles are restricted to arc-related sedimentary basins: they will not be of global extent nor globally synchronous.

The development of areally extensive tectonic sequences in southern Central America is discussed briefly in Sections 6.3.2 and 7.4. The formation of regional unconformities that can be traced for distances of 900 km was attributed by Seyfried et al. (1991) to intraplate stresses (Sect. 11.4) reflecting tectonic adjustments during subduction.

Some authors claim to have observed or demonstrated a correlation between episodes of magmatic activity and periods of rapid sea-floor spreading and subduction. Winsemann and Seyfried (1991), who studied forearc sedimentation on the western coast of Nicaragua, stated "the most striking feature is the almost simultaneity of tectonic/volcanic activity and second order cycles of eustatic sea-level fluctuations during Late Cretaceous, Paleocene, and Middle–Late Eocene times." They go on to say:

Late Cretaceous to Late Eocene drift rates calculated from the North Atlantic Ocean correlate closely with global orogenic phases which in turn coincide with crustal movements in Central America. These episodes of major tectonic/volcanic activity occurred at 80–75, 63, 55–53, and 42–38 Ma and agree with major tectonic and volcanic pulses in Cenral America.... This corroborates that global tectonic processes play an important role in generating major sea-level fluctuations and hence in the formation of depositional sequences.

There is a considerable danger of the self-fulfilling prophecy based on circular reasoning in this kind of correlation. Much depends on the accuracy and precision of the dating of the magmatic episodes and the stratigraphic successions on which these correlations are based. Examples of "forced" and unconvincing correlations with the global cycle chart in forearc basins are referred to in Section 7.4. The nature of the problem of correlation and geological time is discussed in detail in Part IV.

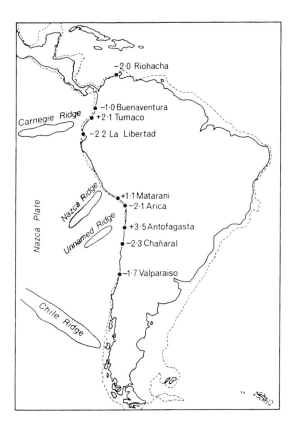

Fig. 11.12. Mean changes in relative sea level (mm/year) during the last 30 years along the western coast of South America, based on tide-guage records. The Andean margin is mostly undergoing subsidence (negative values), except where aseismic ridges have entered the subduction zone. Buoyancy of the subducting ridge is leading to relative uplift of the adjacent coastline. (Flint et al. 1991)

11.3.2 Tectonism Versus Eustasy in Foreland Basins

It has long been known that foreland-basin strata are characterized by the intertonguing of marine and nonmarine strata, indicating episodic regression and transgression. For example, Weimer (1960) recognized four regional regressive-transgressive cycles in his classic work on the Upper Cretaceous of the Western Interior of the United States, plus many minor events that he did not, at that time, attempt to correlate. More recent work on these cycles is described and illustrated in Sections 7.2 and 8.6. Many nonmarine (molasse) successions consist of several separate wedges of coarse detritus that formed by episodic progradation into the basin (Miall 1978; Van Houten 1981; Blair and Bilodeau 1988).

The question of causality of these cycles is particularly difficult to resolve in foreland basins, which, by their very nature, owe their origins to regional tectonic activity. The processes of continental collision, terrane accretion, nappe migration, thrust movement, and imbricate fault propagation, thicken and flexurally load the crust while generating tectonic highlands. Erosional unroofing of the highlands subsequently causes isostatic uplift. These processes lead to episodic cycles of relative changes in sea level on time scales of 10^4–10^7 years. It has long been accepted that these cycles are responsible for large-scale molasse pulses (Miall 1978; Van Houten 1981). The numerical models of Jordan (1981) and Beaumont (1981) provide the necessary theoretical background for explaining subsidence in terms of tectonism. Allen et al. (1986) and Cross (1986) plotted subsidence curves that illustrated the typical subsidence pattern of foreland basins. Subsidence rates tend to increase with time, and the curves commonly are characterized by one or more sharp inflexion points where the rate of subsidence rapidly increases as a result of a specific thrust-loading event (Fig. 11.13). Subsidence rates may also vary over a 10^6–10^7-year time scale because of heterogeneities in the underlying basement. The rate of flexural response to loading is governed largely by the strength and elasticity of the underlying crust, and if this varies laterally, subsidence rates vary as different parts of the crust are loaded during contractional movements, including fold-thrust development and subduction (Waschbusch and Royden 1992).

It has commonly been assumed that wedges of coarse sediment prograding from a basin margin are "syntectonic," and in many earlier studies the dating of such wedges has been used to infer the timing of major orogenic episodes (see summary and references in Miall 1981; Rust and Koster 1984). It can now been seen that this interpretation is simplistic. In fact, the relationship between tectonism and sedimentation is complex, and depends on the balance between a range of controls, including sediment supply and sediment type, and the con-

figuration and rigidity of the flexed basement. Some recent studies have demonstrated that tectonism does not necessarily coincide with the progradation of wedges of coarse sediment, but may precede such progradation by a significant period of time (Blair and Bilodeau 1988; Jordan et al. 1988; Heller et al. 1988). In other cases, sediment input and progradation rates are adequate to keep pace with thrust-sheet loading (Sinclair et al. 1991). Tectonism, in the form of basin-margin fault movement, leads to increased rates of subsidence, particularly at the edge of the basin. Foreland basins, in particular, are developed as a result of crustal loading by overriding thrust sheets, and are characterized by syntectonic subsidence of the proximal regions of the basin. The immediate basinal response may be marine or lacustrine incursions, with ponding of coarse debris against the basin margin. Tectonism eventually leads to increased basin-margin relief and hence to a rejuvenation of the supply of coarse sediment, but this process take time, and it has been suggested that clastic-wedge progradation is largely a posttectonic phenomenon (Blair and Bilodeau 1988; Heller et al. 1988). In this case, areally extensive coarse fluvial deposits may not be deposited until posttectonic uplift of the basin takes place, driven by isostatic rebound following erosional unroofing of the fold-thrust belt. Following this scenario, increased rates of subsidence accompanying tectonism would be recorded by increased rates of sedimentation of fine-grained deposits, particularly shallow-marine and lacustrine sediments, at the basin margins, and at this time river systems may actually flow *towards* the fold belt (Burbank et al. 1988). Only in the most fault-proximal regions is tectonic activity likely to be recorded by the rapid development of a clastic wedge. Heller and Paola (1992) referred to this model of coarse sedimentation as *antitectonic*. There has been considerable debate in the literature regarding the validity of the syntectonic and antitectonic models with respect to particular foreland basins (see summary in Miall 1996, Sect. 11.3.6).

Posamentier and Allen (1993a) developed a useful model for the sequence stratigraphy of foreland ramp-type basins, in which the relationship between eustatic sea-level change and flexural subsidence was explored (Fig. 11.14). They suggested that in many foreland basins there may be a proximal region close to the fold-thrust belt where subsidence is alway faster than the rate of eustatic sea-level change. In this area eustasy does not lead to the generation of subaerial (type 1) uncon-

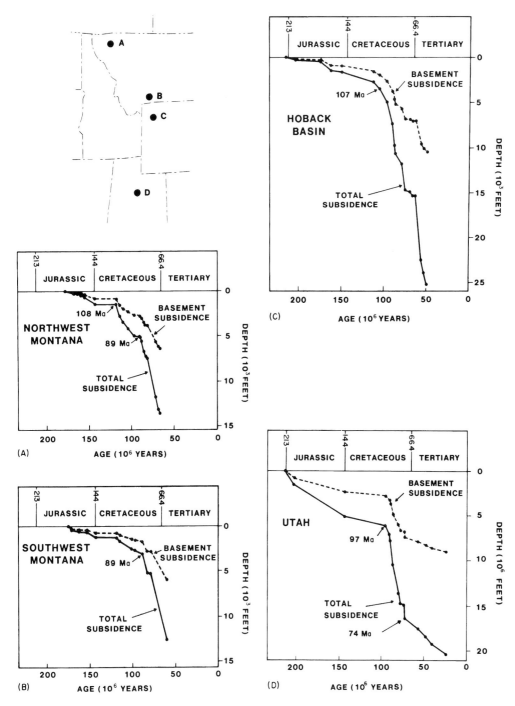

Fig. 11.13. Examples (A–D) of subsidence curves for the Rocky Mountain foreland basin, showing the characteristic convex-up shape, with inflexion points corresponding to thrust-loading events. This shape contrasts with the concave-up shape of subsidence curves derived from extensional basins, such as is shown in Fig. 11.3. (Cross 1986)

formities. Subsidence and sedimentation are more or less continuous, but may vary in rate. This area they term zone A. In more distal regions, the rate of flexural subsidence is less, and in this area, termed zone B, eustatic fall may outpace flexural sub-

sidence, and a type 1 unconformity is the result. The distal margin of the basin is marked by a tectonic hinge, beyond which vertical motion of the forebulge is in the opposite direction to that of the basin. The location within the basin where the rate

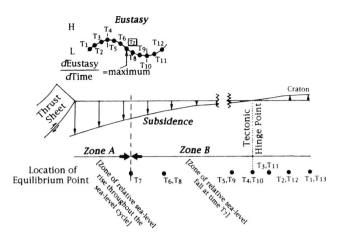

Fig. 11.14. The foreland ramp model of Posamentier and Allen (1993a). The rate of flexural subsidence decreases away from the thrust sheet. The equilibrium point is the point where the rate of flexural subsidence and the rate of sea-level change are equal. It moves across the basin as sea level rises and falls, as shown by the position of points $T1$–$T13$. The location of the most proximal position of the equilibrium point defines the boundary between *zones A* and *B*

of tectonic subsidence and the rate of eustatic fall are equal is termed the equilibrium point. It moves basinward of the tectonic hinge during times of falling sea level, and toward the forebulge during times of rising sea level. The boundary between zone A and zone B is defined as the most proximal position reached by the equilibrium point, which it attains at the inflection point during the falling leg of the sea-level curve (Fig. 11.14). An application of this model to the Western Interior Basin of North America is described in the next section.

Some retroarc foreland basins are wider and deeper than can be acounted for by the flexural-loading model. Supracrustal loading typically can account for a basin up to about 400 km wide, whereas some basins are double this width. The Alberta basin, for example, is more than 1000 km wide in its central part (from the fold-thrust belt to the edge of the Canadian Shield), occupying most of Alberta, plus the southern parts of the adjacent provinces of Saskatchewan and Manitoba. Beaumont (1982) suggested that a regional basinward tilting of the crust occurred at the time of the flexural loading, and proposed that the tilting was a response of the lithosphere to convective mantle flow coupled to a subduction zone. A pattern of secondary mantle flow in the overriding lithospheric plate was suggested by Toksöz and Bird (1977). Flow takes place toward the subduction zone and is drawn down parallel to the cold, descending oceanic plate. The crust is tilted toward the subduction zone by the mechanical drag effects of the downgoing current. When subduction ceases, buoyancy forces, coupled with erosional unroofing of the accreted terrane, combine to reverse the tilting process, leading to uplift of the basin. Mitrovica and Jarvis (1985) and Mitrovica et al. (1989) modeled this process, and showed that the width of

the crust affected by the tilting increases as the subduction angle decreases. At subduction angles of $20°$, the tilt effect extends for more than 1500 km from the thrust front. The degree of tilting is also affected by the flexural rigidity of the overriding lithosphere. The cessation of crustal shortening accompanies the termination of subduction and its associated mantle convection currents, so that the mechanical down-drag effect ceases, and the basin then tends to rebound (Mitrovica et al. 1989). The entire cycle takes a few tens of million years to complete, and would generate a cycle of relative sea-level change on a regional scale. This dependence of basin architecture and subsidence history on mantle thermal processes is an example of dynamic topography, a topic discussed in Section 9.3.2.

Mantle thermal processes and regional flexural loading generate basement movements over a time scale of 10^6–10^7 years. However, the crustal response to contractional movements may generate flexural loading and vertical movements of the basin over much shorter time scales. Individual earthquakes, which may cause vertical movements of several meters, may have recurrence intervals of 10^1–10^2 years, and longer term cycles of movement, over the 10^3–10^5-year time band, may be related to the loading and uplift of large detached crustal sheets (terranes, nappes, thrust complexes) and their individual components, including imbricate slices (Fig. 11.15). Intraplate (in-plane) stress may transmit the effects of such tectonism throughout the basin and beyond (Sect. 11.4). Ancient foreland basins contain many examples of stratigraphic sequences formed over 10^4–10^6-year time scales, and an examination of these processes is therefore essential. The relationship between the time scale of sequences and possible tectonic mechanisms was examined in Pyrenean foreland basins by Dera-

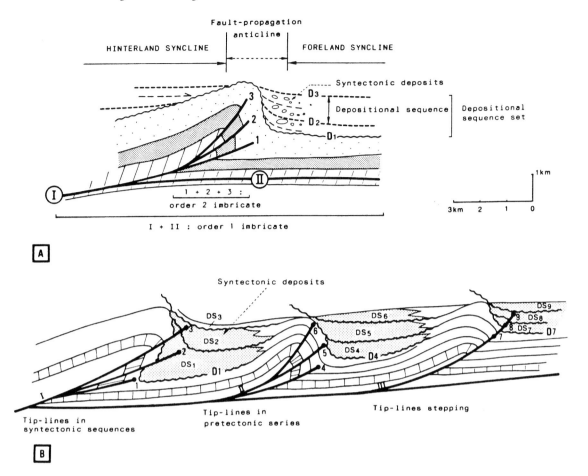

Fig. 11.15. The relationship between the development of a fold-thrust belt and the stratigraphy of the adjacent foreland basin. Unconformities, numbered *D1–D9*, develop as imbricate thrust slices develop. Uplift of each slice is recorded by a corresponding numbered unconformity in the basin. (Deramond et al. 1993, reproduced by permission of The Geological Society, London)

mond et al. (1993), from whose paper Table 11.1 is adapted.

It is becoming possible to suggest approximate correlations of individual clastic pulses with specific tectonic events, such as times of terrane accretion (Cant and Stockmal 1989; Stockmal et al. 1992), or thrust faulting events (Burbank and Raynolds 1988; Deramond et al. 1993). More difficult to resolve is the question of causality for intertonguing fine-grained marine and nonmarine clastic successions far from the sediment source, where the influence of basin-margin tectonism is not obvious. The relative importance of tectonism and eustasy in generating these sequences is by no means clear, particularly where the occurrence of contemporaneous eustasy cannot be independently demonstrated.

These problems can be exemplified by a brief review of two North American basins where the stratigraphic record is particularly well know, the Cretaceous cyclicity of the North American Western Interior Seaway, and the Late Paleozoic succession of the Appalachian foreland basin. Pyrenean basins and the Himalayan foredeep are also discussed briefly.

11.3.2.1 The North American Western Interior Basin

The sequence stratigraphy of this basin is described and illustrated in Section 7.2 (Figs. 7.17–7.25), and Section 8.6 (Figs. 8.45–8.60) Ten major transgressive-regressive cycles have been recognized in the Rocky Mountain foreland basin (Fig. 7.20; Weimer 1960, 1986; Kauffman 1984). Transgressions are characterized by the development of thick and areally extensive mudstone units (e.g., Mancos Shale), and by fine-grained limestones (including chalk) in areas distant from sediment sources (e.g.,

Table 11.1. The relationship between tectonic processes and stratigraphic signatures in foreland basins, at different time scales

Duration (m.y.)	Scale	Tectonic process	Stratigraphic signature
>50	Entire tectonic belt	Regional flexural loading, imbricate stacking	Regional foredeep basin
10–50	Regional	Terrane docking and accretion	multiple "molasse" pulses
10–50	Regional	Effects of basement heterogeneities during crustal shortening	Local variations in subsidence rate; may lead to local transgressions/regressions
>5	Regional	Fault-propagation anticline and foreland syncline	Subbasin filled by sequence sets bounded by major enhanced unconformities
5–0.5	Local	Thrust overstep branches developing inside fault-propagation anticline	Enhanced sequence boundaries; structural truncation and rotation; decreasing upward dips; sharp onlaps; thick lowstands, syntectonic facies
<0.5	Local	Movement of individual thrust plates; normal listric faults; minor folds	Depositional systems and bedsets geometrically controlled by tectonism and bounded by unconformable bedding-plane surfaces; maximum flooding surfaces superimposed on growthfault scarps; shelf-perched lowstand deposits.

This table was adapted mainly from Deramond et al. (1993), with additional data from Waschbusch and Royden (1992), and Stockmal et al. (1992).

Niobrara Formation). Regressions gave rise to extensive clastic wedges, in which nonmarine sandstones and conglomerates pass basinward into shoreline and shelf sandbodies. The Indianola and Mesaverde groups of Utah and Colorado are good examples of major regressive stratigraphic packages, which contain numerous minor transgressive-regressive cycles nested within them (Fig. 7.19; Weimer 1960; Fouch et al. 1983; Lawton 1986a,b; Swift et al. 1987; Van Wagoner et al. 1990, 1991; Olsen et al. 1995; Yoshida et al. 1996).

Are the broad patterns of sedimentation in the Western Interior Basin the result of regional flexural subsidence or eustatic sea-level change? Recent work by Pang and Nummedal (1995) has demonstrated the importance of basement heterogeneity as a control on subsidence patterns, and thereby confirmed the importance of flexural subsidence as a primary control of basin architecture. Pang and Nummedal (1995) constructed flexurally back stripped subsidence profiles for six transects through Upper Cretaceous strata in the Western Interior Basin between Montana and New Mexico (Figs. 11.16, 11.17). Variations in subsidence patterns reveal two important tectonic controls: regional differences in basement rigidity, the effects of which were discussed by Waschbusch and Royden (1992; see previous section), and the presence of

"buttresses" in the basement that slowed or prevented crustal shortening during thrust movements, resulting in smaller flexural effects. The second point is exemplified by profiles (a) in Montana, (b) in Wyoming, and (f) in Utah-Colorado (Fig. 11.17), along which the rate of subsidence during the Late Cretaceous did not vary much, indicating very weak flexural effects. Reactivated basement structures and the presence of rigid basement blocks are cited as the reasons. Note also the effect of the Douglas Creek Arch (DCA in Fig. 11.17, profile e), near the Utah-Colorado border.

Stockmal et al. (1992) suggested a relationship between the docking history of terranes on the western coast of Canada, and the generation of clastic pulses in the foreland basin in Alberta (Fig. 11.18). The accretion process caused lithospheric delamination, which resulted in the terranes being emplaced onto the continental margin above a major detachment surface. This increased their flexural reach, and led to the generation of multiple pulses of contractional thrusting and flexural subsidence in the foreland basin, some hundreds of kilometers inboard from the continental margin.

Some of the cycles with 10^6-year periodicities in the Western Interior Basin may be global in origin, and related to eustatic changes in sea level. Partial correlation between the sea-level curve for the

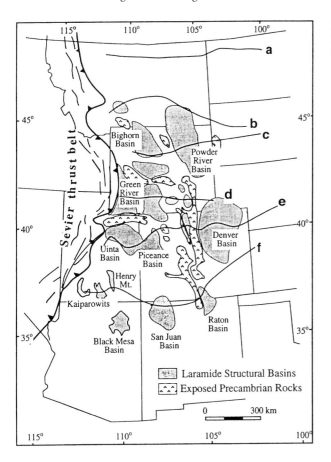

Fig. 11.16. Location of transects *a-f* across the Western Interior Basin of the Rocky Mountain states, shown in Fig. 11.17. (Pang and Nummedal 1995)

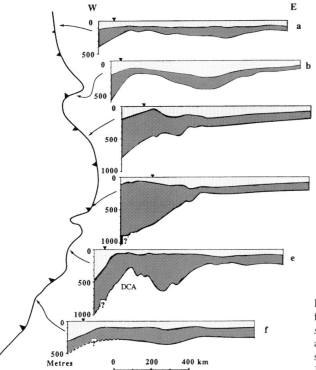

Fig. 11.17. Flexurally backstripped subsidence profiles along the transects shown in Fig. 11.16. *Light shading* indicates subsidence during the 97 (94 on **e** and **f**) to 90-Ma interval, *dark shading* shows subsidence during 90–80 Ma. (Pang and Nummedal 1995)

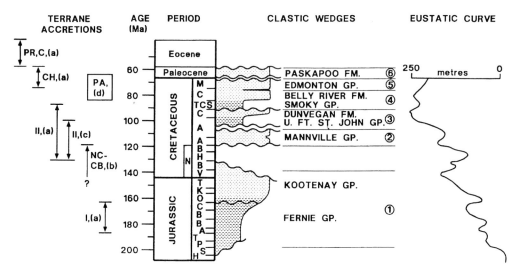

Fig. 11.18. Clastic pulses *1–6* in the Alberta basin, plotted against the global cycle curve and the estimated times of terrane accretion on the Pacific margin of North America. (Stockmal et al. 1992, reproduced by permission)

western United States and that for northern Europe has been claimed (Hancock and Kauffman 1979; Weimer 1986), although correlations are bedeviled by the difficulties in reconciling various chronostratigraphic time scales. Arguments regarding the validity of this type of correlation, and the precision of the dating on which such correlation depends, are considered in Part IV. Kauffman (1984) claimed that there is a consistent correlation among transgression, thrusting (in Wyoming and Utah), and volcanism in the Western Interior (Fig. 11.19). He also claimed that his 10^6-year cycles are correlatable between North America and Europe (Hancock and Kauffman 1979). The regional tectonic associations suggest that the cycles of sea-level change are regional in origin, related to episodic loading in the foreland fold-thrust belt and consequent basin subsidence effects. However, if the cycles can be correlated into Europe this would imply eustasy as the main control. All the tectonic events in the Cordillera are presumably tied to convergent plate movements between the paleo-Pacific Ocean and the North American continent. Yet Kauffman argued that changes in rates of seafloor spreading, which caused episodic tectonism, were also widespread enough to generate cycles of eustatic sea-level change, as demonstrated by the intercontinental correlation of his cycles in the Western Interior. One possibility is that 10^6-year cyclicity represents the effects of regional plate rifting and convergence superimposed on 10^7-year cycles of ridge length and volume changes. However, the initial correlations between stratigraphic cyclicity,

volcanism and thrust faulting on which these arguments are based (Fig. 11.19) seem weak.

There is no doubt that some of the transgressions are contemporaneous with thrust-faulting episodes within the Sevier orogen of Utah, possibly including episodes in the Albian, Santonian, and Maastrichtian (compare Kauffman 1984; Lawton 1986a,b; Weimer 1986). Other thrusting events within the Sevier orogen are not clearly correlated with regional changes in sea level in the basin, but are correlated with the development of major clastic wedges that prograded across the basin margins. Villien and Kligfield (1986), Lawton (1986a,b), and DeCelles et al. (1995) indicated a link between thrusting and clastic-wedge formation in Utah between the Middle Albian and the Late Eocene, although it is not possible to provide the tight correlation between individual tectonic and stratigraphic events that is now available for other foreland basins, such as parts of the Himalayan foredeep and Pyrenean basins (Sect. 11.3.2.3), and on these grounds we cannot yet distinguish between the syntectonic and antitectonic model for all the units in this clastic wedge.

Are individual clastic tongues within major wedges (those discussed in Sect. 8.6) tectonic or eustatic in origin? There is no doubting an overriding tectonic control of sedimentation for many of these successions. Paleocurrent and petrographic evidence indicate shifting sediment sources and changes in regional paleoslope during deposition of the Mesaverde Group and associated units in Utah. Lawton (1986a,b) documented unroofing of in-

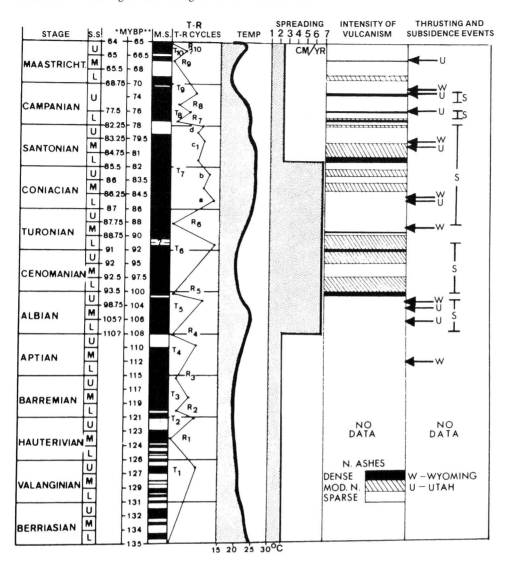

Fig. 11.19. Suggested correlation of 10^6-year stratigraphic cycles in the Rocky Mountains foreland basin with episodes of volcanism and thrust faulting. Spreading rates refer to average rates in the Pacific and Atlantic oceans. (Kauffman 1984, reproduced by permission)

trabasin uplifts from which early basin-fill sediments were cannibalized, and changes in dispersal patterns related to basin tilting (Fig. 11.20). The volume of sediment within the wedge is too great to have been controlled by passive changes in sea level (see also Galloway 1989b). However, the question remains whether tectonically induced sediment input was modulated by sea-level control, leading to high-frequency (10^4–10^5-year) cyclicity along the fringes of the clastic wedge (such as in the Book Cliffs of Utah: see Swift et al. 1987; Van Wagoner et al. 1990). Posamentier and Vail (1988) argued that fluvial coastal-plain progradation is switched on during initial sea-level fall after a time of highstand,

as a result of the lateral shift in stream profiles and the creation of sedimentary accommodation space. However, Miall (1991a) argued that this concept is faulty on several grounds (e.g., base-level fall normally leads to incision), and maintained that considerations of sediment supply, driven by tectonism, are the major causes of coastal-plain progradation (see Sect. 15.3.1 and a more extended discussion of this problem in Miall 1996).

The sequence architecture of the Castlegate Sandstone and equivalent units in the Book Cliffs of Utah (Fig. 8.58) was interpreted by Yoshida et al. (1996) in terms of the simultaneous action of two types of tectonic control acting over different time

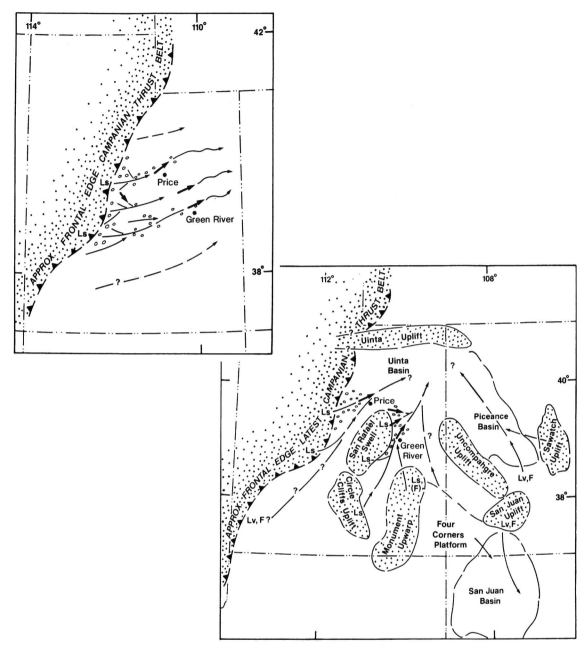

Fig. 11.20. Evolution of fluvial dispersal systems in Utah from the Middle Campanian (Castlegate Formation, *left*) to the Late Campanian (Price River, Tuscher Formations, *right*), based on analysis of source terranes and paleocurrents (*heavy arrows*). The Middle Campanian pattern is typical of transverse-oblique drainage from a foreland-fold-thrust belt. The Late Campanian pattern indicates the development of basement uplifts within the basin, and major reorganization of drainage patterns. Stratigraphic sequences formed in the basin clearly were strongly influenced by tectonism. (Lawton 1986b)

scales. The single Castlegate sequence at Price Canyon represents one of a series of such cycles (Olsen et al. 1995) that are interpreted as the product of cyclic variations in the rate of subsidence on an approximately 5 m.y. time scale. During periods of slow subsidence basinward sediment transport was facilitated, following the antitectonic model of Heller et al. (1988), leading to the development of extensive, sheetlike depositional units. The Lower Castlegate Sandstone and the Bluecastle Tongue are examples. These deposits represent tectonically generated nonmarine analogs of lowstand systems tracts. They extend more than 150 km down depositional dip from the Price Canyon area. The sequence boundaries at their base may represent intervals of considerable erosion. An increase in the rate of subsidence led to higher rates of generation of accommodation space, with greater preservation potential for floodplain deposits, looser channel stacking patterns (Wright and Marriott 1993), and possibly subtle changes in fluvial style (Schumm 1993). The Upper Castlegate Sandstone, which contains evidence of tidal invasion, is an example of this phase, corresponding to a transgressive systems tract.

This interpretation represents a modification of the model of Posamentier and Allen (1993a), in which Yoshida et al. (1996) suggested that the long-term rate of flexural subsidence underwent cyclic variations. The amalgamated channel sandstones of the Lower Castlegate Sandstone, the Bluecastle Tongue, and the other "lowstand" sandstones of Olsen et al. (1995), were attributed to slower rates of long-term subsidence in zone A, not to a geomorphic adjustment by the rivers to a short-term fall in baselevel in zone B, as in the original model of Posamentier and Allen (1993a). Studies of the timing of thrusting in the Sevier orogen (DeCelles et al. 1995) indicate a pause in thrusting during the Middle Campanian, which may correspond to one or more of these times of uplift. In fact it is possible that considerable uplift and erosion took place throughout the project area, generating the type 1 sequence boundaries that bracket the Castlegate sequence. This may in part explain why the sequence does not thicken westward toward the proximal part of the basin. The high frequency sequences mapped by Van Wagoner et al. (1990, 1991) in the downdip part of the Book Cliffs (Fig. 8.58) were attributed by Yoshida et al. (1996) to short-term tectonism (Table 11.1), such as flexural response to the movement of individual thrust sheets, the transmission of the flexural effects across the basin by in-plane stress, and subsequent erosional unroofing and isostatic uplift (Sect. 11.4).

Other researchers have speculated about the possible tectonic control of high-frequency sequence architecture. Leithold (1994) pointed out that the Upper Cretaceous succession of southern Utah contains sequences with a frequency in the Milankovitch band and speculated about possible glacioeustatic mechanisms, but also discussed episodic thrust loading. Similarly, Kamola and Huntoon (1995) documented the pattern of repetitive parasequence progradation in the Blackhawk For-

mation, the unit below the Castlegate Sandstone in the Book Cliffs, and suggested episodic movement of imbricate slices in thrust sheets as a cause of loading and stratigraphic cyclicity.

Other examples of high-frequency cycles dominated by shelf and shoreline depositional systems are abundant farther north within the Western Interior Seaway, particularly in Alberta and British Columbia (Figs. 8.48–8.55; Leckie 1986; Plint et al. 1986; Leggitt et al. 1990). In all these cases, appeal has been made to tectonic mechanisms to explain short-term changes in relative sea level. Leggitt et al. (1990) described beveling of an erosion surface, the geometry and orientation of which appears to indicate intrabasin tilting between intervals of sedimentation. Hart and Plint (1993) and Plint et al. (1993) discussed tectonic mechanisms that governed the construction of the Smoky Group on the Alberta–British Columbia border. The overall architecture of the Smoky Group was interpreted as a product of a varying rate of flexural subsidence and associated forebulge movement on a 10^6-year time scale (Plint et al. 1993). Bevels on certain erosion surfaces within the Cardium Formation were interpreted as the result of flexure over reactivated basement faults and resulting shoreline incision (Hart and Plint 1993).

Catuneanu et al. (in press) examined the sequence stratigraphy of the Bearpaw Formation across the Western Interior Basin from Alberta to Manitoba, and documented a succession of cyles with frequencies of 10^4–10^6 years through this predominantly marine unit that represents the last marine transgression in the basin. Correlation of the sequences was carried out mainly with the use of wireline logs, and was facilitated by the presence of numerous bentonites. Good biostratigraphic control was also available, based on ammonites, microfossils and palynomorphs. The main Bearpaw transgression occurred in response to long-term flexural subsidence, and this was modulated by 10^5-year cycles of subsidence. Bentonite correlation and biostratigraphic evidence clearly indicate that the sequences are diachronous, becoming younger away from the fold-thrust belt. This architecture is interpreted as the response of the basin to a flexural "wave" rolling out from the fold-thrust belt in response to each pulse of thrust-induced subsidence. The basin rotated about a hingeline corresponding to the forebulge of the basin, generating a reciprocal stratigraphy in which subsidence in the basin can be correlated with uplift east of the hingeline, and vice versa.

As discussed in Section 11.3.3 there is ample evidence from modern areas of active tectonism (e.g., New Zealand Alps, Alpine orogen of Europe, Himalayan peripheral basin, Banda Arc) that rates of tectonism and the episodicity of uplift and subsidence in modern foreland basins are of the right order of magnitude to accommodate the high-frequency cycles documented in Utah, Alberta and elsewhere within the Western Interior of North America. It remains for future research to provide the tight chronostratigraphic correlation between tectonic episodes and the sequence stratigraphy that would confirm tectonism as the major depositional control.

11.3.2.2 The Appalachian Foreland Basin

Ettensohn (1994) reviewed what used to be termed the geosynclinal cycle, in which stages of basin development are related to the evolving structural geology, including unconformities, and the sedimentary evolution, which tends to show a repetition of the succession from deep-water shale up through turbidite sandstone (flysch) to shallow marine-nonmarine sandstone (molasse). This model may be compared with that for the Alberta Basin by Cant and Stockmal (1989). Ettensohn (1994) went on to provide a detailed review of the Paleozoic unconformities in the Appalachian basin. He noted:

The recurrence of unconformities, many accompanied by similar overlying sedimentary sequences, strongly suggests some type of cyclicity, even though recurrence intervals are irregular. Ten of the [13 interregional] unconformities ... are interpreted to be primarily tectonic in origin. Interpretation of tectonic origin is based on the presence of a distinctive, overlying, flexural stratigraphic sequence, the coincidence of unconformity formation with the inception of established orogenies or tectophases therein, and the distribution of unconformities relative to probable loci of tectonism. In contrast, the absence of an overlying flexural sequence and no coincidence with orogeny suggests that unconformities were predominantly eustatic in origin. Only the unconformity at the Ordovician-Silurian boundary seems to be of this type.

Washington and Chisick (1994) examined the Taconic (Upper Ordovician) unconformities between Newfoundland and Virginia in detail and claimed that the pattern of repeated sedimentary breaks makes more sense in terms of a eustatic model. However, they did not provide detailed evidence of correlation between the various unconformities along strike, so their case remains unproven. A very widespread unconformity mark-ing the end of the Ordovician has now been interpreted as the product of a short-lived glacial episode (Sect. 10.5).

The Mississippian-Permian stratigraphy of the Appalachian Foreland Basin is a particularly interesting case, because this was a tectonically active basin in which cycles of undisputed glacioeustatic origin, the cyclothems, were deposited. Unraveling the complexities of tectonism and eustasy in this basin has been a major challenge, and many questions and disagreements between different specialists still remain. A major contribution to this debate is a set of papers edited by Dennison and Ettensohn (1994) dealing with the Upper Paleozoic cyclothems, in which tectonism and eustasy have been approached from different points of view by a wide variety of authors. The following discussion of this basin is based mainly on papers from this book.

The glacioeustatic control of the Pennsylvanian cyclothems in the Appalachian basin is now well established, as discussed in Section 10.4. However, various lines of evidence suggest that contemporaneous tectonism was also important. Chesnut (1994) documented three scales of cycle in the Pennsylvanian of the central Appalachians: (a) a large-scale coarsening-upward succession corresponding to the Breathitt Group, which represents 9–34 m.y., depending on which estimate of Pennsylvian time is used; (b) within this are eight major-transgression cycles which average 1.1–4.3 m.y. (again, depending on time scale); (c) Within these are coal-clastic cycles averaging 0.2–0.7 m.y. The large-scale cycle of the Breathitt Group represents long-term regional-scale flexural subsidence of the basin. The second scale of cyclicity is interpreted as tectonic, and in fact probably corresponds to a 10^6-year cycle of tectonism, of the type Yoshida et al. (1996) suggested for the Castlegate Sandstone (Sect. 11.3.2.1). The smallest scale of cyclicity almost certainly represents Milankovitch glacioeustatic control.

Pashin (1994) constructed isopach maps for coal-bearing cyclothems of the Lower Pennsylvanian Pottsville Formation in the Black Warrior Basin of Alabama. Each is estimated to represent 0.2–0.5 m.y. He demonstrated that these cycles each show different isopach patterns, and stated that this demonstrates,

[T]hat major changes in basin geometry occurred as deformation loading of the Alabama promontory proceeded. Hence, significant spatial and temporal variations of subsidence rate occurred in the same time frame as deposition of each coarsening- and coaling-upward cycle....

Thus, flexure of the lithosphere below the Black Warrior Basin represents multiple events that occurred as elements of different tectonic terranes collided with and were thrust onto the Alabama promontory.

However, Pashin (1994) also stated, "although flexural subsidence may have amplified marine transgression, no tectonic causes of regional marine regression were identified that operated at the time scale of deposition of a single Pottsville cycle. For this reason, glacial eustasy is considered to have been the dominant cause of cyclicity in the study interval." In another study, Beuthin (1994) documented a fluvial paleovalley trunk-tributary system that appears to map out the slope of the Appalachian foreland basin and its flanks, indicating control by tectonic slope. However, a eustatic fall in sea level may have occurred to expose the deeper parts of the basin to fluvial erosion.

Both the studies noted in the previous paragraph emphasized glacioeustasy and failed to make use of the concept of high-frequency tectonic control that is now being proposed for some stratigraphic successions in the Western Interior Basin and the Pyrenean basins. Klein and Kupperman (1992) and Klein (1994) attempted to devise a quantitative method for distinguishing between tectonic and eustatic causes of sedimentary cyclicity. They used backstripping calculations to determine the amount of subsidence that can be attributed to tectonism during the accumulation of each cyclothem, and they used sedimentological methods, such as depths at which typical facies are deposited, to determine water-depth changes during cyclothem accumulation. The difference betwen these estimates then indicates the changes in water depth that can be atttributed to eustasy for each cyclothem. Although these calculations are instructive, there are many sources of error in the estimates, and the results should probably be regarded as qualitative rather than precise in their indications. Undoubtedly tectonism was important in generating clastic sediment supply to the Appalachian Basin, and in increasing the rate of subsidence relative to the cratonic interior but, if Heckel (1994) is correct, high-frequency glacioeustasy can be detected in the Appalachian Basin. Careful biostratigraphic correlation between the Midcontinent, the Illinois Basin, and the Appalachian Basin suggests that individual cyclothems can be traced between the three regions. However, a bundling of the cyclothems is detectable in the Appalachian Basin. Bundles of cyclothems dominated by marine units are interbedded with bundles dominated by terrestrial deposits, and

containing major paleosol intervals. The cyclicity responsible for the bundling is attributed by Heckel (1994) to tectonic flexure and unloading with a periodicity of about 3 m.y.

11.3.2.3 Pyrenean and Himalayan Basins

In these basins the use of magnetostratigraphic data to supplement biostratigraphy has provided tight constraints on models of tectonism and sequence development. The technique may permit a precision of dating to within the nearest 10^5 years. For example, Burbank and Raynolds (1984, 1988) used magnetostratigraphic techniques to date the fluvial deposits of the Himalayan foredeep of Pakistan. These results showed that, in the Himalayan foredeep, thrusting and uplift episodes were rapid and spasmodic, and that they did not occur in sequence into the basin, as the classic model (e.g., Dahlstrom 1970) would predict (Fig. 11.21). In one case, Burbank and Raynolds (1988) were able to demonstrate the uplift and removal of 3 km of sediment over the crest of an anticline within the basin over a period of 200,000 years, an average uplift and erosion rate of 1.5 cm/year. In most basins, the precision of dating obtained by these authors cannot be attempted, and this therefore suggests that caution should be used in assessing published reconstructions of the rates of convergence, crustal shortening, and subsidence, except where these are offered as long-term (>1 m.y.) averages. These results confirm the importance of high-frequency tectonism in the development of the fold-thrust belt and the potential for corresponding high-frequency sequence development in the adjacent deposits (Fig. 11.15).

Burbank et al. (1992) used magnetostratigraphic control to document timing of thickness and facies changes in the eastern Pyrenean foreland basin (the succession of sequences illustrated in Fig. 7.30). They demonstrated that in most cases coarse-grained progradation occurs during times of slow subsidence, and rapid subsidence is generally reflected by thick fine-grained deposits, thus supporting the Heller et al. (1988) two-phase model for foreland-basin development. Structural relationships, including faulted cross-cutting of sequences, enabled the history of deformation to be reconstructed and related to that of sequence development. The results (Fig. 11.22) show a complex pattern of thrusting, including the development of imbricates, and out-of-sequence thrusts. The se-

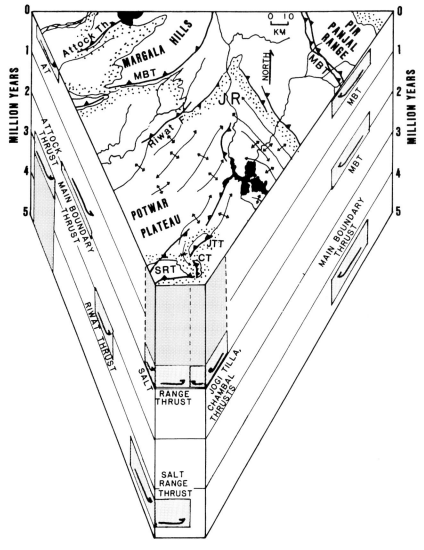

Fig. 11.21. Chronology of fault motions within part of the Himalayan foredeep of Pakistan. *Shaded boxes* indicate the time and space domain over which each thrust is interpreted to have been active. (Burbank and Raynolds 1988)

quences and their intervening unconformities therefore developed in response to a shifting pattern of flexural loading, with continual changes in the location of sediment sources and the rate of sediment input, and changes in areas of erosional unroofing and unloading.

11.3.3 Rates of Uplift and Subsidence

One of the most impressive features of basins that develop in regions of convergent tectonism, including arc-related basins and those within and adjacent to suture zones, is the extremely rapid

rates of tectonic and sedimentary processes that occur in these basins. Figure 11.23 shows the history of subsidence and uplift of various locations along the Banda Arc of Indonesia, which is an area of convergence and collision between Eurasia and the Australian and Pacific plates. At the bottom of this diagram is shown the global cycle chart of Haq et al. (1987, 1988a) at the same scale. Clearly the fluctuations indicated in this chart would be swamped by tectonic effects in an area undergoing a tectonic evolution comparable to that of the Banda Arc.

Rapid rates of uplift and subsidence have been calculated from many regions of active tectonism

MPTS DATED SEQUENCES AND FORMATIONS TECTONIC HISTORY

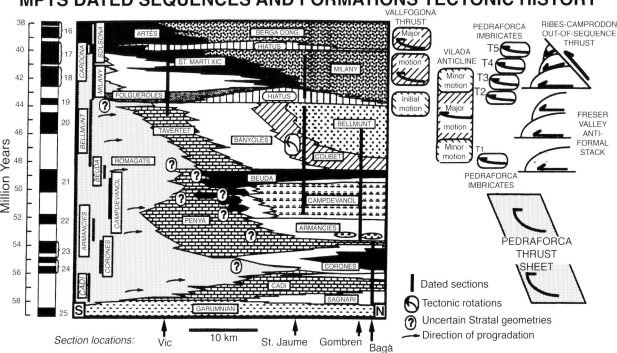

Fig. 11.22. Schematic Wheeler diagram showing age, sedimentary facies, sequence development, and chronology of thrusting in the eastern Pyrenean foreland basin. The sequences named in the *vertical boxes at left* correspond to the sequences shown in Fig. 7.30. (Burbank et al. 1992)

(rates are summarized in Table 9.1). Pedley and Grasso (1991) documented uplift rates in a subduction-accretion zone in southern Sicily, indicating localized Quaternary uplift of up to 1 km. Bishop (1991) studied raised terraces in the New Zealand Alps, a region characterized by transpressional uplift. Calculated uplift rates range from 5 to nearly 8 m/ka over periods of about 10 ka, with longer term rates an order of magnitude smaller (1 m/ka for 100 ka). Data on rates of movement along the San Andreas transform system were summarized by May and Warme (1987), who noted measured modern subsidence rates of 1.5 m/ka in the Imperial Valley, and up to 2 m/ka in the Eocene Ventura Basin. Uplift rates in the Banda Arc are illustrated in Fig. 11.23, and range from 0.5–10 m/ka over periods of less than 2 m.y. (Fortuin and de Smet 1991). Subsidence rates are not as rapid. In the Banda Arc they range from 0.08 to 1 m/ka for periods of up to 15 m.y., resulting in a total recorded subsidence of up to 4 km. In the Himalayan peripheral basin of Pakistan Johnson et al. (1986)

and Burbank et al. (1986) demonstrated subsidence (actually sediment-accumulation) rates of up to 0.6 m/ka persisting for 2–4 m.y. The same overall rate of sediment accumulation, 0.6 m/ka averaged over 2.5 m.y., was obtained for an Upper Cretaceous forearc-basin succession in Baja California (excluding possible but unknown eustatic effects) by Fulford and Busby (1993), and Eocene forearc basins in California subsided at rates between 0.1 and 0.7 m/ka for periods measured over millions of years. Hiroki (1994) determined vertical changes in relative sea level during the Quaternary in a forearc setting in coastal Japan and differentiated glacioeustatic from tectonic changes. He showed changes between subsidence and uplift over a 10^4–10^5-year interval, with considerable local variation in rates, probably reflecting local structural controls. The maximum uplift rate, neasured at one locality, was 0.29 m/ka, and the maximum subsidence rate was 0.45 m/ka.

The evidence indicates that uplift and subsidence are spasmodic and localized, with movement rates

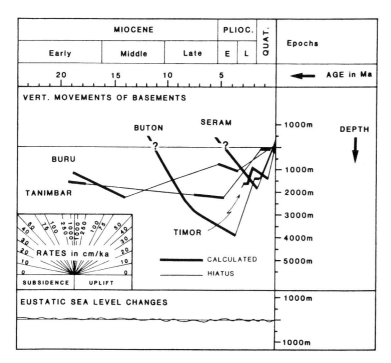

Fig. 11.23. Selection of age-versus-depth plots for various locations along the Banda Arc of Indonesia. *Thick lines* indicate plots derived from geohistory analysis. These are linked by *thin lines*, representing the average of vertical motions during erosional intervals. In each case uplift must have occurred to bring about erosion, but vertical amplitude is difficult or impossible to ascertain. At the *base of the diagram* is plotted the global cycle chart of Haq et al. (1987, 1988) at the same vertical and horizontal scales. (Fortuin and de Smet 1991)

and the total amount of movement varying over distances of only a few kilometers. Uplift on Timor is described as "regional arching over distances of 10 km to more than 100 km" (Fortuin and de Smet 1991), and this is probably one of the more extensive areas to be affected simultaneously by a single episode of convergent tectonism. Based on measurements along a 40-km stretch of coastline Hiroki (1994) deduced that crustal bending had a wavelength of 50–100 km.

11.3.4 Discussion

When examined in a regional context it would seem unlikely that tectonism could ever be confused with eustasy in tectonically active basins such as foreland or forearc basins. However, the paucity of stratigraphic data in many basins may tempt the regional geologist to attempt comparisons with the global cycle chart. Sengör (1992) argued that convergent margins are characterized by the successive development of numerous local tectonic unconformities. Poor dating control and limited outcrop may tempt the mapper to assign these to a single regional event, and even to correlate it with one of the events on the global cycle chart. In the case of old and no-longer active basins much of the evidence of active tectonism may have been lost to erosion. The difficulties are likely to be acute where only one or two exploration holes with limited biostratigraphic data

are available for analysis. In Section 13.4.1 several examples are given where careful structural and stratigraphic mapping clearly indicates tectonic control of sequence development, yet correlations with the global cycle chart have been attempted. Fortuin and de Smet (1991, p. 87) offered this critical appraisal of such correlations based on their observations in Timor:

[U]p to four sequence boundaries might be preserved according to the Haq et al. (1987) sea-level chart, notably at 3.0, 2.4, 1.6, and 0.8 Ma. We can indicate two distinct sequence boundaries, one around 2.1 Ma, when pelagic sedimentation was suddenly replaced by channel and fan deposits, and one around 0.2 Ma, which is the second uplift unconformity separating the marine sediments from overlying terrestrial conglomerates.... The 2.1-Ma level has been interpreted to be of tectonic origin.... Around the 2.4-Ma eustatic low, bathyal pelagic chalks were being deposited, far from clastic sources, so that this and the 3.0-Ma lowstand period probably cannot be reflected in the sedimentary record. After the first pulse of uplift waned, quiet deposition, dominantly of marls, followed and the newly emerged hinterland of proto-Timor cannot have been very far away. The timing around 2.0 Ma of this facies change might well indicate control by a eustatic rise. In that case one would also expect to find a sequence boundary around 1.6 Ma, due to the next eustatic fall, but no facies change was observed there. Apart from possible age correlation errors these results seem an illustration of our conclusion that prediction of chronostratigraphic relations using the cycle chart (Haq et al. 1987) in basinal settings in tectonically active areas is uncertain, at least before some information concerning the ages of the main (tectonic) unconformities is available.

11.4 Intraplate Stress

11.4.1 The Pattern of Global Stress

"Ridge-push" and "slab-pull" are informal terms that have been used for some time in the plate-tectonics literature to refer to the horizontal (in-plane) forces associated with, respectively, the horizontal compressional effects resulting from the elevation differences between a spreading center and the deep ocean basin, and the tensional effects on an oceanic plate generated by the downward movement under gravity into a subduction zone of a cold slab of oceanic crust. In-plane (horizontal) stress is a central theme of the paper by Molnar and Taponnier (1975) describing a model of Himalayan collision, in which it was demonstrated that most of the Cenozoic structural geology of western China, extending for 3000 km north of the Indus suture to

the edge of the Siberian craton, could be explained as the product of deformation resulting from in-traplate stresses transmitted into the continental interior from the India-Asia collision zone. It came to be recognized that tectonic plates can store and transmit horizontal forces many thousands of kilometers from zones of plate-margin stress. The presence of residual stress fields in continental interiors has long been known from such evidence as the development of active joints ("break-outs") in exploration holes (summary in Cloetingh 1988). Stress data for northwestern Europe are shown in Fig. 11.24. Cloetingh (1988) provided similar data for the India-Australia plate, which revealed high compressive stresses, particularly in the north-eastern Indian Ocean, associated with the north-ward collision of this plate with Eurasia. There the seafloor has been deformed into broad flexural folds as a result of intraplate compressive stress. Cloe-tingh (1988, p. 206) suggested that:

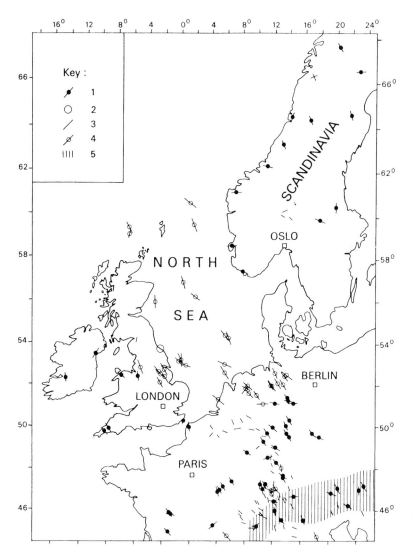

Fig. 11.24. Compilation of observed maximum horizontal stress directions in the northwestern European platform. *1* In situ measurements; *2* horizontal stresses equal in all directions, as determined from in situ measurements; *3* determinations from earthquake focal-mechanism studies; *4* well break-outs; *5* location of Alpine fold-belt. (Cloetingh 1988)

[T]he observed modern stress orientations show a re-markaby consistent pattern [in northwestern Europe], especially considering the heterogeneity in lithospheric structure in this area. These stress-orientation data indicate a propagation of stresses away from the Alpine collision front over large distances in the platform region.

In the late 1980s the study of in-plane stress evolved into the World Stress Map Project, under the auspices of the International Lithosphere Program. Zoback (1992) provided a report on this project, as one of a series of papers discussing intraplate stress. She compiled over 7300 in situ stress orientation measurements, and provided a series of maps documenting the results. Zoback (1992) and Richardson (1992) confirmed the observation that stress orientations are remarkably consistent over large continental areas, including areas that are characterized by considerable crustal heterogeneity. Richardson (1992) pointed out that on a global scale orientation measurements are most readily interpreted with respect to the location and orientation of active spreading centers, and suggested that ridge-push forces are the most important in determining in-plane stress. Most stresses are compressional, with extensional stresses having been recorded mainly in areas of high topography. Stresses associated with plate collision are locally important, but the patterns of stress indicate a complex relationship between intraplate stress and tectonic deformation.

11.4.2 In-Plane Stress as a Control of Sequence Architecture

Cloetingh et al. (1985) were the first to recognize the significance of in-plane stress as a control on basin architecture. They argued, "that variations in regional stress fields acting within inhomogeneous lithospheric plates are capable of producing vertical movements of the Earth's surface or the apparent sealevel changes ... of a magnitude equal to those deduced from the stratigraphic record."

The important contribution which Cloetingh et al. (1985) made was to demonstrate by numerical modeling that horizontal stresses modify the effects of existing, known, vertical stresses on sedimentary basins (thermal and flexural subsidence, sediment loading), enlarging or reducing the amplitude of the resulting flexural deformation. The principles are illustrated in Fig. 11.25. They demonstrated that a horizontal stress of 1–2 kbar, well within the range of calculated and observed stresses resulting from

plate motions, may result in a local uplift or subsidence of up to 100 m, at a rate of up to 0.1 m/ka (Table 9.1). Compressional stresses generate uplift of the flanks of a sedimentary basin and increased subsidence at the center. Extensional stresses have the reverse effect (Fig. 11.25). The magnitude of the effect varies with the flexural age (rigidity) of the crust, as well as the magnitude of the stress itself.

The stratigraphic results of this process are extremely important. Figure 11.26 models the effects of imposing a horizontal stress of 500 bars on a continental margin undergoing long-term thermally induced subsidence. Cloetingh (1988, p. 214) stated, with reference to this model:

When horizontal compression occurs, the peripheral bulge is magnified while simultaneously migrating in a seaward direction, uplift of the basement takes place, an offlap develops, and an apparent fall in sea level results, possibly exposing the sediments to produce an erosional or weathering horizon. Simultaneously, the basin center undergoes deepening [Fig. 11.26b], resulting in a steeper basin slope. For a horizontal tensional intraplate stress field, the flanks of the basin subside with its landward migration producing an apparent rise in sea level so that renewed deposition, with a corresponding facies change, is possible. In this case the center of the basin shallows [Fig. 11.26c], and the basin slope is reduced.

These architectural features are locally identical to the onlap and offlap patterns from which Vail and his coworkers have derived their global cycle charts, although on a basinal scale sequence architectures show important differences. For example, build-up of compressional stresses over an extensional margin results in uplift of the continental shelf and offlap, and may lead to the generation of an erosional unconformity, at the same time as the basin center undergoes deepening. Increased tilting of the continental margin may increase the tendency for slope failure and mass wasting, with increased potential for large-scale submarine-fan deposition at the base of the slope. Conversely, an increase in tensional stress may result in flooding of the continental shelf and uplift and enhanced erosion of the basin floor, with the potential for the development of submarine unconformities. Changes of stress regime result in the typical indicators of relative sea-level change that are used in sequence analysis (unconformities, transgressions, progradation, retrogradation, etc.) being out of phase by a half cycle between the basin margin and the basin center (although this may be impossible to demonstrate from the limited chronostratigraphic evidence that is commonly available). The kind of local, outcrop-scale analysis that is used as the basis for many

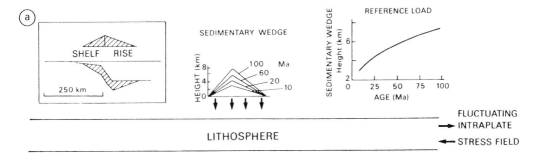

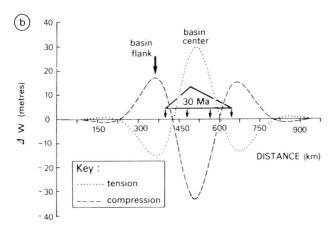

Fig. 11.25. a Intraplate stress: principles of the geophysical model. At *top left* a simple shelf-slope-rise sedimentary wedge at a continental margin is shown. The thickness and age of this wedge are shown at *top, center* and *right*.

b The resulting flexural deformation attributable to a horizontal stress of 1 kbar acting on this continental margin. Compressional and tensional stresses yield equal but opposite effects. (Cloetingh 1988)

sequence anlyses (e.g., Van Wagoner et al. 1990) could potentially therefore give very misleading result. As Karner (1986) pointed out, "basin margin and interior regions should experience opposite baselevel movements. The nature and distribution of sediment cyclicity across either a passive margin or intracratonic basin therefore offers an excellent opportunity to test the concept and importance of lateral stress-induced baselevel variations."

Figure 11.27 illustrates simplified, hypothetical subsidence curves for three positions within a basin undergoing thermally induced subsidence in which the subsidence has been affected by changes in intraplate compressive stress. As pointed out by Cathless and Hallam (1991) in opposition to Cloetingh's hypotheses, the elevation changes produced within any given sedimentary basin are small in areal extent, limited by the flexural wavelength of the basement underlying the basin. However, the point they misssed themselves is that the actual intraplates stresses are much more widespread – they may be transmitted for thousands of kilo-

meters and simultaneously affect all basins within that plate.

The initial hypothesis of Cloetingh et al. (1985) has been elaborated by Karner (1986), Cloetingh (1986, 1988), Lambeck et al. (1987), Cloetingh et al. (1990) and Cloetingh and Kooi (1990). Karner (1986), in particular, noted that "the simple relationship between in-plane stress and plate rigidity is likely to be complicated by the real rheological properties of the lithosphere," and that "on the application of an in-plane stress, the rigidity of the lithosphere changes and in so doing will modify any preexisting deformation." Karner (1986) carried out analysis and modeling of several aspects of this complex interrelationship of tectonic forces and processes. For example, he studied the change from in-plane extension to compression that characterizes the basin history in many complex orogens, such as the Tertiary basins of western Europe. Many of these basins are characterized by inversion, leading to uplift. The Wessex Basin is often quoted as the typical example of this process. Karner (1986)

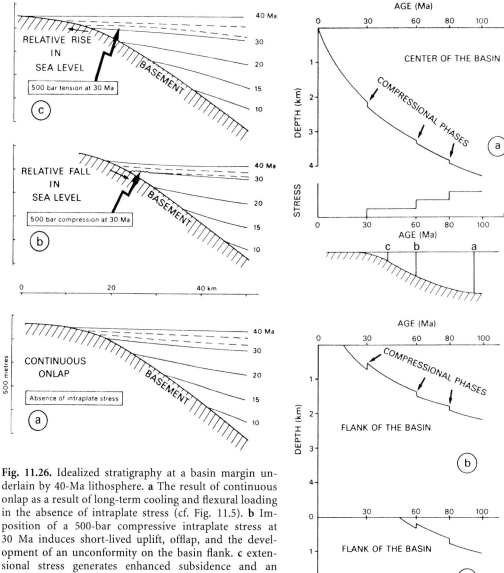

Fig. 11.26. Idealized stratigraphy at a basin margin underlain by 40-Ma lithosphere. **a** The result of continuous onlap as a result of long-term cooling and flexural loading in the absence of intraplate stress (cf. Fig. 11.5). **b** Imposition of a 500-bar compressive intraplate stress at 30 Ma induces short-lived uplift, offlap, and the development of an unconformity on the basin flank. **c** extensional stress generates enhanced subsidence and an increase in the rate of onlap. The stratigraphy in **b** and **c** indicate a relative fall and rise in sea level respectively. (Cloetingh 1988)

Fig. 11.27. Effect of intraplate compressive stress on subsidence curves at three different positions within a sedimentary basin. In the basin center (**a**) the effect of compression enhance subsidence, whereas on the flanks of the basin (**b,c**) enhanced subsidence changes to uplift as the flexural node migrates during long-term flexural widening of the basin. (Cloetingh 1988)

demonstrated that fault reactivation leads to uplift and inversion, whereas unfaulted basins, or those whose basin-margin faults are locked, experience enhanced subsidence under compressive in-plane stresses.

The importance of tectonism as a control of sea-level change, as evaluated by those who have studied it in detail, has led to a suggestion that Vail's work be virtually turned on its head. Karner (1986) stated in the conclusions to his work:

Knowing the origin of the Vail curve puts its correlative powers onto a strong theoretical basis and helps to define where and when the curve is of use in seismic strati-

graphy. Quite independently however, if this in-plane stress mechanism is right, it implies that the Vail curve can be used to measure the absolute value of paleo in-plane stress variations and also as a pointer to the timing of past plate boundary reconfigurations.

Elsewhere Karner (1986) stated:

Because lateral force variations will be globally balanced (that is, an increase in relative compressive force at one plate boundary must be balanced by a relative increase in tensile forces at another), then in-plane stress-induced unconformities cannot be global. If there exists a causal (but undoubtedly complex) relationship between plate interactions and in-plane stress variations, then transgressive/regressive sequences from a variety of basins will be correlatable if they share the same in-plane stress system, which potentially may span a number of plates. In essence the Vail et al. (1977) coastal onlap curve can be correctly applied to widely spaced basins only when the above criteria are established.

Cloetingh and his colleagues have now carried out a considerable number of modeling experiments, in which they have applied the concepts of intraplate stress to extensional (Kooi and Cloetingh 1992a,b) and contractional (Peper 1994; Peper et al. 1992, 1995; Heller et al. 1993) tectonic settings.

Kooi and Cloetingh (1992b) examined the effect of the depth at which crustal attenuation ("necking") occurs during crustal extension, and its implications for sequence architecture, and they discussed the importance of basinal tilting induced by flexural effects (Fig. 11.25). They showed that the stratigraphic effects of changes in intraplate stress regime are particularly pronounced and different from simple models of subsidence-with-eustasy in the case of extensional margins that undergo crustal thinning and "necking" at relatively shallow levels. Modeling experiments demonstrated that where necking occurs at deep levels, the resulting stratigraphic architecture is more similar to that of the eustatic model. Kooi and Cloetingh (1992b) showed that the angular unconformities that result from the warping movements driven by changes in stress regime are very subtle, typically $0.5°$ or less, and therefore difficult to detect in most data sets. They confirmed that the observations suggested by Embry (1990) for the detection of tectonic influences in sequence generation (Sect. 11.1) are all consistent with tectonism driven by changes in the regional in-plane stress regime.

The response of foreland basins to flexural loading is discussed in Section 11.3.2. Most recent work has been concerned with tectonism on a 10^6-year time scale. Recent studies of intraplate stress, aided by numerical modeling, have suggested that the crust may respond on much shorter time scales, possibly less than 10^4 years (Cloetingh 1988; Peper et al. 1992, 1995; Heller et al. 1993). Peper et al. (1992) modeled responses of periodic tectonic movement at time scales down to that corresponding to the spacing of individual earthquakes (10^0–10^1 years), and made a convincing case for tectonic cyclicity that could explain cycles with frequencies of 10^5–10^4 years. Heller et al. (1993) considered the

effects of compressional tectonism in areas containing crustal heterogeneities, including active faults and inherited basement structures. They concluded that vertical motions of meters to a few tens of meters extending over distances of tens of kilometers could be attributed to such stresses. This is quite enough to overprint the effects of modest eustatic sea-level changes, and Heller et al. (1993) pointed to several subtle structural and stratigraphic features in the Rocky Mountain states that they suggested could be attributed to in-plane stress occurring as part of compressional tectonism during the Jurassic to Middle Cretaceous.

11.4.3 In-Plane Stress and Regional Histories of Sea-Level Change

Lambeck et al. (1987), and Cloetingh (1988) compiled stratigraphic and tectonic data data for the North Sea Basin as a starting point for the evaluation of the importance of intraplate stress in this area, and Cloetingh et al. (1990) expanded the data-base to the entire North Atlantic region. They chose this region in part because of its importance in the original establishment of the global cycle chart by Vail in the 1970s. Many workers (e.g., Glennie 1984; Miall 1986; Hallam 1988) had independently noted the importance of regional tectonism in controlling the stratigraphy of this area. Hallam (1988), in particular, noted "that a significant number of Jurassic unconformities are confined to the flanks of North Sea Basins, consistent with the predictions of [intraplate stress]." (Cloetingh 1988). The chart compiled by Cloetingh and his coworkers is shown in Fig. 11.28. It is of interest that the paleostress curve compiled for comparison to dated tectonic events is based in part on the onlap-offlap curves of Vail et al. (1977) and Vail and Todd (1981; Fig. 11.29). It seems strange that Cloetingh and his colleagues would rely on these curves, given that they have argued that the curves are a reflection of tectonic processes rather than eustatic sea-level change, and that therefore the methods of regional compilation used by Vail and his coworkers might be regarded as suspect. The "averaged" curve in Fig. 11.29 has been calibrated by the incorporation of various assumptions about eustatic sea-level change, including long-term rises in sea level during the Jurassic and Cretaceous, based on the work of Hallam (1981) and Hancock (1984), and the sea-level fall due to glacioeustasy in the Middle Oligocene.

Compilations such as that in Fig. 11.28 (see also Schwan 1980; Ziegler 1982) are difficult to evaluate

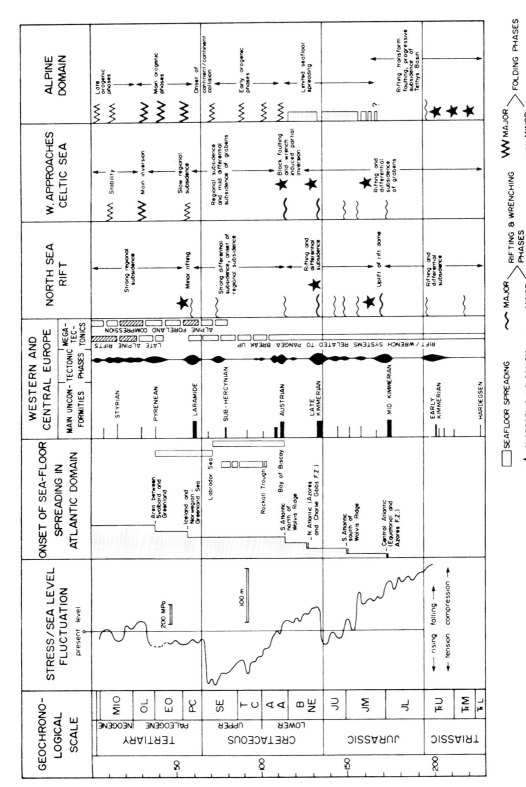

Fig. 11.28. Relationship between interpreted paleostress and tectonic events in Europe and the North Atlantic region. The paleostress curve is based on sea-level curves compiled from various sources, as given in Fig. 11.29. (Lambeck et al. 1987)

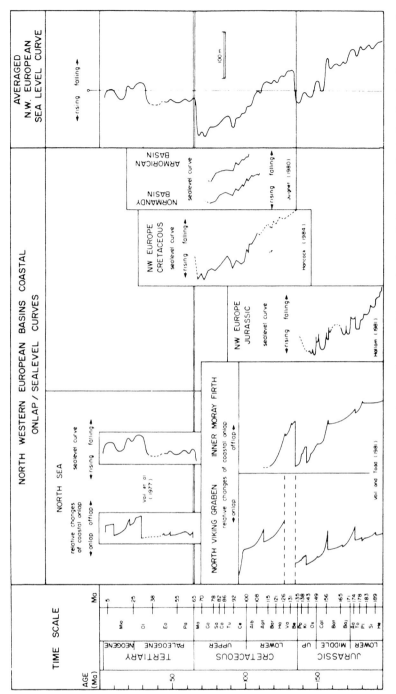

Fig. 11.29. Onlap-offlap curves compiled by various authors and arranged by Lambeck et al. (1987) as a basis for the development of a regional paleostress curve. The averaged curve at *right* is based on calibration of various events in the component curves using local stratigraphic data. The assumption is that these onlap-offlap curves are largely a reflection of thermal and flexural subsidence modified by intraplate stress, rather than a product solely of eustatic sea-level change. (Lambeck et al. 1987)

because of the uncertainties inherent in the positioning and relative importance of each event. The analysis which Cloetingh (1988) and Lambeck et al. (1987) offered of their own compilation is little more than "permissive," in the sense that they show that all relative sea-level events in Europe and the North Atlantic could be explained by changes in intraplate stress, but are unable to offer quantitative proof, in the form of detailed numerical models.

In another attempt to model paleostress Cloetingh and Kooi (1990) provided a curve of estimated changes in the paleostress field of the United States Atlantic margins, based on a modification of the approach of Watts and his colleagues (which is discussed in Sect. 11.2), in which they superimposed short-term intraplate stress on the long-term flexural subsidence of the continental margin (Fig. 11.30; following discussion taken largely from Miall 1991a).

Changes in the stress regime of the United States Atlantic margin occurred because of the continual change in the plate kinematics of the Atlantic region as the ocean opened and the spreading center extended and underwent various changes in config-

uration. Important events, such as the initiation of spreading in Labrador Sea, the extension of spreading between Greenland and Europe, and various jumps in ridge position, brought about significant changes in plate trajectories, which would have been accompanied by changes in the direction and intensity of intraplate stresses.

Sea-floor spreading patterns in the northern and central Atlantic Ocean (Srivastava and Tapscott 1986; Klitgord and Schouten 1986) provide a detailed history of plate motions. The reconstruction of a continuous succession of matched plate margins as the Atlantic Ocean opened indicates that the rotation poles of the North American plate relative to Europe and Africa changed at intervals of about 2–16 m.y. This episodicity is comparable in magnitude to the duration of 10^6-year stratigraphic cycles, a fact that is highly suggestive. Furthermore, some of the changes in rotation pole can be correlated in time with changes in the position of depocenters in the Mesozoic-Cenozoic stratigraphic record of the United States Atlantic continental margin. Poag and Sevon (1989) compiled isopach maps of the postrift sedimentary record, and in-

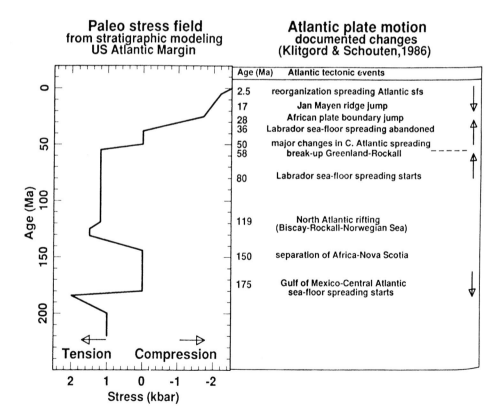

Fig. 11.30. Paleostress curve inferred from a stratigraphic model of the United States Atlantic margins. Timing of tectonic events in the region is shown at right. (Cloetingh and Kooi 1990)

terpreted these in terms of the shifting pattern of denudation in and sediment transport from the Appalachian and Adirondack Mountains. It seems likely that many of the major existing rivers that presently carry sediment into this region have been in existence since the Mesozoic. Poag and Sevon (1989) were able to show many changes with time in the relative importance of these various sediment sources, and it is suggestive that several of these changes occurred at times of change in Atlantic plate kinematics. The sea-floor spreading record reveals at least fourteen changes in plate configuration in the central Atlantic since the Middle Jurassic (Klitgord and Schouten 1986). Seven of these changes, at 2.5, 10, 17, 50, 59, 67 and 150 Ma, correspond to times when the major sediment dispersal routes from the Appalachian Mountains to the continental shelf underwent a major shift (data from Poag and Sevon 1989).

As noted in Section 11.2.2, Janssen et al. (1995) were able to relate Late Jurassic to present-day subsidence histories of basins within and on the flanks of Africa to to the timing of sea-floor spreading events in the ocens that surround Africa. This indicates that changes in intraplate stress regimes were a major control on the rate of generation of accommodation space in these basins.

Tectonism is diachronous and, in the case of the intraplate stresses discussed by Cloetingh (1986, 1988), the effects vary across the plate. Adjacent plates may be expected to have dissimilar tectonic histories, except where major extensional or collisional events affect adjoining plates. However, there is no mechanism for generating globally simultaneous tectonic events of similar style and magnitude. If, as suggested by Cloetingh (1986, 1988), many 10^6-year cycles are tectonic in origin, it should not be possible to construct a *global* cycle chart (but see below). The fact that such a chart has been constructed suggests that we should examine the basis on which it has been made. This examination forms the basis for Part IV of this book.

It is ironic that Karner and Cloetingh should suggest using the Vail curve to prove a hypothesis that opposes Vail's own basic premise – the preeminence of eustasy. It is doubly ironic that Vail's so-called "global" curve should be used as a test of regional paleostress patterns (in the North Sea and North Atlantic regions). Such a proposal incorporates the circular reasoning of the original curve as well as the inversion of Vail's central geological hypothesis! It points to a continuing lack of appreciation of the methods used by Vail and his

coworkers to construct the Vail curves, especially the methods of interregional correlation and calibration. These critical areas are examined in Part IV of the book.

The fact that the earth's crust consists of a finite series of interracting plates suggests the possibility for widespread, even global tectonic episodes that are synchronous but of different style and magnitude in different areas, as noted in Section 11.1. A possible example of this was noted by Hiroki (1994), who documented tectonic and glacioeustatic movements in a forearc basin on the eastern coast of Japan during the Quaternary. He noted that a change from subsidence to uplift between 331 and 122 ka B.P. occurred at about the same time as an increase in uplift rates in New Zealand, and a decrease in uplift rates in New Guinea. He stated that, "the synchronous change in vertical crustal movement may be explained by a change in the regional horizontal stress field due to rotation of the Pacific plate." Likewise, Embry (1993), in an analysis of Jurassic sea-level changes in the Canadian Arctic Islands, documented several that are clearly at least in part tectonic in origin, as they occurred at times of significant changes in thickness, facies or sediment dispersal patterns; yet they correlate remarkably well with the "global" events in Hallam's (1988) synthesis (Fig. 14.9). Embry (1993) suggested that these events may be tectono-eustatic in origin. Careful documentation and correlation of deformation and sea-level histories in adjacent but separate regions may reveal more such genetically related episodes of tectonism in widely separated parts of the earth's surface.

11.5 Basement Control

The rate and magnitude of all the tectonic processes discussed in this chapter depend to a considerable extent on the nature of the basement underlying a sedimentary basin. The importance of ancient lines of weakness in Precambrian basement has long been discussed as a possible cause of the localization of cratonic basins (Quinlan 1987; Miall 1990; Sect. 9.3.6.1), and many authors have documented the system of uplifts and lineaments within large cratonic areas that are thought to have modified the effects of epeirogeny (e.g., Sanford et al. 1985). However, it is only relatively recently that sequence-stratigraphic analyses and numerical models of basin development have begun to taken into ac-

count the importance of local and regional variations in elasticity, and the effects of basement heterogeneities, such as deep-seated faults and buried
sutures between tectonically dissimilar terranes
(with different crustal thickness, strengths, and
anisotropies). Within regions of the crust that have
undergone a long history of repeated plate movements, with the overprinting of many separate tectonic episodes, the effects of basement
heterogeneity on vertical movements of the crust
are likely to be very complex, whether the major
driving forces are flexural loading, crustal stretching, or mantle thermal effects (dynamic topography). Such regions of the earth would include
most of northwestern Europe, which has been located astride major plate junctions for most of the
Phanerozoic, and the southwestern United States,
which has been affected by repeated collisional and
extensional events on its southern and western
margins throughout the Paleozoic and Mesozoic.
Even the cratonic sedimentary cover above Precambrian basement may be affected by subtle
basement heterogeneities during epeirogenic warping or during changes in the in-plane stress regime.

Several examples of this important type of local
tectonic control have been referred to above. Most
cited examples are located in foreland basins, and
include the importance of buried faults in the Alberta Basin (Hart and Plint 1993), and the influence
of regional heterogeneities and basement "buttresses" during flexural loading of the western
continental margin of the United States (Pang and
Nummedal 1995). Yoshida et al. (1996) suggested
that regional variations in sequence styles in the
Upper Cretaceous Mesaverde Group of the Book
Cliffs, Utah, could be explained in part as the differential response of Paleozoic-Mesozoic basement
cover to flexural loading during the Sevier orogeny.
Part of the study area is located above the Paradox
Basin, an area of thinned crust overlain by Upper
Paleozoic evaporites and cut by numerous faults.
Overlying this area the sequence stratigraphy of one
interval, equivalent to the Castlegate Sandstone, is
characterized by high-frequency sequences with
well-developed trangressive tidal deposits and
highstand marine units (Fig. 8.58), whereas beyond
the edge of the Paradox Basin these sequences appear to merge into a single sequence spanning
about 5 m.y. The relative changes in sea level that
generated this stratigraphy are attributed to vertical
basement movements driven by in-plane stress
transmitted from the Sevier orogen – the process
discussed by Heller et al. (1993; see Sect. 11.4.2),

and it is suggested that the area of the Paradox
Basin had a lesser structural strength and integrity
during the imposition of these stresses, and therefore demonstrated a more sensitive response to
changes in the stress pattern. Similar basement
tectonic control on older Mesozoic units overlying
the Paradox Basin were described by Baars and
Stevenson (1982) and Baars and Watney (1991).

11.6 Other Speculative Tectonic Hypotheses

Cathless and Hallam (1991) developed a hypothesis
that elaborated on the ideas about intraplate stress,
and suggested a mechanism whereby this stress may
lead to elevation changes in the continents that
could have true global eustatic effects. They suggested that the stress involved in rifting leads to an
increase in crustal density, which propogates across
the lithosphere and affects entire plates in a few tens
of thousands of years. The result is that the continent affected undergoes a transient subsidence (an
actual transgression), and this generates an increase
in the ocean-basin volume, which leads to a global
fall in sea level. Mantle flow set up by isostasy
eventually compensates for the density changes, and
the cycle of sea-level change is completed within
about 0.05 m.y. Cathless and Hallam (1991) proposed this idea in the form of a hypothetical numerical geophysical model. They suggested that the
occurrence of significant sudden biotic extinction
events during the Phanerozoic could be
explained by this model. Some evidence
suggests these extinctions were accompanied
by short term regressive-transgressive couplets of possible worldwide extent. This model to
explain extinctions is intended as an alternative to
the bolide-impact hypothesis, which has received
considerable prominence in recent years. Its possible importance in the history of sea-level changes
remains to be evaluated.

Another idea that is still speculation is that of
rapid epeirogenic movement induced by instabilities
in near-surface mantle convection. Officer and
Drake (1985) proposed the idea to explain observations of warping of modern coastal regions based on
leveling data, and variations in the depth of the
continental shelves, after filtering out the effects of
reef growth, marine erosion, etc. The authors pointed to various types of cycle in the geological record
and suggested mantle instabilities as an alternative
to the more widely accepted tectonic and orbital-

forcing mechanisms. However, their hypothesis is based only on ideas generated by the need to explain the observations of crustal warping, and no numerical geophysical model nor hard evidence from the geological record was offered.

It has been suggested that the Mediterranean Sea became isolated from the world ocean and dried out during the Late Miocene as a result of convergent plate movements (Hsü et al. 1973; Adams et al. 1977). The Mediterranean contains 0.28% of the volume of the world ocean, and Berger and Winterer (1974) calculated that when this water was added to the main body of the ocean it would have raised sea level by about 10–15 m. Did this effect occur at any other time in geological history? Berger and Winterer (1974) suggested that the incipient South Atlantic Ocean might have been similarly isolated, before separation between Africa and South America was complete in the Early Cretaceous. This is a much larger, deeper basin, and its desiccation could have caused a eustatic rise of about 60 m. There is no evidence that this did, indeed occur but it is an idea worth exploring. Thick evaporite deposits occur on the continental margins of western Africa and Brazil (Rona 1982), which is consistent with this possibility. Many oceans develop evaporite basins during the incipient seafloor spreading stage, suggesting a possible desiccation phase, and therefore eustatic rises might be sought during times of continental breakup. However, this immediately preceeds the time of most active seafloor spreading, when eustatic rises are to be expected in response to expansions in the world midoceanic ridge system. Separation of various causes of eustasy might be difficult.

It seems likely that this mechanism is not one that can be called upon to explain regular, frequent eustatic sea-level changes, but might have added its effects spasmodically during times of continental collision or breakup. Rowley and Markwick (1992) suggested that only the following intervals of oceanic isolation and desiccation seem likely: the Mediterranean and the Red Sea during the Miocene, the South Atlantic during the Aptian, and the Gulf of Mexico during the Callovian.

Hodell et al. (1986) argued that the first desiccation event to have been suggested, that of the Mediterranean in the Miocene, may have been triggered by a glacioeustatic fall not by tectonic isolation of the ocean basin. The global reality of the event has now been confirmed by Aharon et al. (1993), who reported evidence from the South Pacific for eight episodes of sea-level fall over a 1-m.y.

period that correlate with the Messinian events. The evidence was derived from a core through a carbonate platform. Strontium isotope geochronology was used to correlate the events, which are indicated by ^{18}O and ^{13}C data that reveal times of fresh-water diagenesis during low sea-level stands.

11.7 Sediment Supply and the Importance of Big Rivers

Sediment supply is controlled primarily by tectonics and climate. In geologically simple areas, where the basin is fed directly from the adjacent margins and source-area uplift is related to basin subsidence, supply considerations are likely to be directly correlated to basin subsidence and eustasy as the major controls of basin architecture. Such is the case where subsidence is yoked to peripheral upwarps, or in proximal regions of foreland basins adjacent to fold-thrust belts. However, where the basin is supplied by long-distance fluvial transportation, complications are likely to arise.

Many sedimentary basins were filled by river systems whose drainage area has been subsequently remodeled by tectonism, and it may take considerable geological imagination to reconstruct their possible past position. For example, McMillan (1973) envisaged a Tertiary river system draining from the continental interior of North America into Hudson Bay. Potter (1978) discussed several similar examples, and pointed out that major river systems may cross major tectonic boundaries, feeding sediment of a petrographic type unrelated to the receiving basin, into the basin at a rate unconnected in any way with the subsidence history of the basin itself. The modern Amazon river is a good example. It derives from the Andean Mountains, flows across and between and is fed from several Precambrian shields, and debouches onto a major extensional continental margin. From the point of view of sequence stratigraphy, the important point is that large sediment supplies delivered to a shoreline may overwhelm the stratigraphic effects of variations in sea level. A region undergoing a relative or eustatic rise in sea level may still experience a major stratigraphic regression if large delta complexes are being built by major sediment-laden rivers.

A significant example of this long-distance sedimentary control is the Cenozoic stratigraphic history of the Gulf of Mexico. This continental margin is fed with sediment by rivers that have occupied essentially the same position since the Early Tertiary (Fisher and McGowen 1967). The rivers feed into the Gulf Coast from huge drainage basins occupying large areas of the North American Interior (Fig. 11.31). Progradation has extended the continental margin of the Gulf by up to 350 km. This has taken place episodically in both time and space, developing a series of major clastic wedges, some hundreds of meters in thickness (Figs. 11.32, 11.33). According to the Exxon sequence-stratigraphy models these clastic wedges would be interpreted as highstand deposits, but Galloway (1989b) showed that their age distribution shows few correlations with the global cycle chart (Fig. 11.34). The major changes along strike of the thickness of these clastic wedges (Fig. 11.33) is also evidence against a control by passive sea-level change. Highly suggestive are the correlations with the tectonic events of the North American Interior, for example, the timing of the Lower and Upper Wilcox Group wedges relative to the timing of the Laramide orogenic pulses along the Cordillera (Fig. 11.34). It seems likely that sediment supply, driven by source-area tectonism, is the major control on the location, timing and thickness of the Gulf Coast clastic wedges. A secondary control is the nature of local tectonism on the continental margin itself, including growth faulting, evaporite diapirism and gravity sliding. As shown by Shaub et al. (1984) and as reemphasized by Schlager (1993), variations in deep-marine sediment dispersal in the Gulf of Mexico show very similar patterns to the coastal and fluvial variations illustrated by Galloway (1989b). Large-scale submarine-fan systems are therefore dependent, also, on considerations of long-term sediment supply variation, which may be controlled by plate-margin tectonism, in-plane stress regime and dynamic topography.

In arc-related basins volcanic control of the sediment supply may overprint the effects of sea-level change. For example, Winsemann and Seyfried (1991) stated:

Sediment supply and tectonic activity overprinted the eustatic effects and enhanced or lessened them. If large supplies of clastics or uplift overcame the eustatic effects, deep marine sands were also deposited during highstand of sea level, whereas under conditions of low sediment input, thin-bedded turbidites were deposited even during lowstands of sea level.

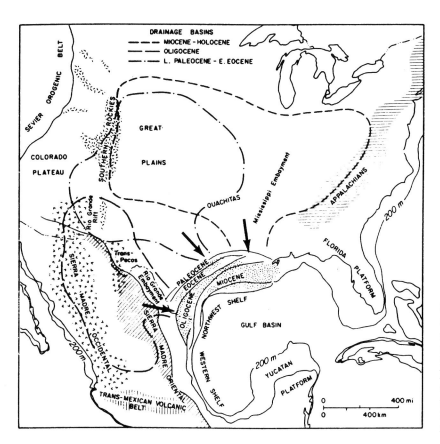

Fig. 11.31. The major drainage basins which fed delta complexes on the Gulf Coast during the Cenozoic. *Arrows* indicate the position of three long-lasting "embayments" through which rivers entered the coastal region. (Galloway 1989b, reproduced by permission)

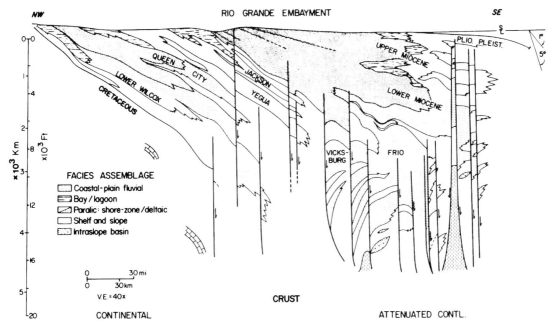

Fig. 11.32. Generalized dip-oriented stratigraphic cross section through the Rio Grande depocenter, in the northwestern Gulf Coast (location shown in Fig. 11.33), indicating principal Cenozoic clastic wedges. Many of these thicken southward across contemporary growth faults. (Galloway 1989b, reproduced by permission)

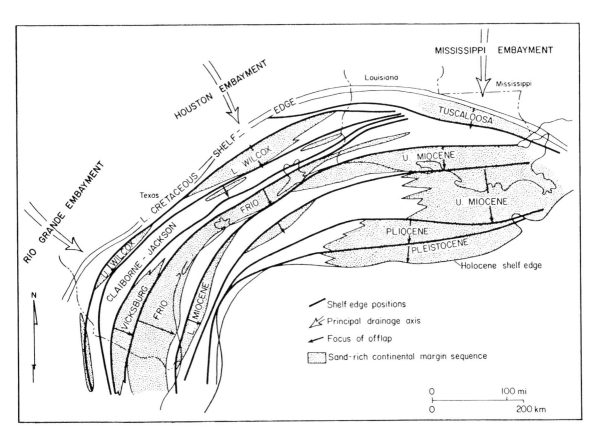

Fig. 11.33. Progradation of the Gulf Coast continental margin by the development of clastic wedges during the Cenozoic. Locations of the three principal embayments through which rivers entered the coastal plain are also shown. (Galloway 1989b, reproduced by permission)

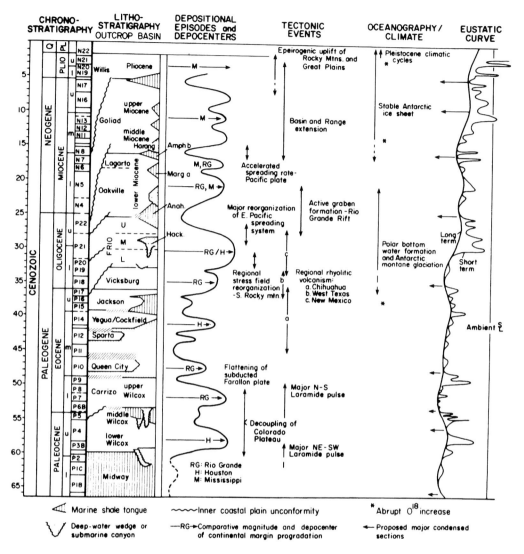

Fig. 11.34. Age of the major clastic wedges along the Gulf Coast, compared with the age range of tectonic events in the North American Interior. The global cycle chart of Haq et al. (1987, 1988) is shown for comparative purposes at the *right*. (Galloway 1989b)

Other examples of the tectonic control of major sedimentary units are provided by the basins within and adjacent to the Alpine and Himalayan orogens. Sediments shed by the rising mountains drain into foreland basins, remnant ocean basins, strike-slip basins, and other internal basins. However, the sediment supply is controlled entirely by uplift and by the tectonic control of dispersal routes. For example, Van Houten (1981) showed how the Oligocene Molasse of the Swiss peripheral basin was deposited by rivers flowing axially along the basin, and that these underwent reversal in transport directions as a result of changes in the configuration of the basin and the collision zone during orogen-

esis. Brookfield (1992, 1993) discussed the shifting of dispersal routes through basins and fault valleys within the Himalayan orogen of central and southeastern Asia. Some of the major rivers in the area (Tsangpo, Salween, Mekong) are known to have entirely switched to different basins during the evolution of the orogen. Much work remains to be done to relate the details of the stratigraphy in these various basins to the different controls of tectonic subsidence, tectonic control of sediment supply, and eustatic sea-level changes.

11.8 Environmental Change

Some sequence boundaries in carbonate sediments are not due to sea-level changes, but to environmental changes that ultimately are related to the tectonic evolution of the area. Two main processes have been documented. The first of these results in what Schlager (1989) termed drowning unconformities. Changes in nutrient supply, water chemistry, or the clastic content of the sea may cause a termination of carbonate-producing biogenic growth. Secondly, surfaces having the same character as sequence boundaries can be generated erosively by submarine currents, such as the Gulf Stream. Schlager (1992a) provided several examples of both these types of process, which are discussed in more detail and illustrated in Section 2.2.1 of this book. He argued that several events in the Exxon global cycle chart were generated by these noneustatic processes.

11.9 Main Conclusions

Regarding extensional basins and continental margins:

1. The stratigraphic architectures that have been used by Vail and his coworkers to interpret sea-level changes on a 10^7-year time scale could be generated by other processes. Coastal onlap (relative rises in sea level) may be caused by flexural downwarp of extensional continental margins. Relative falls in sea level may be caused by thermal doming accompanying rifting.
2. The processes which lead to long-term eustatic sea-level changes, such as changing rates of sea-floor spreading (Chap. 9) also lead to tectonic adjustents of the continents (e.g., initiation of the rift-thermal subsidence cycle), so that relative changes in sea level may have multiple causes which are not necessarily simply interpretable in terms of eustasy.
3. The North Sea Basin, one of the type areas for the Vail curves, was affected by almost continuous tectonism during the Mesozoic and Cenozoic, and many of the sequence-boundary events and stratigraphic architectures there can now be interpreted in terms of specific local or regional episodes of rift faulting and thermal uplift and subsidence.

Regarding convergent plate margins and collision zones, and their associated foreland basins:

4. Arcs are characterized by active regional tectonism as a result of variations in subduction history, arc migration, and the filling and deflation of magma chambers. The rates, magnitudes and episodicities of the relative changes in sea level caused by this tectonism have been determined by detailed mapping in modern arcs and other regions of convergent tectonism, and are consistent with the sea-level changes interpreted to have generated stratigraphic sequences with 10^4- to 10^7-year frequencies that have been mapped in arc-related basins.
5. The various elements of convergent tectonism that generate foreland basins by flexural loading (continental collision, terrane accretion, nappe migration, thrust movement, imbricate fault propogation) can cause relative changes in sea level on a 10^4- to 10^7-year time scale.
6. Stratigraphic sequences of 10^6- to 10^7-year duration in foreland basins, such as the Western Interior of North America and Pyrenean basins, may commonly be correlated with specific tectonic episodes in the adjacent orogen, and there is increasing evidence that high-frequency sequences, of 10^4- to 10^5-year duration, may be correlated with local tectonic events. The correlation of any of these sequences with the global cycle chart may therefore be fortuitous.

Regarding intraplate stress:

7. Stresses generated by plate-margin extension or compression, and transmitted horizontally, have an important modifying effect on the flexural behavior of sedimentary basins, acting to amplify or subdue the flexural deformation caused by regional tectonics. Changes in paleostress fields change these upwarps and downwarps, resulting in offlap and onlap patterns comparable in stratigraphic duration and magnitude to the "third-order" (10^6-year) stratigraphic cyclic changes in the Exxon global cycle chart.
8. High-frequency sequences (of 10^4- to 10^5-year duration), of limited, regional extent, may be generated by the transmission of stresses induced by flexural loading of individual structures and their structural components, as thust faults and their imbricates are propagated during contractional tectonism.
9. Stratigraphic patterns induced by intraplate stress are of opposite sign in basin margins and

basin centers. The proposition of intraplate stress is therefore a testable hypothesis, requiring detailed correlation of unconformities and their correlative conformities across basins.

Regarding the importance of sediment supply:

10. The sediment supply to a basin may not be governed by local tectonics but by tectonic events in distant, structurally unrelated areas, with sediment transported across cratons and through orogens by major rivers. In such cases the distribution and age of fluvial, coastal and submarine clastic wedges does not relate to the sea-level cycle in the basin, but to the tectonic evolution of the sediment source area.

Regarding the importance of environmental change:

11. Carbonate sedimentation may be interrupted by changes in nutrients, water chemistry or clastic content, which inhibit carbonate production. Carbonate platforms are also susceptible to erosion by shifting submarine currents. The result, in all these cases, may be breaks in sedimentation that appear similar to but are unrelated to sequence boundaries produced by sea-level fall.

General conclusions:

12. If tectonic subsidence varies in rate within and between basins, as is typically the case, but is comparable to the rate of eustatic sea-level change, sequence boundaries and flooding surfaces are markedly diachronous. No synchronous eustatic signal can be generated in such cases.

13. The availability of tectonic mechanisms to explain stratigraphic cyclicity of all types and at all geological time scales removes the need for global eustasy as a primary mechanism for the generation of stratigraphic architectures. This being the case, the onus is on the supporters of eustasy to prove their case by quantitative documentation of eustatic processes, and by rigorous global correlation of supposed eustatic events.

14. Given the finite size of the earth, major plate-tectonic events (e.g., collisions, ridge reordering events, changes in rotation vectors), and their stratigraphic responses, may be globally synchronous. The potential therefore exists for global stratigraphic correlation. However, the structural and stratigraphic signature would vary from region to region, such that, for example, a relative sea-level rise in one location may be genetically related to a fall elsewhere.

15. The invocation of tectonic mechanisms to explain stratigraphic architecture has led to the suggestion that Vail's charts be "inverted," to provide a source of data from which tectonic episodicity could be extracted. While this is an interesting idea, there remains the problem that the global accuracy and precision of the Vail curves is in question, and tectonic arguments are in as much danger as eustatic arguments of falling into the trap of false correlation and circular reasoning.

IV Chronostratigraphy and Correlation: Why the Global Cycle Chart Should Be Abandoned

Are sequences in a given basin regional or global in distribution? Or are both types present? If so, how do we tell them apart? The basic premise of the Exxon approach is that there exists a globally correlatable suite of eustatic cycles, and that all stratigraphic data may be interpreted in keeping with this concept. However, this basic premise remains unproven, and it sidesteps the issue of regional tectonic control, of the type demonstrated in Parts II and III of this book. Unfortunately, the global cycle chart is commonly presented as its own proof, with many researchers using the chart as a guide to correlation, and then citing the result as successful test of the chart. One of the critical tests of the Exxon chart is to demonstrate that successions of cycles of the same age do indeed exist in many tectonically independent basins around the world. The chronostratigraphic accuracy and precision of the chart and of the field sections on which it is based are therefore of critical importance. This requires independent studies of sequence stratigraphies in tectonically unrelated basins.

The global cycle chart has become accepted by many as a new standard of geological time. It appears on many key synthesis charts, despite the absence of a rigorous, published examination of the supporting data. Such a casual approach to an apparently important new tool for correlation contrasts dramatically with the enormous effort that has been ongoing since the work of William Smith in the late eighteenth century to refine and calibrate biostratigraphic, radiometric, and magnetostratigraphic times scales. Revisions of these scales appear frequently, whereas the Exxon global cycle chart continues to be accepted as a finished product. Changes in the Exxon chart have been made over the years, and different versions of parts of the chart, published at the same time, actually differ from one another, but the topic of potential error and of changes due to revisions and improvements is not discussed in the supporting literature by Exxon personnel.

The purpose of this part of the book is to examine the nature of the chronostratigraphic record of cycle correlation, to examine the potential for error, and to review the various sea-level curves that have been prepared by Exxon and by other workers, to demonstrate the present level of uncertainty.

- A sea-level curve may be compared with other regional data by plotting it on a chrono-stratigraphic correlation chart for the region.... Such a combination shows the relations of sea-level changes to geological age, distribution of depositional sequences, unconformities, facies and environment, and other information. (Vail et al. 1977, p. 77)

- It is ... important to obtain precise ages of the sequence boundaries and rate and magnitude of apparent sea-level change to establish their global or regional nature. In this context, critical stratigraphic examination of seismic sections across the inner edge of passive margins to the seaward part is of particular importance. (Cloetingh 1988, p. 219)

- Most refinement in correlation derives from a fuller understanding of the properties of the characters employed. There is always room for improvement, however there are several reasons why an apparent age differs more or less from the true age. Most obvious is nonavailability of appropriate characters and gaps in the record which lead to indeterminate results. In other cases human skill or error affect the uncertainty as in observational and experimental errors. These can in some degree be measured as standard errors, or as systematic errors due to mistaken assumptions (e.g., decay constant) or due to inadequate understanding of the characters used, taxonomic lumping, splitting, or mistakes. (Harland 1978, p. 18)

- Biostratigraphy is not without its problem areas.... Variations in the inferred relationship between absolute time and biostratigraphic zones comprise a major area of controversy. (Kauffman and Hazel 1977b, p. iv)

- Without a precise time control the depositional mechanisms forming beds and sequences cannot be sufficiently understood.... timing has remained an elusive problem. Too many inaccuracies are involved in resolving stratigraphic durations, including a large range of error in radiometric age determinations, poor biostratigraphic as well as magnetostratigraphic resolution, and an incompleteness of sedimentary sections. As a result, time estimates are commonly imprecise, and the range of error is often larger than the actual time span considered.... (Ricken 1991, p. 773).

12 Time in Sequence Stratigraphy

12.1 Introduction

One of the features of the existing geological time scale about which Vail et al. (1977) expressed concern is that "the boundaries of the global cycles in several cases do not match the standard epoch and period boundaries." The cycles in the Exxon global cycle chart are bounded by unconformities, and traditional stratigraphic methods also made use of marked breaks in sedimentation, including angular unconformities, as boundaries between many stratigraphic units. Hancock (1977) has described the historical development of the time scale, including the biostratigraphic methods developed by Oppel and D'Orbigny during the nineteenth and early twentieth century. Many of the major boundaries in the Phanerozoic time scale were established in the nineteenth century on the basis of gross lithostratigraphic or paleontological changes, not on detailed zonal studies. For example, the Silurian-Devonian boundary was initially defined in Britain, at a major unconformity and facies change from marine to nonmarine deposits. We can now recognize that these characteristics are a reflection of regional tectonics (the Caledonian Orogeny), and that in other areas (e.g., eastern Europe) sedimentation continued through the lengthy time period represented by this regional unconformity. Accounting for and classifying this "missing" time and deciding how to redefine the Silurian-Devonian boundary constituted a major research project in the 1960s (McLaren 1973). The modern approach to the construction of the time scale recognizes that time is continuous, and that the use of breaks in sedimentation as major chronostratigraphic boundaries does not allow for the calibration of the time missing at such breaks. Hedberg (1976) described a different technique, in which stratigraphers search for continuous sections and, as far as possible, define zone, stage and other boundaries at points in the section where "nothing happened," fixing such boundaries by the use of *golden spikes*. These modern techniques are summarized in many recent textbooks (including Miall 1990, Chap. 3).

One of the major problems with Vail's approach is that it carries the danger of circular reasoning in placing undue emphasis on significant "events" that are presumed to be synchronous in different areas. This presumption makes for difficulties in performing meaningful tests of the global cycle chart, as shown in Chapter 13. Secondly, there is no reason why changes in sea level, which are the events defining stratigraphic sequences, should show any relationship to zone and stage boundaries that are based on biological evolution. Changes in sea level affect the ecology of marine environments, but the relationship of such changes to biotic evolution and extinction is a complex one. We should not therefore expect any relationship between sequence boundaries and biostratigraphically based boundaries in the standard geological time scale.

The purpose of this chapter is to demonstrate the importance of time breaks in the sedimentary record. The next chapter discusses the difficulty in carrying out global correlation of stratigraphic sections, including their contained hiatuses.

12.2 Hierarchies of Time and the Completeness of the Stratigraphic Record

It has become a geological truism that many sedimentary units accumulate as a result of short intervals of rapid sedimentation separated by long intervals of time when little or no sediment is deposited (Ager 1981). Therefore, although time is continuous, the stratigraphic record of time is not. Time, as recorded in the stratigraphic record, is discontinuous on several time scales (Fig. 12.1). Breaks in the record range from such trivial events as the nondeposition or erosion that takes place in front of an advancing bedform (a few seconds to minutes), to the nondeposition due to drying out at

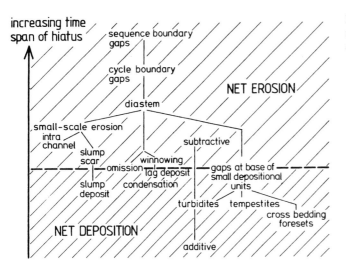

Fig. 12.1. The types of stratigraphic gap, arranged in a hierarchy to illustrate the relationship to time spans. (Ricken 1991)

ebb tide (a few hours), to the summer dry periods following spring run-off events (several months), to the surfaces of erosion corresponding to sequence boundaries (tens to at least hundreds of thousands of years), to the longer breaks caused by tectonism, up to the major regional unconformities generated by orogeny (millions of years). There is a similarly wide variation in actual rates of continuous accumulation, from the rapid sandflow or grainfall accumulation of a cross-bed foreset lamina (time measured in seconds), and the dumping of graded beds from a turbidity current (time measured in hours to days), to the slow pelagic fill of an oceanic abyssal plain (undisturbed for hundreds or thousands of years, or more; Fig. 12.2).

It is now widely realized that rates of sedimentation measured in modern depositional environments or the ancient record vary in inverse proportion to the time scale over which they are measured. Sadler (1981) documented this in detail, using 25,000 records of accumulation rates (Fig. 12.3). His synthesis showed that measured sedimentation rates vary by eleven orders of magnitude, from 10^{-4}–10^7 m/ka. This huge range of values reflects the increasing number and length of intervals of nondeposition or erosion factored into the measurements as the length of the measured stratigraphic record increases. Miall (1991b) suggested that the sedimentary time scale constitutes a natural hierarchy corresponding to the natural hierarchy of temporal processes (diurnal, lunar, seasonal, geomorphic threshold, tectonic, etc.). Crowley (1984) determined by modeling experiments that as sedimentation rate decreases the number of time lines preserved decreases exponentially, and the completeness of the record of depositional events decreases linearly. Low-magnitude depositional events are progressively eliminated from the record.

Many workers, including Berggren and Van Couvering (1978), Ager (1981), Sadler (1981), McShea and Raup (1986), Einsele et al. (1991b), and

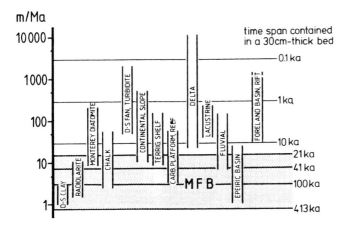

Fig. 12.2. Long-term sedimentation rates for various types of deposit. Time spans at *right* indicate time required for accumulation of a 30-cm bed at the given sedimentation rate. *Shaded area* corresponds to time scales within the Milankovitch frequency band (*MFB*). (Ricken 1991)

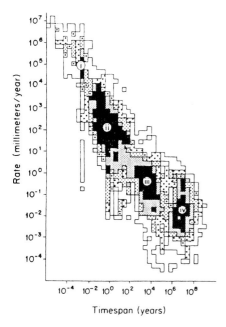

Fig. 12.3. A plot of sedimentary accumulation rates against time span, compiled by Sadler (1981). The data group into four clusters, labeled *i–iv*. Sadler (1981) explained these clusters as follows: *i* continuous observation of a modern environment; *ii* reoccupation of survey stations after a lapse of time; *iii* data points determined by radiocarbon dating; *iv* data points determined by biostratigraphy, calibrated by radiometric dating, magnetostratigraphy, etc. (Reproduced by permission of the University of Chicago Press)

Ricken (1991) have been aware of the hierarchical nature of stratigraphic events, and the problem this poses for evaluating the correlation of events of very different time spans. Algeo (1993) proposed a method for estimating stratigraphic completeness based on preservation of magnetic reversal events. The method is most suitable for time intervals such as the Late Jurassic-Early Cretaceous, and the latest Cretaceous-present, during which periods reversal frequency was in the range 1–5 m.y. At other times, reversal frequency was much lower, and the sensitivity of magnetostratigraphic methods correspondingly less. Miall (1991b) developed a table which classifies depositional units into a hierarchy of physical scales, and compiled data on sedimentation rates for the units at the various scales (Table 12.1). It is important to note that sedimentation rates vary by up to an order of magnitude between one rank of this hierarchy and the next.

A simple illustration of the hierarchy of sedimentation rates, and the important consequences this has for correlation is shown in Fig. 12.4. Se-

dimentation rates used in this exercise and quoted below are based on the compilation of Miall (1991b). The total elapsed time for the succession of four sequences in this diagram is 1 m.y., based on conventional geological dating methods, such as the use of biostratigraphy. However, in modern environments where similar sedimentary successions are accumulating, such as on prograding shorelines, short-term sedimentation rates typically are much higher. Elapsed time calculated on the basis of continuous sedimentation of an individual sequence amounts to considerably less (total of 400 ka for the four sequences), indicating a significant amount of "missing" time (600 ka). This time is represented by the sedimentary breaks between the sequences. Algeo and Wilkinson (1988) concluded, following a similar discussion of sedimentation rates, that in most stratigraphic sections only about one thirtieth of elapsed time is represented by sediment. Even in the deep oceans it has been demonstrated that in some cases "there is almost as much geological time represented by unconformities as there is time represented by sediments" (Aubry 1995).

As discussed below, a further analysis could take into account the rapid sedimentation of individual subenvironments within the shoreline (tidal channels, beaches, washover fans, etc.), and this would demonstrate the presence of missing time at a smaller scale, within the 100 ka represented by each sequence (e.g., Swift and Thorne 1991, Fig. 16, p. 23). Thus, facies successions ("parasequences," in the Exxon terminology; e.g., Van Wagoner et al. 1990) that constitute the components of sequences, such as delta lobes and regressive beaches, represent 10^3–10^4 years and have short-term sedimentation rates up to an order of magnitude higher than high-frequency sequences, in the 1–10 m/ka range.

The important point to emerge from this simple exercise is that the sedimentary breaks between supposedly continuous successions may represent significant lengths of time, much longer than is suggested by calculations of long-term sedimentation rates. This opens the possibility that the sequences that were deposited during the time span between the breaks may not actually correlate in time at all. Physical tracing of sequences by use of marker horizons, mapping of erosion surfaces and sequence boundaries, etc., may confirm the existence of a regional sequence framework, but these sequences could, in principal, be markedly diachronous, and correlation between basins, where no such physical tracing is possible, should be viewed with extreme caution.

Table 12.1. Hierarchies of architectural units in clastic deposits (Miall 1991b)

Group	Time scale of process (years)	Examples of processes	Instanta-neous sed. rate (m/ka)	Fluvial, Deltaic *Miall*	Eolian *Brookfield, Kocurek*	Coastal, Estuarine *Allen[a] Nio and Yang[b]*	Shelf *Dott and Bourgeois[c], Shurr[d]*	Submarine Fan *Mutti and Normark*
1	10^{-6}	burst-sweep cycle		Lamina	Grain flow, grain fall	Lamina	Lamina	
2	10^{-5}–10^{-4}	bedform migration	10^5	Ripple (microform) [1st-order surface]	Ripple	Ripple [E3 surface[a]] [A[b]]	[3-surf in HCS[c]]	
3	10^{-3}	diurnal tidal cycle	10^5	Diurnal dune incr., react. surf. [1st-order surface]	Daily cycle [3rd-order surface]	Tidal bundle [E2 surface[a]] [A[b]]	[2-surf in HCS[c]]	
4	10^{-2} 10^{-1}	neap-spring tidal cycle	10^4	Dune (mesoform) [2nd-order surface]	Dune [3rd-order surface]	Neap-spring bundle, storm, layer [E2[a], B[b]]	HCS sequence [1-surf[c]]	
5	10^0–10^1	seasonal to 10 yr flood	10^{2-3}	Macroform growth increment [3rd-order surface]	Reactivation [2nd-, 3rd-order] surfaces, annual cycle	Sand wave, major storm layer [E1[a], C[b]]	HCS sequence [1-surface[c]]	
6	10^2–10^3	100 year flood	10^{2-3}	Macroform, e.g. point bar levee, splay [4th-order surface]	Dune, draa [1st-, 2nd-order surfaces]	Sand wave field, washover fan [D[b]]	[facies pack-age (V)[d]]	Macroform [5]
7	10^3–10^4	long term geomorphic processes	10^0–10^1	Channel, delta lobe [5th-order surface]	Draa, erg [1st-order, super surface]	Sand-ridge, barrier island, tidal channel [E[b]]	[elongated lens (IV)[d]]	Minor lobe, channel levee [4]
8	10^4–10^5	5th-order (Milankovitch) cycles	10^{-1}	Channel belt [6th-order surface]	Erg [super surface]	Sand-ridge field, c-u cycle [F[b]]	[Regional lentil (III)[d]]	Major lobe [turb. stage: 3]
9	10^5–10^6	4th-order (Milankovitch) cycles	10^{-1}–10^{-2}	Depo. system, alluvial fan, major delta	Erg [super surface]	c-u cycle [G[b]]	[ss sheet (II)[d]]	Depo. system [2]
10	10^6–10^7	3rd-order cycles	10^{-1}–10^{-2}	Basin-fill complex	Basin-fill complex	Coastal-plain complex [H[b]]	[Lithosome (I)[d]]	Fan complex [1]

Hierarchical subdivisions of other authors are given in square brackets. Names of authors are at head of each column, with corresponding superscript letters in the columns. The hierarchy of depositional units and bounding surfaces is discussed in Chap. 15.

Figure 12.5 makes the same point regarding missing time in a different way. Many detailed chronostratigraphic compilations have shown that marine stratigraphic successions commonly consist of intervals of "continuous" section representing up to a few million years of sedimentation, separated by disconformities spanning a few hundred thousand years to more than one million years (e.g., MacLeod and Keller 1991, Fig. 15; Aubry 1991, Fig. 6). The first column of Fig. 12.5, labeled MC (for cycles in the million-year range), illustrates an example of such a succession. Each such cycle may be composed of a suite of high-frequency cycles, such as those in the hundred-thousand-year range, labeled HC in Fig. 12.5. Detailed chronostratigraphic compilations for many successions demonstrate that the hiatuses between the cycles represent as much or more missing time than is recorded by actual sediment (e.g., Ramsbottom 1979; Heckel 1986; Kamp and Turner 1990; for example, see Figs. 7.3, 7.37, 8.18, 10.15). Sedimentation rates calculated for such sequences (Table 12.1) confirm this, and the second column of Fig. 12.5 indicates a possible chronostratigraphic breakdown of the third-order cycles into component Milankovitch-band cycles. Each of these cycles consists of su-

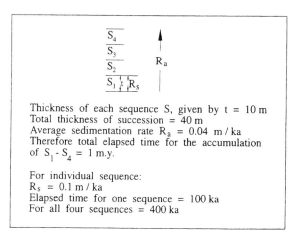

Thickness of each sequence S, given by t = 10 m
Total thickness of succession = 40 m
Average sedimentation rate R_a = 0.04 m / ka
Therefore total elapsed time for the accumulation
of S_1 - S_4 = 1 m.y.

For individual sequence:
R_s = 0.1 m / ka
Elapsed time for one sequence = 100 ka
For all four sequences = 400 ka

Fig. 12.4. Comparison of sedimentation rates measured at two different scales. The typical sedimentation rate for long-term sedimentation at a geomorphic scale, comparable to that which can be measured for many fifth-order and some fourth-order stratigraphic sequences, is 0.1 m/ka (groups 8 and 9 of Table 12.1). Longer-term sedimentation rates, such as those estimated from geological, chronostratigraphic data, are in the order of 0.01 m/ka (group 10, 3rd-order cycles, Table 12.1). Calculations of total elapsed time using these two rates give results that are an order of magnitude different, indicating considerable "missing" time corresponding to nondeposition and erosion

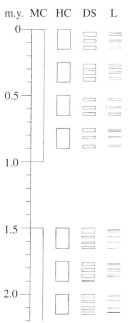

Fig. 12.5. A demonstration of the predominance of missing time in the sedimentary record. Two cycles with frequencies in the million-year range are plotted on a chronostratigraphic scale (*column MC*), and successively broken down into components that reflect an increasingly fine scale of chronostratigraphic subdivision. The second column shows hundred-thousand-year cycles (*HC*), followed by depositional systems (*DS*) and individual lithosomes (*L*), such as channels, deltas, beaches, etc. At this scale chronostratigraphic subdivision is at the limit of line thickness, and is therefore generalized, but does not represent the limit of subdivision that should be indicated, based on the control of deposition by events of shorter duration and recurrence interval (e.g., infrequent hurricanes, seasonal dynamic events, etc.)

perimposed depositional systems (column DS) such as delta or barrier-strandplain complexes, and each of these, in turn, is made up of individual lithosomes (column L), including fluvial and tidal channels, beaches, delta lobes, etc. According to the hierarchical breakdown of Table 12.1 the four columns correspond to sediment groups 10, 9, 8, and 7, in order from left to right. In each case, moving (from left to right) to a smaller scale of depositional unit focuses attention on a finer scale of depositional subdivision, including contained discontinuities. The evidence clearly confirms Ager's (1981) assertion that the sedimentary record consists of "more gap than record."

Devine's (1991) lithostratigraphic and chronostratigraphic model of a typical marginal-marine sequence (Fig. 8.57) demonstrates the importance of missing time at the sequence boundary (his subaerial hiatus). Shorter breaks in his model, such as the estuarine scours, correspond to breaks between depositional systems, but more are present in such a succession than Devine (1991) has indicated. His chronostratigraphic diagram is redrawn in Fig. 12.6 to emphasize sedimentary breaks, and numerous additional discontinuities have been indicated, corresponding to the types of breaks in the record

introduced by switches in depositional systems, channel avulsions, storms and hurricanes, etc. Cartwright et al. (1993) made a similar point regarding the complexity of the preserved record, particularly in marginal-marine deposits, and commented on the difficulty of "forcing-through" meaningful stratigraphic correlations using seismic-reflection data.

The conclusion is that the sedimentary record is extremely fragmentary. A time scale that focuses on continuity is to be preferred over one that is built on unconformities. The modern method of refining the geological time scale, uses "continuous" sections for the definition of chronostratigraphic boundaries and encompasses a method for the incorporation of missing time by defining only the base of chronostratigraphic units, not their tops. If missing time is subsequently documented in the

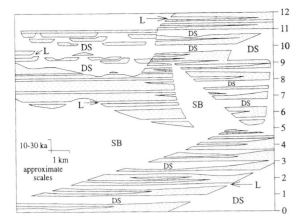

Fig. 12.6. A redrawing of Fig. 8.57B (based on Devine 1991) to emphasize the gaps in the stratigraphic record. The most significant break in sedimentation, labeled *SB*, represents a minor local disconformity between parasequences in Devine's original work. In larger-scale systems this would be the position of the sequence boundary, and would correspond to the kind of gap shown in column HC of Fig. 12.5. Intermediate-scale gaps are those between individual depositional systems (*DS*), and correspond to some of the time-line "events" in Devine's original model (Fig. 8.57A). The smallest gaps (*L*) are those between individual lithosome, caused by storm erosion, unusually low tides, etc. Only a few of these are labeled

stratigraphic record by careful chronostratigraphic interpretation it is assigned to the underlying chronostratigraphic unit, thereby avoiding the need for a redefinition of the unit (Ager 1964; McLaren 1970; Bassett 1985).

In the Exxon work much use is made of the term "correlative conformities," as in their original definition of a sequence as "a relatively conformable succession of genetically related strata bounded at its top and base by unconformities or their correlative conformities" (Vail et al. 1977, p. 210). However, sequence boundaries are diachronous. The transgression and regression that constitute a sea-level cycle generate breaks in sedimentation at different times in different parts of a basin margin. Kidwell (1988) demonstrated that the major break in sedimentation on the open shelf occurs as a result of erosion during lowstand and sediment bypass or starvation during transgression, whereas in marginal-marine environments the major break occurs during regression. The sequence-boundary unconformities are therefore offset by as much as a half cycle between basin-margin and basin-center locations. Kidwell (1988) referred to this process as reciprocal sedimentation. Variations in the balance between subsidence rate, sediment supply and sea-

level change also generate substantial diachronism in sequence boundaries (Sects. 11.1, 16.3).

Given the diachronous nature of sequence boundaries, the accuracy of sequence correlation could be improved by dating of the correlative conformities in deep-marine settings, where breaks in sedimentation are likely to be at a minimum, but it is doubtful if this is often possible. It requires that sequences be physically traced from basin margins into deep-water environments, introducing problems of physical correlation in areas of limited data, and problems of chronostratigraphic correlation between different sedimentary environments in which zonal assemblages are likely to be of different type. As discussed in Chapter 13, problems of correlation across environmental and faunal-province boundaries are often significant. There is also a theoretical debate regarding the extension of subaerial erosion surfaces into the shallow-marine environment (Sect. 15.2.2). In view of the ubiquity of breaks in sedimentation in the stratigraphic record it is arguable whether in fact the concept of the correlative conformity is realistic.

The significance of missing section and the ambiguity surrounding the correlation of unconformities is discussed in Chapter 13.

Correlation of sequence frameworks that cannot be related to each other by physical tracing must be based on methods that provide an adequate degree of accuracy and precision. What is required is correlation to a precision corresponding to a fraction of the smallest possible time span of an individual sequence, at the scale under consideration. Cycles of 10^7- to 10^8-year duration require dating with an accuracy to within a few millions of years. This is not a difficult requirement, at least within marine rocks of Phanerozoic age, and a global framework of second-order cycles is within our grasp. In fact, Sloss began the process of establishing this framework many years ago, as described earlier in this book. Whether the Exxon framework of "second-order" sequences is to be considered reliable is another question, which is addressed in Chapter 14.

Cycles with million-year periodicities require a precision of ±0.5 m.y., or better for reliable correlation. As is demonstrated in Chapters 13 and 14, conventional geological techniques, such as biostratigraphy calibrated with radiometric, chemostratigraphic, and/or magnetostratigraphic data, can rarely attain this degree of precision (except in the youngest Cenozoic), and it is therefore highly questionable whether the data currently permit us to attempt the erection of an interbasinal framework of "third-order" cycles. In special cases the use of high-resolution correlation techniques, including graphic correlation, permit precision to within about ±100 ka in rocks of Early Cenozoic and Mesozoic age (Kauffman et al. 1991). In such cases a framework of regional correlation can be built with some degree of confidence. However, it is doubtful if we are anywhere near ready to extend such frameworks to other basins on different continents, in different tectonic settings, and in different climatic zones.

High-frequency sequences (those with periodicities in the Milankovitch band) cannot yet be reliably correlated interregionally, except for the glacioeustatic sequences of the Late Cenozoic, as documented in Chapter 8. For this most recent interval of geological time the rapid changes in sea level have climatic causes which have left a reliable chronostratigraphic signature in the form of the oxygen-isotope record.

12.3 Main Conclusions

1. Only a small fraction of elapsed time normally is represented by sediment preserved in the geological record.
2. The hierarchy of sequences represent time spans and sedimentation rates that differ from each other by orders of magnitude.
3. Accurate and precise estimates of time spans of the component sequences and hiatuses in a succession cannot be achieved by determining bracketing ages from stratigraphically widely spaced marker beds that enclose several or many sequences.
4. Correlation of sequences requires that they be dated with a precision, or potential error range, equal to a fraction of the range of the shortest sequence in the succession. For example, sequences with 10^6-year periodicities (those comprising the main basis of the Exxon global cycle chart) require a correlation precision of ±0.5 m.y., or less.

13 Correlation, and the Potential for Error

13.1 Introduction

The global cycle model represented a new paradigm in geology when it was proposed by Vail et al. (1977). It was proposed that stratigraphic sequences are chronostratigraphic indicators superior to all other forms of stratigraphic data, because they were generated by synchronous global eustatic processes. According to this view they therefore comprise the ideal basis for a superior standard of geological time. While considerable use is made of biostratigraphic and other conventional data for the dating and correlating of sequences, the new paradigm explicitly subordinated these data to the sequence framework, thereby downplaying the efforts of more than 200 years of stratigraphic research to develop and refine a geological time scale based primarily on painstaking biostratigraphic research (e.g., see Harland et al. 1990; Berggren et al. 1995b). It was on the basis of the new paradigm that the global cycle chart of Vail et al. (1977) and its subsequent revised version (Haq et al. 1987, 1988a; the "Vail curve" or the "Exxon global cycle chart") was built. The main theoretical basis for the paradigm is the supposition that global stratigraphic architecture is controlled primarily by eustatic sea-level changes with an episodicity in the range of 1–10 m.y. – the so-called "third-order cycles" (Chap. 7).

It is commonly forgotten by geologists that the basic premise of the paradigm remains unproven. There is as yet no convincing, independent evidence that a suite of globally correlatable eustatic cycles on this scale exists. The critical test of the Exxon chart is therefore to demonstrate that successions of cycles of the same age do indeed exist in many tectonically independent basins around the world (Miller and Kent 1987; Gradstein et al. 1988; Miall 1992). The chronostratigraphic accuracy and precision of the chart and of the sections on which it is based are therefore of critical importance. This requires independent studies of sequence stratigraphies in many different basins. Unfortunately,

this is often not what is done. While biostratigraphers continue to refine the stratigraphic framework in individual basins, many sequence studies begin by using the Exxon chart as a template for stratigraphic correlation (e.g., Olsson 1988; Baum and Vail 1988; most of the papers in Ross and Haman 1987). Successful correlations are then presented as confirmation that the cycle chart is correct (e.g., Baum and Vail 1988), and excellent local biostratigraphic data may even be distorted or ignored in favor of a correlation with the Exxon curve (Hancock 1993a). The dangers of circular reasoning are obvious.

Virtually none of the events in the Exxon chart has received independent *global* confirmation by careful chronostratigraphic work. A few exceptions might include such major events as the Middle Oligocene eustatic drop related to rapid build-up of Antarctic ice (Pitman 1978; Miller and Kent 1987; but see the discussion of this event in Sect. 10.3). A few specific events in the Cretaceous were discussed by Hancock (1993b), and the topic of possible global eustatic events received lengthy treatment in Hallam (1992). Most of the careful local independent studies that have been carried out (e.g., Hubbard 1988; Carter et al. 1991; Underhill 1991; Hancock 1993a,b) indicate significant differences from the succession of events in the Exxon curve (Figs. 6.15, 14.3). How, then, is the basic premise of the Exxon chart, that globally correlatable cycles actually exist, to receive an independent test and confirmation? I suggest that the Exxon methods, as exemplified by Haq et al. (1988a), are seriously flawed.

It is the main objective of this chapter to argue that current chronostratigraphic dating techniques do not permit the level of accuracy and precision in sequence correlation claimed for the global cycle charts that have been published by Peter Vail and his former Exxon colleagues and coworkers. Regional cyclicity of relative sea level on a 1–10 m.y. ("third-order") time scale can be amply demonstrated from the stratigraphic record (Chap. 8), but we cannot yet convincingly isolate any global eustatic signal. Until this has been done it is premature to construct a "global" cycle chart.

13.2 The New Paradigm of Geological Time?

Peter Vail and his coworkers have made the theoretical basis for their global-eustasy model quite clear. For example, in their first major publication they stated:

One of the greatest potential applications of the global cycle chart is its use as an instrument of geochronology. Global cycles are geochronological units defined by a single criterion – the global change in the relative position of sea level through time. Determination of these cycles is dependent on a synthesis of data from many branches of geology. As seen on the Phanerozoic chart..., the boundaries of the global cycles in several cases do not match the standard epoch and period boundaries, but several of the standard boundaries have been placed arbitrarily and remain controversial. Using global cycles with their natural and significant boundaries, an international system of geochronology can be developed on a rational basis. If geologists combine their efforts to prepare more accurate charts of regional cycles, and use them to improve the global chart, it can become a more accurate and meaningful standard for Phanerozoic time. (Vail et al. 1977, p. 96).

This approach has been used throughout the Exxon work. For example, with reference to the Jurassic of the North Sea, Vail and Todd (1981, p. 217) stated that "several unconformities cannot be dated precisely; in these cases their ages are based on our global cycle chart, with age assignment made on the basis of a best fit with the data." An example of this approach is given later in this same paper (p. 230) where Vail and Todd stated that, "the Late Pliensbachian hiatus described by Linsley et al. (1979) fits the basal Early Pliensbachian sequence boundary on our global cycle chart." In other words, the age assignment of the earlier workers is subordinated to the sequence framework. The Pliensbachian stage is now estimated to span approximately 7 m.y., which provides an indication of the magnitude of the revision Vail and Todd (1981) are willing to make based on their sequence analysis. Vail et al. (1984, p. 143) stated that "interpretations [of stratigraphic sequences] based on lithofacies and biostratigraphy could be misleading unless they are placed within a context of detailed stratal chronostratigraphic correlations." The context of the word "chronostratigraphic" in this reference implies correlation by tracing seismic reflections.

Baum and Vail (1988, p. 322) stated that "sequence stratigraphy offers a unifying concept to divide the rock record into chronostratigraphic units, avoids the weaknesses and incorporates the strengths of other methodologies, and provides a global framework for geochemical, geochronological, paleontological, and facies analyses." Baum and Vail commented on the inconsistent placement of stage boundaries in the Cenozoic section of the Gulf Coast. Some are at sequence boundaries, others are at major surfaces within sequences, such as transgressive surfaces. They recommended the use of a sequence framework for redefining the stages, and defining the stage boundaries at the correlative conformities of the sequences. This approach indicates a misunderstanding of the modern concept of a chronostratigraphic time scale, in which boundaries are defined within continuous sections independent of tectonic, eustatic or other "events" that may or may not be global in scope.

An example of a standard Exxon-type sequence analysis was provided by Mitchum and Uliana (1988). Their correlation of a carbonate basin-margin section in a backarc setting with the global cycle chart was done on the basis of a general positioning of the stratigraphy within the Tithonian-Valanginian interval, by comparison (not detailed correlation) of the subsurface with nearby outcrops, where ammonite zonation had been carried out. No faunal data were available from the wells used to correlate the seismic section! However, the pattern of seismic sequence boundaries was said to match the global pattern for this interval.

In one of his more recent overview papers Vail et al. (1991) stated that sequences "can be used as chronostratigraphic units if the bounding unconformities are traced to the minimal hiatus at their conformable position and age dated with biostratigraphy." (p. 622) and "Sequence cycles provide the means to subdivide sedimentary strata into genetic chronostratigraphic intervals.... Sequences, systems tract, and parasequence surfaces provide a framework for correlation and mapping" (p. 659).

As is made clear by these quotes, the Exxon approach subordinates biostratigraphic and other data to the sequence framework, where conflicts arise. It fails to recognize the fundamentally independent nature of these data. The Exxon method is essentially that summarized in Fig. 13.1A. The assumption is made that an important sea-level event in any given stratigraphic section represents a eustatic event. From the paradigm of global eustasy it then

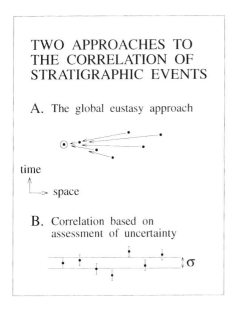

Fig. 13.1. Two approaches to the correlation of the same set of stratigraphic events (*dots*). **A** A sea-level event in one section is assumed to represent global eustasy, and it is assumed that all other events correlate with it. **B** Age, with *error bar*, is determined independently for each location, with the degree of correlation determined by statistical means

follows that a comparable pattern of sea-level events in other sections is by definition correlated with the original event, even when the chronostratigraphic data may not support such correlations. The application of this approach to correlation to construct a "global master curve" is illustrated in Fig. 13.2. Figure 13.1B illustrates schematically the quantitative approach to correlation, which is to attach error bars to assigned ages, based on numerical estimates of age, calculations from sedimentation rates, etc. A correlation band may then be erected based on standard error expressions, such as standard deviations. The accuracy and precision of correlations can readily be assessed from such diagrams.

Miller and Kent (1987), while arguing for the need for careful chronostratigraphic correlation, pointed out that "the durations of the third-order cycles are at the limit of biostratigraphic resolution." They went on to state:

We agree that to test the validity of the third-order cycles it is not necessary to establish that *every* [their emphasis] third-order cycle is precisely the same age on different margins. Haq et al. (1987) utilized a sequence approach to recognize third-order events above known datum levels.

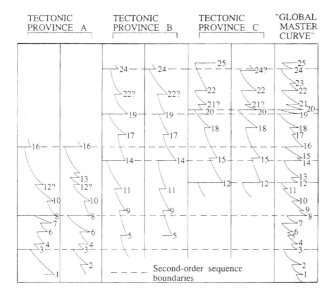

Fig. 13.2. The construction of a "global master curve" of stratigraphic events from local sequences of events in three separate tectonic provinces, based on the assumption that each event is globally synchronous. The "master list" of events (numbered *1–25*) is not present in complete form in any single locality, and very few individual events may be correlated between tectonic provinces. Events indicated with *question marks* may or may not be correlated with those in other tectonic provinces, depending on assumptions that must be made about acceptable im

precision in chronostratigraphic correlation. The curves are drawn in this diagram to simulate the Exxon method of plotting coastal onlap. Some of the larger onlap events are labeled as "second-order" sequence boundaries, and may reflect regional plate-tectonic events. The other events reflect regional tectonic events related to in-plane stress, basin subsidence, etc. In the Exxon method all the events are interpreted to be eustatic (global) in origin. "Missing" events at any given locality are attributed to local tectonism, removal by erosion, etc.

Assuming that they observed the same patterns on different margins, their observation of the same ordinal hierarchy of events within a given time window on different margins argues against a local cause and points to eustatic control.... However, the simple matching of third-order cycles between locations is complicated by gaps in the records, uncertainties in establishing datum planes, and the ability to discriminate between these cycles at the outcrop level.

Miller and Kent (1987), Christie-Blick et al. (1988), Gradstein et al. (1988), and Miall (1991a) have pointed out some of the problems and imprecisions in chronostratigraphic correlation. Ricken's (1991) comment regarding chronostratigraphic imprecision and error is quoted in the introduction to Part IV of this book.

The dangers inherent in simple pattern recognition, of the type alluded to by Miller and Kent (1987) are that, given the density of stratigraphic events present in the Vail curve, there is literally an "event for every occasion" (Miall 1992). Practically any stratigraphic succession can be made to correlate with the Vail curve, even synthetic sections constructed from tables of random numbers (Miall 1992; Sect. 13.4.3). Dickinson (1993), in a discussion of Miall (1992), demonstrated that the average duration of the main ("third-order") cycles in the Exxon chart increases with age, and suggested that this reflects a decrease in the quality of the sequence data in older sections – in other words, the event spacing is at least in part an artefact of the data quality and the analytical methodology used to construct the chart. Haq et al. (1988b) referred to a procedure of "rigorous pattern matching" of sequences and systems tracts, but have nowhere described their methodology or attempted to quantify the degree of "rigor."

What if Vail and his colleagues are wrong? What if there is in fact no suite of global cycles dependent on and accurately reflecting a history of eustatic sea-level change? Vail's basic premise has not yet been proved. Part III of this book demonstrates in detail that many alternative mechanisms exist for the generation of regional suites of stratigraphic sequences, at comparable rates to the major eustatic processes, and producing stratigraphic results of identical character. There is therefore no reason for favoring Vail's interpretation of the cycle charts over, say, regional tectonic mechanisms, unless very accurate and precise global correlations between the cycles can be demonstrated. *This is the only possible independent proof of Vail's paradigm.* The demonstration of global correlation of sequence frameworks between distant and tectonically unrelated

basins is the essential prerequisite before Vail's paradigm can be taken seriously as a testable geological hypothesis. The Exxon workers have not even begun to address the serious problems that must be dealt with before such a proof can be developed. It is the purpose of Part IV of this book to demonstrate that existing methodologies are at present not capable of providing such precise correlations for cycles with periodicities of 10^6 years and less, except in certain exceptional situations. We are many years away from even knowing whether it will ever be possible to construct a chart of global cycles in the million-year range. *We do not yet know that such a framework exists.*

It is important to remember that where tectonic subsidence and eustatic sea-level change have comparable rates, a synchronous eustatic signal may not be preserved in the geological record (Sect. 11.1). Therefore, even where eustasy was an important process it may be difficult to demonstrate this from the stratigraphic record. Conversely, the absence of a synchronous record does not disprove the occurrence of eustatic sea-level change.

13.3 The Dating and Correlation of Stratigraphic Events: Potential Sources of Uncertainty

The dating and correlation of stratigraphic events between basins, where physical tracing-out of beds cannot be performed, involves the use of biostratigraphy and a variety of other chronostratigraphic methods. The process is a complex one, fraught with many possible sources of error. Many textbooks and review articles have dealt with various aspects of this subject, but practical reviews for the working geologist have not been developed. What follows is an attempt to break the process down into a series of discrete "steps," although in practice, dating successions and erecting local and global time scales is an iterative process, and no individual stratigrapher follows the entire procedural order as set out here. The geologist is able to draw on the accumulated knowledge of the geological time scale that has (as noted earlier) been undergoing improvements for more than 200 years, but each new case study presents its own unique problems.

There are two related but distinct problems to be addressed in the construction of a global stratigraphic framework. On the one hand we need to test whether similar stratigraphic events (such as sequence boundaries) and successions of events (e.g., succesions of sequences) have the same ages in different parts of the earth, to test models of global causality (e.g., eustasy). From this perspective precise correlation is critical, but determination of absolute age is not, although all the Exxon publications on the global cycle chart appear to suggest that absolute age is of primary importance by the way the cycle charts have been annotated. On the other hand there is the ongoing effort to determine the precise age of stratigraphic events, leading to continuing refinements in the global time scale. In either case the potential error, or uncertainty, involved with one or more of the "steps" described above must be acknowledged and addressed as part of a determination of the level of refinement to be expected in stratigraphic interpretation.

Standard correlation methods are discussed by Miall (1990; Chap. 3) and in several standard textbooks on stratigraphy. Harland et al. (1990) provided what is probably the most thorough and

scholarly discussion of the development of the geological time scale. Kauffman and Hazel (1977a) edited a valuable collection of papers containing many different types of biostratigraphic study, and including a useful historical article by Hancock (1977). Agterberg (1990) and Guex (1991) presented detailed descriptions of quantitative methods for correlation. Numerous books and papers have addressed the subject of biogeography and faunal provincialism. Review articles by Ludvigsen et al. (1986) and Smith (1988) provide useful summaries.

Six main "steps" are involved in the dating and correlation of stratigraphic events (Miall 1991a; 1994). Figure 13.3 summarizes these steps and provides generalized estimates of the magnitude of the uncertainty associated with each aspect of the correlation and dating of the stratigraphic record. Some of these errors may be cumulative, as discussed in the subsequent sections. The assignment of ages and of correlations with global frameworks is an iterative process that, in some areas, has been

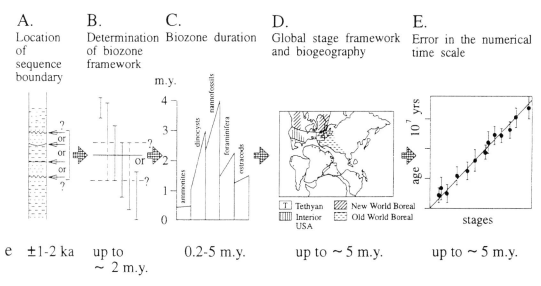

Fig. 13.3. Steps in the correlation and dating of stratigraphic events. *e* Typical range of error associated with each step. **A** In the case of the sequence framework, location of sequence boundaries may not be a simple matter, but depends on interpretation of the rock record using sequence principles. **B** Assignment of the boundary event to the biozone framework. An incomplete record of preserved taxa (almost always the case) may lead to ambiguity in the placement of biozone boundaries. **C** The precision of biozone correlation depends on biozone duration. Shown here is a simplification of Cox's (1990) summary of the duration of zones in Jurassic sediments of the North Sea Basin. **D** The building of a global stage framework is fundamental to the development of a global time scale. However, global correlation is hampered by faunal provincialism. Shown here is a simplification of the

faunal provinces of Cretaceous ammonites, shown on a Middle Cretaceous plate-tectonic reconstruction. Based on Kennedy and Cobban (1977) and Kauffman (1984). **E** The assignment of numerical ages to stage boundaries and other stratigraphic events contains inherent experimental error and also the error involved in the original correlation of the datable horizon(s) to the stratigraphic event in question. Diagrams of this type are a standard feature of any discussion of the global time scale (e.g., Haq et al. 1988a; Harland et al. 1990). The establishment of a global biostratigraphically based sequence framework involves the accumulation of uncertainty over steps *A–D*. Potential error may be reduced by the application of radiometric, magnetostratigraphic or chemostratigraphic techniques which, nonetheless, contain their own inherent uncertainties (step **E**)

underway for many years. There is much feed-back and cross-checking from one step to another. What follows should be viewed therefore as an attempt to break down the practical business of dating and correlation into more readily understandable pieces, all of which may be employed at one time or another in the unraveling of regional and global stratigraphies. The main steps are as follows:

1. Identification of sequence boundaries. Determining the position of the sequence boundary may or may not be a straightforward procedure. There are several potential sources of error and confusion (Sect. 13.3.1).
2. Determining the chronostratigraphic significance of unconformities. Unconformities, such as sequence boundaries, represent finite time spans which vary in duration from place to place. In any given location this time span could encompass the time span represented by several different sedimentary breaks at other locations (Sect. 13.3.2).
3. Determination of the biostratigraphic framework. One or more fossil groups is used to assign the selected event to a biozone framework. Error and uncertainty may be introduced because of the incompleteness of the fossil record (Sect. 13.3.3.1).
4. Assessment of relative biostratigraphic precision. The length of time represented by biozones depends on such factors as faunal diversity and rates of evolution. It varies considerably through geological time and between different fossil groups (Sects. 13.3.3.2, 13.3.4, 13.3.5).
5. Correlation of biozones with the global stage framework. The existing stage framework was, with notable exceptions, built from the study of macrofossils in European type sections. Correlation with this framework raises questions of environmental limitations on biozone extent, our ability to interrrelate zonal schemes built from different fossil groups, and problems of global faunal and floral provinciality (Sect. 13.3.6).
6. Assignment of absolute ages. The use of radiometric and magnetostratigraphic dating methods, plus the increasing use of chemostratigraphy (oxygen and strontium isotope concentrations) permits the assignment of absolute ages in years to the biostratigraphic framework. Such techniques also constitute methods of correlation in their own right, especially where fossils are sparse (Sect. 13.3.7).

13.3.1 Identification of Sequence Boundaries

The first step is that a well section or outcrop profile is analyzed and the positions of sequence boundaries are determined from the vertical succession of lithofacies. The possible errors in this procedure arise from the potential for confusion between allogenic and autogenic causes for the breaks in the stratigraphic record. Autogenic causes include condensed sequences, ravinement surfaces, and channel scours. Various other processes also generate sedimentary breaks that may be confused with sequence boundaries, such as environmental change in carbonate settings. The subject is discussed at some length in Section 2.2.1. These errors could lead to possible errors in boundary placement of at least several meters. Given a typical sedimentation rate of 0.1 m/ka, a 10-m error is equal to 100 ka.

Armentrout et al. (1993) used petrophysical logs tied to seismic cross sections to develop a correlation grid in Paleogene deposits in the North Sea. In this basin, where wells are up to tens of kilometers apart, they determined that tracing markers around correlation loops could result in mismatches of up to 30 m. However, errors of this magnitude would not be expected in mature areas, such as the Alberta Basin or the Gulf Coast.

13.3.2 Chronostratigraphic Meaning of Unconformities

Assigning an age to an unconformity surface is not necessarily a simple matter, as illustrated in the useful theoretical discussion by Aubry (1991). An unconformity represents a finite time span; it may have a complex genesis, representing amalgamation of more than one event. It may also be markedly diachronous, because the transgressions and regressions that occur during the genesis of a stratigraphic sequence could span the entire duration of the sequence. As noted in Section 12.2, Kidwell (1988) demonstrated that this results in an offset in sequence boundary unconformities by as much as one half of a cycle between basin center and basin margin.

Unconformities actually represent amalgamations of two surfaces, the surface of truncation of older strata and the surface of transgression of younger strata. These two surfaces may vary in age considerably from location to location, as indicated by chronostratigraphic diagrams (e.g., Vail et al. 1977, Fig. 13, p. 78). Even if the two surfaces have

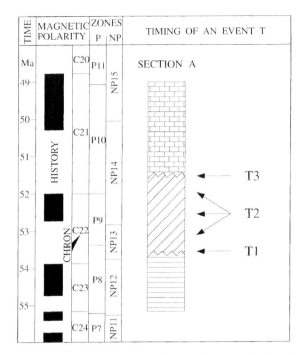

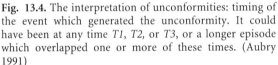

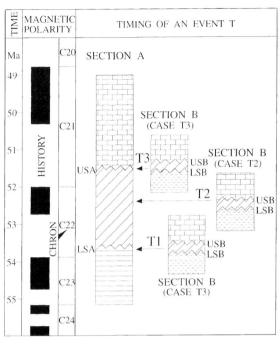

Fig. 13.4. The interpretation of unconformities: timing of the event which generated the unconformity. It could have been at any time *T1*, *T2,* or *T3*, or a longer episode which overlapped one or more of these times. (Aubry 1991)

Fig. 13.5. The interpretation of unconformities: The generation of a single major unconformity by the amalgamation of more than one event. (Aubry 1991)

been dated this still does not provide an accurate estimate of the timing of the event or events that generated the unconformity. As shown in Fig. 13.4, a fall in relative sea level may have occurred at time T1, in which case no erosion or deposition took place prior to the deposition of the overlying sequence. An alternative is that sedimentation continued until time T3, followed by a rapid erosional event and transgression. Or the sea-level event may have occurred at any time T2.

A given major unconformity may represent the combined effects of two or more unrelated sea-level events, of eustatic and tectonic origin (Fig. 13.5). Recognizing the occurrence of more than one event requires the location and dating of sections where the hiatus is short, as in the various possible sections B in Fig. 13.5. As explained by Aubry (1991, p. 6646);

If an unconformity Y with a short hiatus on the shelves of basin A can be shown to be exactly correlative (=isochronous=synchronous) in the stratigraphic sense with an unconformity y with a short hiatus on the shelves on basin B (i.e., if the two hiatuses overlap almost exactly), it is probable, although not certain, that both unconformities Y and y are correlative, in the genetic sense, with a unique event T. Overlap between hiatuses of stra-

tigraphically correlatable unconformities in two widely separated basins fulfills a condition required but insufficient to establish global eustasy. Unconformities Y and y become a global eustatic signal if other correlative (=synchronous) unconformities ... with short hiatus can be recognized on as many shelves as possible of widely separated basins.

Even this procedure begs the question of the interpreted duration of sea-level events. Given the rapid events that characterize glacioeustasy (frequency in the 10^4–10^5-year range) this is not a trivial question. Aubry's figures (Figs. 13.4, 13.5) were drawn to illustrate actual problems which arose in attempts to assign ages to unconformities in Eocene sections. The difference between times T1 and T3 in Fig. 13.4 is 2 m.y., which represents a significant potential range of error. It could correspond to one entire million-year sequence.

Vail and his coworkers assert that sequence boundaries are the same age everwhere, despite the influence of tectonism, sediment supply or other factors (e.g., Vail et al. 1991). They believe that sediment-supply factors may affect shoreline position but not the timing of sequence boundaries, based on the models of Jervey (1988). However, this conclusion is not supported by later modeling work.

Theoretical studies by Pitman and Golovchenko (1988) demonstrated a phase-lag between sea-level change and the stratigraphic response, particularly in areas of slow subsidence and sea-level change, and rapid sediment supply. This was confirmed by the computer modeling experiments of Jordan and Flemings (1991), who stated that, "the sequence boundary for an identical sea-level history could be of different ages and the ages could differ by as much as 1/4 cycle." Schlager (1993) and Martinsen and Helland-Hansen (1994) pointed out that the deltas of big rivers such as the Mississippi and the Rhône have prograded while the nearby coast is still retreating as a result of the postglacial sea-level rise. In effect, this is equivalent to a 1/4-phase shift in the sequence cycle, a difference between "highstand" deltas and "transgressive" shorelines. The lag may vary from basin to basin depending on variations in sediment flux, subsidence rates, and the amplitude of the sea-level cycle. In the case of the 10^6-year ("third-order") cycles considered in these experiments, this amounts to a variation of up to several millions of years. This work by Pitman, Jordan and their coworkers is important because it indicates that sequence boundaries are inherently imprecise recorders of sea-level change (see also Fig. 11.1 and discussion thereof). Practical examples of this were illustrated by Leckie and Krystinik (1993). Additional results of computer models are described in Chapter 16.

13.3.3 Determination of the Biostratigraphic Framework

Sequence boundaries are dated and correlated by biostratigraphic, radiometric or magnetostratigraphic techniques. The potential for error here stems from the imperfection of the stratigraphic record, including its fossil content. For example, first and last taxon occurrences could vary in position by several meters to even a few tens of meters among sections of the same age. Errors in age assignment of complete biozones or half zones are not uncommon.

The biostratigraphic data base for the global time scale has experienced an orders-of-magnitude expansion since microfossils began to take on increasing importance in subsurface petroleum exploration in the postwar years, and following the commencement of the DSDP in 1968. However, the main basis for the global time scale remains the classical macrofossil assemblages (e.g., ammonites, graptolites) that have been in use, in many cases, since the nineteenth century. Supplementing macrofossil and microfossil collections from outcrops, drilling operations on land and beneath the oceans have contributed a vast biostratigraphic sample base over the last few decades, permitting the evaluation of a wide range of microfossils for biostratigraphic purposes in all the world's tectonic and climatic zones. The result has been a considerable improvement in the flexibility and precision of dating methods, and much incremental improvement in the global time scale. Several useful reviews of this area of research are contained in the synthesis of the first decade of DSDP edited by Warme et al. (1981).

As noted by Cope (1993) detailed, meticulous taxonomic work still holds the potential for much improved biostratigraphic resolution and flexibility at all levels of the geological column. In older Mesozoic and Paleozoic strata such fossil groups as conodonts and graptolites have yielded very refined biostratigraphic zonal systems. However, this potential has yet to be fully tapped by sequence stratigraphers.

The various types of zones, and the methods for the erection of zonal schemes, are subjects dealt with in standard textbooks and reviews (e.g., Miall 1990, Chap. 3; Kauffman and Hazel 1977a) and are not discussed here. The purpose of this section is to focus on two major problems that affect the accuracy and precision of biostratigraphic correlation.

13.3.3.1 The Problem of Incomplete Biostratigraphic Recovery

Because of incomplete preservation, poor recovery, or environmental factors, the rocks rarely yield a complete record through time of each biozone fauna or flora. Practical, measured biostratigraphic time, as indicated by the imperfect fossil record, may be different from hypothetical "real" time, which can rarely be perfectly defined (Murphy 1977; Johnson 1992). This is illustrated in Fig. 13.6, in which the difference between observed lines of correlation and hypothetical but invisible and unmeasurable time lines is indicated by gaps labeled "r."

How important are the "r" gaps shown in Fig. 13.6? Many researchers have attempted to sidestep this problem, by treating fossil occurrences quantitatively, and applying statistical treatments to assessments of preservation and correlation (e.g., Riedel 1981; McKinney 1986; Agterberg 1990; Guex 1991). However, the method of graphic correlation

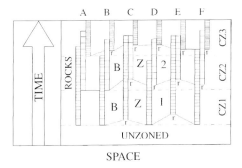

Fig. 13.6. Hypothetical time-rock diagram, showing the ranges of three species as recorded in six stratigraphic sections. The first-appearances of the three species in section *A* are used to define biozones *BZ1* and *BZ2*, the recorded extent of which is shown by the *heavy lines* connecting the six sections. *Dashed lines* extending from section *A* indicate ideal chronostratigraphic time lines. In most sections these differ from the biozone boundaries by a time increment *r*, resulting from failure of the fossilization process, sample spacing, or environmental factors. (Johnson 1992)

(Sect. 13.3.4), which focuses attention on the incompleteness of the fossil record, allows us to examine this question. Results presented by Edwards (1989) indicate that differences of up to 10 m between biostratigraphic and hypothetical "real" time lines are not uncommon. Doyle (1977) illustrated in detail an attempt to correlate two wells using palynological data (Fig. 13.7). Ranges of correlation error up to 30 m are apparent from his data, and result from incomplete preservation or wide sample spacing. At a sedimentation rate of 0.1 m/ka 30 m of section is equivalent to 300 ka, more than enough to lead to miscorrelation of sequences in the Milankovitch band.

Conventional wisdom has it that pelagic organisms are the best biostratigraphic indicators because they are widely distributed by ocean currents and tend to be less environmentally sensitive. However, recent detailed studies of nektonic and planktonic forms have indicated a wide range of factors that lead to uneven distribution and preservation even of these more desirable forms. Thierstein (1981) and Roth and Bowdler (1981), in studies of Cretaceous nannoplankton, discussed a variety of mechanisms that affected distribution of these organisms. Among the most important environmental factors are changes in ocean currents, which affect water temperatures and the concentrations of oxygen and carbon dioxide and nutrients in the seas. Changes in temperature and sea level affect the position of the carbonate compensation depth in the oceans, and hence affect the rates of dissolution and preservability of calcareous forms in deep-sea sediments. Conversely, bottom dwelling faunas, which are specifically limited in their geographical distribution because of ecological factors, may exhibit a diversity and rapid evolutionary turn-over that makes them ideal as biostratigraphic indicators in specific stratigraphic settings. For example, many corals have been found to be of great biostratigraphic utility in the study of reef limestones of all ages.

13.3.3.2 Diachroneity of the Biostratigraphic Record

Another common item of conventional wisdom is that evolutionary changes in faunal assemblages are dispersed so rapidly that, on geological time scales, they can essentially be regarded as instantaneous. This argument is used, in particular, to justify the interpretation of what biostratigraphers call "first-appearance datums" (FADs) as time-stratigraphic events (setting aside the problems of preservation discussed in Sect. 13.3.3.1). However, this is not always the case. Some examples of detailed work have demonstrated considerable diachroneity in important pelagic fossil groups.

MacLeod and Keller (1991) explored the completeness of the stratigraphic sections that span the Cretaceous-Tertiary boundary, as a basis for an examination of the various hypotheses that have been proposed to explain the dramatic global extinction occurring at that time. They used graphic correlation methods, and were able to demonstrate that many foraminiferal FADs and LADs (last-appearance datums) are diachronous. Maximum diachroneity at this time is indicated by the species *Subbotina pseudobulloides*, the FAD of which may vary by up to 250 ka between Texas and North Africa. However, it is not clear how much of this apparent diachroneity is due to preservational factors.

An even more startling example of diachroneity is that reported by Jenkins and Gamson (1993). The FAD of the Late Cenozoic foraminifera *Globorotalia truncatulinoides* differs by 600 ka between the southeastern Pacific Ocean and the North Atlantic Ocean, based on analysis of much DSDP material. This is interpreted as indicating the time taken for the organism to migrate northward from the South Pacific following its first evolutionary appearance there. As Jenkins and Gamson (1993) concluded "the implications are that some of the well documented evolutionary lineages in the Cenozoic may show similar patters of evolution being limited to

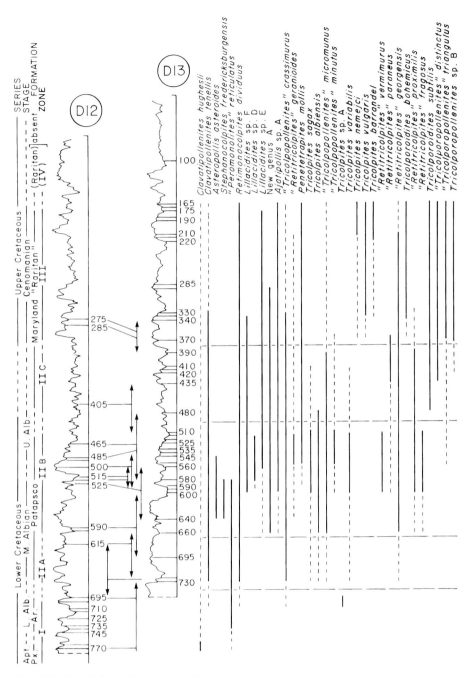

Fig. 13.7. Correlation of two wells through Cretaceous strata in Delaware, using palynological data. Correlation brackets (*double-headed arrows*) terminate just above and below samples in well *D13* that bracket the age of the indicated sample in well *D12*. (Doyle 1977)

discrete ocean water masses followed by later migration into other oceans.... If this is true, then some of these so-called 'datum planes' are diachronous." This conclusion is of considerable importance, because the result is derived from excellent data, and can therefore be regarded as highly reliable, and deals with one of the most universally preferred

fossil groups for Mesozoic-Cenozoic biostratigraphic purposes, the foraminifera. It would appear to suggest a limit of up to about one half million years on the precision that can be expected of any biostratigraphic event.

The two cases reported here may or may not be a fair representation of the magnitude of diachroneity in general. After a great deal of study, experienced biostratigraphers commonly determine that some species are more reliable or consistent in their occurrence than others. Such forms may be termed "index fossils," and receive a prominence reflecting their usefulness in stratigraphic studies. Studies may indicate that some groups are more reliable than others as biostratigraphic indicators. For example, Ziegler et al. (1968) demonstrated that brachiopod successions in the Welsh Paleozoic record were facies controlled and markedly diachronous, based on the use of the zonal scheme provided by graptolites as the primary indicator of relative time. Armentrout (1981) used diatom zones to demonstrate that molluscan stages are time transgressive in the Cenozoic rocks of the northwestern United States. Wignall (1991) demonstrated the diachroneity of Jurassic ostracod zones.

13.3.4 The Value of Quantitative Biostratigraphic Methods

Much work has been carried out in attempts to apply quantitative, statistical methods to biostratigraphic data, to refine stratigraphic correlations and to permit these correlations to be evaluated in probabilistic terms. Excellent syntheses have been provided by Agterberg (1990) and Guex (1991). However, many problems remain, because the biostratigraphic data base does not necessarily meet some of the necessary assumptions required for statistical work. Guex (1991, p. 179), in discussing the use of multivariate methods, quoted Millendorf and Heffner (1978, p. 313), who stated:

This approach ignores the effects of faunal gradation within an isochronous unit with respect to geographic position. Thus, if the faunal composition of such an isochronous unit changes across the study area, samples taken from distant points in the unit might be dissimilar enough not to cluster. Simply, the greater the lateral variation and the larger the distance between them, the less similar are the two isochronous samples.

Guex (1991, p. 180) himself stated: "In one way or another, all methods based on global resemblance between fossil samples end in fixing the boundaries of statistical biofacies, and they do not make it possible to find, within a fossil assemblage, the species that are characteristic of the relative age of the deposits under study."

Biofacies distributions are determined in part by environment. The definition of "statistical biofacies" is therefore not a very uesful contribution to problems of correlation. Not only biofacies, but sampling methods and questions of preservation

also affect the distribution of fossils (e.g., Agterberg 1990, Chap. 2). It is questionable therefore whether statistical methods can assist directly with solving the problem of assessing error in global correlation.

There is one important exception to this generalization, and that is a technique known as graphic correlation. The method was proposed by Shaw (1964), and has been developed by Miller (1977) and Edwards (1984, 1989). The procedure is summarized by Miall (1990, p. 115–118). Mann and Lane (1995) edited a collection of papers describing modern methods and case studies. The data on foraminiferal diachroneity at the Cretaceous-Tertiary boundary quoted above from MacLeod and Keller (1991) were obtained using the graphic correlation method. The statistical methods that have evolved for ranking, scaling and correlation (Agterberg 1990; Gradstein et al. 1990) are a form of quantified graphic correlation.

As with conventional biostratigraphy, the graphic method relies on the careful field or laboratory recording of occurrence data, and focuses on the collation and interpretation of FADs and LADs. The objective is to define the local ranges for many taxa in at least three complete sections through the succession of interest. The more sections that are used, the more nearly these ranges correspond to the total (true) ranges of the taxa. To compare the sections, a simple graphical method is used.

One particularly complete and well-sampled section is chosen as a standard reference section. Eventually, data from several other good sections are amalgamated with it to produce a composite standard reference section. A particularly thorough paleontological study should be carried out on the standard reference section, as this enables later sections, for example, those produced by exploration drilling, to be correlated with it rapidly and accurately.

The graphic technique is used both to amalgamate data for the production of the composite standard and for correlating the standard with new sections. Figure 13.8 shows a two-dimensional graph in which the thicknesses of two sections X and Y have been marked off on the corresponding axes. The FADs and LADs are marked on the sections by circles and crosses, respectively. If the taxon occurs in both sections, points can be drawn within the graph corresponding to FADs and LADs by tracing lines perpendicular to the X and Y axes until they intersect. For example, the plot for the top of fossil 7 is the coincidence of points X=350 and Y=355.

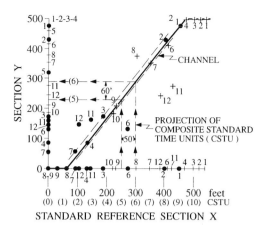

Fig. 13.8. Typical data plot used in the graphic correlation method. *Section X* is chosen as the standard reference section, and *section Y* is any other section to be correlated with it. FADs are shown by *circles* and LADs by *crosses*, plotted along the axis of each section. Data points within the graph indicate correlations of FADs and LADs, and are the basis for defining the line of correlation (*diagonal line*). Points off the line reflect incomplete sampling or absence as a result of ecological factors. Progressive correlations of other sections to the standard enables "true" ranges of each taxon to be refined in the standard section, resulting in a detailed basis for further correlation and the erection of standard time units. (Miller 1977)

If all the taxa occur over their total range in both sections and if sedimentation rates are constant (but not necessarily the same) in both sections, the points on the graph fall on a straight line, called the line of correlation. In most cases, however, there is a scatter of points. The X section is chosen as the standard reference section, with the expectation that ranges are more complete there. The line of correlation is then drawn so that it falls below most of the FADs and above most of the LADs. FADs to the left of the line indicate late first appearance of the taxon in section Y. Those to the right of the line indicate late first appearance in section X. If X is the composite standard, it can be corrected by using the occurrence in section Y to determine where the taxon should have first appeared in the standard.

If the average, long-term rate of sedimentation changes in one or other of the sections, the line of correlation bends If there is a hiatus (or a fault) in the new, untested sections (sections Y), the line shows a horizontal terrace. Obviously, the standard reference section should be chosen so as to avoid these problems as far as possible. Harper and Crowley (1985) pointed out that sedimentation rates are in fact never constant and that stratigraphic sections are full of gaps of varying lengths (Sect. 12.2). For this reason, they questioned the

value of the graphic correlation method. However, Edwards (1985) responded that when due regard is paid to the scale of intraformational stratigraphic gaps, versus the (usually) much coarser scale of biostratigraphic correlation, the presence of gaps is not of critical importance. Longer gaps, of the scale that can be detected in biostratigraphic data (e.g., missing biozones), give rise to obvious hiatuses in the line of correlation, as noted previously.

The advantage of the graphic method is that once a reliable composite standard reference section has been drawn up it facilitates chronostratigraphic correlation between any point within it and the correct point on any comparison section. Correlation points may simply be read off the line of correlation. The range of error arising from such correlation depends on the accuracy with which the line of correlation can be drawn. Hay and Southam (1978) recommended using linear regression techniques to determine the correlation line, but this approach assigns equal weight to all data points instead of using one standard section as a basis for a continuing process of improvement. As Edwards (1984) noted, all data points do not necessarily have equal value; the judgment and experience of the biostratigrapher are essential in evaluating the input data. For this reason, statistical treatment of the data is inappropriate.

Figure 13.9 illustrates an example of the use of the graphic method in correlating an Upper Cretaceous succession in the Green River Basin, Wyoming, using palynological data (from Miller 1977). The composite standard reference section has been converted from thickness into composite standard time units, by dividing it up arbitrarily into units of equal thickness. As long as the rate of sedimentation in the reference section is constant, these time units are of constant duration, although we cannot determine by this method alone what their duration is in years. Isochrons may be drawn to connect stratigraphic sections at any selected level of the composite standard time scale. These isochrons assist in defining the architecture of the succession. For example, time unit 30 in Fig. 13.9 is truncated, indicating the presence of an unconformity.

An important difference between the graphical method and conventional zoning schemes is that zoning methods provide little more than an ordinal level of correlation (biozones, as expressed in the rock record, have a finite thickness which commonly cannot be further subdivided), whereas the graphic method provides interval data (the ability to

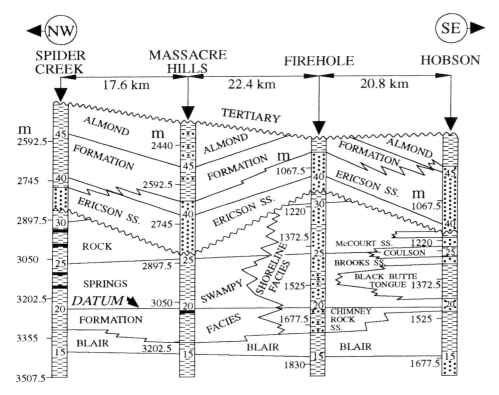

Fig. 13.9. An example of the correlation of four wells through Cretaceous sections in Wyoming. *Numbers* within each log are composite standard time units derived from graphic correlation with the standard reference section. They can be used to generate correlations on an interval scale rather than the ordinal scale obtainable using conventional biostratigraphic zonations. (Miller 1977)

make graduated subdivisions of relative time). Given appropriate ties to the global time frame, the composite standard time units can be correlated to absolute ages in years, and used to make interpolations of the age of any given horizon (such as a sequence boundary) between fossil occurrences and tie points. The precision of these estimates is limited solely by the accuracy and precision obtainable during the correlation to the global standard. Mac-Leod and Keller (1991) provided excellent examples of this procedure, and their results suggest an obtainable precision of less than ±100 ka. Other examples of the use of graphic correlation are given by Scott et al. (1988), although no data plots are presented. In a later paper Scott et al. (1993) used graphic correlation methods in a study of core data to demonstrate diachroneity of some Cretaceous sequence boundaries of more than 0.5 m.y. Additional examples of graphic correlation were given by Mann and Lane (1995).

13.3.5 Assessment of Relative Biostratigraphic Precision

The precision of biostratigraphic zonation is partly a reflection of the diversity and rate of evolution of the fossil group used to define the zonal scheme. This varies considerably over time and between different fossil groups. For example, Fig. 13.3C is a simplified version of a chart provided by Cox (1990) to illustrate the time resolution of various fossil groups used in the subsurface correlation of Jurassic strata in the North Sea Basin. The best time resolution is obtained from ammonites, which have been subdivided into biozones representing about 0.5 m.y. Hallam (1992) and Cope (1993) quoted detailed studies of British ammonites that yield a local resolution estimated at less than 200 ka, but such a level of accuracy cannot yet be extended globally for the purpose of testing the global cyclicity model, and ammonites are rarely obtainable in subsurface work. The microfossil groups that are typically used provide time resolution ranging from 1 to 4 m.y. The sloping caps to each bar in Fig. 13.3C illustrate the varying length of biozones for each fossil group through the Jurassic. For example, dinocyst zones each represent about 1 m.y. in the Late Jurassic but are about 3 m.y. long in the Early Jurassic.

Moore and Romine (1981), in a study of the contributions to stratigraphy of the DSDP, examined the question of biostratigraphic resolution in detail. In 1975 (the latest data examined in this paper) the resolution of foraminiferal zonation, as expressed by the average duration of biozones, varied from 4 m.y. during much of the Cretaceous, to 1 m.y. during parts of the Neogene. Srinivasan and Kennett (1981) found that foraminiferal zones in the Neogene ranged from 0.4 to 2.0 m.y. They suggested that,

[E]xperience shows that this resolution seems to have reached its practical limit. This ... is largely constrained by the evolution of important new species within distinctive and useful lineages. Further subdivision of the existing zones is of course possible when additional criteria are employed, but further subdivision of zones into shorter time-intervals does not guarantee a practical scheme for biostratigraphic subdivision; that is, such zones may not be widely applicable.

Combinations of foraminiferal zones with other microorganisms occurring in the same sediments, such as calcareous nannofossils and radiolaria, increase biostratigraphic precision (Moore and Ro-

mine 1981; Srinivasan and Kennett 1981; Cope 1993), but not necessarily by a large amount. Figure 13.10 shows the biostratigraphic resolution that was achievable in 1975 based on combinations of all three fossil groups. The combination of three fossil groups does increases accuracy and precision (to a 0.3- to 2.0-m.y. range), but commonly zonal boundaries of more than one group coincide in time, so that no additional precision is provided. It is not thought likely that precision will increase by very much. In fact, the system of numbered microfossil zones that was established early during the DSDP still forms the basis for the Exxon global cycle charts of the late 1980s. As noted earlier in this book, the analysis of different fossil groups from the same stratigraphic sections may also indicate that some groups are more facies controlled than others, and demonstrate diachronism. Cross-checking between these different groups may therefore be important in the reduction of biostratigraphic uncertainty (e.g., Ziegler et al. 1968; Armentrout 1981).

It has been suggested that because sequence boundaries are dated primarily by biostratigraphic data, they should be referred to and correlated on this basis, without reference to the absolute time

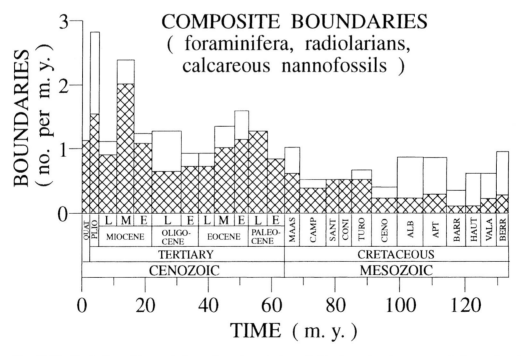

Fig. 13.10. Biostratigraphic resolution of marine strata based on combined zonation of foraminifera, radiolaria and calcareous nannofossils, expressed as the number of biozone boundaries per million years, averaged over individual epochs. Average resolution as it existed in 1969, at the beginning of the DSDP, is shown by the *cross-hatched columns*. Improvements up to 1975 are indicated by the *blank top to each column*. (Moore and Romine 1981)

scale, to circumvent the imprecisions associated with this scale (as discussed in the next sections). However, even where no attempt is made to provide absolute ages for sequence boundaries, this discussion has shown that a built-in biostratigraphic imprecision of between about one half million years and (in the worst case) several million years must be accepted for the ages of sequence boundaries. Errors are larger for earlier parts of the Phanerozoic time scale. A potential error in this range, which is typical of most regional biostratigraphic frameworks, is already too great to permit the interregional correlation of sequences that are less than a few million years in duration (most of the "third-order" cycles that constitute the main basis for the global cycle chart). Biozones in the Exxon synthesis of Mesozoic-Cenozoic time have durations of 1–5 m.y. Uncertainties of this magnitude are therefore to be anticipated.

13.3.6 Correlation of Biozones with the Global Stage Framework

Hedberg (1976) suggested that the stage be regarded as the basic working unit of chronostratigraphy but, as Hancock (1977) pointed out, stages were originally defined as groups of biozones, and stages are therefore biostratigraphic entities. However, modern work continues to focus on the stage as the most practical basis for subdividing the stratigraphic record, and for standardizing the geological time scale. This is facilitated by the gradual incorporation into the stage framework of different faunal and floral biozone systems, by the use of radiometric, chemostratigraphic and magnetic methods to assign numerical ages to the stages, and by the use of stratotypes defined by golden spikes to establish global reference sections for unambiguous time correlation (Miall 1990, Chap. 3). For some systems and series the same stage framework is now used globally, although the types of organisms and the assemblages of biozones used to define the stages varies from region to region.

Published global time scales, such as the wall chart accompanying Harland et al. (1990), the Elsevier charts, and the bio- and chronostratigraphic framework accompanying the Exxon chart (Haq et al. 1987, 1988a) convey an impression of completeness, precision and certitude. However, an examination of the evidence that is being used to build these time scales reveals numerous gaps, generalizations and inconsistencies. It has proved difficult to extend the stage framework globally because most organisms are limited in their distribution by tectonic, physiographic and climatic barriers. Broad ecological differences from region to region and continent to continent have generated distinct faunal and floral provinces (e.g., Kennedy and Cobban 1977; Gray and Boucot 1979; Kauffman 1984; Smith 1988; Hancock 1993a,b). Time correlation across provincial boundaries is typically fraught with error or uncertainty because of the limited number of taxa that cross the province boundaries. Commonly such correlations depend on the fact that faunal boundaries shift to and fro in response to climatic changes or plate-tectonic events, permitting faunas and floras from adjacent provinces to become interbedded. Uncertainties of the magnitude of one or more biozones are possible in this process of cross-correlation.

Hancock (1993a, p. 8) stated that, "it is seldom realized by geologists at large how insecure most zonal schemes are, and how few are the regions in which any one scheme has been successfully tested." A synthesis with which he was involved (Birkelund et al. 1984; see also Hancock 1993b) noted as many as nine different possible biostratigraphic standards that could be used to define specific stage boundaries in the Cretaceous. Many are partly in conflict with one another.

Figure 13.11 illustrates the process of collating chronostratigraphic data from a variety of sources around the world to construct a global stage framework calibrated with absolute ages. During much of the Mesozoic the presence of the wide east-west-oriented Tethyan Ocean, separation of the North and South American continents, and the broad latitudinal extent of the American continents led to faunal provincialism in benthic and some pelagic fossil groups. The Tethyan, Boreal and other provinces indicated in Fig. 13.11 are based on the distribution of Cretaceous ammonites. As Hancock (1993a, p. 8) stated "every wide-ranging stratigrapher working on the Mesozoic meets the difficulty of correlations between boreal and tethyan realms." Faunal provincialism is also a problem with microfossils, including most of those favored by biostratigraphers for intercontinental correlation. Theirstein (1981) and Roth and Bowdler (1981) described latitudinally controlled biogeographic distribution of Cretaceous nannoplankton. The latter authors also described neritic-oceanic biogeographic gradients.

As a single example of the magnitude of the potential for error at this stage of the analysis Sur-

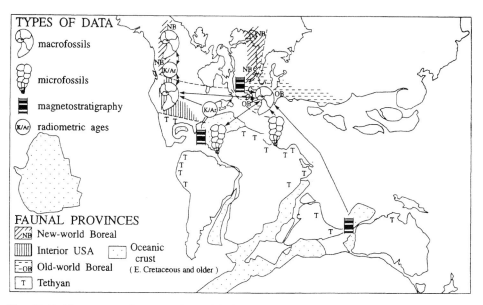

TYPES OF DATA

Fig. 13.11. The process of constructing a global time scale based on the collation of biostratigraphic, magneto-stratigraphic and radiometric data. This diagram is loosely based on the construction of the Cretaceous time scale by Van Hinte (1976b) but, of necessity, shows only a few of the key data points used in the compilation. Macrofossil data from the original nineteenth-century type sections in Europe comprise a critical base for the time scale. The Western Interior of North America is also a key area because of the wealth of well-studied sections there (e.g., Obradovich and Cobban 1975). A few other data points, many representing DSDP data, are indicated. Biostratigraphic correlation is complicated by the problem of faunal provincialism. A very simplified distribution of Cretaceous ammonite faunal provinces (from Kennedy and Cobban 1977; Kauffman 1984) is shown

lyk's (1990, 1991) study of the sequence stratigraphy of the Jurassic section of eastern Greenland may be mentioned. He derived a sea-level curve for this area, and compared it to the curves of Hallam (1988) and Haq et al. (1987, 1988a). In doing so he noted "that the correlation between the Boreal and Tethyan stages across the Jurassic-Cretaceous boundary is only precise within 1/2 stage, rendering the eustatic ... nature of sea-level curves rather meaningless for this time interval."

An error in correlation of a biozone to the standard by as little as one zone as a result of faunal provincialism could add an error of up to some 5 m.y., although this should be considered a pessimistic maximum. The degree of uncertainty varies considerably, depending on the degree of evolutionary divergence between organisms (provinciality) that has arisen because of long-continued climatic difference or plate separation between regions.

13.3.7 Assignment of Absolute Ages

An important step in the testing of regional and global correlations is to assign absolute ages to stratigraphic events using numerical methods based on radiometric, magnetostratigraphic and chemostratigraphic methods. This may be done in two main ways. Firstly, it may be accomplished by the dating of the sequence framework itself, for example, by the radiometric analysis of interbedded volcanic horizons or glauconitic beds, or by the establishment of a magnetostratigraphic or chemostratigraphic framework (which, nonetheless, requires supplementary biostratigraphic or radiometric dates for purposes of calibration). Secondly, the biostratigraphic framework may be related to one or other of the published global time scales, making use of the accumulated evidence for the age of, say, the Campanian-Maastrichtian boundary to fix the age of local biostratigraphically dated horizons. These two approaches are not, of course, mutually exclusive, and much use may be made of local, regional, and global correlation networks and data bases to achieve the best result.

The accuracy and precision of time resolution varies with the chronostratigraphic methods used and the level of the stratigraphic column. Various authors have estimated the dating precision over various intervals of geological time, and have assessed the incremental improvements in the global time scale that have been built from local, regional, and inter-regional correlation programs. Kidd and Hailwood (1993) estimated the resolution for various time slices back to the Triassic (Table 13.1).

Kauffman et al. (1991) described a method of high-resolution correlation that utilizes biostratigraphy (graphic correlation methods), magnetostratigraphy, chemostratigraphy and the correlation of event beds. Numerous datable ash beds are present in his sections. A potential uncertainty of only ±100 ka is claimed for Cretaceous beds of the Western Interior of the United States. However, this provides only a regional framework, and it is unlikely that it could be extended to other unrelated basins with the same degree of precision.

Miller (1990) indicated that chronostratigraphic resolution of Cenozoic sections could ideally attain accuracies of ±100 ka, based on the use of modern chronostratigraphic techniques and biostratigraphic data bases, but he conceded that uncertainties of 0.5–2.0 m.y. are common in many actual case studies. Aubry (1991), in a discussion of the Early Eocene record of sea-level change, stated that under ideal conditions combinations of biostratigraphic (mainly microfossil) and magnetostratigraphic data should permit dating to within 0.2–0.3 m.y. Yet commonly, according to her, the data from specific locations are inadequate to permit the use of all available tools.

In the absence of a magnetostratigraphic signature that can be calibrated against biostratigraphy – which is the case for some Upper Cretaceous and Cenozoic strata, the fossil record is still by far the most important tool for precise age determination, although some new techniques may challenge the supremacy of biostratigraphy in the forseeable future. Chemostratigraphy is now providing valuable supplementary data beyond the oxygen isotope record of the Late Cenozoic. For example, Jones et al. (1994a,b) examined the strontium isotope signature in Jurassic and Cretaceous strata, and suggested

that at least parts of the time scale could be calibrated to within one or two ammonite zones (±0.5–1 Ma). House (1985) proposed that the record of orbital frequencies preserved in the rock record could assist in the refinement of a geological time scale by providing calibration of biostratigraphic data with a 10^4- to 10^5-year precision. This has become a fruitful area of research (House and Gale 1995), but much remains to be done to clarify possible changes with time in orbital frequencies (Sect. 10.2.3), and the broad framework of absolute ages within which refined cyclostratigraphic determinations can be carried out. So far, a reliable cyclostratigraphic (*astrochronological*) time scale is only available for the youngest Cenozoic strata, back to 5 Ma (Hilgen 1991; Berggren et al. 1995a).

Unfortunately there is still no universal agreement on the ages of many of the major chronostratigraphic boundaries (Menning 1989; Kent and Gradstein 1985; Gradstein et al. 1995). Many authoritative scales have been published, such as that used in the Decade of North American Geology Project (Palmer 1983), the synthesis of Harland et al. (1990), and the new Mesozoic and Cenozoic time scales provided in Berggren et al. (1995b). However, residual differences between these scales exist, as illustrated in Fig. 13.12.

Harland et al. (1990) indicated that the range of possible ages for stage boundaries in the Cretaceous varies by as little as ±4 m.y. for the Albian-Cenomanian boundary (the difference between the likely minimum and maximum possible ages) and as much as ±25 m.y. for the Aptian-Barremian boundary. They assigned an overall average 2% uncertainty to the calibration of the Phanerozoic scale (±2 m.y. at 100 Ma). These error values relate to the best available global data calibrated by several independent means, yet they reveal a residual imprecision that would not permit the dating of any given stratigraphic event, even in one of the global stratotypes, to better than ±2 m.y. In their new Mesozoic time scale Gradstein et al. (1995) provided revised estimates of Mesozoic stage boundaries with uncertainties expressed in million years to two standard deviations. These increase from ±0.5 m.y. for some of the Late Cretaceous stage boundaries, to ±2.6 m.y. for the Jurassic-Cretaceous boundary, to errors increasing from ±3 m.y. to ±4.8 m.y. for stages extending back from the Late Jurassic to the Early Triassic.

Table 13.1. Achievable resolution for integrated stratigraphy in marine successions

Quaternary	<1–3 ka
Late Cenozoic	5–10 ka
Early Cenozoic	10 ka–1 m.y.
Late Cretaceous	100 ka–to 1 m.y.
Early Cretaceous	~10 m.y.
Jurassic	50–150 ka
Triassic	225 ka–2 m.y.

Simplified from Kidd and Hailwood (1993).

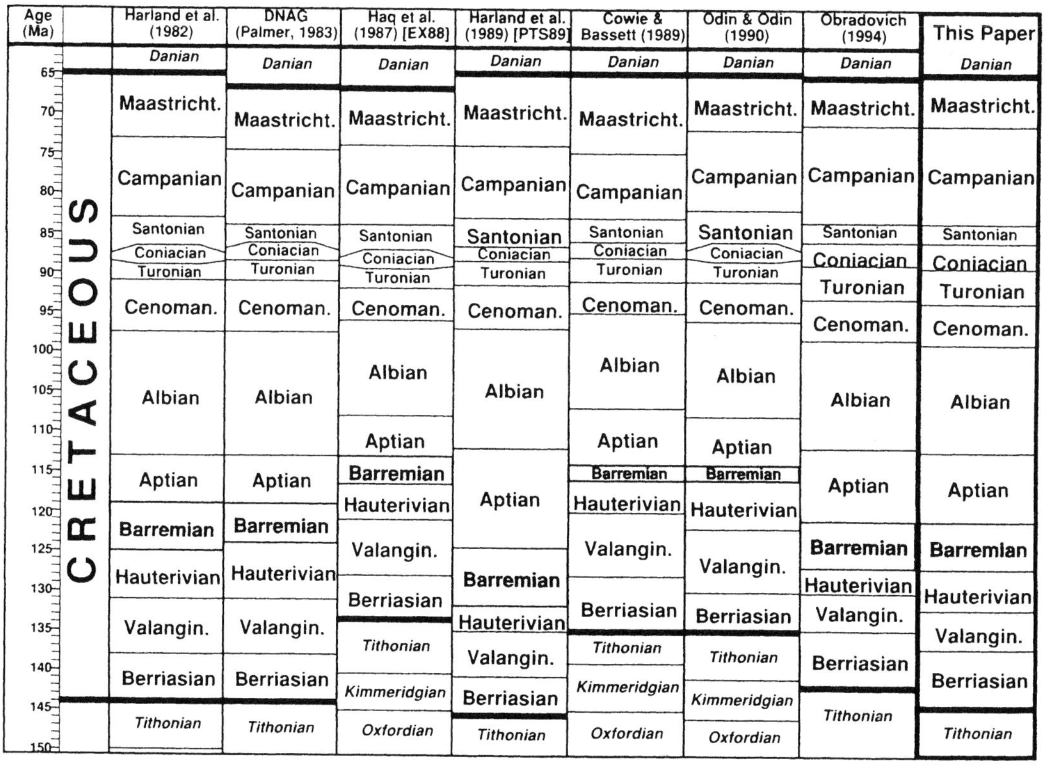

Age (Ma)		Harland et al. (1982)	DNAG (Palmer, 1983)	Haq et al. (1987) [EX88]	Harland et al. (1989) [PTS89]	Cowie & Bassett (1989)	Odin & Odin (1990)	Obradovich (1994)	This Paper
65		Danian	Danian	Danian	Danian	Danian	Danian	Danian	Danian
70		Maastricht.	Maastricht.	Maastricht.	Maastricht.	Maastricht.	Maastricht.	Maastricht.	Maastricht.
75–80		Campanian	Campanian	Campanian	Campanian	Campanian	Campanian	Campanian	Campanian
85	C R E T A C E O U S	Santonian	Santonian	Santonian	Santonian	Santonian	Santonian	Santonian	Santonian
90		Coniacian / Turonian	Coniacian / Turonian	Coniacian / Turonian	Coniacian / Turonian	Coniacian / Turonian	Coniacian / Turonian	Coniacian / Turonian	Coniacian / Turonian
95		Cenoman.	Cenoman.	Cenoman.	Cenoman.	Cenoman.	Cenoman.	Cenoman.	Cenoman.
100–105		Albian	Albian	Albian	Albian	Albian	Albian	Albian	Albian
110		Aptian	Aptian	Aptian	Aptian	Aptian	Aptian	Aptian	Aptian
115–120		Aptian	Aptian	Barremian / Hauterivian	Aptian	Barremian / Hauterivian	Barremian / Hauterivian	Aptian	Aptian
125		Barremian	Barremian	Valangin.	Barremian	Valangin.	Valangin.	Barremian	Barremian
130		Hauterivian	Hauterivian	Berriasian	Hauterivian	Berriasian	Berriasian	Hauterivian	Hauterivian
135		Valangin.	Valangin.	Tithonian	Valangin.	Tithonian	Tithonian	Valangin.	Valangin.
140		Berriasian	Berriasian	Kimmeridgian	Berriasian	Kimmeridgian	Kimmeridgian	Berriasian	Berriasian
145–150		Tithonian	Tithonian	Oxfordian	Tithonian	Oxfordian	Oxfordian	Tithonian	Tithonian

Fig. 13.12. A comparison of Cretaceous stage-boundary ages in eight recent time scales. "This paper" refers to the source of this diagram, a revised Mesozoic time scale by Gradstein et al. (1995)

13.3.8 Implications for the Exxon Global Cycle Chart

The Exxon curves are not the first curves that have been compiled in an attempt to provide a global standard. Early work by Stille, Schuchert, Grabau and others was reviewed by Hallam (1992) who, himself, has been compiling and revising sea-level curves for many years. However, it is the Exxon curves that have received so much attention, and that have appeared in global syntheses of geological time, such as that by Harland et al. (1990). The Exxon work therefore needs to be examined carefully.

Potential error in the global time scale is discussed in some detail by Haq et al. (1988a), but the relationship between this error and the dating and correlation of sequence boundaries is not discussed. Sequence boundaries are not shown with attached error bars in the global cycle chart. Haq et al. (1987, 1988a) showed sequence boundaries spaced 1 m.y. or less apart and, in some cases, dated them to within ±0.5 m.y., even though this is less than the stated potential age error. Given the potential error summarized in the previous section what this means is that in most cases the error bar for the age

of any one sequence boundary encompasses the assigned age of at least one sequence above and below. Distinguishing the ages of such adjacent sequence boundaries is thus not possible on chronostratigraphic grounds. The implied precision for the Mesozoic part of the Exxon global cycle chart, and the density of event spacing, are therefore simply impossible, and none of the "events" shown in this chart can be relied upon as proven. (It must be emphasized that this relates to the problem of *global* correlation. As noted above, much greater precision can be achieved locally, but this does not help in the testing of global cyclicity). This inconsistency, plus residual disagreements between various chronostratigraphers involved in the compilation of the global cycle chart, has led to numerous errors and inconsistencies in the papers published in support of the chart. These were summarized by Miall (1991a; Sect. 14.2). The correlation and dating of stacked sequences and the use of pattern-matching techniques may reduce the potential for error, but this procedure can introduce its own problems, as discussed in Section 13.4.1 with reference to Cretaceous clastic wedges in Alberta.

The practice of assigning radiometric ages to sequence boundaries even where the absolute ages are subject to revision is a source of considerable confusion. Baum and Vail (1988, p. 314) stated:

Although sequence boundaries are named by a radiometric age (corresponding to the age of the sequence boundary where it becomes conformable), the sequence boundaries are dated paleontologically.... Currently there are numerous composite radiometric-time scales, all varying slightly from one another. Thus, some unnecessary confusion exists because different authors prefer different time scales. The radiometric age of a sequence boundary may vary from author to author but the paleontological age is the same.

Comparisons between the sea-level curves published in the same book by Haq et al. (1988a) and Baum and Vail (1988) show that they are not the same (Fig. 14.1). Both versions are clearly labeled as "eustatic," indicating that they are presenting the end product of the Exxon sequence-stratigraphic analytical methodology. This being the case how can they be different? It is explicit in the methodology that tectonics and other complicating factors may be ignored in the *placement* (age) of sequence boundaries, although it is allowed that the *amplitude* of the sea-level deflections comprising the eustatic curve may be modified by tectonics (Vail et al. 1991). Underhill (1991) has also demonstrated major inconsistencies in the Jurassic curves derived from Exxon research in the North Sea Basin (Fig. 14.3). Apparent conflicts between the two curves coauthored by P.R. Vail can readily be understood as a reflection of the kinds of error or uncertainty discussed in this chapter, but this is not addressed by the authors of these curves, who (as noted elsewhere) do not deal effectively with the significance of error in any of their papers. These points are addressed in more detail in Section 14.2.

Hancock (1993a,b) carried out an examination of the Cretaceous part of the Exxon curve. He pointed out that the biostratigraphic framework consists of an amalgam of regional biozones, many of which do not occur together in a given location. Serious problems of Tethyan-Boreal provinciality were not addressed in the Exxon work, and Hancock (1993a) had difficulty correlating his own detailed stratigraphic synthesis with the Exxon curve using either biostratigraphic comparisons or radiometric ages. He concluded that "It is perhaps inevitable that discussion of the Exxon chart should seem to be a catalogue of complaints."

In some cases residual error in the geological time scale has left room for debate regarding sequence correlation and duration. An example is the recent differences of opinion that have been expressed regarding the duration of the Pennsylvanian period. Although this particular problem primarily concerns high-frequency cycles of glacioeustatic origin, and the consequences this has for estimating periodicities of cyclothems driven by Milankovitch processes, the problems exemplify in an extreme form the difficulties that are present throughout the geological time scale. The debate was opened by Klein (1990), who reviewed recent work on the assignment of absolute ages to the major boundaries within the Pennsylvanian. There is substantial disagreement in the literature, with the estimated duration of the period ranging from 19 to 40 m.y. Discussions of Klein's article (Langenheim et al. 1991) raised the usual problems of the limits of the available data and the difficulty of correlating faunas across tectonic and faunal-province boundaries. The age range of the Pennsylvanian is of concern to those attempting to demonstrate Milankovitch control of cyclothems of this age, because absolute ages of stage boundaries are required to provide brackets on the age range of the cyclic successions from which periodicities are calculated. Attempted correlations with the periodicities of the modern earth's orbit are meaningless in the absence of reliable age-range determinations, particularly because of the possible variations in these periodicities during the geological past (Sect. 10.2.3).

As Kauffman et al. (1991) showed, dating with a precision of ±100 ka can potentially be carried out on a regional scale in rocks of Mesozoic age, given good biostratigraphic, magnetostratigraphic, and radiometric data. However, this is only the beginning in the effort to construct a global cycle chart, for which the establishment of global correlations is a necessary condition. I have suggested elsewhere (Miall 1992) that without such global correlations there is no independent proof that the paradigm of "third-order" global eustasy has any validity whatsoever. It is misleading to state, as do Posamentier and Weimer (1993, p. 737) that, on the one hand chronostratigraphic error is a serious problem, while on the other hand arguing that the global cycle chart is most useful to stratigraphers where there is no age dating available! I respectfully submit that "They don't get it!"

It is simply not enough for the Exxon workers, and their colleagues, to present isolated sections with their assigned ages and correlations to the global standard (several of the papers in Special Publication 42). No presentation of the potential error is given. It is no help to be told that the Mesozoic-Cenozoic cycle chart has been assembled with the use of data from scores of sections around the world (the locations are listed by Haq et al. 1988a, Appendices C–F, but no actual data are provided). How has the potential for the four types of error listed above been accounted for? Where are the plots showing this error, with allowance made for alternative interpretations?

13.4 Correlating Regional Sequence Frameworks with the Global Cycle Chart

13.4.1 Circular Reasoning from Regional Data

One of the most well-known of Vail's diagrams is that reproduced here as Fig. 5.1. This chart purports to show correlation between four Cretaceous to Recent basins. A series of major and minor sea-level changes has been correlated among these basins, producing what is, at first sight, a convincing proof of the eustatic sea-level model. The evidence as presented here is suspect for two reasons. Firstly, in Fig. 5.1, correlation lines are all based on sudden sea-level falls. The questionable validity of this interpretation is discussed in Sections 2.2.3 and 15.3.1. Second, it is rare in geology for the evidence to permit such precise correlation of events that they can be indicated by the straight lines encompassing the globe shown in Fig. 5.1 (Miall 1986). As discussed in Section 13.3, a considerable margin of potential error must be allowed for, which is not indicated on this chart. If such data are not provided, it is difficult to allow for revisions and refinements.

One of the most serious problems that has arisen as a result of the introduction of the Exxon method

is the tendency to regard the Vail curves as some kind of approved global chronostratigraphic standard. Vail and his coworkers encourage this trend and even favor seismic correlation over biostratigraphic correlation when the latter does not appear to fit their models, as shown by several quotations from their work in Section 13.2. Many other workers have adopted this practice, and it is common for excellent biostratigraphic evidence to be ignored in favor of correlation with the global cycle chart, even where this is not supported by the actual age data (Hancock 1993a). This approach does not lend itself to independent tests of the global cycle chart (Miall 1986).

Numerous examples of self-fulfilling correlation could be provided to illustrate the dangers of circular reasoning. A good example is the attempt by Plint et al. (1992) to correlate the Cretaceous clastic pulses in the Alberta Basin, Canada, with the global cycle chart (Fig. 8.47). Table 13.2 reproduces Plint's (1991) table of possible stage ages for the Middle Cretaceous, showing the ages assigned in four recent time scales. Plint (1991) used these data as the basis for his correlation of one of the clastic pulses, the Marshybank Formation. This unit is assigned a latest Coniacian to Early Santonian age on biostratigraphic grounds. Plint's table indicates that the Coniacian-Santonian boundary is between 86 and 88 Ma, depending on the time scale used. The arithmetic average of these values is 86.87, and Plint (1991) is therefore confident in assigning the Marshybank clastic pulse to the 87.5-Ma eustatic event of the Exxon global cycle chart. Given the magnitudes of the errors discussed in this book, a correlation with the 88.5-Ma event, or even the 85-Ma event on the global cycle chart would not be unreasonable. However, as can be seen from Fig. 8.47, this would require a realignment of many of the other clastic pulses in the Alberta Basin, none of

Table 13.2. Boundaries and durations of the Coniacian and Santonian stages. (Plint 1991)

	Obradovitch and Cobban (1975)	Palmer (1983)	Haq et al. (1988)	IUGS (1989)	Average
Campanian					
	82	84	84	83	83.25
Santonian	(4)	(3.5)	(4)	(3)	(3.62)
	86	87.5	88	86	86.87
Coniacian	(1)	(1)	(1)	(2)	(1.25)
	87	88.5	89	88	88.12
Turonian					

which has been dated with any greater degree of precision. The best that can be said for this correlation exercise is that it is "permissive" – the data do not negate it as a possibility. The alternative possibility is that these clastic pulses are tectonic in origin – in other words, typical "molasse," in the sense of Van Houten (1981). Plint and his co-workers have in fact examined various tectonic mechanisms for the generation of sequence architectures in the Alberta Basin (Plint et al. 1993; Hart and Plint 1993). Miall (1991a, 1992) suggested that many of the "events" in the Exxon global cycle chart are of regional tectonic origin, and the example of the Alberta clastic pulses given here is typical of the difficulties involved in assessing the alternatives.

Dixon (1993) addressed the tectonics-versus-eustasy question in the case of regional unconformities present in the Cretaceous succession of northwestern Canada. Imprecision and uncertainty in biostratigraphic correlation with the Exxon curve exist in part because of the difficulty of precise correlation across faunal-province boundaries into the Boreal realm. Dixon (1993) demonstrated that each of the eight sequence boundaries in the succession could be correlated to two or three of the "events" in the Exxon curve (Fig. 13.13), confirming Miall's (1992) suggestion that the curve contains "an event for every occasion." In several cases, however, regional evidence clearly indicates a tectonic origin for the sequence boundary. For example, a Late Albian–Early Cenomanian sequence boundary corresponds to a break-up unconformity preceding the sea-floor spreading that generated the Canada Basin.

Another example of self-fulfilling correlation is the Eocene sequence of the Hampshire Basin, England, illustrated in Fig. 3.5. Note that the 48.5 and 49.5-Ma events in the Haq et al. (1987, 1988a) global cycle chart are indicated as missing in this succession. Plint (1988b), who compiled this diagram, stated:

The eustatic curves of Haq et al. (1987) were constructed through the integration of seismic, well-log and outcrop data, the latter including the Eocene sections of the Hampshire Basin. Although this may be construed as introducing a certain circularity to the correlation suggested here [Fig. 3.5], the Palaeogene sea-level curves of Haq et al. (1987) are stated to be broadly based on widely separated sequences in Europe, the eastern seaboard of the United States, and in Australia and New Zealand. I am reasonably confident that the "local" to "global" sequence correlation described here retains its validity.

In fact the circularity of correlating a succession to the "global" chart that was partly or largely developed from the same sediments is clear. Plint (1988b) argued that the absence of marginal marine deposits on the Atlantic Coastal Plain of the United States at 49.5 Ma, the same age as one of the missing sequences in the Hampshire Basin, supports the correlation. Both the Hampshire Basin and the United States Atlantic margin are part of the same oceanic system, and regional tectonics might be

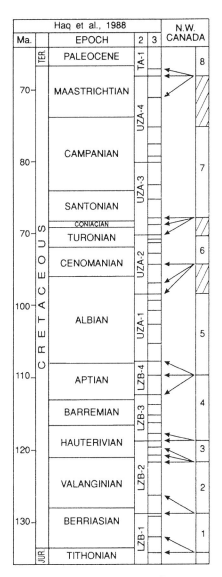

Fig. 13.13. Cretaceous unconformities in northwestern Canada (*hachured areas* between numbered sequences), and the possible choices of correlative unconformities in the global cycle chart of Haq et al. (1987, 1988a). The columns headed with the numerals *2* and *3* provide the "second-" and "third-order" sequences from this chart. (Dixon 1993, reproduced by permission of Academic Press)

expected to play a major role in developing synchronous events, as demonstrated in Chapter 11. Figure 6.15 illustrates that this event is not present in other basins bordering the Atlantic ocean.

Many studies offer correlations with the Exxon global cycle chart without providing rigorous tests of sequence age. Successions of sequences are simply lined up against the events shown in the chart (e.g., Mitchum and Uliana 1988; Brink et al. 1993; Muntingh and Brown 1993). In some cases supposed correlations are not at all clear to the reader. For example, Kendall et al. (1992) asserted that the Cretaceous sections of the United Arab Emirates and Gulf Coast correlate with each other and with the global cycle chart, but in another study of a nearby section in Oman Pratt and Smewing (1993) noted that there is really no similarity with the global cycle chart, and stated that the available biostratigraphic control in the Arabian peninsula is too crude to substantiate any such correlations. They suggested that regional tectonism can readily explain the observed succession in the area.

Some studies appear to be specifically undertaken to explore the relationship between sedimentation and active basin-margin tectonism, and succeed in developing tectonic models for sequence development, but then offer correlations with the global cycle chart and go on to discuss eustatic control. The research by Millan et al. (1994) and Deramond et al. (1993) in the tectonically highly active Pyrenean foreland basins are good examples of this. The latter authors refer to "tectonically enhanced" unconformities and stated: "The apparent correlation between the two groups of independent phenomena is an artefact of the method which calibrates the tectonic evolution by comparison with eustatic fluctuations." This form of "calibration" is, of course, circular reasoning.

The presence of missing or extra sequences relative to the Exxon global cycle chart does not pose a problem for those using the chart. Such sequences are readily attributed to local tectonics. For example, Mancini and Tew (1993), discussing the Lower Paleocene stratigraphy of southern Alabama, stated that "the global sea-level and coastal-onlap changes for Paleocene strata as published by Haq et al. (1988a) are partly recognisable in the Paleocene ... of southern Alabama." In fact, Haq et al. (1987, 1988a) showed five sequence boundaries in the interval under study, of which Mancini and Tew (1993) claimed to have documented three, plus three others that are not in the global cycle chart. This amount to a 50% correlation. However, to be

rigorously objective, if sequence X is the result of tectonics, why not the identical sequence Y, which happens to correlate with the global cycle chart? The only evidence of eustatic control is this correlatability, which brings into sharp focus once again the importance of chronostratigraphic precision and error.

Correlations with the Exxon curve as good as those illustrated in Figure 3.5 are almost always possible, because the spacing of events in the Exxon chart (cycles of sea-level change, sequence boundaries, etc.) is less than the potential error involved in dating those events, given the acccuracy and precision of present-day dating techniques. Therefore, when the best available chronostratigraphic techniques are used, almost all stratigraphic events can be found to correlate successfully with this chart when allowance is made for error. Such correlations are therefore of questionable value, as they prove little. Therefore no independent checks of the chart are yet possible, and the chart itself may be no better than one constructed from spurious noise.

13.4.2 A Rigorous Test of the Global Cycle Chart

The work of Aubry (1991) has been referred to above, in Section 13.3.2, where the problem of assigning accurate dates to unconformities was discussed. The main purpose of her paper was to use the method of rigorous evaluation of unconformities to examine the framework of Lower-Middle Eocene sedimentation in Europe, North Africa, the Atlantic margins, and North America, and including data from several DSDP sites. A summary of some of the main results is given in Fig. 13.14, and the paper contains a detailed discussion of the biostratigraphic and magnetostratigraphic evidence, including an evaluation of several zonal schemes and their relationships in various locations.

Among the results of Aubry's analysis was the discovery of more than one unconformity at the Lower-Middle Eocene boundary, within the CP12A and NP14 nannofossil zones, which Haq et al. (1987, 1988a) date as 49 Ma, but which Aubry (1991) assigned an age of 52 Ma (a time difference corresponding to four sequences in the Exxon global cycle chart). Aubry (1991, p. 6672) stated: "Assuming that sequence stratigraphy reflects global changes in sea level, hiatuses around the Lower Middle Eocene boundary on the North Atlantic margins, in Cyrenaica and in California would be

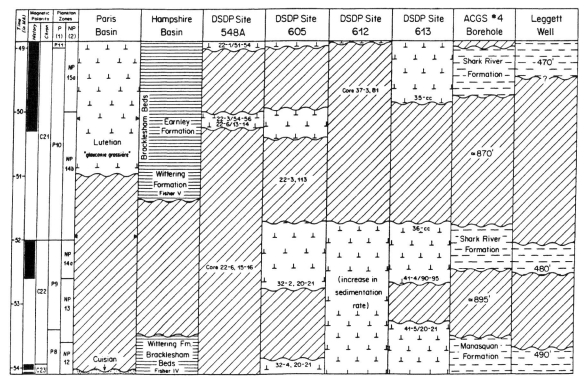

Fig. 13.14. Interpretations of the stratigraphy of part of the Lower-Middle Eocene interval at selected sites around the North Atlantic basin. (Aubry 1991)

indicative of at least two global eustatic events." The oldest of these unconformities would have occurred between 52.8 and 53.3 Ma or between 52.8 and 52.9 Ma, depending on two different interpretations.

Correlative unconformities would correspond to the TA2.9/TA3.1 (the 49.5 Ma) sequence boundary. A younger event would have occurred between 50.3 and 51.7 Ma as deduced from the stratigraphic record on the North Atlantic margins, and the correlative unconformities would represent the TA3.1/TA3.2 (the 48.5-Ma) sequence boundary. That hiatuses overlap is a condition required for their physical expressions, the unconformities, to be genetically related. It is not however a sufficient condition, unless the hiatuses are so short that the timing of the causative event can be precisely determined and correlated with a genetic mechanism. The hiatuses associated with the uppermost Lower Eocene unconformities are short (less than 1 m.y.).... However, these unconformities appear to have limited regional extent ... which suggests that they may be unrelated. In contrast the Lower Middle Eocene unconformities, which have a broader geographic extent ... are associated with long hiatuses (over 2 m.y.), and may be polygenetic.

The discussion continues in this vein for several additional paragraphs. The sense that emerges is that no single clear unconformable signal can be extracted for any level within this short pile of strata. Aubry (1991) developed four possible scenarios to explain the observed data (Fig. 13.15). The first alternative is that both unconformities result from a single eustatic event, within the NP14 zone, which would seem to correspond to the 48.5-Ma sequence boundary. However, this raises problems of different interpretations of Aubry (1991) and Haq et al. (1987, 1988a) regarding the relationships between the specific zones, stage boundaries, and sequence boundaries, a discussion of which is beyond the scope of this book. An interpretation involving two successive eustatic events represents Aubry's second possible interpretation (Fig. 13.15). However, without the framework of sequence stratigraphy to tie the events in the various locations together their differing character and time duration would suggest alternative mechanisms, including regional tectonism. For example, Aubry (1991) stated that "the formation of a nodular bed in Cyrenaica and the development of an unconformity on the New Jersey continental margin at apparently the same time (within the limits of biochronological resolution) may or may not be related." This type of argument accounts for the other two interpretations given in Fig. 13.15.

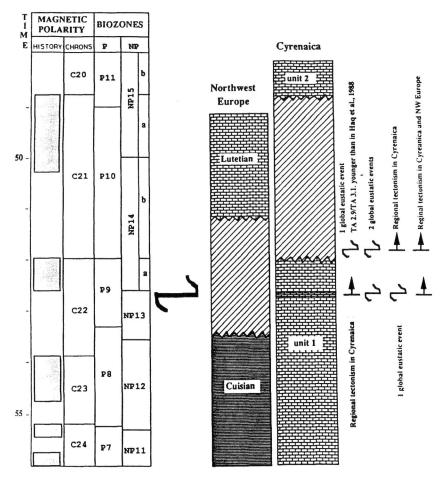

Fig. 13.15. Correlation of Lower-Middle Eocene stratigraphy in northwestern Europe and Cyrenaica, and four possible interpretations of the observed breaks in sedimentation. Explained in text. (Aubry 1991)

In her conclusions, Aubry (1991) argued that:

Since it cannot be demonstrated that Lower Middle Eocene unconformities (and probably no other Paleogene unconformities) formed simultaneously on the margins of a basin, there is no a priori reason to assume that they result from a single event, i.e., a global sea-level fall. It may be that unconformities on the slope and rise ... form independently from one another at different times, the difference in timing between two erosional events at two different localities being so little (a few tens to a few hundreds of thousands of years perhaps) that the resulting unconformities are apparently synchronous.

In a later paper, Aubry (1995) extended her analysis of Lower and Middle Eocene unconformities to the entire central and southern Atlantic Ocean, examining over one hundred deep-sea sites. She demonstrated the presence of multiple unconformities, and no clear pattern that would support the construction of a single summary section that could serve to summarize all the data.

13.4.3 A Correlation Experiment

I conducted a simple experiment to illustrate the questionable value of the Exxon cycle chart (Miall 1992). The experiment replicates the correlation exercise performed by basin analysts attempting to compare data from a new outcrop section or well to an existing regional or global standard. It is basic geological reasoning that if such a correlation exercise results in the definition of numerous tie lines between the new data and the standard, two conclusions may be drawn: (a) the events documented in the regional or global standard also occurred in the area of the new section; (b) the new data confirm and strengthen the validity of the standard. There is, of course, a danger of circular reasoning in such an analytical procedure.

Figure 13.16 shows the 40 Cretaceous sequence boundaries in the Exxon chart compared with the event boundaries recognized in four other sections.

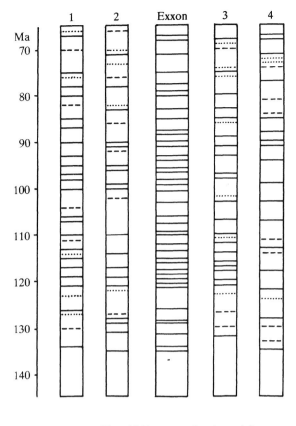

Ties with Exxon chart in columns 1-4:
—— Events correlated to within ±0.5 m.y.
- - - Events differing by >0.5 m.y., <1 m.y.
······· Events differing by >1 m.y.

Fig. 13.16. Event boundaries in the Exxon global cycle chart (total of 40, center column). Columns *1–4* are independent data sets used in the correlation experiment. (Miall 1992)

In these four sections the events have been classified into three kinds: (a) events dated to within ±0.5 m.y. of an event in the Exxon chart, (b) events outside this range but falling within ±1 m.y. of an event in the Exxon chart, and (c) events that differ by >1 m.y. from any event in the Exxon chart. Table 13.3 lists the number of such events, and demonstrates a high

degree of correlation of all four sections with the Exxon chart. In fact, the lowest correlation success rate is 77%, from section 3. By normal geological standards, considering the fuzziness and messiness of most geological data, this would be regarded by most experienced geologists as indicating a high degree of correlation. This is especially the case when we bear in mind the normal types of imprecision inherent in chronostratigraphic data. An error range of ±1 m.y. is unrealistic in most cases. With an error range of ±2 m.y. the correlation success rate would be almost 100%. The data in Fig. 13.16 would therefore appear to provide a successful confirmation of the Exxon chart. The catch is that all four of the test sections were constructed by random-number generation!

Columns 1 and 2 were constructed by selecting 40 values from a table of two-digit random numbers ranging from 0 to 99. These were normalized to the range 66.5–131, this being the range (Ma) of the Cretaceous Period. The resulting numbers were then treated as stratigraphic events and were plotted to produce the columns seen in Fig. 13.16, their "ages" rounded to the nearest 0.5 m.y. The symbol used to plot each "event" indicates the proximity in age to one of the events on the Exxon chart (see key for Fig. 13.16). Upon rounding, several of the events overlapped. Therefore the number of events plotted in each case is less than 40. Columns 3 and 4 were constructed by a different method. Event spacing in the Exxon chart for the Cretaceous ranges from 0.5 to 4.5 m.y. Suites of 40 random numbers were normalized to this range, rounded to the nearest 0.5 m.y., and then used to construct a stratigraphic sequence. The first (oldest) event shown in columns 3 and 4 of Fig. 13.16 represents 131 minus the first normalized random-number value. The second event is the age of the first plotted event, minus the second normalized value, and so on. Again, some events overlapped, and are not shown, so the resulting columns contain fewer than 40 events.

Table 13.3. Results of Correlation Experiment (Miall 1992)

Section	No. of events	±0.5 m.y. ties		±1 m.y. ties (no.)	Total ties		Mismatches
		No.	% Fit		No.	% Fit	
1	32	22	69	5	27	84	5
2	28	18	64	6	24	86	4
3	31	21	68	3	24	77	7
4	27	17	63	7	24	89	3

Convincing correlations have been achieved by both random positioning of stratigraphic events and random sequencing of events. Such a high degree of correlation must therefore be expected to exist, as a kind of background noise, in any actual field test of the Exxon chart. How can we be sure that any given event on the chart is not the product of local processes and compounded error, if random processes can produce such convincing correlations? The answer is that we cannot.

This exercise was offered by Miall (1992) in the same spirit as the classic paper "Cycles and Psychology" by Zeller (1964). In that paper Zeller succeeded in convincing his colleagues to construct stratigraphic correlations between sections whose rock types and thicknesses had been compiled entirely from random numbers. As Zeller stated, his colleagues "should have found it completely impossible to make any correlation ... with plotted stratigraphic sections." However, they did, thus demonstrating the training geologists receive in pattern-recognition techniques.

13.4.4 Discussion

Research into global stratigraphy and global stratigraphic controls has received an enormous impetus from the work of Peter Vail and his colleagues. The sequence-architecture model is now firmly established, and providing a basis for much exciting research and synthesis. However, as this chapter has attempted to demonstrate, the question of global synchroneity of stratigraphic events remains unresolved. It is still beyond our powers to test the global synchroneity of the "third-order" cycles (those ranging between about 1 and 10 m.y. in duration) that constitute the main basis of the Exxon global cycle chart because, as demonstrated in this chapter, the accuracy and precision of our methods of dating and correlation are in most cases associated with potential uncertainties of up to a few millions of years. Much research is now demonstrating the potential for tectonic processes to develop cycles with 10^6-year episodicities ("third-order" type; Macdonald 1991; Williams and Dobb 1993; Chap. 11). This currently constitutes one of the most vigorous and exciting areas of stratigraphic research. Cycles of tectonic origin may have areal extents of regional or continental scope, but are very unlikely to be global. For this reason the test of global synchroneity remains central to a resolution of the tectonic-versus-eustatic debate, and

for the forseeable future improvements in biostratigraphic correlation represent our best hope for resolving this question.

13.5 Main Conclusions

1. The proposed supremacy of stratigraphic sequences as chronostratigraphic indicators and their suitability to form the basis for a superior standard of time represents a new paradigm in geology. *However, at present the paradigm remains unproven.*
2. The recommendation that sequence boundaries be used for the redefinition of stage boundaries indicates a misunderstanding of the purpose of the geological time scale, which is to provide a practical subdivision of continuous time *independent* of any geological process.
3. The main theoretical basis for the paradigm is the supposition that stratigraphic sequences are controlled primarily by eustatic sea-level changes.
4. Following the eustatic-sequence paradigm, additional or missing sequences in a succession, relative to the global cycle chart, are attributed to local tectonic processes.
5. Given the availability of many other processes for generating sequences (Part III of this book) the main test of eustasy is global correlation of the sequence framework.
6. Currently available chronostratigraphic techniques (biostratigraphy, radiometric dating, magnetostratigraphy, chemostratigraphy) are inherently imprecise, which must be allowed for in sequence correlation.
7. Higher order sequences in the Exxon global cycle chart (those with episodicities of 10^6 years and less) have spacings less than the range of error involved in their dating and correlation (except for Late Cenozoic glacioeustatic cycles). *The precise ages for sequence boundaries provided in the Exxon global cycle chart cannot therefore be supported.*
8. Given current levels of chronostratigraphic imprecision and event spacing, almost any stratigraphic succession can be made to correlate with the global cycle chart, including synthetic successions constructed from random data (subject to points 3 and 4, above).
9. Rigorous correlation exercises carried out where detailed data are available reveal a complexity of

regional hiatuses of varying duration and areal extent. Clear eustatic signals may not necessarily be present.

10. Many published sequence analyses include "forced" correlations to the global cycle chart

even where individual stratigraphic events can be interpreted in terms of local or regional tectonism.

11. Given points 1–10, above, the Exxon global cycle chart should be abandoned.

14 Sea-Level Curves Compared

14.1 Introduction

The purpose of this chapter is to describe some of the internal inconsistencies and errors within the various versions of the Exxon global cycle chart, and to compare these curves with other published curves of sea-level change.

14.2 The Exxon Curves: Revisions, Errors, and Uncertainties

Two main versions of the Vail curves have been published, the original versions in AAPG Memoir 26 (Vail et al. 1977), and the revised curves prepared by Haq et al. (1987, 1988a). The later sets of curves are quite different. Some of the revisions that went into the revised curves are summarized in Chapter 5.

Haq et al. (1987, 1988a) described the methods used to construct a time scale by which the global cycle chart was calibrated. Biostratigraphic, radiometric, and magnetostratigraphic data were employed. The stated ranges of error of stage boundaries vary from epoch to epoch, as follows:

- Early Tertiary: ± 1.4 m.y
- Late Cretaceous: ± 1.75
- Early Cretaceous: ± 3.0
- Jurassic: ± 3.5
- Triassic: ± 5.0

These ranges of error relate to the precision of radiometric dating and the degree of confidence in the relation between radiometric and magnetic events. However, the assessment of these correlations has changed with the acquisition of new data. The following are changes in the assigned ages of stage boundaries from earlier (Vail et al. 1977) versions of the cycle chart. The figures indicate the maximum indicated change for some of the stages during the period of time indicated.

- Early Tertiary: 1.5 m.y.
- Late Cretaceous: 6.0 m.y.
- Early Cretaceous: 5.0 m.y.
- Jurassic: 12.0 m.y.
- Triassic: 24.0 m.y.

Revisions of age assignments such as those listed immediately above are understandable, but what is not understandable is the claim that sequence boundaries may be spaced 1 m.y. or less apart and, in some cases, dated to within ±0.5 m.y., even though this is less than the stated potential age error. To put it another way, the potential dating error of any one sequence boundary in most cases encompasses the assigned age of at least the sequence above and below. Distinguishing the ages of such adjacent sequence boundaries is thus not possible. This discussion, of course, does not take into account the nature of other chronostratigraphic errors discussed in Chapter 13.

The potential for error can be illustrated by a careful reading of the papers in SEPM Special Publication 42. There are numerous internal inconsistencies, not all of which can be dismissed as drafting errors, but must reflect internal uncertainties and unresolved conflicts in the research. For example, in different figures in Haq et al. (1988a) there are the following discrepancies:

- *Dispar-inflatum* zone boundary
    - In Fig. 12: 96.5 Ma
    - In Fig. 15: 97.8 Ma
- Top NP16 zone
    - In Fig. 11: 44 Ma
    - In Fig. 14: 41 Ma
- Top NP15 zone
    - In Fig. 11: 46 Ma
    - In Fig. 14: 43.1 Ma

These are internal errors of up to 3 m.y., representing potential miscorrelation of an entire sequence.

There are similar errors or differences in assigned ages in other papers in this book. For example, the following zone and stage boundaries are given different ages in the paper by Haq et al. (1988a) and that by Olsson (1988): top P1 zone, Campanian/Maastrichtian stage boundary, top *G. elevata* zone, base *G. elevata* zone, Albian/Cenomanian boundary, base TA2 zone, base UZA-4 zone, base UZA-3 zone, base UZA-2 zone. The differences are as large as 4 m.y. Similar discrepancies exist between the Haq et al. (1988a) paper and that by Baum and Vail (1988). In these two papers the Paleocene-Oligocene stage boundaries and their contained biozones vary in indicated age by up to 2 m.y.

One of Olsson's specific purposes was to relate the Upper Cretaceous stratigraphy of New Jersey to the Vail and Haq sequence curves (Olsson 1988, Fig. 7). Baum and Vail (1988) carried out the same type of research on Paleogene sections of the Gulf Coast and Atlantic margin of the United States. Whose assigned ages do we trust, and how reliable are the sequence correlations? No discussion of these points was given, nor were we presented with estimates of possible error in these papers.

Another unexplained feature is that two different papers in the same volume, both with Vail as a coauthor, actually contain different versions of part of the Eocene eustatic curve (Fig. 14.1). Both versions, that of Haq et al. (1988a) and that of Baum and Vail (1988) are clearly labeled as " eustatic," indicating that they are presenting the end product of the Exxon sequence-stratigraphic analytical methodology. This being the case how can they be different? It is explicit in the methodology that tectonics and other complicating factors may be ignored in the *placement* (age) of sequence boundaries, although it is allowed that the *amplitude* of the sea-level deflections comprising the eustatic curve may be modified by tectonics (Vail et al. 1991; see Sect. 11.1).

Dickinson (1993) pointed out that the average duration of the main (" third-order") cycles in the Haq et al. (1987, 1988a) chart decreases through Mesozoic and Cenozoic time (Fig. 14.2). Setting aside some fundamental geological cause, which seems unlikely, he suggested that the reason lies in the methodology of compilation of the chart, and is therefore an artefact of the chart itself, and not a true reflection of geological events. Among the factors that might have contributed to the observed secular trend he suggested:

(a) [T]he tendency for younger events to be more readily detected than older events in the geological record, (b) loss of resolution with depth on seismic reflection lines as seismic velocities increase owing to loss of sediment

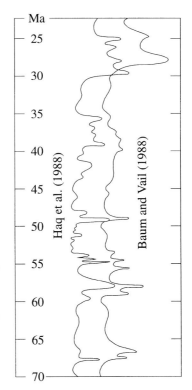

Fig. 14.1. Comparison of Paleogene portions of the eustatic sea-level curve from Haq et al. (1988a) and Baum and Vail (1988). These are not rival curves. Both papers contain P.R. Vail as a coauthor. How, then, can they be different?

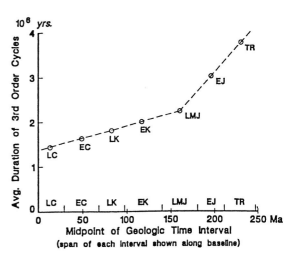

Fig. 14.2. Relationship between the duration of inferred "third-order" cycles in the Haq et al. (1987, 1988a) global cycle chart, and their absolute age. *LC* Late Cenozoic; *EC* Early Cenozoic; *LK* Late Cretaceous; *EK* Early Cretaceous; *LMJ* Middle-Late Jurassic; *EJ* Early Jurassic; *TR* Triassic. (Dickinson 1993)

porosity by compaction and cementation, (c) degradation of seismic signal with depth owing to progressive scatter and attenuation, (d) incorporation of more younger stratal sections than older sections in the dataset used to construct the chart. All these factors might promote recognition of more presumed cycles in younger strata, lying generally at shallower depths in sedimentary basins, than in older strata lying generally at greater depths.

All these possibilities seem likely, and this confirms that the global cycle chart cannot be trusted to tell us anything useful.

Underhill (1991) drew attention to revisions in the Upper Jurassic portion of the global cycle chart. He demonstrated that the original version of this part of the chart was based largely on data from the Inner Moray Firth, Scotland, which he was able to show was an area dominated by the effects of local tectonism (Sect. 11.2.2). The closely spaced sequence boundaries in this part of the section were shown to be the result of repeated block faulting and rotation, generating local onlap-offlap patterns (Fig. 11.10). Figure 14.3 compares the old and new versions of the Upper Jurassic portion of the global cycle chart. As Underhill (1991, p. 94) stated of the Haq et al. (1987, 1988a) chart:

Footnotes attached to these papers state that determination of the new Jurassic chart has been based on the sequence stratigraphy of outcrops exposed between the Isle of Portland and Swanage in southern England and the Montsalvens area in Switzerland. The revision of the chart by reference to outcrops in these areas was intended to provide refinement rather than replacement of original areas. However, it remains unclear whether the Inner Moray Firth still remains a key area in the construction of the newer cycle charts or whether the original seismostratigraphic determinations have been discredited, superceded or refined and hence, have been omitted from the analysis. Although it should be borne in mind that different methodologies have been used to derive both charts, a simple comparison interestingly shows that the curve resulting from work on the sequence stratigraphy of Dorset and Switzerland shows few detailed similarities to that derived from the previous Inner Moray Firth seismic-based studies [Fig. 14.3]. Despite this the chart still preserves a surprising rapidity of Late Jurassic sea-level fluctuations relative to the rest of the curve.

The curves from the two areas should, of course, be similar, subject to refinement. Eustasy is eustasy, wherever it is measured, according to the Exxon model and, again according to the model, local tectonism may be ignored in the dating of sequence boundaries. The fact that the curves from the Inner Moray Firth and Dorset – two areas within the general North Atlantic-North Sea Basin tectonic province – are almost completely different, requires explanation by the authors. None of the peaks and troughs in the two eustatic curves match (Fig. 14.3). In part the differences probably relate to changes in age assignments of the sequence boundaries, but this is not acknowledged or dealt with in the relevant publications.

In conclusion, it seems clear that the implied precision of the Mesozoic-Cenozoic Global Cycle Chart (Haq et al. 1988a) is not justified. The built-in potential error in the chart itself has not been acknowledged, the potential error in correlating new sections with the chart has not been dealt with, and revisions and updates to the chart have not been explained.

14.3 Other Sea-Level Curves

Various methods have been used to compile and calibrate sea-level curves, by reference to measureable changes in some stratigraphic parameter, or by calculation of the sea-level response to a presumed eustatic or tectonic process. These methods are reviewed in Chapter 2, and are summarizd in Table 14.1. Good reviews of the difficulty of achieving accurate estimates of relative and absolute sea-level change have been given by Burton et al. (1987), Kendall and Lerche (1988), Sahagian and Watts (1991), and McDonough and Cross (1991). Most methods require the determination of numerous corrections, such as complex calculations of subsidence, which in turn depend on several assumptions regarding isostasy and tectonic mechanisms. Also, as explained by Burton et al. (1987) and Kendall and Lerche (1988), none of the methods is fully satisfactory, because there are no stationary datums on earth from which measurements of sea level can be made. The earth's surface, which records the changing position of shorelines, is itself subject to vertical movements in response to tectonism, loading by sediment and water, and changes in the thermal structure of the mantle (dynamic topography). Essentially three main parameters control local sea-level change, eustasy, sediment supply, and tectonism. Unfortunately, calculations of the magnitude of any one of these three is typically dependent on a knowledge of the other two. One solution is to use numerical methods to model a variety of possible combinations of change in the three input parameters, and choose those solutions that best fit the observed stratigraphy for further testing. This approach is discussed further in Chapter 16.

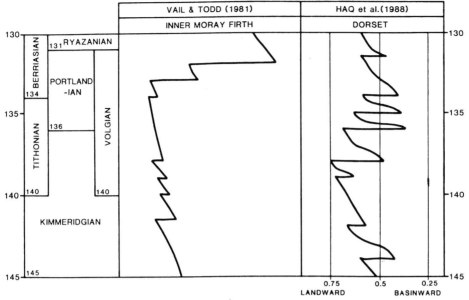

EXXON'S RELATIVE CHANGES OF COASTAL ONLAP

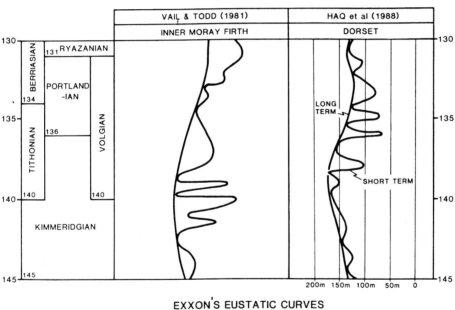

EXXON'S EUSTATIC CURVES

Fig. 14.3. Comparison of old and new versions of the Upper Jurassic portion of the Exxon global cycle chart. (Underhill 1991, reproduced by permission of Blackwell Scientific Publications)

In their original work, Vail et al. (1977) used the calculations of Pitman (1978) and Sleep (1976) to calibrate their coastal onlap chart, as discussed in Section 9.3.1. However, as shown by the discussion of the work of Kominz (1984) in this same section, only a very crude calibration is available using this method (compare Figs. 9.9 and 9.10). Lambeck et al. (1987) also attempted to calibrate the Exxon curves

(the updated versions of Haq et al.) using various estimates, such as the work of Hallam (1984), and then suggested that these curves could be used as paleostress indicators (Figs. 11.28, 11.29).

In this section other sea-level curves, compiled by a variety of techniques (Chap. 2) for the Cretaceous and Jurassic, are briefly reviewed. A recent book by Hallam (1992) discusses this subject at

Table 14.1. Methods for estimating the size of eustatic sea-level changes (Kendall and Lerche 1988)

Method	Measured variable	Assumptions	Problems
Hypsometric	Area of continent covered by marine sediments for time interval on equal area projection, measured with planimeter	The relationship between continental relief and area of continent at that elevation today was the same in the past	1. Time interval may be too long 2. Paleogeographic maps are inaccurate 3. Tectonic behavior unknown 4. Continental thickness unknown
Vail sediment onlap	1. Distance of onlap of seismic reflectors on unconformities, perpendicular to shore 2. Height of onlap	Onlap not a product of: (1) tectonic subsidence; (2) compaction; or (3) isostatic response	Cannot put dimensions on tectonic subsidence, compaction, or isostatic response
Paleo-bathymetric markers 1. Shoaling cycles	Thickness of cycle between high-water-mark indicators	1. Thickness a result of eustasy 2. Tectonic subsidence, compaction, and isostasy negligible	Effects of tectonic subsidence, compaction, and isostasy are unknown
2. Strand-line markers (beaches, reefs, notches, peats, etc.)	Elevation above present sea level and between markers	1. Result of eustasy and constant rate of tectonic uplift	Tectonic uplift rate unknown. Constant behavior unknown
Crustal subsidence curves 1. Divergence from thermotectonic curves	Difference between crustal subsidence curve for a well and predicted thermo-tectonic curve for same location	1. Depth of average 1% porosity and/or basement 2. Compaction history, response to sediment load on crust 3. Thermotectonic model.	Depth to 1% porosity and basement may not be known. Compaction history, isostatic response of crust unknown. Thermo-tectonic behavior unknown. Cannot determine the assumptions
2. Perturbations on stacked crustal subsidence curves	Size of perturbations from integrated stacked subsidence curves	1. Depth of average 1% porosity and/or basement 2. Compaction history 3. Isostatic response to sediment and water load on crust 4. Lithospheric rigidity and thermotectonic model	
Oxygen isotopes	$\delta^{18}O$ values	1. Variation in $\delta^{18}O$ value is a result of ocean volume 2. Isostatic response of crust to weight of water the same everywhere 3. Can estimate volume of continental ice and volume of ocean ice as function of time 4. No diagenetic effect	Cannot prove any of the assumptions

considerable length, and the purpose of this section is not to repeat Hallam's review, but to illustrate the various approaches used by different authors, to point to some of the difficulties inherent in attempts to compile such curves, and to illustrate the difficulties in arriving at unambiguous conclusions.

14.3.1 Cretaceous Sea-Level Curves

Numerous researchers have compiled local and regional sea-level curves for the Cretaceous Period. Figures 14.4-14.6 illustrate some of these.

Harris et al. (1984) developed a regional sea-level curve based on their studies of the dominantly carbonate section of the Arabian craton and platform margin. This might be considered an ideal area for the study of eustasy because it was, at that time, a tectonically stable area in which sedimentation was not affected by a large clastic sediment supply. Their curve is compared with two other standard curves of the late 1970s, those of Kauffman (1977) and Vail et al. (1977). The differences between the curves are considerable

(Fig. 14.4). Harris et al. (1984, p. 77) explained them as follows:

It is not surprising that differences exist between the various curves considering that the data was collected from different geographic locations and different geological foundations and reflect discrepancies in absolute age curves, position of various biostratigraphic markers, and overall degree of biostratigraphic resolution.

Scott et al. (1988) compiled curves from eight diferent sources, and compared them with one developed by themselves (Fig. 14.5). According to these authors " Some of the discrepancies in the timing of various peaks among these curves may result from different criteria used in identifying stadial boundaries." This point has been examined at length in Chapter 13. Scott et al. (1988) also referred to the effects of local tectonics. Their own curve was assembled by comparing stratigraphic sections in Oman and Texas, using graphic correlation methods (Sect. 13.3.4) to provide tight biostratigraphic correlations. The cycles of sea-level change that appear on their curve are of much lower frequency than those appearing on the Exxon global cycle chart. Because of the more refined method of

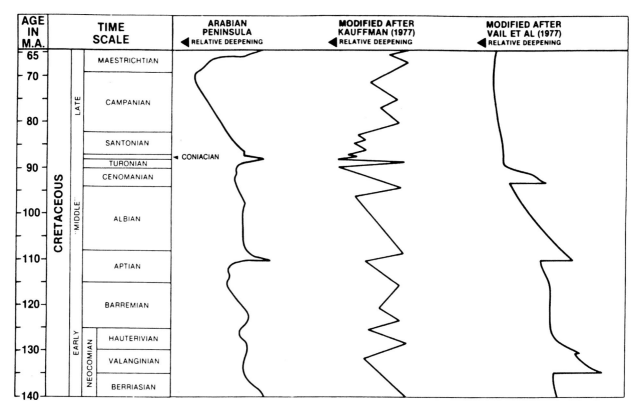

Fig. 14.4. A comparison of paleodepth curves for eastern Arabia with two other " standard" curves from the 1970s. (Harris et al. 1984, reproduced by permission)

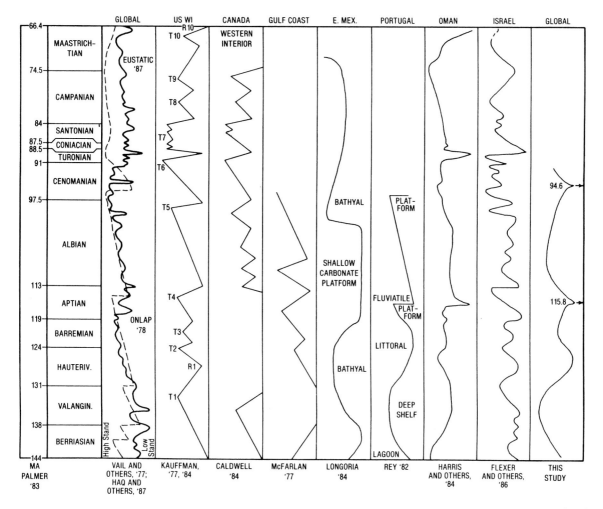

Fig. 14.5. A comparison of sea-level or paleodepth curves for the Cretaceous, prepared by various authors. The diagram was compiled by Scott et al. (1988), whose curve resulting from their own work is given in the *right-hand column*, labeled " this study"

correlation used in this study, it was possible to eliminate the effects of local tectonics, and the correlations that remain therefore seem far more realistic. It is therefore of interest that the times of low sea level that appear on their curve do not correlate very closely with those in any of the other curves, except the Exxon curve, which, as discussed in Section 13.4.3, is capable of being correlated with anything, even random data, because of the frequency of sea-level events therein (see also the comment on Cretaceous sea-level changes in the Arabian Peninsula by Pratt and Smewing 1993; Sect. 13.4.1).

A useful discussion by Schlager (1991) described several means by which sequence boundaries could be generated by processes other than sea-level change. Submarine erosion is one such process, as discussed in Section 2.2.1. Erosion by the Gulf Stream

on Blake Plateau, off the eastern United States, is interpreted to have generated several regional unconformities that do not correlate with the global cycle chart (Fig. 2.2). Drowning unconformities were also discussed by Schlager (Sect. 2.2.1). According to him, many sequence boundaries that have been identified in carbonate sections may represent this process, not sea-level change. He suggested that two Valanginian events in the Exxon global cycle chart may be of this type. It is significant that they are not present in other curves for this time interval, as shown in Fig. 14.6.

The Cretaceous system is considered in detail, along with the rest of the Phanerozoic record, in Hallam's (1992) compilation. This review is replete with examples of where the local stratigraphy confirms the Exxon global cycle chart and examples

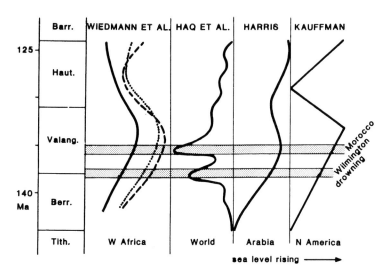

Fig. 14.6. Various curves for the Early Cretaceous, compared with the global cycle chart of Haq et al. (1987, 1988a). Note the presence of two short-term events in the Exxon curve, that do not appear in the other curves. (Schlager 1991, reproduced by kind permission of Elsevier Science-NL, Sara Burgerhartstraat 25, 1055 KV Amsterdam, The Netherlands)

where local tectonics or a high sediment supply have yielded a different curve, which does not correlate with the Exxon chart. The closest correlations appear to be in areas that provided a strong influence on the construction of the Exxon chart, namely the Western Interior of the United States, and the North Sea Basin. Circular reasoning again? It is, perhaps, significant, that Hallam did not attempt to assemble his own global cycle chart based on his examination of the published record – a more thorough discussion of a more complete data base than that used by the Exxon workers. Perhaps the reason is that it is impossible to construct a global chart because very few true global eustatic signals can be detected through the local and regional "noise."

14.3.2 Jurassic Sea-Level Curves

One of the world's acknowledged authorities on Jurassic stratigraphy is Hallam, who has included a sea-level curve for this period in many of his publications (Hallam 1978, 1981, 1984, 1988, 1992). He has been steadily enlarging his own area of field activity over the years to encompass more of the world's major Jurassic sections in an attempt to refine a synthesis of eustasy and tectonics. For this reason, most other workers researching Jurassic stratigraphy make reference to his standard curve, of which two versions have been published.

Figure 14.7 illustrates a family of curves compiled by Poulton (1988) based on his stratigraphic

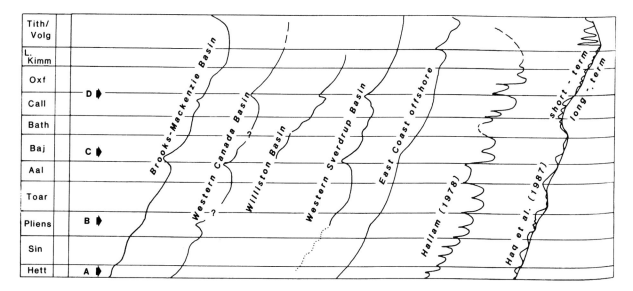

Fig. 14.7. Curves of " relative subsidence" prepared by Poulton (1988) for various cratonic and continental-margin basins in Canada, compared with the original Jurassic curve of Hallam (1978)

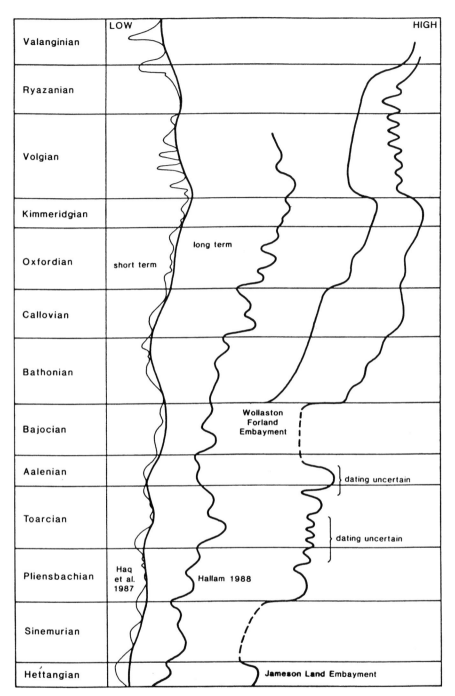

Fig. 14.8. Sea-level curves for two areas within eastern Greenland, compared with the global cycle chart of Haq et al. (1987, 1988a) and Hallam's (1988) revised curve.

(Surlyk 1990, reproduced by kind permission of Elsevier Science-NL, Sara Burgerhartstraat 25, 1055 KV Amsterdam, The Netherlands)

work in various regions of Canada, compared with the original curve of Hallam (1978). The curves are illustrated in mirror-image relative to the usual form of display, with rising sea level or relative subsidence increasing to the right. All the curves are in agreement that sea level was rising throughout the Jurassic, probably in response to the breakup of Pangea (Sect. 3.2, 9.2). Beyond this, however, there are few correlations, those that can be made indicating the effects of regional tectonics.

Surlyk has worked in eastern Greenland for many years, and his regional studies permitted him to compile regional sea-level curves for two separate basins (Surlyk 1990), which are shown in Fig. 14.8. Again, an overall sea-level rise during the Jurassic is indicated by his data, but beyond this the comparisons are few. That there are not more similarities between his curve and those of Hallam (1988, the revised Hallam curve) and Haq et al. (1987, 1988a) is, perhaps, surprising, because eastern Greenland is part of the North Atlantic region that underwent regional extension during the Jurassic, and was the source of some of the critical data used by both sets of authors in the compilation of their curves. Part of the problem may be difficulties in carrying out biostratigraphic correlation across faunal bound-

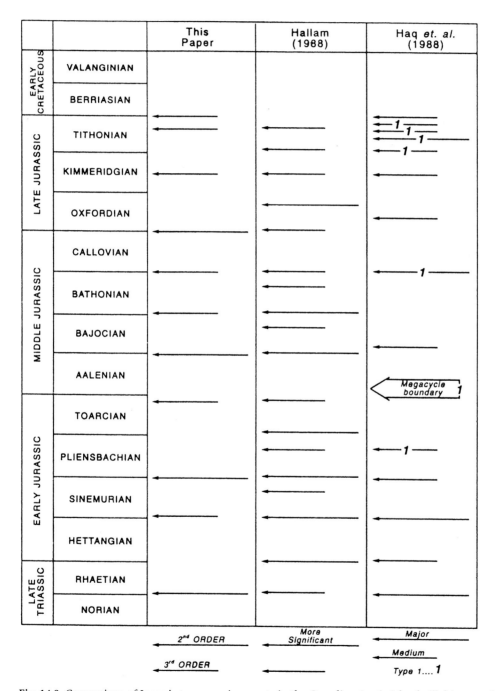

Fig. 14.9. Comparison of Jurassic transgressive events in the Canadian Arctic Islands (" this paper") with those in two other syntheses. (Embry 1993)

aries. As noted in Section 13.3.6, Surlyk estimated that a potential error of about one half of a stage must be allowed for in correlations between the European standard sections and Greenland. Further comments on Surlyk's work and that of others in the North Atlantic area are given in Section 11.2.2.

Stratigraphic analysis of the thick Mesozoic clastic section in the Sverdrup basin, Canadian Arctic Islands, led Embry (1993) to make several important contributions to the study of sequence stratigraphy. Figure 14.9 illustrates his synthesis of the major sea-level events in the Arctic, compared with those derived from the work of Hallam (1988) and Haq et al. (1988a). The history of transgressive events in the Arctic is very similar to the succession of events determined by Hallam (1988), but there is

very little similarity with those of Haq et al. (1988a). As Embry (1993) stated, it is difficult to evaluate this lack of similarity with the Exxon chart, because none of the data on which it was built has been made available. Embry (1993) found the close comparison with Hallam's "global" events significant, because many of these events (his "second order events"; see Sect. 15.2.1 for a discussion of his event classification) are clearly associated with regional tectonism in the Arctic, in that they are times when significant changes occurred in thickness and facies distributions and in sediment dispersal patterns. Embry (1993) concluded that these sea-level events were "tectono-eustatic" in origin, and referred to the intraplate stress model of Cloetingh (1988) as the basis for a possible explanation. The

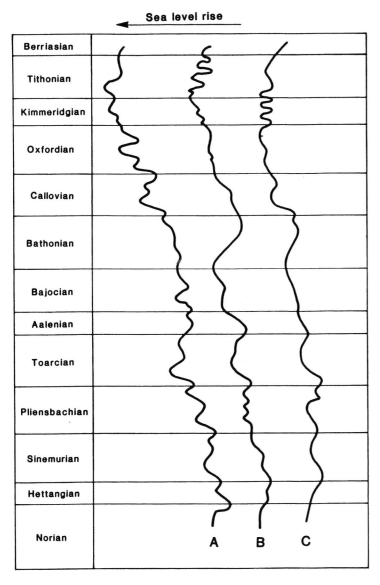

Fig. 14.10. Comparison of the three most well-known Jurassic curves. A Hallam (1988). B Haq et al. (1987, 1988a). C Vail and Todd (1981), Vail et al. (1984). This diagram was prepared by Hallam (1992). (From "Phanerzozic Sea-Level Changes," copyright by Columbia University Press; reproduced with permission of the publisher)

possibility of globally synchronous tectonic events is discussed in Section 11.4.

Finally, Fig. 14.10 compares the three most well-known sea-level curves for the Jurassic, that of Hallam (1988) and the two versions of the Exxon chart. In only a few places are there genuinely close similarities between these curves, indicating the possibility of a eustatic event that shows through the different methods of data compilation and analysis, and has survived the various revisions of the Exxon curve. For example, sea-level rises in the Early Hettangian and Late Toarcian, and falls in the Early Toarcian and Early Callovian, appear on all three curves.

14.3.3 Why Does the Exxon Global Cycle Chart Contain So Many More Events Than Other Sea-Level Curves?

With the exception of the Hallam curves, the other curves illustrated in this section do not show the ubiquitous cycles of 1–5 m.y. frequency that characterize the Exxon global cycle chart. Cycles of 5–20 m.y. duration are more typical. The reason for this seems likely to reflect the different modes of data synthesis (Dickinson 1993).

It is suggested that the paradigm of stratigraphic eustasy followed by the Exxon group allows them to interpret virtually all sea-level events as eustatic in origin. Hallam (1992, p. 92) said, of the Vail and Todd (1981) article on the North Sea: "The very title of the Vail and Todd paper implies that, at that time at least, seismic sequence analysis of only one region was believed to be sufficient to obtain a global picture." Therefore every new sea-level event that is discovered, can be added to the global chart. Elsewhere I have stated:

What does the Exxon chart represent? It undoubtedly constitutes a synthesis of real stratigraphic data from around the world. I suggest that most of the sequences are the product of regional events that originated as a result of plate-margin and intraplate tectonism, including basin loading and relaxation and in-plane stresses. These may have been well documented in one or more basins in areas of similar kinematic history, but their promotion to global events should be questioned.... Some events may be duplicates of other events that have been miscorrelated because of ... chronostratigraphic error. With so many events to choose from, such miscorrelation is almost impossible to avoid." (Miall 1992, p. 789)

My interpretation of the Exxon method of data synthesis is discussed in Section 13.2 (see Figs. 13.1, 13.2).

14.4 Main Conclusions

1. The margin of error in the assigned ages of sequence boundaries in the Exxon global cycle curves is, in most cases, greater than the spacing of these boundaries. Distinguishing these boundaries on the basis of chronostratigraphic data is therefore not possible.
2. The Exxon curves and the supporting work published in SEPM Special Publication 42 and elsewhere, contain errors and revisions that are not acknowledged or explained.
3. If the Exxon global cycle charts are a measure of eustasy, as claimed, the positions of sequence boundaries should not vary from location to location, or from curve to curve, although the amplitude of the curve may vary in response to local tectonics. However, different versions of the Exxon curves are in some cases markedly different, and these differences have not been explained or discussed by the authors of the curves.
4. Many other sea-level curves have been prepared by other authors, based on outcrop and subsurface stratigraphic data. In only a few cases are there genuinely close similarities between these curves, indicating the possibility of eustatic events that show through the different methods of data compilation and analysis.
5. Broad similarities between some of the curves, such as the overall rise in sea level through the Jurassic, reveal long-term trends corresponding to 10^7- to 10^8-year stratigraphic cyclicity, which is probably of eustatic origin (Chap. 9).
6. The almost total lack of correspondence between the high-frequency (1–5 m.y.) events ("third-order" cycles) revealed in the various sea-level curves confirms that global signals on this time scale are not globally correlatable.
7. The close spacing of sea-level events in the Exxon global cycle chart is not matched in other charts. This is interpreted as an outcome of the application of the paradigm of stratigraphic eustasy by the Exxon group, which allows them to interpret virtually all sea-level events as eustatic in origin. On this basis, every new sea-level event that is discovered, can be added to the global chart (Sect. 13.2).
8. However, the lack of proof for a global eustatic signal on the 10^6-year ("third-order") time scale (the lack of correspondence between the various other curves at that time scale) indicates that

either or both of the following statements is correct: (a) The imprecision and inaccuracy of current chronostratigraphic correlation methods do not permit a reliable test of global eustasy of this frequency (Chap. 13). (b) If a 10^6-year eustatic signal exists, it is largely or entirely masked by local to regional variations in sediment supply and tectonically driven uplift and subsidence of comparable rate and magnitude (Chap. 11).

9. In a few cases there is growing evidence for widespread sea-level events that are simultaneous (within the limits of chronostratigrahic precision) but also occur at times of tectonism, as indicated by changes in thickness, facies or dispersal patterns. These may be exmples of a newly recognized class of tectono-eustatic sea-level changes, with in-plane stress as the major driving force (Sect. 11.4).

V Approaches to a Modern Sequence-Stratigraphic Framework

Having demonstrated, in the earlier parts of this book, that stratigraphic sequences are widespread, and caused by many mechanisms of relative sea-level change, and that it is premature to attempt to define a global cycle framework, the question now arises, "where do we go from here?" Sequence-stratigraphic concepts are proving to comprise a powerful set of interpretive tools, and the science of stratigraphy has undergone a revitalization as a result of their application. The techniques of observation and interpretation that underlie our investigations of of sea-level change, are improving. It is now realized that sea level has been rising and falling (at least in a relative sense) continuously, throughout the Phanerozoic and possibly the Precambrian.

Modern work is proceeding along two main lines, which are the subject of the two chapters in this part of the book. Firstly, the integration of sequence-stratigraphic concepts with the methods of facies analysis is providing new insights into stratigraphic architecture and depositional processes. These are finding wide application in the area of petroleum exploration (Part V). Secondly, the relationships between sedimentation, subsidence and sea-level change are an ideal subject for the application of the new techniques of numerical and graphical simulation by computer, and this is an area that is currently exploding with new ideas.

- Two different conceptual models underlie the application of sequence stratigraphy by Vail and his coworkers: one model relates to presumed sea-level behavior through time; the other model relates to the stratigraphic record produced during a single sea-level cycle. Though the two models are interrelated they are logically distinct, and we believe that it is important to test them separately.... Our studies lead us to have considerable confidence in the correctness and power of the Exxon sequence-stratigraphic model as applied to sea level controlled, cyclothemic sequences.... At the same time, we suspect that the Exxon "Global" sea-level curve, in general, represents a patchwork through time of many different local relative sea-level curves. (Carter et al. 1991, p. 42 and 60)

- If stratigraphy is the study of layered rocks, incorporating many topics traditionally considered the domains of individual subdisciplines within the earth sciences, then we are in the midst of an evolving form of stratigraphic analysis which has been termed quantitative dynamic stratigraphy. Quantitative dynamic stratigraphy is the application of mathematical, quantitative procedures to the analysis of geodynamic, stratigraphic, sedimentological and hydraulic attributes of sedimentary basins, treating them as features produced by the interactions of dynamic processes operating on physical configurations of the earth at specific times and places. (Cross 1990, p. ix)

15 Elaboration of the Basic Sequence Model

15.1 Introduction

The sequence models established by Peter Vail and his coworkers in such publications as Vail (1987), Van Wagoner et al. (1987), Loutit et al. (1988), Posamentier et al. (1988), Posamentier and Vail (1988), Van Wagoner et al. (1990), and Vail et al. (1991) have served to stimulate a revitalization of the science of stratigraphy. With sea-level change as a model for the reinterpretation of complex clastic and carbonate stratigraphies a great deal of new research has been carried out, and many complex problems of stratigraphic correlation and facies analysis have been resolved. For much of this research the global cycle chart is irrelevant to the problems at hand, which bear on local correlation, paleogeographic evolution, and the search for generative mechanisms. An excellent example is the evolution of thought about the Cardium Formation, a conglomerate-sandstone unit of some importance in the Alberta Basin because of the numerous oil fields for which the Cardium is the reservoir. A brief discussion of the Cardium is provided in Section 8.6, and reference is made to the major publications in which Cardium sequence stratigraphy has been discussed (see, in particular, Plint et al. 1986, and the discussions and replies relating to this paper by Rine et al. 1987). Sedimentological techniques in general have undergone a major revision, as can be seen by a comparison between the second (Walker 1984) and third (Walker and James 1992) editions of the well-known *Facies Models* publication of the Geological Association of Canada. The third edition has the phrase *Response to sea level change* as its subtitle, and the book contains many references to the sequence-stratigraphic literature.

The purpose of this chapter is to sketch out some of the recent improvements in our understanding of depositional systems, drawing particularly from Van Wagoner et al. (1990, 1991), Walker and James (1992), Schlager (1992a), Hunt and Tucker (1992,

1995), Posamentier et al. (1992), Posamentier and Weimer (1993), Posamentier and Allen (1993b), and Kolla et al. (1995), plus many other sources. Three recent edited collections of research papers in the field of sequence stratigraphy (Posamentier et al. 1993; Weimer and Posamentier 1993; Loucks and Sarg 1993) contain many important ideas. As noted by H.W. Posamentier, P. Weimer, and their colleagues in their recent review papers, much controversy has arisen in the sequence stratigraphy literature during the 1990s because of attempts to apply the early, generalized Exxon models to specific case examples where conditions of subsidence style, sediment supply, accumulation rate, climatic control, or some other important parameter, are markedly different from those in the the initial models. No model should be used as a rigid template for interpretation, merely as a guide. It is also true that some valuable new concepts have emerged from these controversies. A few are touched upon in this chapter.

15.2 Definitions

15.2.1 The Hierarchy of Units and Bounding Surfaces

Depositional units constitute a hierarchy of scales, from the small-scale ripple cross-laminae through the various units constituting depositional systems, to the scale of the entire basin fill. This concept of "hierarchy" is discussed in the context of geological time in Chapter 12, and the ideas are summarized in Table 12.1. Important recent papers that deal with the subject of depositional hierarchy are those of Van Wagoner et al. (1990), Miall (1991b), Mitchum and Van Wagoner (1991), Nio and Yang (1991) and Drummond and Wilkinson (1996). There is a need to relate the hierarchies based primarily on outcrop studies of depositional units and bounding surfaces, to the sequence-stratigraphic system of nomenclature.

Figure 15.1 illustrates Miall's (1988, 1991b) hierarchy developed for fluvial deposits. A very similar field scheme was devised for tidal deposits by Nio and Yang (1991), based on that of Miall. Bounding surface codes for the two systems are given in Table 12.1. However there are substantial disagreements between the two systems in terms of the interpreted time duration of the various units. Sequence nomenclature of the Exxon group of workers has been concerned primarily with "third-order" cycles (group 10 of Table 12.1). These represent depositional units and time-spans at least an order of magnitude greater than most of the units with which Miall (1991b) was concerned. However, Van Wagoner et al. (1990) and Mitchum

and Van Wagoner (1991) claimed that they can trace widespread high-frequency sequences nested within the third-order cycles through the Gulf of Mexico. Their stratigraphy (and the terminology they have developed to describe it) involves depositional units of the same scale as those classified by Miall (1988, 1991b) and Nio and Yang (1991). Some attempt at a reconciliation of the terminologies is therefore necessary.

Figure 15.2 and Tables 15.1 and 15.2 illustrate the nomenclature of the sequence stratigraphers. High-frequency ("fourth-order") sequences in Fig. 15.2 are equated by Mitchum and Van Wagoner (1991) with the parasequences (facies successions) of "third-order" sequences (see definition of para-

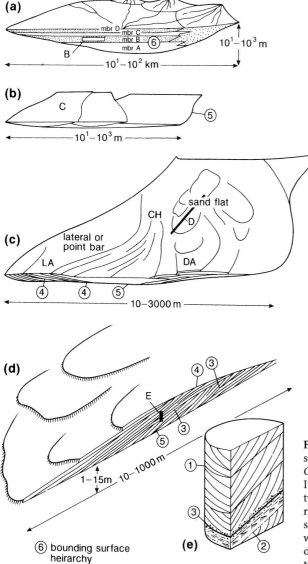

Fig. 15.1a–e. Scales of depositional elements in a fluvial system, showing a hierarchy of bounding surfaces. *Circled numbers* indicate ranks of bounding surfaces. In **c** two-letter codes indicate architectural element types: *CH* channel-fill; *DA* downstream-accreted macroform; *LA* lateral accretion deposit (macroform). In **d** sand flat is shown as being built up by migrating sand waves. Foreset terminations of these are shown at *top* of the diagram, but resultant internal cross-stratification has been omitted for clarity. (Miall 1988, 1991b)

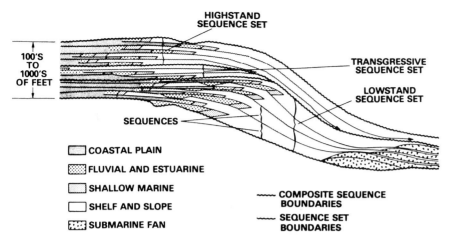

Fig. 15.2. Diagram of sequences, sequence sets, and composite sequences. Individual sequences, composed of parasequences, stack into lowstand, transgressive, and highstand sequence sets to form composite sequences.

(Mitchum and Van Wagoner 1991, reproduced by kind permission of Elsevier Science-NL, Sara Burgerhartstraat 25, 1055 KV Amsterdam, The Netherlands)

sequence and facies-succession in Sect. 4.2). However, the relationship is not necessarily as simple as this. In some studies high-frequency (10^5-year) sequences constitute the basic building-blocks of the stratigraphy, and may themselves be subdivided into parasequences. For example, Fig. 8.51 illustrates the 10^5-year sequences of the Dunvegan Delta, Alberta. Each of the sequences, named informally A–G, has been defined as an allomember, and each may itself be subdivided into suites of offlapping shingles representing 10^4 years of sedimentation. Each of the shingles is an upward coarsening facies succession (Fig. 8.52), which therefore itself fits the definition of a parasequence. Shingle boundaries would be classified as sixth-order bounding surfaces in the Miall (1988) classification, or as "F" surfaces in the Nio and Yang (1991) system. At least three scales of succession therefore need to be accommodated in the nomenclature.

Not acknowledged by Mitchum and Van Wagoner (1991) and the other Exxon workers is the importance of autogenic processes in the development of facies successions (parasequences). Those successions constituting the "shingles" in Fig. 8.51, and the the various lobes of the postglacial Mississippi Delta (Figs. 8.4–8.6) are interpreted to be of autogenic origin. They developed by delta-lobe switching as a result of major channel crevassing and consequent shortening of the transport path to the sea (Bhattacharya and Walker 1992). The resulting vertical succession is very similar to that of stratigraphic sequences, including the presence of prograding ("constructive") clinoforms similar to

highstand systems tracts, and delta-lobe abandonment surfaces followed by "destructive" deposits comparable to the sequence boundaries and transgressive systems tracts of the sequence models. The successions are comparable in vertical succession and time span to 10^4-year ("fifth-order") sequences. However, no sea-level change need be invoked to explain delta-lobe switching. Very careful delta mapping is required to distinguish autogenic delta shingles and lobes from "true" high-frequency cycles (e.g., Bhattacharya 1991; Bhattacharya and Walker 1991).

Table 15.3 is an attempt to set out the relationships between three types of hierarchical stratigraphic terminology that have been described in the literature. Sequence nomenclature is that of Van Wagoner et al. (1990) and Mitchum and Van Wagoner (1991). These authors defined third-order cycles as those having durations of 1–2 m.y., which represents only the shorter end of the time spans traditionally classified as third-order. Fourth- and fifth-order sequences, in their classification, are those with periodicities or episodicities of 10^5 and 10^4 years, respectively. The bounding-surface classifications of Miall (1988, 1991b) and Nio and Yang (1991) are based on outcrop architectural descriptions. These classifications were built from the small-scale units upward, by mapping the way in which facies units combine into macroforms (e.g., bars), channels, and larger features. This contrasts with the sequence nomenclature, which represents an attempt to develop an ever more refined subdivision of the larger-scale (regional or global)

Table 15.1. Definitions and descriptions of terms used in sequence stratigraphy (Van Wagoner et al. 1990) [p.6]

Stratal units	Definitions	Range of thicknesses (feet)	Range of lateral extents (sq.miles)	Range of times for formation (years)	Tool Resolution
Sequence	A relatively conformable succession of genetically related strata bounded by unconformities and their correlative conformities (Mitchum and others 1977)	1000–100	10 000–1000	10^6–10^5	
Parasequence set	A succession of genetically related parasequences forming a distinctive stacking pattern and commonly bounded by major marine-flooding surfaces and their correlative surfaces	100–10	1000–100	10^5–10^4	
Parasequence	A relatively conformable succession of genetically related beds or bedsets bounded by marine-flooding surfaces and their correlative surfaces	100–10	100–10	10^4–10^3	
Bedset	See Table 15.2	10–1	10–1	10^3–1	
Bed	See Table 15.2	1–Inches	10–1	10^2–1	
Lamina set	See Table 15.2	1–Inches	1	1	
Lamina	See Table 15.2	1–Inches	1	1	

Tool Resolution (horizontal range indicators):
Core and outcrop
Well log
Exploration Seismic
Paleo

Table 15.2. Detailed characteristics of lamina, bed, and bedset (Van Wagoner et al. 1990; based on Campbell 1967)

Stratal unit	Definition	Characteristics of constituent stratal units	Depositional processes	Characteristics of bounding surfaces
Bedset	A relatively conformable succession of genetically related beds bounded by surfaces (called bedset surfaces) of erosion, non-deposition, or their correlative conformities	Beds above and below bedset always differ in composition, texture, or sedimentary structure from those composing the bedset	Episodic or periodic.(same as bed below)	(Same as bed below) plus • Bedsets and bedset surfaces form over a longer period of time than beds • Commonly have a greater lateral extent than bedding surfaces
Bed	A relatively conformable succession of genetically related laminae or laminasets bounded by surfaces (called bedding surfaces) of erosion non-deposition, or their correlative conformities	Not all beds contain laminasets	Episodic or periodic. Episodic deposition includes deposition from storms, floods, debris flows, turbidity currents. Periodic deposition includes deposition from seasonal or climatic changes	• Form rapidly, minutes to years • Separate all younger strata from all older strata over the extent of the surfaces • Facies changes are bounded by bedding surfaces • Useful for chronostratigraphy under certain circumstances • Time represented by bedding surfaces probably greater than time represented by beds • Areal extents vary widely from square feet to 1000's square miles
Laminaset	A relatively conformable succession of genetically related laminae bounded by surfaces (called laminaset surface) of erosion non-deposition or their correlative conformities	Consists of a group or set of conformable laminae that compose distinctive structures in a bed	Episodic, commonly found in wave or current-rippled beds, turbidites, wave-rippled intervals in hummocky bedsets, or cross beds as reverse flow ripples or rippled toes of foresets	• Form rapidly, minutes to days • Smaller areal extent than encompassing bed
Lamina	The smallest megascopic layer	Uniform in composition/texture Never internally layered	Episodic	• Forms very rapidly, minutes to hours • Smaller areal extent than encompassing bed

Table 15.3. Suggested relationships between various hierarchical systems of stratigraphic classification

Sequence duration (years)	Sequence nomenclature	Bounding surface classifications Miall[a], Nio and Yang[b]	Proposed allostratigraphic terminology
10^6	3rd-order sequence	8^a	Alloformation
10^5	4th-order sequence (or regional parasequence set)	7^a	Allomember
10^4	5th-order sequence (or regional parasequence set)	$6^a, F^b$	Allomember or submember
10^{3-4}	Parasequence	5 or 6^a, E or F^b	(Facies succession)

Bounding-surface lettering of Nio and Yang (1991) is inserted here on the basis of a comparison between the architectural features described by these authors and Miall (1988), but does not correspond to their sequence designation. The hierarchical sequence nomenclature shown in column 2 is based on Vail et al. (1977) but is no longer recommended.

stratigraphic units. The major overlap and possible source of confusion in the nomenclature systems is at the level of Miall's (1991b) groups 8 and 9 (Table 12.1). High-frequency ("fourth-" and "fifth-order") sequences and large-scale autogenic facies successions may have very similar vertical thicknesses and lithological profiles, and represent similar time spans, but they are quite different in three-dimensional distribution and have quite different origins. It should be borne in mind that modern work is now suggesting that although natural hierarchies of sequences are common in the rock record, the durations of the constituent sequences are variable, and there is no genetic basis for a formal rank ordering (Sect. 3.1).

An appropriate system of nongenetic stratigraphic nomenclature might be the solution to the confusion. A perfect system does not, at present, exist, but the system of allostratigraphy devised by the North American Commission on Stratigraphic Nomenclature (1983) is a useful first approach. Allostratigraphic nomenclature is discussed in Section 1.4, and some of the terms and suggested relationships to sequence and bounding-surface nomenclature are shown in Table 15.3.

Embry (1993, 1995) objected to the rank-order classification of sequences (first- to sixth-order) proposed by Vail et al. (1977). This classification is not based on any descriptive characteristics of the sequences themselves except their duration, which may be poorly known, and has led to classifications of sequences in the global cycle chart that Embry (1993, 1995) regards as arbitrary. Embry's (1993, 1995) proposed solution is to define a five-fold classification of sequences based on six descriptive criteria:

- The areal extent over which the sequence can be recognized
- The areal extent of the unconformable portion of the boundary
- The degree of deformation that strata underlying the unconformable portion of the boundary underwent during boundary generation
- The magnitude of the deepening of the sea and the flooding of the basin margin as represented by the nature and extent of the transgressive strata overlying the boundary
- The degree of change in the sedimentary regime across the boundary
- The degree of change in the tectonic setting of the basin and surrounding areas across the boundary.

The application of these criteria to a sequence classification is illustrated in Fig. 15.3. There are two problems with this classification. One is that it implies tectonic control in sequence generation. Sequences generated by glacioeustasy, such as the

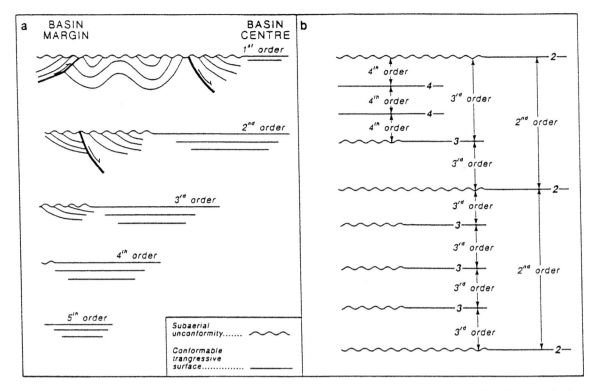

Fig. 15.3. a Schematic classification of stratigraphic sequences, according to the proposal by Embry (1993, 1995). As shown in the hypothetical sequence diagram **b**, sequences cannot contain a sequence boundary with the same or lower order than its highest order boundary, and the order of a sequence is equal to the order of its highest order boundary. (Embry 1993, 1995, reproduced by kind permission of Elsevier Science-NL, Sara Burgerhartstraat 25, 1055 KV Amsterdam, The Netherlands)

Upper Paleozoic cyclothems and those of Late Cenozoic age on modern continental margins, would be first-order sequences in this classification on the basis of their areal distribution (they are potentially global in extent), but fifth-order on the basis of the nature of their bounding surfaces (because of the absence of tectonism in the generation of the sequence boundary). The second problem is that the classification requires good preservation of the basin margin in order for deformation at the sequence boundary to be properly assessed. This may not always be available, and it is possible that sequences of first to third order in this scheme could be mistakenly assigned to the fourth order if their uplifted and eroded marginal portions are not preserved. However, this classification has certainly clarified relationships in basins such as the Sverdrup Basin of the Canadian Arctic, and there are undoubtedly other areas where this particular type of objective descriptive approach could be useful.

In this book a formal classification of sequences according to duration or level of deformation is not used because such classifications are increasingly being shown to be arbitrary (Sect. 3.1). However, it is useful to refer to sequences in a simple descriptive sense as being of low- or high-frequency, or to make reference to their periodicity or episodicity in terms of the order of magnitude of the cycle frequency (e.g., 10^7-year cycles for what were earlier termed "second-order" cycles).

15.2.2 Systems Tracts and Sequence Boundaries

With some important exceptions that are discussed here, the suite of systems tracts defined in the original sequence models (Chap. 4) has served sequence stratigraphers well. However, a few important modifications must now be made to these early models on the basis of a more sophisticated understanding of depositional processes during changing relative sea level. In addition, some common misconceptions must be clarified.

First of all, it is important to fully comprehend the chronostratigraphic limitations of systems-tract and sequence boundaries. Dalrymple and Zaitlin

(1994, p. 1085) commented on the diachronous nature of the fluvial-estuarine contact in incised-valley-fill deposits, and stated that "its use as the transgressive surface would cause the LST and TST to overlap in age, and the TST would pinch out landward, causing the HST to lie directly on the LST in area landward of the limit of estuarine sedimentation.... This would violate the desired chronostratigraphic significance of systems tracts." In fact this is a misreading of the concept of systems tracts. Posamentier and Allen (1993b) stated that "in general, eustasy and sea-floor subsidence/uplift determine the timing of sequence bounding surfaces, whereas sediment flux and physiography are most effective in determining the stratal architecture between those bounding surfaces." Systems-tract boundaries typically are diachronous because of progradation and aggradation and variations in sediment supply. Ito and O'Hara (1994) and Ito (1995) provided particularly good examples of systems-tract diachroneity, as proved by ash dating and correlation.

The diachroneity of sequence boundaries has been discussed elsewhere in this book, for example,

in Section 13.3.2 we discuss the chronostratigraphic significance of unconformities. An important debate has, however, been underway in recent years regarding the correlation of the sequence boundary across the major marine to nonmarine facies transitions. A sequence boundary is, in principal, easy to recognize in the nonmarine environment where it is represented by an erosional unconformity, although it may be difficult to distinguish such a surface from other erosion surfaces formed by autogenic channel scouring. It is much less obvious where the sequence boundary should be placed in correlative marine deposits, and serious disagreements have emerged between the original team of Exxon sequence workers (Van Wagoner et al. 1990; Posamentier et al. 1992; Posamentier and Allen 1993b; Kolla et al. 1995) and some other workers (notably Hunt and Tucker 1992, 1995; Helland-Hansen and Gjelberg 1994; Embry 1995).

Hunt and Tucker (1992) drew attention to the importance within the sequence framework of shoreline deposits formed during falling sea level. Although they did not refer to the earlier work on this topic by Plint (1988a, 1991; see Sect. 15.3.3),

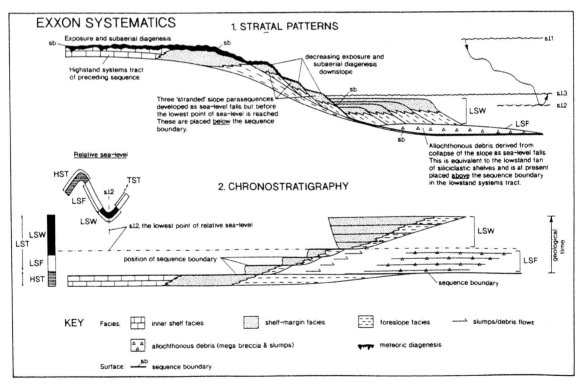

Fig. 15.4. Cross section and "Wheeler" diagram of a carbonate sand-shoal rimmed shelf, showing facies, systems tracts and positioning of the sequence boundary, according to the Exxon terminology. (Hunt and Tucker 1992, reproduced by kind permission of Elsevier Science-NL, Sara Burgerhartstraat 25, 1055 KV Amsterdam, The Netherlands)

their paper is important in drawing together a range of observations and problems bearing on the significance of falling-stage deposition (Figs. 15.4, 15.5). These deposits were not originally recognized in the Exxon sequence scheme, but are now assigned to their own systems tract, which Hunt and Tucker (1992) termed the *forced regressive wedge systems tract* (FRWST). Recognition of the importance of this systems tract raised the question, where to draw the sequence boundary? Posamentier et al. (1988) and Posamentier and Vail (1988) had related changing accommodation rates and facies shifts to the inflection points in the sea-level curve. Following this approach, Posamentier et al. (1992) suggested that the sequence boundary represents the beginning of the sea-level fall, and proposed that the boundary be drawn below the deposits formed during the falling stage. The problem with this approach is that this contact may in part be conformable, it is truncated by the regressive surface of marine erosion formed during the falling stage (Figs. 8.55, 15.6), and it assigns to the overlying sequence what may constitute a significant thick-

ness of regressive shoreface deposits formed during the falling leg (Hunt and Tucker 1992; Helland-Hansen and Gjelberg 1994; Embry 1995). The sub-aerial unconformity commences formation at the beginning of the falling leg, but erosion and downcutting continue until the end of the falling stage. Because this tends to be a prominent surface at the basin margin it is a good place to define the sequence boundary, which therefore corresponds in time to the beginning of the lowstand rather than the end of the highstand. The question remains how to correlate this surface into deeper parts of the basin.

Embry (1995) demonstrated that six genetically discriminated types of surface may be recognized within a shelf-slope succession (Figs. 15.6, 15.7). The slope onlap surface, although commonly prominent in seismic-reflection cross sections, is not a good choice of sequence boundary, because it develops during sea-level fall, when sediment is funnelled down submarine canyons and the slope becomes starved. This surface is, in turn, incised by the regressive surface of marine erosion, and by the

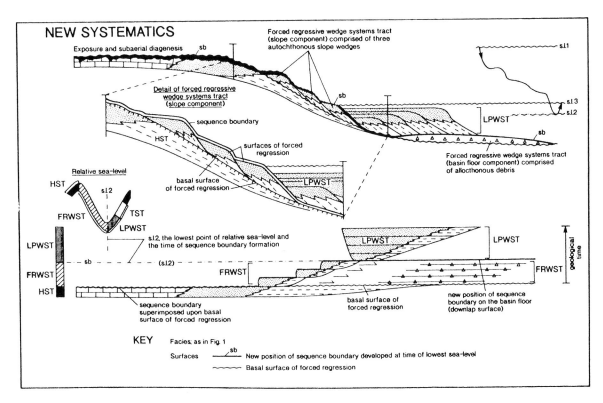

Fig. 15.5. A revised version of Fig. 15.4 showing a re-interpretation of the position of the sequence boundary relative to the lowstand fan deposits – what is now termed the lowstand prograding-wedge systems tract. (Hunt and Tucker 1992, reproduced by kind permission of Elsevier Science-NL, Sara Burgerhartstraat 25, 1055 KV Amsterdam, The Netherlands)

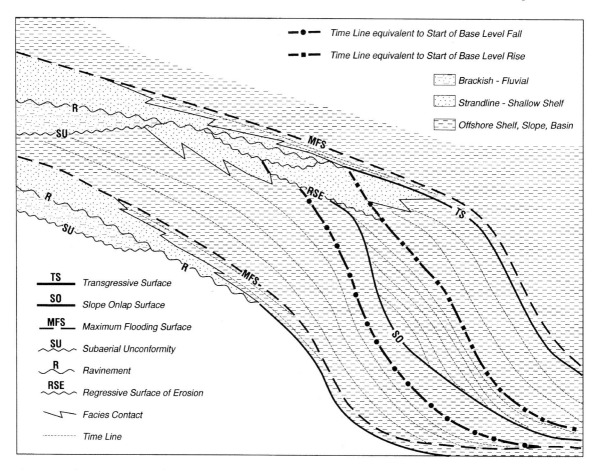

Legend:

— ● — Time Line equivalent to Start of Base Level Fall

— ■ — Time Line equivalent to Start of Base Level Rise

Brackish - Fluvial

Strandline - Shallow Shelf

Offshore Shelf, Slope, Basin

TS — Transgressive Surface

SO — Slope Onlap Surface

MFS — Maximum Flooding Surface

SU — Subaerial Unconformity

R — Ravinement

RSE — Regressive Surface of Erosion

— Facies Contact

------- Time Line

Fig. 15.6. Schematic stratigraphic cross section through a basin margin with a distinct shelf and shelf-slope break, depicting the six types of surface that may be distinguished. (Embry 1995, reproduced by kind permission of Elsevier Science-NL, Sara Burgerhartstraat 25, 1055 KV Amsterdam, The Netherlands)

subaerial unconformity, so that the latter surface would be in part at the base of the sequence and, in its more basinward extent, within the sequence (Fig. 15.6). This is not satisfactory. In ramp-type basins a slope-onlap surface is not present (Fig. 15.7). Embry (1995) recommended defining the sequence boundary in the basinward part of the succession at the surface of transgression. This is a surface that is normally readily mappable because it occurs at a distinct change in facies. In most cases the surface of transgression merges landward with the subaerial unconformity. Posamentier and Allen (1993b) demonstrated that in some instances this transgressive surface, near the base of the continental slope, at the toe of the "last clinoform" of the older regressive deposits, may be covered by a wedge of fine-grained sediment shed seaward by ravinement erosion. They referred to this as the *healing phase* of the transgression, "because the area seaward of the relatively steeply dipping

shoreface or delta front is partly infilled or healed over' as a lower-gradient shelf equilibrium profile is reestablished."

Hunt and Tucker (1992) pointed out an additional problem, the Exxon practice of drawing the sequence boundary below the lowstand fan deposits (Fig. 15.4). Walker (1992b) has also pointed out some problems with the Exxon systems tract terminology for deep-sea sediments. There may or may not be a continuous surface from the shelf to the base of the continental slope, as suggested in the cross section of Fig. 15.4, but it is certainly chronostratigraphically inconsistent to place the sequence boundary at the base of these deposits (see the Wheeler diagram in Fig. 15.4). Just as the FRWST is formed during the falling stage, so are the so-called lowstand fan deposits. In fact the term lowstand fan is incorrect, because the deposits are formed as a result of sediment bypass and reworking as the continental shelf is exposed to erosion during the

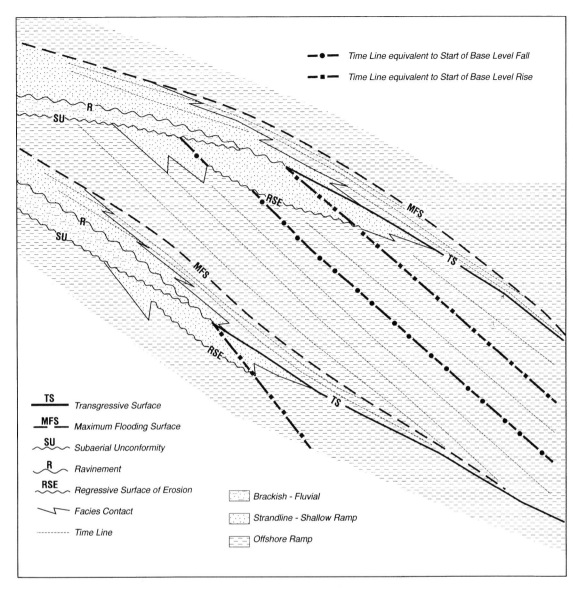

falling stage. Hunt and Tucker (1992) did not use the term lowstand fan, and recommended classifying these deposits with the FRWST (Fig. 15.5). They drew the sequence boundary at the top of these deposits. They argued that the sequence boundary as redefined is then consistent in that it is "everywhere coincident with the lowest point of relative sea level." (Hunt and Tucker 1992, p. 5). Kolla et al. (1995), in their discussion of this proposal, argued that fan sedimentation follows sea-level drop, and that therefore the placement of the sequence boundary below the so-called lowstand fan is the correct procedure; but this ignores the source of the sediment transported down to the base of the slope, and the timing of its accumulation, as described above. Kolla et al. (1995) pointed out, correctly, that placing the sequence boundary at the top of the fan deposits means, in many cases, placing the boundary within a conformable succession, although in many cases this is at a major facies change, possibly, within a condensed succession. This they found problematic, preferring to follow the historical precedent of drawing sequence boundaries at unconformities, even though this may disregard the elementary sedimentology of the falling stage systems tract (as described briefly above, and in greater detail in subsequent sections), and even though the term "correlative conformity" was introduced by Mitchum et al. (1977b) to include the type of situation where an unconformable sequence boundary passes laterally into a conformable succession. Other former members of the Exxon team (Mitchum et al. 1993) have correctly noted that the basin floor fan is formed during the "early phase of a relative fall in sea level," yet persisted in placing the sequence boundary below it.

Kolla et al. (1995) also objected to the introduction of the new term forced regressive wedge systems tract, although they use the term *forced regressive deposits*, and the acronym FRD, throughout their own discussion. Hunt and Tucker (1995), in their reply to the discussion of Kolla et al. (1995), reiterated their original points with regard to the placement of the sequence boundary and the need for an additional systems tract, and illustrated them with several examples. This discussion and reply provides a very useful summary of some of the differences of opinion that have emerged in the sequence stratigraphic literature in recent years, and is recommended to the reader.

Another collection of research articles on sequence stratigraphy (Van Wagoner and Bertram 1995) contains an updated discussion of systems tracts and offers new definitions of sequence terms (Van Wagoner 1995) which generate a whole new class of confusions. In earlier papers by the Exxon group systems tracts were defined and named with explicit reference to their interpreted relationship to the position on the sea-level curve (Van Wagoner et al. 1987; Posamentier et al. 1988). How else is one to interpret such terms as "highstand systems tract" except that it is interpreted to have been deposited during a sea-level highstand? However, in this new book it is stated that "erroneous interpretations of the block diagrams in Posamentier and Vail (1988) – that systems tracts are defined by their relationship to the eustatic curve – is still influencing stratigraphic thinking" (Van Wagoner 1995, p. xi). This is unfair. The initial definitions of the systems tracts (Posamentier et al. 1988, p. 110) stated that each is *interpreted* to be associated with a specific segment of the eustatic curve ... although not *defined* on the basis of this association" (italics added). However, all the text and diagrams in these earlier papers made the link between systems tracts and the eustatic sea-level curve, and geologists making systems-tracts interpretations can hardly be

faulted for overlooking this qualification. In recent years it has become increasingly clear that the relationships among systems tracts, sequence boundaries and relative sea-level change is far from straightforward. In fact, the use of such interpretive terminology as "highstand systems tract" could be questioned. Van Wagoner's (1995) new definitions of systems tracts make no reference to sea level. As Van Wagoner (1995) pointed out, incised valley-fills, which are typically assigned to the lowstand to transgressive systems tracts, may also contain fluvial deposits formed during the falling leg of the sea-level curve. Another example of confusion is the Pleistocene shelf-margin deltas in the Gulf of Mexico described by Suter and Berryhill (1985). These have been correlated with the falling leg of the sea-level curve on the basis of chronostratigraphic data, yet would, according to Van Wagoner (1995), be termed highstand deposits based on their stacking patterns, types of bounding surface and position on the shelf. Van Wagoner's (1995) proposed solution to this dilemma is to suggest "whether one chooses to define systems tracts using rock properties or the relationship to a sea-level curve is currently a matter of personal preference." I submit that this proposal can only lead to further confusion, and that we are in need of simple, nongenetic, descriptive terms for systems tracts that avoid this problem.

It has always been assumed that deep-marine sediments would contain fewer hiatuses than those deposited in shallow environments, and it is this assumption that lies behind the widespread acceptance of the idea of the deep-basin "correlative conformity" as the place ultimately to resolve the age of a sequence boundary. Recent work by Aubry (1995) is demonstrating that the pattern of unconformities in deep-marine sediments is surprisingly complex, and that sedimentation in the deep oceans is not nearly as continuous as has been assumed. The reason for this is not yet known, but may relate to meandering of erosive ocean currents, or to tectonism, including the triggering of widespread slumping (M.-P. Aubry 1996, personal communication).

Most stratigraphic research is now being carried out with sequence concepts in mind, and further developments, of the type discussed here, are to be expected. However, the application of sequence models is not always straightforward, as illustrated by the case of the British Millstone Grit. Read (1991) attempted to apply Exxon sequence concepts to these Carboniferous deposits of the Pennine district, including the recognition of systems tracts and sequence boundaries. However, Collinson et al. (1992) pointed out the many complications in Millstone Grit stratigraphy arising from autogenic processes and tectonic effects that make the simple application of the Exxon terminology suspect. Controversies of the type exemplified by this debate can only be resolved by additional detailed stratigraphic documentation.

Given the ongoing controversies described in this section regarding the definition of sequence boundaries and systems tracts, it is incumbent on working geologists to document their stratigraphic data as carefully as possible, paying particular regard to the recognition and correlation of major sedimentary breaks. Researchers should not be too dependent upon any particular predictive model of sequence architecture or sequence boundary correlation, but should let the rocks be their guide. In this way it is to be hoped that the misconceptions and errors that arise from model-driven research can gradually be eliminated.

15.3 The Sequence Stratigraphy of Clastic Depositional Systems

The purpose of this section is to examine some of the recent literature dealing with the architecture and environments of depositional systems that are undergoing relative sea-level change.

15.3.1 Fluvial Deposits and Their Relationship to Sea-Level Change

The early sequence model of Posamentier et al. (1988) and Posamentier and Vail (1988) emphasized the accumulation of fluvial deposits during the late highstand phase of the sea-level cycle, based on the graphical models of Jervey (1988). The model suggested that the longitudinal profile of rivers that are graded to sea level would shift seaward during a fall in base level, and that this would generate accommodation space for the accumulation of nonmarine sediments. This idea was examined critically by Miall (1991a), who described scenarios where this may and may not occur. The response of fluvial systems to changes in base level was examined from a geomorphic perspective in greater detail by Wescott (1993) and Schumm (1993), and the sequence stratigraphy of nonmarine deposits was critically reviewed in an important paper by Shanley and McCabe (1994). A recent extensive discussion of the sequence stratigraphy of fluvial deposits was given by Miall (1996, Sect. 11.2.2 and Chap. 13), and only a brief summary of this material is given here.

Shanley and McCabe (1994) discussed the relative importance of downstream base-level controls versus upstream tectonic controls in the development of fluvial architecture (Fig. 15.8). In general, the importance of baselevel change diminishes upstream. In large rivers, such as the Mississippi, the evidence from the Quaternary record indicates that sea-level changes affect aggradation and degrada-

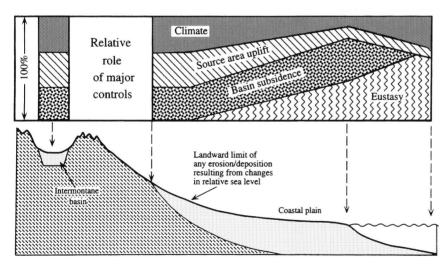

Fig. 15.8. The relative importance of the major controls on fluvial sequence architecture, showing how the balance of control varies between source area and shoreline. Inter-montane basins (*left*) are, of course, unaffected by sea-level change. (Shanley and McCabe 1994, reproduced by permission)

tion as far upstream as the region of Natchez, Mississippi, about 220 km upstream from the present mouth. Farther upstream than this source-area effects, including changes in discharge and sediment supply, resulting from tectonism and climate change, are much more important. In the Colorado River of Texas baselevel influence extends about 90 km upstream, beyond which point the river has been affected primarily by the climate changes of the Late Cenozoic glaciations. Blum (1994, p. 275), based on his detailed work on the Gulf Coast rivers, stated that "At some point upstream rivers become completely independent of higher order relative changes in base level, and are responding to a tectonically controlled long-term average base level of erosion." The response of river systems to climate change is complex. As summarized by Miall (1996, Sect. 12.12.2), cycles of aggradation and degradation in inland areas may be driven by changes in discharge and sediment load, which are in part climate dependent. These cycles may be completely out of phase with those driven purely by baselevel change.

The influence of tectonism in the development of fluvial clastic wedges is better understood, and has been discussed in many reviews (e.g., see Sect. 11.3.2).

The elements of a generalized sequence model for fluvial deposits are shown in Figs. 15.9 and 15.10. The sequence boundary is commonly an incised valley eroded during the falling stage of the relative sea-level cycle. This is filled with fluvial or estuarine deposits during the lowstand to transgressive part of the cycle, with the thickness and facies composition of these beds determined by the balance between the rates of subsidence, baselevel change and sediment supply. Away from the incised valley, on interfluve areas, the sequence boundary may be marked by well-developed paleosol horizons. It is a matter of debate whether the fluvial fill of an incised valley should be assigned to the lowstand or the transgressive systems tract. The shape of the sea-level curve and the timing of these deposits relative to this curve is usually not knowable, and so this is a somewhat hypothetical argument.

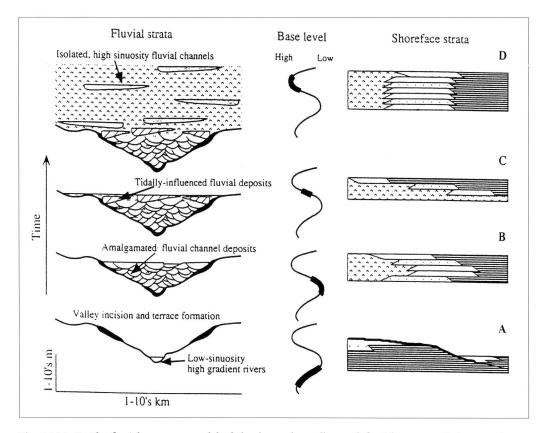

Fig. 15.9A–D. The fluvial sequence model of Shanley and McCabe (1991, 1994), showing the relationship between shoreface and fluvial architecture and base-level change. **A** Falling stage systems tract, with development of incised valley and fluvial terraces. **B** Lowstand systems tract. **C** Tidal influence indicates the beginning of the transgressive systems tract. **D** Highstand systems tract. (Reproduced by permission)

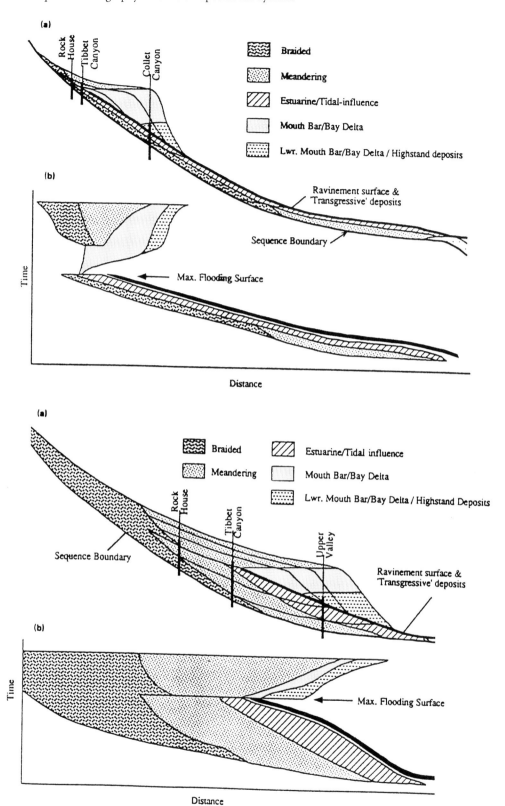

Fig. 15.10. a Longitudinal stratigraphic cross section and **b** chronostratigraphic diagram (Wheeler chart) of fluvial-coastal sequences developed during conditions of rapid rise in base level (*above*) and slow rise in base level (*below*). Based on studies of Upper Cretaceous sequences in southern Utah. (Shanley et al. 1992)

Transgression is commonly indicated by the appearance of tidal influence in the fluvial succession. Sigmoidal crossbedding, tidal bedding (wavy, flaser and lenticular bedding), oyster beds and brackish to marine trace fossils are all typical indicators of rising base level. The transition from fluvial to tidal is typically diachronous, as the rate of baselevel rise increases, and the filling of the incised vally changes from aggradational to retrogradational (Fig. 15.10). Inland from tidal influence the change from the lowstand to the transgressive phase may be marked by a change in fluvial style or by the development of coal beds. Coal commonly occurs during an initial increase in accommodation, before this is balanced by an increase in clastic supply. Within the valleys of major rivers the increase in accommodation can result in more loose stacking of channel sand bodies and greater preservation of overbank fines. Changes in fluvial style are also common, with braided rivers typifying lowstand systems and anastomosed or meandering rivers common during times of high rate of generation of accommodation, as during the transgressive phase of the base-level cycle.

The highstand systems tract develops when baselevel rise slows down, and the rate of generation of accommodation space decreases to a minimum. There are two possible depositional scenarios for this phase of sequence development. Retrogradation of the river systems during transgression will have led to reduced slopes, and a low-energy landscape undergoing slow accumulation of floodplain deposits, limited channel agrgadation, and closely spaced, well-developed soil profiles (Shanley and McCabe 1994). Given no change in source-area conditions, however, the sediment supply into the basin continues, and vigorous channel systems are eventually reestablished, Under these conditions channel bodies form that show reduced vertical separation relative to the TST, leading to lateral amalgamation of sandstone units and high net-to-gross sandstone ratios (Wright and Marriott 1993; Olsen et al. 1995; Yoshida et al. 1996). Basinward progradation of coastal depositional systems leads to downlap of deltaic and barrier-strandplain deposits onto the maximum flooding surface (Fig. 15.10). A good nonmarine example of this was described by Ray (1982), who mapped the progradation of an alluvial plain and deltaic system into lake deposits.

It seems likee that the HST is poorly represented in most nonmarine basins, because the highstand is immediately followed by the next cycle of falling baselevel, which may result in the removal of much or all of the just-formed HST deposits by subaerial erosion. A minor increase in the sand-shale ratio immediately below the sequence boundary may be the only indication of the highstand phase, as in the Castlegate Sandstone of Utah (Olsen et al. 1995; Yoshida et al. 1996).

Care must be taken to evaluate all the evidence in interpreting such data as net-to-gross sandstone ratios. Changes in this parameter may not always be attributable to changes in the rate of generation of accommodation space. Smith (1994) described a case where an increase in the proportion of channel sandstones in a section seems to be related not to changes in the rate of generation of accommodation space, but to increased sediment runoff resulting from increased rainfall. In the case of sequences driven by orbital forcing mechanisms, where both baselevel change and climate change may be involved, unraveling the complexity of causes and effects is likely to be a continuing challenge. In the model of Shanley and McCabe (1994) a greater degree of channel amalgamation is shown in the TST than in the HST (Fig. 15.9), the opposite of that shown in the model of Wright and Marriott (1993). Shanley and McCabe (1994) suggested that where rising baselevel is the main control on the rate and style of channel stacking, the rate of generation of accommodation is small during transgression in inland areas while the coastline is still distant, and increases only once transgression has brought the coastline farther inland where the effect of baselevel rise on the lower reaches of the river produces a more rapid increase in accommodation. In this model the rate of generation of accommodation is greater during the highstand than during transgression, and results in low net-to-gross sandstone ratios. However, this line of reasoning omits the influence of upstream factors, and must therefore not be followed dogmatically. One must also be cautious in using systems-tract terminology derived from marine processes for the labeling of nonmarine events. There may be a considerable lag in the transmission of a transgression upstream to inland positions by the process of slope reduction, aggradation and tidal invasion (Fig. 15.10). The inland reaches of the river does not "know" that a transgression is occurring, and it is questionable therefore whether the deposits formed inland during the initial stages of the marine transgression should be included with the TST.

Fluvial influence may be important even where no fluvial deposits are preserved. The high-frequency Cardium Sandstone cycles described by

Plint et al. (1986) and illustrated in Fig. 8.53 contain conglomerate bodies that constitute important local oil reservoirs. These are interpreted to be of shoreface origin, but there has been debate in the literature as to the ultimate origin of the gravels. Plint et al. (1986) proposed that they were of fluvial origin, but, as they pointed out here and in later papers (e.g., Walker and Plint 1992), no evidence of the fluvial feeder channels has been preserved. Arnott (1992) examined the evidence, and showed how fluvial incision during static sea-level at a time of highstand, followed by the next period of fall, could have developed incised channels that delivered gravel to the shoreline. This was reworked into shoreline conglomerate bodies during the next phase of sea-level rise, while most of the evidence for the feeder channels was stripped away by erosion during transgression.

15.3.2 The Concept of the Bayline

Posamentier et al. (1988, p. 118) defined the bayline as the demarcation line between fluvial and paralic/delta plain environments. They stated that this, rather than the shoreline, is the base-level point to which stream profiles are adjusted. All of the models discussed by Posamentier et al. (1988) and Posamentier and Vail (1988) assume an area of bays or lagoons along a shoreline, with rivers graded to the landward side of this area. Further, they defined the equilibrium point as the point on the continental margin at which subsidence and sea-level change are in balance. As sea-level rises and falls this equilibrium point migrates landward or seaward. A series of block diagram models were presented by Posamentier and Vail (1988) showing the response of the bayline and the coastal environments landward and seaward of it, to movement of the equilibrium point as sea-level rises and falls. For example, Fig. 15.11 (Posamentier and Vail 1988, p. 136, Fig. 16) shows the response of a shoreline

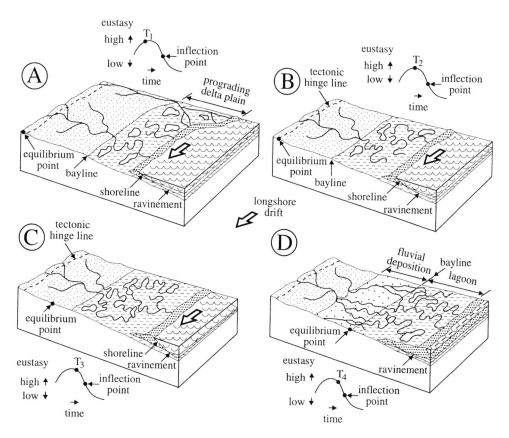

Fig. 15.11. Response of a fluvial and barrier-lagoon system to sea-level change, according to Posamentier and Vail (1988)

characterized by a barrier-lagoon system to changes in relative sea level. Fig. 15.12 (their Fig. 17, p. 137) illustrates the evolution of a delta system.

The delta models illustrate what seem to me some fundamental logical flaws in their sequence concepts. As Posamentier et al. (1988, p. 135) correctly stated "deltaic deposition occurs when a stream encounters a standing body of water and flow velocity abruptly decreases." However, this encounter occurs at the mouth of the stream, not at a hypothetical bayline somewhere near the head of the delta. This is why distributaries of the Mississippi delta have extended 3 km during the last century (Gould 1970), with the deposition of mouth bars at the *mouths* of the distributaries, where the flow expands into the Gulf of Mexico, and competence abruptly decreases (Coleman and Gagliano 1965). It is only possible for a river to grade itself to this point, which is why the entire Lower Mississippi is subject to flooding, and why crevassing occurs throughout the delta plain (subject to engineering control works). How could a delta build

seaward during a rapid sea-level fall (as required by the Exxon sequence models) if the source river has graded itself to a point at the delta head, implying no seaward (downslope) flow to maintain sediment transport?

The assumption throughout the Exxon models that rivers grade to the bayline ignores factors of sediment supply and grade, as in the Mississippi example discussed above. This is well illustrated by the barrier-lagoon model of Posamentier and Vail (1988; Fig. 15.11). This series of block diagrams illustrates the response of this type of coastline to a slow rise, followed by a fall in sea level. Throughout, the assumption is made that an ample supply of sediment reaches the barrier system by longshore drift to maintain the barrier in response to sea-level change, whereas a careful interpretation of their model indicates the built-in but unstated assumption that fluvial sediment supply is negligible. Why? This is not explained. The results are curious. Thus, three of the four block models show the equilibrium point landward of the bayline, a situation in which

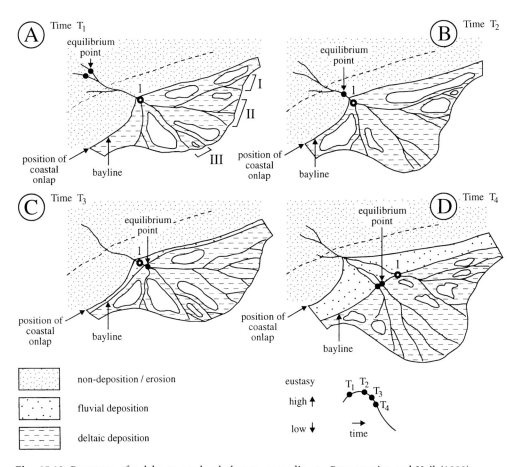

Fig. 15.12. Response of a delta to sea-level change, according to Posamentier and Vail (1988)

the barrier-lagoon system experiences a relative rise in sea level. Studies of Quaternary sea-level change (e.g., Nummedal et al. 1987) indicate that under these conditions the coast retreats, commonly with the development of a prominent ravinement surface (Nummedal and Swift 1987, Fig. 4; Fig. 2.4 of this book). Yet Posamentier and Vail (1988) showed the barrier-lagoon system aggrading and regressing throughout this period, based on their assumption of the maintenance of a sediment supply by long-shore drift.

Assuming that the barrier system does regress, and the lagoon system does aggrade, why does not the river keep pace? This is not made clear. Given an adequate sediment supply the rivers would extend their courses, fill the bays, and shift their position of grade seaward, keeping them at the shoreline (e.g., at the mouth of a tidal inlet) at all times. In this way, accommodation space would be generated, and fluvial aggradation and coastal onlap would take place during sea-level rise, as was actually shown by Jervey (1988, Figs. 10, 11; Fig. 15.13 of this book). Stratigraphic studies of the Gulf Coast Cenozoic (e.g., Winker 1979; Suter et al. 1987) show that only the smaller rivers are graded to a bayline, that coastal plains aggrade during sea-level rise and degrade during sea-level fall. Yet Posamentier and Vail (1988) maintained that fluvial aggradation and coastal onlap occur only during relatively rapid sea-level fall, when the equilibrium point moves seaward to the bayline.

Deltaic sedimentation depends on a host of factors (Coleman and Wright 1975). The slope of the continental margin, discharge, sediment supply and the efficacy of marine reworking processes are among the most important. Posamentier and Vail (1988, pp. 138–139) cited the Tigris-Euphrates flu-vial-deltaic system as an example of one that has extended its profile 300 km during the last 5000 years. They stated that "in this way the points to which the stream profiles are adjusted shift basin-ward, resulting in a basinward shift of equilibrium profiles and concomitant fluvial deposition. Thus, fluvial deposition occurs despite the fact that the position of the equilibrium point is landward of the bayline the entire time." These are a series of as-sertions unsupported by any data regarding tectonic activity, sediment supply and local relative sea-level change, nor do they document their assumptions regarding the shifting positions of the equilibrium point and the bayline, a particularly important omission in the case of the Tigris-Euphrates ex-ample because of its active tectonic setting. In fact, the area of the Tigris-Euphrates delta has been undergoing a relative rise in sea level since pre-historic time, as evidenced by archeological evi-dence indicating subsidence, drowning, and burial of ancient settlements by fluvial-deltaic sediment (Lees and Falcon 1952). Postglacial sea-level rise and tectonic subsidence have continued to generate accommodation space during the last few thousand years, and the fact that the delta has advanced in-dicates conditions of high sediment supply gener-ated by vigorous source-area uplift.

15.3.3 Deltas, Beach-Barrier Systems, and Estuaries

In the standard sequence models of Posamentier et al. (1988) and Posamentier and Vail (1988), that are described in some detail in Chapter 4, large-scale coastal depositional systems, including beach-bar-rier-lagoon systems and deltas, are characteristic features of highstand systems tracts. When sea level

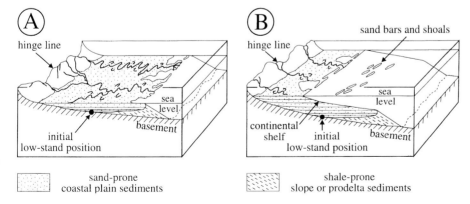

Fig. 15.13. Early Exxon models of the response of a continental shelf to a rise in sea level, and consequent increase in accommodation. (Jervey 1988)

344 Elaboration of the Basic Sequence Model

is stabilized at a highstand, sediment-supply factors may dominate, leading to extensive seaward progradation, the deposition of progradational sequence sets, and the production of clinoform architectures, which typically downlap onto the maximum flooding surface. Examples of this architecture are illustrated in Figs. 7.13–7.14, 7.21, 7.23, 7.37 (10^6-year cycles), and Figs. 8.6, 8.15–8.16, 8.31–8.33, 8.48–8.52, 8.57 (10^4–10^5-year cycles).

Coastal depositional systems occur in three other stages of the sea-level cycle: (a) beach-barrier systems may be deposited during the falling leg of the relative sea-level curve – the product of what have come to be termed *forced regressions;* (b) shelf-margin depositional systems (deltas and barrier systems) form during lowstands; (c) retrogradational deposits may be preserved as part of the transgressive systems tract, at the base of a sequence, between the sequence boundary and an overlying ravinement surface. These are predominantly strandplain (shoreface) deposits and estuarine deposits.

Sequence stratigraphy provides new insights into the popular tripartite delta classification of Galloway (1975), as used in all the popular sedimentology textbooks (Walker 1984; Reading 1986). Although coastal configuration, tide range, wave fetch, and other characteristics are of importance in determining deltaic style (Coleman and Wright 1975), sea-level change can now be seen to also be an important control, other factors being equal. River-dominated deltas are more likely to be formed as part of highstand systems tracts, at the inner margins of broad shelves protected to some extent from wave action. Wave-dominated deltas may be characteristic of shelf-margin deltas developed at lowstands of sea level. Tide-dominated deltas, as argued by Bhattacharya and Walker (1992) are actually sand-dominated estuaries and, as such, are most typical of transgressive systems tracts. The internal architecture of coarse-grained deltas (fan deltas) may yield useful information about the shape of a high-frequency sea-level curve (Postma 1995). For example, the *rollover point*, which is the transition between delta topsets and foresets, is developed close to sea level, and tracking its stratigraphic migration can indicate the occurrence of long-term trends.

Coastal strandplain (shoreface) deposits formed during falling sea level are commonly distinguished from those occurring during highstands by the fact that they are "sharp-based," and rest on regional erosion surfaces (Fig. 15.14b). Examples occur in

the Cardium, Viking, and other Cretaceous formations of the Alberta Basin, as illustrated by Plint (1988a, 1991). This contrasts with the "normal" progradational character of coastal deposits, which grade up from fine-grained offshore deposits, forming coarsening-upward facies successions (Fig. 15.14a).

Plint (1988a, 1991) was the first to recognize the importance of marginal-marine sedimentation during the falling leg of the relative sea-level curve. His model for the development of these sharp-based successions is illustrated in Figs. 8.55 and 15.15. The basal erosion surface develops as a result of wave scour during a relative fall in sea level. This is termed the *regressive surface of marine erosion*. It is overlain by downlapping, prograding shoreface deposits as sea-level fall continues. The formation and preservation of the shoreface deposits depends on the changing balance between sea-level change and subsidence (Fig. 15.15). As shown by Plint (1988a; Fig. 15.15B, C) a temporary increase in the rate of relative sea-level fall may lead to the isolation of shoreface deposits, which then occupy pockets on the basal erosion surface. Hunt and Tucker (1992) referred to these as *stranded parasequences*. As pointed out by Plint (1988a) they have been mistakenly interpreted as "offshore bars" in some older literature. It is important to note that these successions develop as sea level is falling, and before sea level reaches its lowest point. They therefore occur below the sequence boundary, and may be planed off by erosion during the subsequent transgression (Fig. 15.15D).

Subsequent workers have recognized the local importance of these deposits and various proposals for terminology have been made. Posamentier et al. (1992) discussed additional subsurface examples. Hunt and Tucker (1992) proposed the term *forced regressive wedge systems tract*, as discussed in Section 15.2.2 (Fig. 15.5). They listed several terrigenous-clastic and carbonate examples. Posamentier et al. (1992) demonstrated how a recognition of the presence of this systems tract could make subsurface correlation of well logs more explicable. They offered two alternative correlations through such a succession, pointing out that the correct interpretation, based on the appropriate sequence model, requires that the section be "hung" on the appropriate datum (Fig. 15.16). Dominguez et al. (1992) illustrated a slightly different stratigraphic configuration, based on their observations of the modern coast of Brazil, and mapping of Late Cenozoic deltaic and coastal depositional systems.

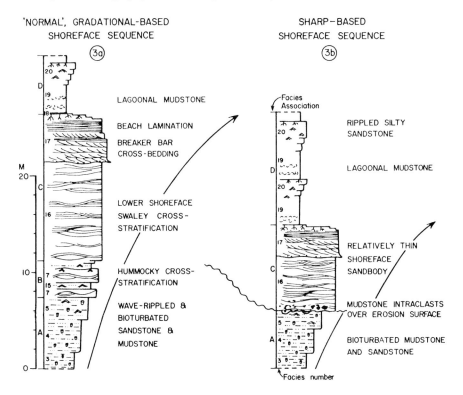

Fig. 15.14. Schematic lithofacies logs illustrating the contrast between "normal" progradational shoreface sucessions and "sharp-based" successions, which rest on erosion surfaces. (Plint 1988a)

In this interpretation (Fig. 15.17) stacked, off-lapping beach complexes rest on ravinement surfaces, The latter formed during sea-level rise, and correspond to marine flooding surfaces. The first beach lens represents the highstand deposits, and these are succeeded by the offlapping, and laterally downdropping lenses formed during regression. As these prograded seaward Dominguez et al. (1992) indicate that they interfinger with offshore marine muds, suggesting a higher rate of subsidence than in the Alberta examples described by Plint (1988a, 1991), where wave scour in front of the prograding beach maintains an erosion surface everywhere at the base of the beach deposits.

Shelf-margin depositional systems are referred to by Posamentier et al. (1988) and Posamentier and Vail (1988), and good examples occur in the Cenozoic deposits of the Gulf Coast (Figs. 8.1, 8.3). There, shelf-margin deltas developed during the Wisconsinan (last glacial) sea-level lowstand, and pass up depositional dip into widespread lowstand fluvial sheets on the continental shelf (Suter and Berryhill 1985; Suter et al. 1987). It seems likely that detailed work would demonstrate the existence of forced-regressive shoreface wedges flanking the deltas along the Cenozoic shelf margin.

The results of Edwards (1981), Winker and Edwards (1983) and Suter et al. (1987) suggest that shelf-margin deltas that form during lowstands are distinctly different in architecture from highstand deltas formed on gently sloping shelves. The latter tend to be "lobe dominated" and characterized by avulsive lobe switching generated by normal auto-genic processes, as noted above. Coarsening-upward facies successions are the result, the architecture of which is similar to that of high-order stratigraphic sequences. In plan view these lobes form ovate or elongate sand bodies oriented at a high angle to the shoreline. The architecture of shelf-margin deltas may be quite different. Shelf margins commonly are an area of gravity-induced lateral extensional stress, which is expressed by the development of growth faults. The result may be the thickening of deltaic deposits across the faults, which are oriented parallel to the shoreline. Sand bodies may then have a distinct shore-parallel thickness distribution (Fig. 15.18).

An interesting theoretical analysis by Muto and Steel (1992) reexamined the backstepping that characterizes the tops of many major deltaic complexes (e.g., the Brent Group, Viking Graben; Fig. 15.19). The conventional interpretation of such

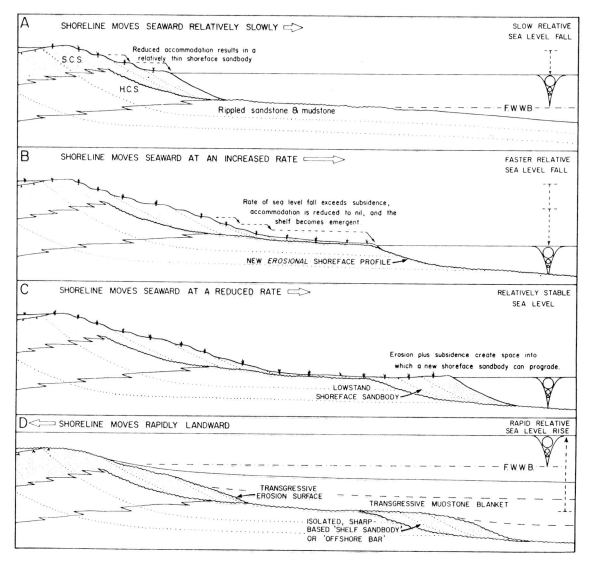

Fig. 15.15. General model for the development of sharp-based shoreface successions during a "forced regression." Preservation of the shoreface deposits depends on the continuous availability of accommodation space during sea-level fall, which, in turn, depends on the balance betwen subsidence and sea-level change. (Plint 1988a)

backstepping is that the sediment supply decreases or that sea level rises. However, Muto and Steel (1992) demonstrated that such backstepping, or retrogradation, is a natural consequence of sediment distribution under conditions of steady sediment supply and steady sea-level rise. The reason is simply that as the delta is extended seaward the surface area of the delta plain and delta front increase, requiring additional volume of sediment deposited to increase the thickness by a constant amount per unit time. If the sediment supply remains constant the response is for the shoreline to retreat landward, and for retrogression to occur.

The third mode of preservation of coastal deposits is during trangression, and constitutes part of the standard transgressive systems tract, as described in Chapter 4. Interpretation of these deposits may hinge on recognition of the ravinement surface, and the transgressive surface. As shown by Demarest and Kraft (1987), details of vertical profiles across the sequence boundary and into the transgressive systems tract vary considerably, and skill in recognition of the various subenvironments (beach, tidal inlet, lagoon, washover fan, estuary, etc.) is required for correct interpretation (Figs. 7.8, 15.20). The reinterpretation of the conglomerates of the

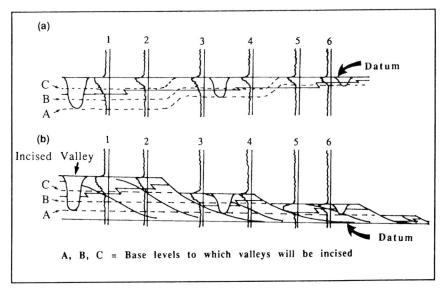

Fig. 15.16. Two alternative interpretive cross sections of a coastal complex. In **a**, a datum corresponding to the top of the sand section is used. This might be interpreted as a flooding surface. The problem with this interpretation is that it "generates" incised fluvial valleys of different depths. In **b**, as the datum a marker is used within the shelf deposits underlying the coastal succession. This lines up the incised valleys better, indicating a common depth of lowstand erosion. (Posamentier et al. 1992, reproduced by permission)

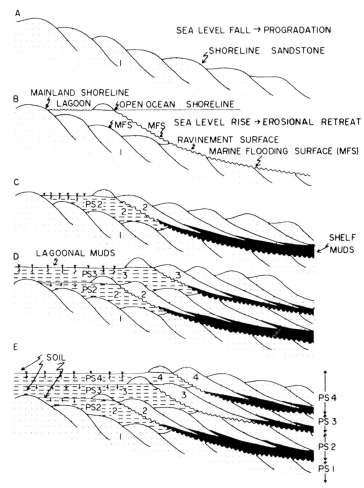

Fig. 15.17A–E. Interpretation of the evolution of beach-barrier-lagoon systems on the coast of Brazil. Shingled, offlapping beach deposits occur above ravinement surfaces. They compare with the sharp-based deposits of Plint. (1988a) but indicate a slightly different sequence of events. (Dominguez et al. 1992, reproduced by kind permission of Elsevier Science-NL, Sara Burgerhartstraat 25, 1055 KV Amsterdam, The Netherlands)

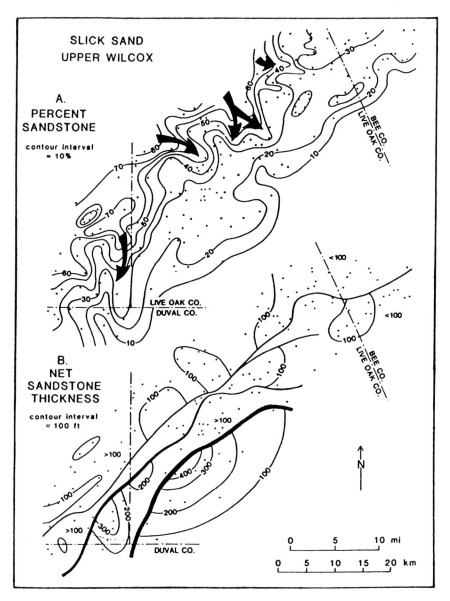

Fig. 15.18. Isolith maps of the Slick Sand (Eocene), the deposit of a shelf-margin delta complex. The percentage-sandstone map (**A**) can be interpreted to indicate the presence of lobate delta bodies and sand entry points. The sand thickness map (**B**) indicates the control of growth faults in generating shore-parallel sand bodies. (Winker and Edwards 1983)

Cardium Formation, Alberta (Fig. 8.53; Sect. 8.6) as shoreface deposits formed during transgression was one of the principal achievments of the work by Plint and his colleagues on these rocks (Plint et al. 1986).

Deltas are rarely formed during transgressions because the river-borne sediment supply tends to be overwhelmed by the rise in base level. Estuaries are the more common result (Reinson 1992). They are characteristic of many valley fills incised into sequence boundaries (Fig. 15.21, 15.22). The vertical successions can be understood in terms of the progressive landward shifting of four major estuarine subenvironments as a result of the rise in sea level (Fig. 15.23).

In recent years the importance of incised-valley-fills in the stratigraphic record has increasingly been recognized. Many small- to medium-scale sandstone petroleum reservoirs occur in this type of stratigraphic setting, such as many of the Mannville reservoirs of Alberta. Most such deposits are fluvial

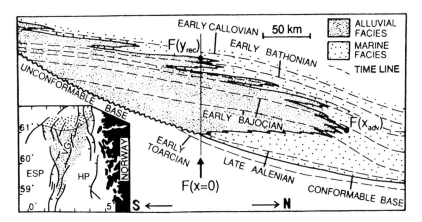

Fig. 15.19. Schematic south-north cross section of the Brent Group delta, Viking Graben. (Muto and Steel 1992)

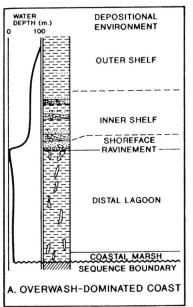

A. OVERWASH-DOMINATED COAST

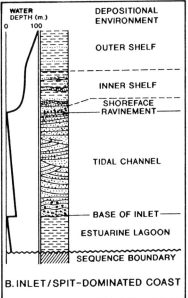

B. INLET/SPIT-DOMINATED COAST

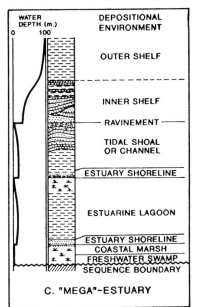

C. "MEGA"-ESTUARY

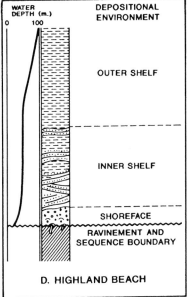

D. HIGHLAND BEACH

Fig. 15.20A–D. Vertical successions representing different types of transgressive systems tract. (Reinson 1992; after Desmarest and Kraft 1987)

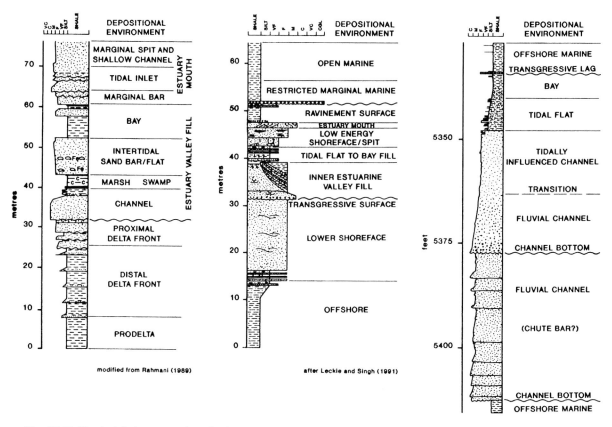

Fig. 15.21. Vertical facies successions in three representative valley-fill estuarine successions. Each profile contains a sequence boundary (*wavy line*) which represents the surface of estuarine transgression. (Reinson 1992)

to estuarine in origin, and the models discussed in this section and Section 15.3.1 are of considerable importance in their analysis. Additional discussion with many case studies, was provided by Dalrymple et al. (1994).

15.3.4 Shelf Systems: Sand Shoals and Condensed Sections

Our knowledge of shelf processes has undergone dramatic advances with the research on the Cardium group in Alberta, led by G. Plint and R.G.

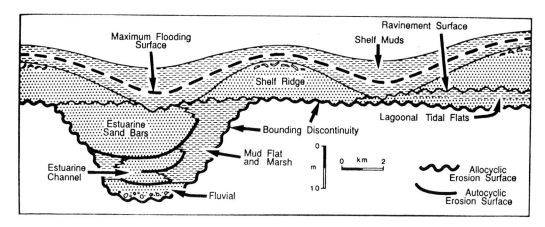

Fig. 15.22. Hypothetical shore-parallel section showing the stratigraphy of a transgressive systems tract. Autogenic erosion surfaces are tidal channel scours, allogenic erosion surfaces represent lowstand and ravinement erosion surfaces. (Dalrymple 1992)

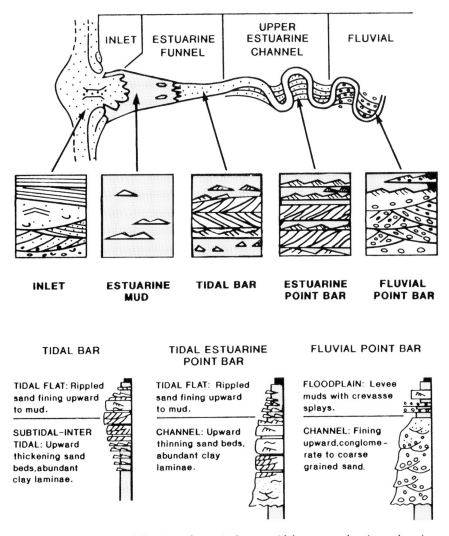

Fig. 15.23. Schematic subdivision of a typical macrotidal estuary, showing subenvironments, suites of sedimentary structures (*boxes in the center*), and vertical profiles. (Reinson 1992)

Walker, and that on the Mesaverde Group, in Wyoming and the Book Cliffs section of Utah, carried out by D.J.P. Swift and his colleagues. Some of this work is referred to and illustrated in Section 8.6. The book by Swift et al. (1991a) contains the most sophisticated approach to the quantification, modeling and interpretation of shelf sedimentation yet attempted. It is, however, concerned primarily with the quantification of sedimentary processes, and provides few large-scale cross sections for use as analogs by practicing petroleum geologists. Up-to-date summaries of tide- and storm-dominated shelves were provided by Dalrymple (1992) and Walker and Plint (1992), respectively. However, most of these articles seem to be concerned predominantly with shoreface deposits, not with the trangressive deposits of the shelf itself – those deposits formed during transgression and after the ravinement surface has been generated and water depths have dropped to below fairweather wave base. Sequence analyses have demonstrated that many so-called "offshore sand bars" of presumed shelf origin are actually shoreface deposits, formed either as the product of forced regression during falling relative sea levels, or as deposits reworked by shoreline processes during transgression (Sect. 15.3.3).

An example of this reinterpretation is the Tocito Sandstone, which comprises a series of lenses in the Gallup Sandstone "third-order" cycle (Figs. 7.23–7.25). The sedimentology and petroleum geology of this succession has been described in detail by Nummedal et al. (1989), and Jennette et al. (in Van Wagoner et al. 1991). Nummedal (1990) interpreted the Tocito as a series of tidally influence shelf sand ridges, but Jennette et al. (in Van Wagoner et al. 1991) carried out detailed subsurface analysis which indicates that the lenses occupy erosional valleys and represent lowstand deposits formed during several successive sea-level lowstands. A brief discussion of this reinterpretation is presented in Section 17.3.1.

Coarsening-upward sandstone successions in the Cardium Formation of Alberta (Fig. 8.53) are truncated by erosion surfaces, and represent highstand, shoaling deposits formed on the continental shelf of the Western Interior foreland basin immediately prior to sea-level fall. Shelf sandstones of the Blackhawk Formation, Utah, display the same general vertical facies succession (Van Wagoner et al. 1990). In other cases, as in the Rancho Rojo Sandstone of Arizona the shelf sandstones fine upward, beacause they were deposited under conditions of increasing water depth during transgression (Kreisa et al. 1986). Hummocky cross-stratification is a distinctive sedimentary structure in such deposits.

Many modern continental shelves are characterized by sand shoals, which have been proposed as analogs for various ancient units. It now seems certain that most of these are relict deposits, first formed during sea-level lowstands, and reworked by coastal and offshore processes during transgression. The modern sand shoals of the Atlantic margin shelf off the United States are a good example (Rine et al. 1991). Dalrymple (1992) summarized work on some other shelf sandstones, including the giant-ridge fields of the North Sea, all of which fall into the category of trangressive deposits (e.g., see Fig. 15.22).

Swift et al. (1991a) discussed numerous case studies of transgressive and regressive shelf-coastal depositional systems from various parts of North America, especially from the Western Interior Basin, and argued that their large-scale architecture is determined primarily by a balance between the rate of generation of accommodation space by relative sea-level change, and the rate of sediment supply. They developed a ratio, ¥, which expresses the general relationship RD/QM, where R is the rate of sea-level rise, D is the rate of sediment transport on the shelf, Q is the rate of sediment input from rivers or along shore, and M represents its character (grain size) They offered five general models (Fig. 15.24). In the first case, rates of transgression are high and preservation potential of retrogradational shoreface deposits is low. Rapid water

deepening leads to fining-upward shelf successions. The modern shelves of the Netherlands and Argentina are examples. The Middle Atlantic Shelf of the United States exemplifies the second case. Here, less rapid trangression leads to reworking of shelf deposits into sand ridges. Where the ratio ¥ is approximately in balance, as in the Gallup Sandstone of New Mexico (Fig. 7.23–7.25) facies successions stack nearly vertically. The remaining two cases represent situations where sediment supply rates were high relative to the rate of creation of accommodation space, resulting in regressive, progradational architectures. The first case is exemplified by the Miocene deposits of Maryland (Figs. 7.9, 7.10), the second by the Mesaverde Group of Utah and the Marshybank Formation of Alberta (Fig. 8.54).

There are three major types of surface associated with the distal, marine extent of continental-margin sequences: (a) the slope onlap surface, (b) the surface of initial trangression, which is a ravinement surface on the shelf, and overlies the lowstand-trangressive deposits of incised valleys, and (c) the surface of maximum marine trangression and highstand downlap (Embry 1995; Figs. 15.6, 15.7). Among the more prominent surfaces within sequences are the flooding surfaces formed during maximum transgression. These are commonly associated with marine hiatuses or condensed sections. Loutit et al. (1988), who provided an excellent treatment of this subject, described condensed sections as follows:

Condensed sections are thin marine stratigraphic units consisting of pelagic to hemipelagic sediments characterized by very low-sedimentation rates. They are areally most extensive at the time of maximum regional transgression of the shoreline. Condensed sections are associated commonly with apparent marine hiatuses and often occur either as thin but continuous zones of burrowed, slightly lithified beds (omission surfaces) or marine hardgrounds.... Condensed sections may also be characterized by abundant and diverse planktonic and benthic microfossil assemblages, authigenic minerals (glauconite, phosphorite, and siderite), organic matter, and bentonites and may possess greater concentrations of platinum elements such as iridium.

Loutit et al. (1988) pointed out that the flooding that leads to the development of condensed successions may serve to bring into continental-margin environments the deeper-water faunas that are normally more reliable for regional and global biostratigraphic correlation than the benthic and shallow-water forms of shallow marine environments. Figure 15.25 illustrates an example of a well-

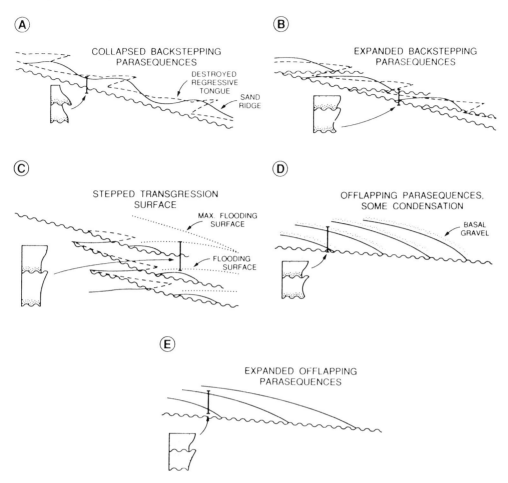

Fig. 15.24A–E. Schematic models of five types of coastal-shelf architecture, reflecting variations in the rate of generation of accommodation space and the rate of sediment supply and its transportation. ¥ is the accommodation/supply ratio. **A** ¥>>1. **B** ¥>1. **C** ¥=1. **D** ¥<1. **E** ¥<<1. (Swift et al. 1991b)

studied condensed section, corresponding to the Cretaceous-Tertiary boundary at an outcrop section in Alabama. Note the abundance of iridium and glauconite immediately above the surface of maximum condensation (indicated by an arrow).

Saito (1991) described the lowstand and transgressive deposits associated with the postglacial sea-level rise in the Sendai shelf area, off northeastern Honshu, Japan. A sand sheet 0.5–1.5 m thick was deposited during transgression. It consists of very poorly sorted muddy and pebbly sand, and bioturbated sand laminae, with a molluscan fauna, shell and plant fragments, and grains of volcanic debris and glauconite. On the inner shelf these sands have now been covered with highstand muds.

The surface of maximum marine flooding commonly serves as a surface of downlap by prograding highstand deposits in the later stage of the cycle of

sea-level rise (e.g., dashed line in Fig. 8.50; allo-member-bounding discontinuities in Fig. 8.51; surface at top of unit A in Fig. 8.54). As Galloway (1989a) noted, these flooding surfaces, whether they are hiatuses or condensed sections, commonly yield high-amplitude reflections on seismic surveys, and also constitute distinctive log markers. permitting reliable regional mapping. Galloway (1989a) recommended using surfaces of maximum flooding, rather than lowstand erosion surfaces, as sequence boundaries, and proposed an alternative definition of sequences, which he termed *genetic stratigraphic sequences*. These are described briefly in Section 4.5. The Exxon workers quickly recognized the significance of these marine surfaces. Vail and Todd (1981) interpreted them as indicators of rapid rise in sea level, leading to temporary sediment starvation in the deeper offshore areas and entrapment of

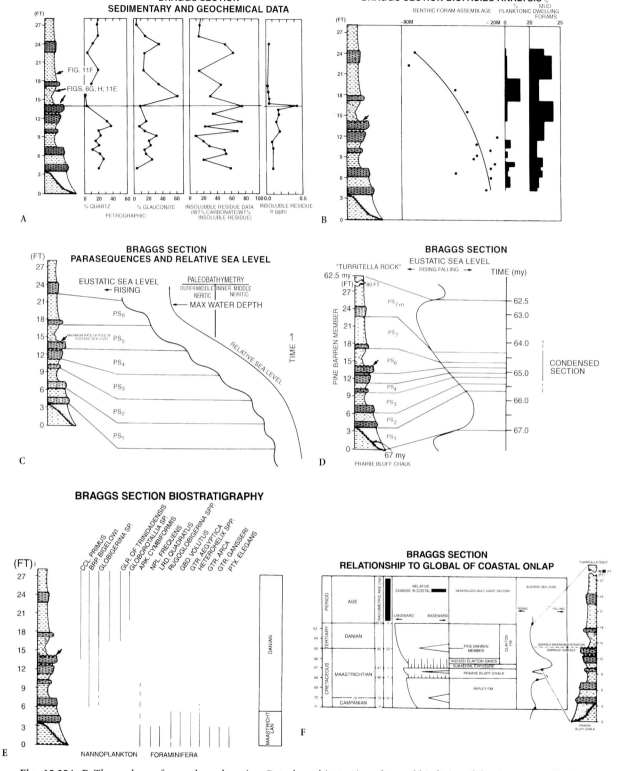

Fig. 15.25A–F. The geology of a condensed section. Petrology, biostratigraphy, and biofacies of the Cretaceous-Tertiary boundary at Braggs, Alabama. (Baum and Vail 1988)

terrigenous detritus in coastal depositional systems. Submarine fans are deprived of their detrital sediment sources, and may be partially or entirely abandoned. Vail and Todd (1981) stated that "such hiatuses are not global. Sometimes they differ in age within the same basin." Yet in a later paper dealing with the same (North Sea) area, this conclusion has been modified. They stated that "the age of a condensed section within a given depositional sequence tends to be synchronous globally but may differ slightly from basin to basin with changes in rates of deposition and subsidence." This revised opinion indicates the growth of a new all-purpose sequence concept, and in subsequent versions of the sequence models condensed sections have been used as invariable indicators of rapid eustatic sea-level rises. Their place in the Exxon sequence framework is shown in Figs. 4.4, 4.5, 7.6 and 11.1A.

As summarized by Miall (1986), evidence from modern and ancient marine deposits does not support the interpretation of every condensed section/marine hiatus as an indication of rapid rise in sea level. Tucholke and Embley (1984) discussed erosional unconformities in the South Atlantic, Indian, and Antarctic Oceans east, south, and west of South Africa. Several have been documented in DSDP cores and, not surprisingly, correlate with the global cycle chart. However, others of Early Oligocene and Late Miocene–Early Pliocene age appear to owe their origin to current erosion arising from increased thermal oceanic circulation associated with the growth of the Antarctic ice cap. The Late Miocene–Early Pliocene unconformity coincides with an episode of low sea level on the Vail et al. (1977) curve. These current-swept areas still exist on the sea floor at the present day. As Pinet and Popenoe (1982), Christie-Blick et al. (1990), and Schlager (1992a) have shown, boundary undercurrents driven by thermal circulation have generated significant regional unconformities in areas such as the Atlantic margins of the United States, which may be mistaken for sequence boundaries, flooding surfaces, or condensed deposits (Sects. 2.2.1).

15.3.5 Slope and Rise Systems

The basic sequence model for deep-marine turbidite and related depositional systems is described in Section 4.3, and examples of Late Cenozoic fan deposits and their relationship to sea-level change are described in Section 8.2. It has long been known that turbidite systems are most active during times of low sea level, when continental shelves are partially or fully exposed, and fluvial systems deliver detritus directly to the continental slope or the heads of submarine canyons (Sect. 6.1).

Hunt and Tucker (1992) pointed out that submarine-fan systems are deposited *during* a sea-level fall, as a result of progressive exposure and erosion of the continental shelf. The time of lowest sea level which is generally used to define sequence boundaries, should therefore be located *above* the lowstand fan deposits, not below them, as in the Exxon sequence model. Their revised sequence model is shown in Fig. 15.5.

The architecture of deep-marine clastic depositional systems, the facies composition of the deposits, and their relationship to the cycle of sea-level change was examined by Mitchum (1985) and Mutti (1985). The latter subdivided turbidite systems into three broad types (Fig. 15.26). Type I deposits consist of thick-bedded lobes of sandstone, formed at times of lowest sea level and highest sediment supply. Type III deposits consist of channel-level complexes formed during higher stands of sea level, when detrital supply is smaller. Type II is an intermediate style, when the submarine fans may contain well-developed, thin-bedded lobe fringes. These three broad types of assemblage form in succession to each other, according to Mutti (1985), with types II and III downlapping onto type I during the rising leg of the sea-level curve. These late-stage turbidite systems also backfill canyons at this time, forming a deep-marine onlap pattern.

Mitchum (1985), in a useful review of the seismic stratigraphy of fans, provided the three-dimensional model reproduced here as Fig. 15.27. The emphasis on downlapping patterns in this diagram should be noted, as it provides the key to the Exxon group of workers for interpreting the sequence-stratigraphic relationship of turbidite systems.

It is important to note, however, that sea level is not the only control on the timing and location of turbidite systems. Kolla and Macurda (1988) pointed to other factors, including the tectonic and climatic setting of the receiving basin. Canyons, which may be the main source of sediment feeding a fan, may indeed be intiated during lowstands by fluvial erosion of the shelf, but they may continue to evolve and deepen by submarine mass wasting during the subsequent sea-level rise. The three-fold subdivision of fan depositional styles defined by Mutti (1985) may not apply to the evolution of very large fans, such as the Indus, Bengal, and Amazon fans, where the depositional slope and sediment supply

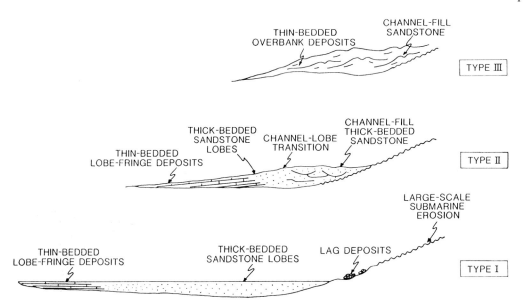

Fig. 15.26. The three main styles of turbidite depositional system. (Kolla and Macurda 1988; based on Mutti 1985)

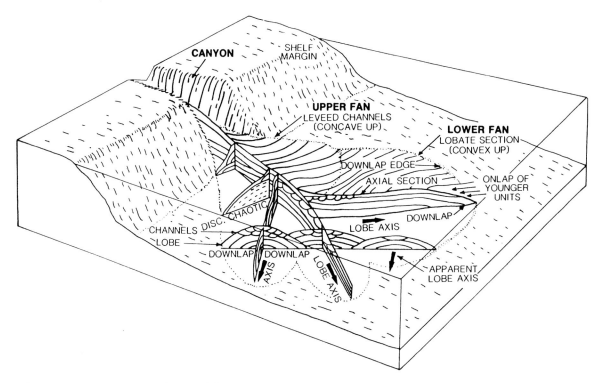

Fig. 15.27. Seismic-stratigraphic model of submarine fans. Downlapping lower fans are interpreted as deposits of the lowest stand of sea level (type I of Mutti 1985). Channel-levee complexes of the upper fan are thought to be the deposits that occurred during rising sea level (type III of Mutti 1985). (Kolla and Macurda 1988; after Mitchum 1985)

are less sensitive to sea-level change than in the small fans that formed the basis of Mutti's classification. Underhill (1991) pointed out the importance of local tectonism in controlling depositional architecture. He related growth patterns in Jurassic fan deposits in the Moray Firth area of the North

Sea basin to tilting of the basin floor by rotational block faulting (Sect. 11.2.2). He argued that Vail and Todd (1981) and Vail et al. (1984) had missed this relationship, and incorrectly attributed downlapping architecture to lowstand progradation. Surlyk (1989) in a review of oilfields in turbidite reservoirs in the North Atlantic–North Sea area also emphasized the importance of tectonic control in determining the style and timing of deep-marine depositional systems.

The control of sediment supply by volcanism is also an important determining factor in controlling the timing and facies architecture of turbidite systems. Winsemann and Seyfried (1991) quoted several cases in the arc-related basins of Central America where high sediment influxes overcame the effects of rising sea level to form major turbidite systems.

Walker (1992b), in a review of deep-marine depositional processes, commented on the systems-tract nomenclature of the Exxon school. He noted that many of the relationships shown in the standard Exxon diagrams (e.g., Fig. 4.4) are not demonstrated by any actual examples of turbidite systems known to him. For example, he could not cite any example where the "lowstand wedge" was buried by highstand deposits, nor examples where the lowstand wedge buries the "slope fan." The latter consists of channel-levee complexes and does not in fact occur on the slope. He pointed out that where data have become available from modern fans, such as the Mississippi fan, the architecture is quite different. For example, there the condensed section recording the sediment starvation that takes place at times of rising sea level rests directly on the basin-floor lowstand fan deposits. Walker (1992b) offered two alternative fan models (Fig. 15.28). Shanmugam et al. (1995) commented on the erroneous assumptions regarding the facies composition of many basin-floor fans. Their detailed core analysis of the Cenozoic fans of the North Sea Basin showed that they are composed dominantly of sandy debris-flow deposits and slump deposits, not turbidites. This would much better explain the mounded seismic facies signature that is a very common signature of such fan deposits (e.g., Mitchum et al. 1993) than if they were composed of turbidites, which would be more likely to generate very flat, tabular seismic facies.

15.4 The Sequence Stratigraphy of Carbonate Depositional Systems

Reviews of carbonate sedimentology and sequence stratigraphy have been provided by Sarg (1988), Walker and James (1992), and Schlager (1992a). A recent book dealing with the sequence stratigraphy of carbonate deposits was prepared by Loucks and Sarg (1993).

Many of the details of carbonate sequence stratigraphy have been discussed elsewhere in this book. In particular, the differences between carbonate and clastic systems in their response to sea-level change is outlined in Section 4.3. The most important single difference is the reciprocal effects sea-level change has on carbonate versus clastic continental margins (Figs. 4.7, 4.8). Carbonate slope and deep-basin deposits are formed most rapidly during times of high sea level, when platform carbonates are actively being deposited, and sedimentation rates may exceed the rate of generation of accommodation space ("highstand shedding"). Such deposits are less significant during times of low sea level, because the platform carbonate factory tends to be shut down at this time, whereas, of course, low sea-level stands are the time when terrigenous detritus is delivered most efficiently to the continental slope and deep-basin environments.

Examples of carbonate sequences are discussed in Sections 7.7 and 8.3.

15.4.1 Platform Carbonates: Catch-Up Versus Keep-Up

The style of carbonate sequence-stratigraphy on the platform is mainly a reflection of the balance between sea-level rise and carbonate production, as summarized in a review by Jones and Desrochers (1992). Where sea-level fall exceeds subsidence rates, exposure occurs, and karst surfaces may develop (Fig. 15.29, 15.30). Rapid trangression, on the other hand, leads to the development of condensed sections, and may shut down the carbonate factory, resulting in a "drowning unconformity" (Sect. 11.8; Fig. 14.6). "Nutrient poisoning" and choking by siliciclastic detritus may also shut down the carbonate factory at times of high sea level, and this can also lead to the development of drowning unconformities, which may be mistaken for sequence boundaries (e.g., Erlich et al. 1993). James and Bourque (1992) argued that poisoning and choking were the processes most likely to cause a shut down

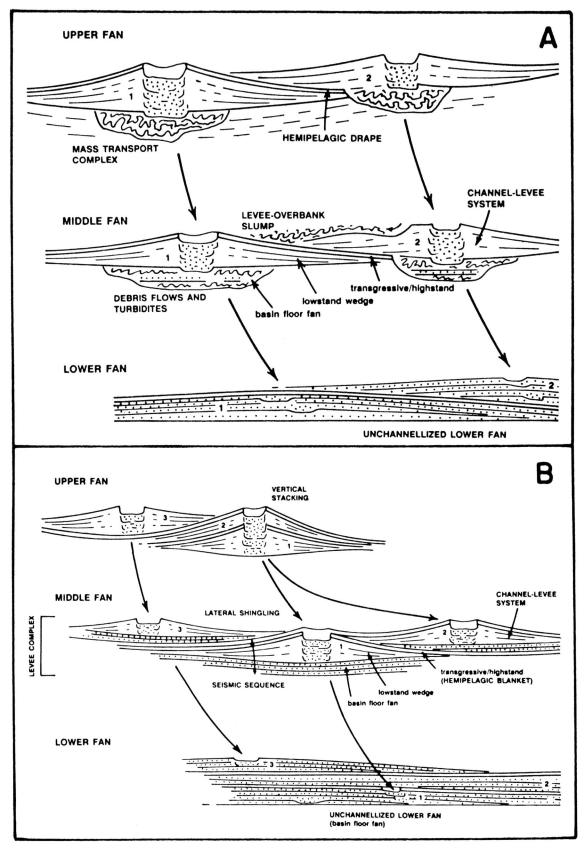

Fig. 15.28. Fan models. A Model commencing with the formation of a "mass-transport complex" and overlain by channel-levee systems. B model commencing with sheetlike basin-floor turbidites. (Walker 1992b)

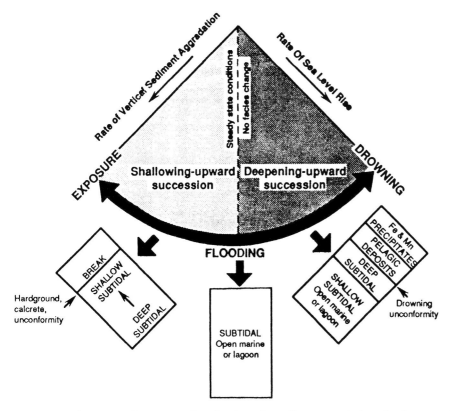

Fig. 15.29. The types of sequence that develop, depending on variations in the rate of relative sea-level rise and carbonate sediment production. (Jones and Desrochers 1992)

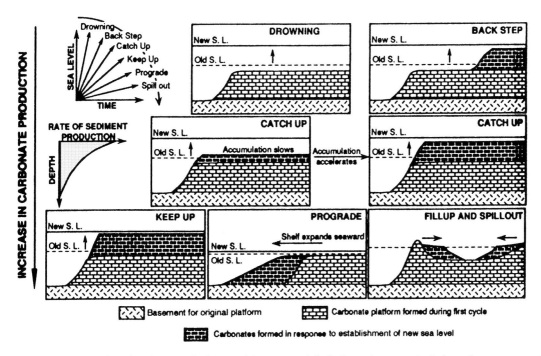

Fig. 15.30. The styles of carbonate-platform architecture and their dependence on the balance between rate of sea-level rise and carbonate productivity. (Jones and Desrochers 1992)

of reef sedimentation, because studies have indicated that under ideal conditions vertical reef growth is capable of keeping pace with the most rapid of sea-level rises.

Figure 15.30 illustrates the main variations in platform architecture that develop in response to changes in the controls noted in the above paragraph. Drowning during a rapid rise in relative sea level is typically followed by backstepping. A slightly less rapid trangression may lead to the so-called "catch-up" architecture. Here the sea-floor remains a site of carbonate production, and as sea-level rise slows, sedimentation is able to catch up to the new sea level. Vertical aggradation characterizes the first stage of the catch-up, but lateral progradation may occur late in the cycle, when the rate of generation of new accommodation space decreases. Shoaling-upward sequences are the result.

A balance between sea-level rise and sediment-production rates leads to "keep-up" successions, in which cyclicity is poorly developed. Eventually, typically at the close of a cycle of sea-level rise, carbonate production may exceed the rate of generation of accommodation space. This can lead to lateral progradation, and "highstand shedding" of carbonate detritus onto the continental slope.

As with coastal detrital sedimentation, autogenic processes may generate successions that are similar to those that are inferred to have formed in direct response to sea-level change. Pratt et al. (1992) summarized the processes of autogenicity in peritidal environments, where shallowing-upward cycles are characteristic, as a result of short-distance transport and accumulation of carbonate detritus. Lateral progradation and vertical aggradation both may occur (Fig. 15.31). Rapid filling of the available accommodation space may lead to a shut-down of the carbonate factory until relative sea-level rises enought to stimulate its reactivation. The alternative to autogenicity is the invocation of a form of rhythmic tectonic movement to generate the required sea-level change, or Milankovitch mechanisms. As Pratt et al. (1992) noted, the scale and periodicity of tectonism required for meter-scale cycles has not yet been demonstrated.

15.4.2 Carbonate Slopes

The consequences of sea-level change for the generation of carbonate sediments on continental slopes has been discussed by Coniglio and Dix (1992).

Sarg (1988) employed a carbonate sequence model comparable to that developed for siliciclastic sediments, including the suggestion of thick lowstand fans and wedges. However, Coniglio and Dix (1992) argued that the sediment supply at times of lowstand would be insufficient to develop such depositional systems. As noted elsewhere in this book (Sect. 4.3) carbonate sedimentary processes respond very differently to changes in sea level, and the Exxon model now seems unrealistic. Coniglio and Dix (1992) stated that "whether sufficient sediments in carbonate systems accumulate to generate recognizable *lowstand systems tracts* analogous to siliciclastic lowstand fans ... is debatable.... From sequence stratigraphic studies of relatively pure carbonate systems, however, the lowstand systems tract appears to be either thin or absent."

Their alternative model is illustrated in Fig. 15.32. A thin lowstand systems tract is shown, consisting of turbidites and pelagic deposits. A condensed sequence (not shown) may develop along the inner slope during transgression. The style of transgressive architecture depends on the balance between sea-level rise and sedimentation, as discussed in the previous section. Slope deposits are likely to be thicker during the trangressive and highstand phases of sea-level change, because of the higher productivity of the carbonate factory (highstand shedding).

Schlager (1992b) commented on specific conditions that would lead to a predominance of platform shedding during lowstands of sea level. This occurs primarily in temperate-water carbonate systems, for several reasons. With reference to the Broken Ridge system of the southeastern Indian Ocean he stated that:

It lacked a protective platform rim that would dam up the sediment, it probably lithified slowly upon exposure because it contained little metastable carbonate minerals, and, very importantly, it hardly lithified at all in the marine environment because of the lack of abiotic submarine precipitation in the temperate zone.... The outer neritic and bathyal sediments that probably contributed very significantly to the reworking have all these attributes to an equal or even greater extent.

Schlager's (1992b) comments were made in the context of a more general debate regarding carbonate depositional models, and the appropriateness of using a sequence framework that was essentially developed for siliciclastic sediments (Vail 1987; Sarg 1988). As pointed out by Driscoll (1992) conditions can vary considerably from carbonate system to system, and generalizations are inadvisable. Gram-

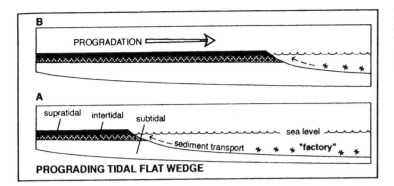

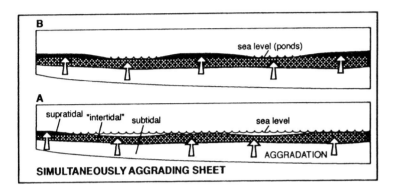

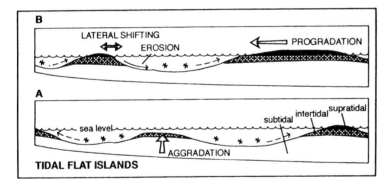

Fig. 15.31. Processes of autogenic sedimentation that can lead to the generation of meter-scale shoaling-upward successions (parasequences) independent of sea-level change. (Pratt et al. 1992)

mer et al. (1993) documented an important phase of shedding and slope construction in the Tongue of the Ocean (Bahamas) during the rapid postglacial rise in sea level.

15.4.3 Pelagic Carbonate Environments

Martire (1992) postulated that condensed sections and hardgrounds commonly occur in pelagic environments during times of low sea level because of enhanced oceanic circulation, erosion, and winnowing that occurs as a result of increased thermohaline circulation. Therefore in this environment

the sedimentary breaks should track eustatic changes, except for the local imprint of tectonic or climatic changes.

15.5 Main Conclusions

Regarding general principles:

1. The application of sequence-stratigraphic concepts to higher-order sequences has raised problems of the hierarchical nature of sequence architectures, requiring revisions in the strati-

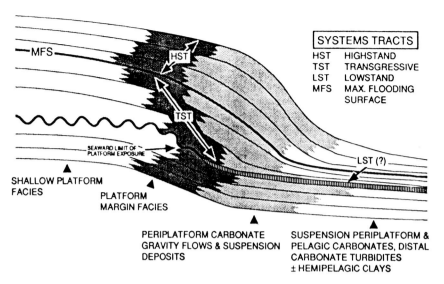

Fig. 15.32. Model of carbonate systems tracts, which differs substantially from that of the Exxon group (e.g., Fig. 4.5). See text for discussion. (Coniglio and Dix 1992)

graphic nomenclature used to name and describe sequences and parasequences.

2. Concepts of hierarchies of architectural elements and bounding surfaces already developed for fluvial and tidal deposits may be reconciled with the sequence-parasequence framework of the Exxon school, but this requires attention to the time scales and physical scales of the processes involved and the units generated.

3. Sequence concepts and nomenclature and the literature of sequence stratigraphy have downplayed the importance of autogenic processes in generating facies successions of the scale of parasequences in both carbonate and siliciclastic depositional systems.

4. Sequence classifications are in a state of flux. A rank-order classification based on duration – the classic Exxon approach – is still widely used, but is not recommended in this book because recent work provides no empirical or genetic justification. A classification by descriptive characteristics has been proposed, based on areal extent, the amount of facies change at the sequence boundary and the degreee of deformation at the boundary, but this classification cannot satisfy all empirical observations. A descriptive terminology referring to the duration or episodicity/periodicity of sequences is useful but does not constitute a formal classification.

Regarding siliciclastic depositional systems:

5. The early Exxon model which purports to explain the development of fluvial and coastal-

plain depositional systems is seriously flawed. Accommodation space for fluvial progradation is not generated during late-highstand seaward shift of the shoreline. Most fluvial deposits are generated during the trangressive and highstand phases of the sea-level cycle, and are also a response to tectonic uplift of source areas.

6. The Exxon model of the bayline also creates some contradictory depositional models that are not supported by observations of modern depositional systems.

7. A need has been expressed for an additional category of systems tracts, corresponding to the "forced regression" of barrier-strandplain systems during falls in sea level: This has been termed the *forced regressive wedge systems tract.* This systems tract lies above a regressive surface of marine erosion, but below the sequence boundary.

8. Deltaic styles, which have traditionally been interpreted in terms of the balance between fluvial and marine processes, also depend on a sea-level control. River-dominated deltas are common in highstand settings; wave-dominated deltas are common shelf-margin systems tracts; tidal deltas are part of the estuarine realm, and are typical of transgressive environments.

9. Many units previously interpreted as shelf deposits, such as many so-called offshore bars, can now be shown to be the product of coastal processes during lowstands. They include forced regressive wedges, reworked trangressive shore-

face deposits, and the fill of incised valleys formed during trangression.

10. Coastal-shelf stratigraphic architecture depends on a balance between the rate of creation of accommodation space and the rate of introduction and transportation of sediment.

11. Condensed sections and marine hiatuses are a characteristic component of trangressive systems tracts, but can also be generated by submarine current erosion. Examples in the Cenozoic sections of the United States Atlantic margin have been mistaken for sequence boundaries.

12. Turbidite depositional systems are a characteristic component of lowstand systems tracts, but the presence of major submarine fan units does not invariably indicate low sea levels. Tectonism and volcanic-dominated sediment supply are other important controls that can lead to submarine fan sedimentation being out-of-phase with and unrelated to the eustatic sea-level curve.

13. In the common situation of sediment supply to the deep sea reaching a maximum during the falling leg of the sea-level curve, because of subaerial erosion and sediment bypass, base-of-slope fans accumulate most rapidly during this falling stage. The sequence boundary, which defines the time of lowest sea level, should therefore be drawn above these deposits, not below them as in the original Exxon model.

Regarding carbonate depositional systems:

14. Exxon sequence models for carbonates are unrealistically similar to those developed for siliciclastics. They do not acknowledge the differences in the response of the two types of system to sea-level change.

15. One of the most important differences is the absence of thick lowstand wedges in carbonate systems. The carbonate factory is at a low level of activity during sea-level lowstands. Deep-water carbonates are more typical of highstand systems tracts, as a result of highstand shedding from carbonate platforms.

16. The styles of platform sedimentation reflect a sensitive balance between sedimentation rates and sea-level change. Rapid sea-level rise may lead to drowning and backstepping of platforms, with subsequent "catch-up" aggradation. Late stages of sea-level rise are characterized by "keep-up" sedimentation of thick tidal deposits, and may be followed by still-stand progradation.

17. Drowning of platforms during rises in sea level by siliciclastic detritus, or influxes of cold, warm, or unusually saline water, or an oversupply of nutrients, may generate "drowning unconformities" that are similar in stratigraphic character to lowstand sequence boundaries.

16 Numerical and Graphical Modeling of Sequences

16.1 Introduction

Geologists have long understood that there are three primary controls on stratigraphic architecture, the rate of tectonic subsidence, the rate of sediment accumulation, and the rate of eustatic sea-level change (e.g., Curray 1964). Burton et al. (1987) demonstrated that we cannot obtain absolute values of any of these variables unless we already know the magnitude of two of them (it is basically a simple problem of algebra: known and unknown variables in equations). Information on sediment-accumulation can normally be derived from stratigraphic data, but the other two variables are elusive, and Burton et al. (1987) concluded, after an extensive review of the various research methods (discussed here in Chap. 2) that "although an accurate eustatic sea-level variation chart would be a boon to stratigraphers ... such a sea-level chart cannot be made."

An alternative that is being pursued with increasing vigor and sophistication is that of computer-based numerical modeling. The values for missing variables can be estimated, and used as input into computer models to simulate stratigraphic architecture. Ranges of values for the three variables may be used as input, to develop families of possible solutions. These are then compared with the actual basin architectures. Numerical modeling of this type permits the application of sensitivity tests to guage the relative importance of the variables. There are two main types of model, *forward models*, that simulate sets of processes and responses, given predetermined input variables, and *inverse models*, that use the structure of a forward model to simulate a specific result, such as an observed basin architecture (Cross and Harbaugh, *in* Cross 1990). Such comparisons do not provide definitive results, but they do help to eliminate unlikely solutions.

Backstripping is a procedure for evaluating subsidence, sedimentation rate and sea-level change using data from individual stratigraphic sections. It is therefore a one-dimensional approach to the problem, dealing with events at one location. The methods and some of the results are discussed in Section 2.3.3, where some of the problems of data input and interpretation are dealt with. For example, it is pointed out how stratigraphic thicknesses, a primary input variable, are dependent in part on compaction as a result of porosity loss with burial, and may also be affected by lithification. Correction for these factors introduces uncertainties into the analysis. Similarly, independent estimates of water depth may be arrived at by using such information as sedimentary facies and the paleoecology of depth-dependent organisms, but such estimates are imprecise, and it is commonly difficult to extract a reliable eustatic signature from the result. The sediment-accumulation rate is in part dependent on the rate of tectonic subsidence, but calculations of this variable are rendered difficult by lack of knowledge of deep crustal structure and its response to stress. Great strides in improving our knowledge in this area have been made by geophysicists (Allen and Allen 1990; or summary in Miall 1990, Chap. 7), but they do not yet permit input of tectonic subsidence data into backstripping calculations that are precise enough to yield accurate output on eustatic sea-level variations. As shown in Section 2.3.3, careful analysis of large volumes of data may lead to some tentative conclusions about sea-level variations for limited intervals of geological time, but this is still a long way from the generation of an all-purpose, independent global cycle chart, which the chart by Haq et al. (1987, 1988a) purports to be.

Geophysical models of basin subsidence, such as the extensional-margin models of Watts (1981), modified to include the effects of in-plane stress by Cloetingh and Kooi (1990), or the foreland-basin model of Beaumont (1981), incorporate the methods and results of backstripping and account for the larger-scale architectural features of basin fills (Chap. 11). The topic under consideration here is the application of modeling to the development of individual stratigraphic sequences.

Most computer models are two-dimensional, simulating cross sections through continental margins. They do this by the development of geometric shapes in which the cross sectional area is dependent on input sediment volumes and the rates of differential subsidence and sea-level change (forward modeling). We are still a very long way from the simulation of the actual physical processes of sediment transport and deposition, although work by J. Syvitski and D.J.P. Swift and their colleagues, not dealt with in this book, is taking important steps in this direction. The models are therefore crude geometric approximations of stratigraphic units.

It also needs to be stressed that, as with all computer models, the results are only as good as the data and ideas that are used to generate them. Although the modeling process has already provided numerous valuable insights, it cannot replace the careful accumulation of data and the interpretation and theoretical development built from such data. There will always be a need for more and better data, collected with improved methods of observation and interpretation in mind.

Many independent groups of modelers are now active. The early work of Jervey (1988), published in SEPM Special Publication 42, but actually carried out much earlier, is of importance in part because it provided the theoretical foundation for the sequence-stratigraphic models of Posamentier et al. (1988) and Posamentier and Vail (1988). Cant (1989b, 1991) discussed some of the simple algebraic and geometrical relationships that are the basis for quantitative modeling. Pitman examined the Exxon sequence concepts in several papers (Pitman 1978, 1986; Pitman and Golochenko 1983, 1988). The University of South Carolina (USC) Group, consisting of C.G.St.C. Kendall and his colleagues, has produced many papers on the topic of computer modeling, and has developed a sophisticated package of computer programs (Burton et al. 1987; Kendall and Lerche 1988; Helland-Hansen et al. 1988; Kendall et al. 1991). Another active team is that at Cornell University, led by T.E. Jordan (Flemings and Jordan 1989; Jordan and Flemings 1990; 1991). Other important contributions have been made by Sinclair et al. (1991), Reynolds et al. (1991), Rivenaes (1992), Steckler et al. (1993), and Lawrence (1993). Several specialized contributions are included in the book edited by Cross (1990), and useful theoretical considerations are presented by Christie-Blick (1991). Kendall et al. (1991) provided a useful review of recent modeling research.

The discussion constituting the remainder of this chapter is brief and selective. Its purpose is to illustrate the kinds of results that are currently being generated, without entering into lengthy treatments of the physical and mathematical justifications of the models, or of the details of the computer programs.

16.2 Model Design

The useful concept of *accommodation space* was explained by Jervey (1988) as follows:

In order for sediments to acumulate, there must be space available below base level (the level above which erosion occurs). On the continental margin, base level is controlled by sea level and, at first approximation, is equivalent to sea level.... This space made available for potential sediment accumulation is referred to as accommodation.

Cant's (1991) simple diagrams set out the basis for models of differential subsidence and sea-level change (Fig. 16.1), and the interaction between coastal nonmarine and marine facies (Figs. 16.2, 16.3). Cant's work was not used in the modeling studies described here, but is presented as an illustration of the processes the modelers are attempting to simulate. Together the diagrams provide a dynamic framework for understanding the concept of accommodation space (diagrams with a comparable purpose were provided by Jervey, but are much more difficult to comprehend. Similar definition diagrams were also provided by Burton et al. (1987), as the basis for their simulations: Fig. 16.4). The first diagram may be compared with Fig. 11.1A, which is the schematic basis for the Exxon sequence models of the interaction between eustasy and subsidence. Figures 16.2 and 16.3 illustrate the principles of facies migration, which depend on the balance between sediment supply and relative sea-level change. The angles shown in the illustrations are exaggerated. In actual continental margins they are only a few degrees from the horizontal, or less.

The actual building of stratigraphic units in the various computer models is carried out in various ways. The physical processes of sedimentation are, of necessity, simplified because of their complexity, and some critical information may be provided as input rather than calculated as part of the simulation, for example, depositional slopes. Burton et al. (1987) added sediment from the sides of the model

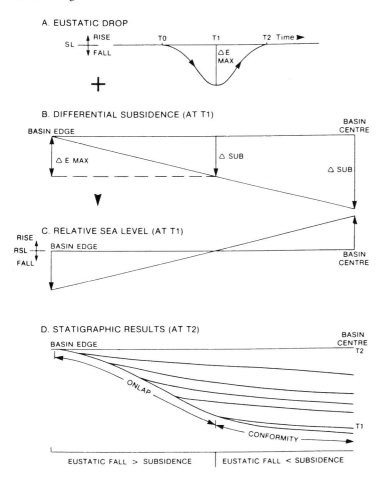

Fig. 16.1. A Diagram showing the integration of a curve of eustatic sea-level change and **B** a plot of differential subsidence to generate a diagram illustrating relative sea-level change. **B** and **C** show the results at time *T1*, and the stratigraphic architecture produced by this process of relative sea-level change is shown in **D**. (Cant 1989b, reproduced by permission of Blackwell Scientific Publications)

using an arbitrary triangular distribution to simulate the downslope decrease in sediment volumes (Fig. 16.5). Different slopes were used for sand and mud to simulate the different transport efficiencies of these two main lithologies. Their first model only simulated marine sedimentation. The procedure for building each unit is as follows. First, locate the "point of greatest onlap" ("Pogo" in Fig. 16.5). Secondly, erode between this point and the previous pogo, then, thirdly, deposit sediment according to the triangular distribution, including the sediment eroded in step 2. The resulting layer is then compacted, and incremental subsidence is carried out, in keeping with the model shown in Fig. 16.4.

The model of Reynolds et al. (1991) builds stratigraphic units by adding a constant volume of sediment with each time step, and distributing it according to predetermined slopes. The model is restricted to marine deposition, differentiating between the continental shelf and slope, but not including nonmarine deposition. Compaction and subsidence are added steps.

The Cornell University model is somewhat more sophisticated than both the approaches described above. Figure 16.6 shows the building of a basic succession of depositional units on an extensional continental margin. Marine and nonmarine deposition are shown. Sediment is moved across the continental margin using a diffusion equation, and the depositional slope is not fixed, but is calculated based on "the diffusional interactions of subsidence and sediment flux" for the two environments, using transport coefficients for marine and nonmarine deposition. Slope is therefore a product of the model, not an input parameter. This makes the model more realistic, although it still represents a crude generalization of reality. For example, it does not take into account along-shore sediment transport, and the transport coefficient is based on deltaic sedimentation, which Jordan and Flemings (1991) argued provided a useful simplification for clastic-dominated coasts. Subsidence and compaction are also built into the model. Note, in Fig. 16.5, the steepening through time of the marine-non-

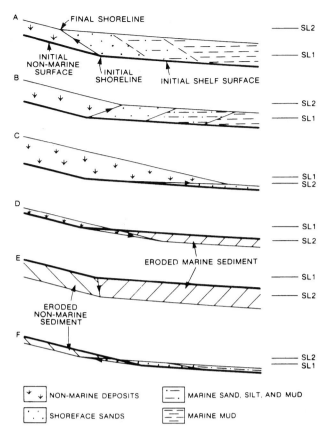

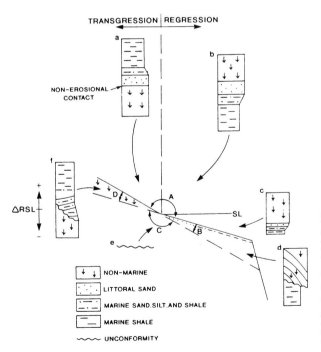

Fig. 16.2A–F. Diagrams illustrating how different directions of shoreline translation generate different facies sequences. Initial sediment surfaces are shown by *heavy lines*, final surfaces by *lighter lines*. *Arrow* within each diagram shows direction of shoreline translation. *Ornamented areas* show relative thicknesses and facies deposited between initial (*SL1*) and final (*SL2*) sea levels. The occurrence of erosion or deposition depends on the angle of shoreline translation in relation to the angles of the depositional surfaces, as shown in Fig. 16.3. **A** Simple trangression. **B** Simple regression. **C** Horizontal regression. **D** Regression with erosion. **E** Regression and erosion. **F** Transgression and erosion. (Cant 1991, reproduced by permission of Blackwell Scientific Publications)

Fig. 16.3a–f. The five facies successions (**a-f**) generated by the five conditions (A-F) illustrated in Fig. 16.2. The angle of shoreline migration in relation to the slopes of depositional/erosional surfaces is the critical differentiation between these diagrams, and can be subdivided into four "sectors" (*A-D*). These slopes are exaggerated for the purpose of illustration. *ΔRSL* Relative change in sea level. (Cant 1991, reproduced by permission of Blackwell Scientific Publications)

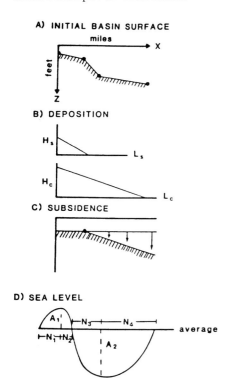

Fig. 16.4A–D. Definitions of major controls for modeling of continental margin sedimentation by the University of South Carolina (USC) group. (Burton et al. 1987, re-produced by kind permission of Elsevier Science-NL, Sara Burgerhartstraat 25, 1055 KV Amsterdam, The Netherlands)

marine facies contact, a process explained by Muto and Steel (1992; Sect. 15.3.3).

16.3 Selected Examples of Model Results

A recent example of the USC modeling output is illustrated in Fig. 16.7. This model run simulated three cycles of eustatic sea-level rise and fall. Sediment input was varied such that it increased with rising sea level, and decreased with falling sea level. Regional subsidence was input for the first two-thirds of the model run, and then switched off. As noted by Kendall et al. (1991) this model shows an output in the form of three sequences. "The first sequence contains only a transgressive and a high-stand system tract which corresponds to the first rise in eustatic sea level, while the later two sequences are complete, reflecting the later two falls and rises in sea level."

The Reynolds et al. (1991) models focused on testing the variations in overall sequence architecture caused by variations in the three major input variables, sediment flux, tectonic subsidence, and the amplitude of the sea-level cycle. The results are shown diagrammatically in Fig. 16.8.

Some of the results of the Cornell work are shown in Fig. 16.9 (from Jordan and Flemings

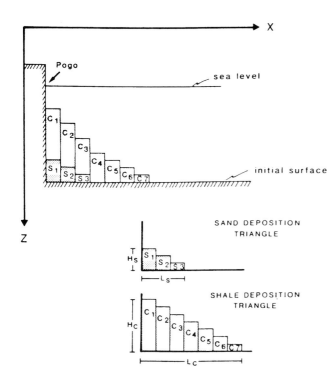

Fig. 16.5. The building of depositional surfaces in the USC model by the arbitrary definition of an array of evenly spaced horizontal points, giving the surface a staircaselike appearance. The scale of the staircase is small enough that the output defines a smooth surface. Diferent slopes are used for sand and shale, as an approximation of the tranport efficiency of these two lithologies. (Burton et al. 1987, reproduced by kind permission of Elsevier Science-NL, Sara Burgerhartstraat 25, 1055 KV Amsterdam, The Netherlands)

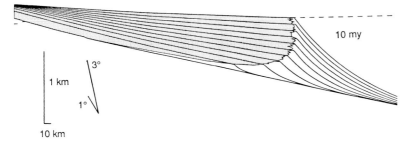

Fig. 16.6. The building of depositional surfaces in the Cornell model. Sediment enters from the side (*left*) and is moved across the surface by application of a diffusion equation. Two major depositional environments are modeled, in contrast to the USC model. *Shaded area* shows nonmarine deposition; *white area* is marine deposition. (Jordan and Flemings 1991)

1991). This figure illustrates the evolution of their extensional-margin model from the starting point shown in Fig. 16.6, through one complete 10^6-year cycle of sea-level change. A chronostratigraphic (Wheeler) diagram of the results is given in Fig. 16.10.

One of the most useful outcomes of these modeling exercises is the insight it has provided on the timing of sequence boundaries. Jordan and Flemings (1991, p. 6687) stated that:

Depending on the particular parameters used in the model, the time of reversal of onlap (ie., the age of the sequence boundary) occurs between the times of maximum rate of fall and the eustatic lowstand.... For example, in the case of a narrower, more rapidly subsiding basin with less sediment supply ... the ages of the sequence boundaries much more closely approximate the times of maximum rate of fall of sea level ... factors that shift the age of the sequence boundary from the age of lowstand toward the time of maximum rate of fall of sea level are ... more rapid subsidence, lower sediment flux, more efficient nonmarine transport ..., longer periodicity

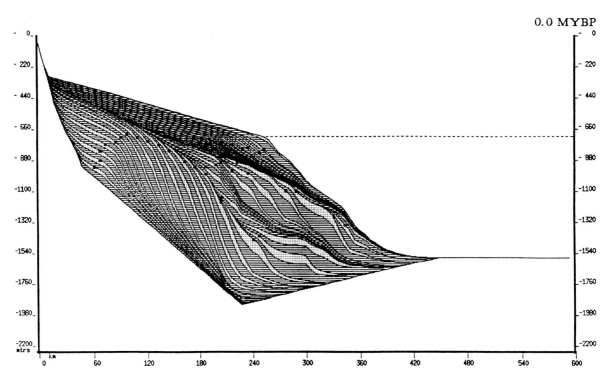

Fig. 16.7. An example of the model output from the USC group. Sand/shale ratio is shown arbitrarily for each time step by separation into two "beds." Sand is shown by *dots* and shale by *horizontal lines*. See text for explanation. (Kendall et al. 1991)

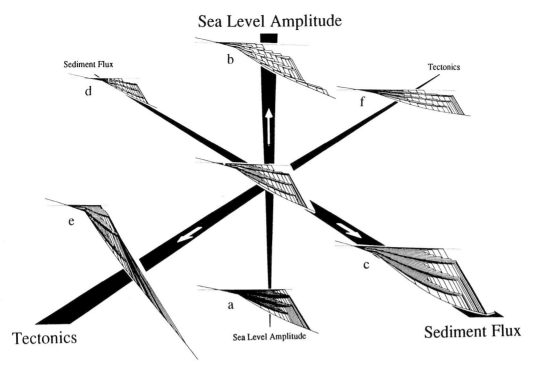

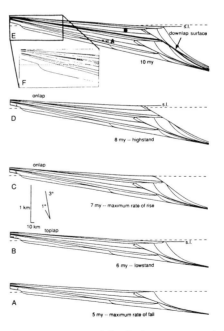

Fig. 16.8a–f. Variations in sequence architecture as a function of variations in the three major controlling parameters. All models represent 100 time steps, totaling 10 m.y., and all are shown at the same scale. The lower three models (**a, c, e**) contain sequence boundaries analogous to the type 2 boundaries of the Exxon terminology (Van Wagoner et al. 1987), while the upper three diagrams (**b, d, f**) contain type 1 boundaries. (Reynolds et al. 1991)

of eustatic change, and smaller amplitude of eustatic change.

The phase lag between sea-level change and stratigraphic response has been pointed out before (e.g., Pitman and Golovchenko 1988), but the experiments of Jordan and Flemings provide what is probably the most thorough computer-model investigation of this phenomenon. Depending on the balance of the various input parameters, Jordan and Flemings (1991) demonstrated that "the sequence boundary for an identical sea level history could be of different ages and the ages could differ by as much as 1/4 cycle." Helland-Hansen et al. (1988) and Steckler et al. (1993) also commented on this point. This directly contradicts the Exxon models, based on the work of Jervey (1988), who asserted that sediment supply affects shoreline position but not the timing of sequence boundaries. It is an important result, because the presumed synchroneity of sequence boundaries is the basis for the Exxon global cycle chart. Yet it has now been demonstrated that the stratigraphic record of a given 10^6-year ("third-order") sea-level cycle could vary by as much as 1/4 of a wavelength, that is, by up to several

Fig. 16.9A–F. Model of Fig. 16.6 extended through a complete 10^6-year ("third-order") cycle of fall and rise. *Dashed line at right* indicates sea level at the time of maximum rate of sea-level fall. Subsequent positions are indicated by the line marked *s.l.* (Jordan and Flemings 1991)

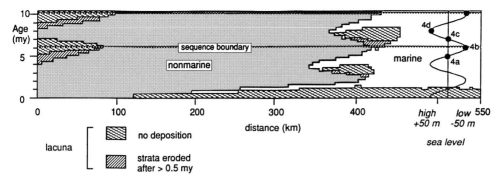

Fig. 16.10. Chronostratigraphic (Wheeler) diagram of the continental-margin architecture simulated in Fig. 16.9. The input sea-level curve is shown at *right*. *4a–4d* refer to the steps A–D in Fig. 16.9. (Jordan and Flemings 1991)

million years. *Sequence boundaries are thus inherently imprecise recorders of sea-level change.*

Jordan and Flemings (1991) also applied their modeling programs to foreland basins, where the basin configurations, the subsidence styles, and the sediment input parameters are all quite different. Similar complex results to those for extensional basins were demonstrated, which do not need to be discussed here.

Attempts are now being made to incorporate the effects of dynamic topography into some models. Burgess and Gurnis (1995) published the results of a preliminary study of Sloss-type sequences in which they combined cycles of eustatic sea-level change with broad cratonic tilts induced by mantle thermal processes.

Johnson and Beaumont (1995) described a new three-dimensional model for foreland basins which incorporates variations in sediment supply and sediment transport.

16.4 Main Conclusions

1. Stratigraphic architecture is a complex response to variations in three major variables, sediment flux, the rate and style of subsidence, and the duration and amplitude of the eustatic sea-level cycle.

2. Computer models of continental-margin stratigraphy cannot yet simulate the physics of sedimentation in all its complex detail, but they can approximate the development of continental margins in two dimensions, by making use of diffusion equations and transport coefficients to provide realistic simulations of sediment dispersal.

3. Stratigraphic architectures can be modeled in various tectonic settings by varying the input of subsidence styles and rates.

4. It has been demonstrated that there can be a phase lag between the sea-level cycle and the stratigraphic response, and that this lag may vary within and between basins by up to 1/4 of a wavelength of the sea-level cycle. Sequence boundaries may therefore vary in age locally and regionally by an amount of time equivalent to 1/4 of the duration of the sequence. In the case of 10^6-year ("third-order") sequences this could amount to several million years.

VI Discussion and Conclusions

In this final part of the book the major conclusions of the analysis are presented. Chapter 17 discusses some practical proposals for the use of sequence-stratigraphic methods in the field of petroleum geology. Chapter 18 draws together the major conclusions reached in the analytical sections of the book (Parts II, III, and IV), and offers an informal analysis of the place of the Exxon methodology in the evolution of sequence stratigraphy.

- Although there are many useful concepts in sequence stratigraphy, [the] models ... must be regarded as preliminary, and in need of an infusion of real data from modern examples. Until some of the problems have been worked out, and until the emphasis on control by global eustatic sea level change has been modified, it is safer to subdivide the geological record by bounding discontinuities of all types (allostratigraphy). (Walker 1992a, p. 12)

- We feel that alternatives to eustasy deserve more attention in sequence stratigraphy. The sedimentological underpinnings of the concept clearly indicate that first, relative and absolute (i.e. eustatic) changes in sea level have the same effect on sequences and, second, sequences can be generated by processes other than sea level. The fascination with, and exclusive emphasis on, eustasy may have reduced rather than improved the role of sequence stratigraphy as a predictive tool. (Schlager 1992a, p. 58)

- There are at least two ways that stratigraphy can contribute directly to more accurate and predictive reservoir characterization. The first is through high-resolution correlation.... A second attribute of stratigraphy useful for reservoir characterization is the relationship between rock properties that are important in controling fluid flow, and the stratigraphic position of genetic sequences within the larger scale stacking pattern of several genetic sequences. (Cross et al. 1993, p. 11)

17 Implications for Petroleum Geology

17.1 Introduction

The purpose of this chapter is to discuss the practicalities of sequence-stratigraphic methods as applied to problems in petroleum geology. Inevitably, the work by Vail and his colleagues receives much attention because this Exxon "school" has dominated research in this area. Section 17.2 comments on research methods, while Section 17.3 discusses two practical examples. An analysis of the origins of the Exxon school of thought is provided in the final chapter (Sect. 18.6), with an attempt to arrive at an understanding of how some of the research seems to have led geologists into blind alleys. There are implications here for how petroleum geologists, particularly those in industry, should carry out comparable research in the future.

17.2 Integrated Tectonic-Stratigraphic Analysis

17.2.1 The Basis of the Methodology

At the end of Chapter 2 reference is made to the recommendations made by the Exxon research group for an integrated basin-analysis approach to problems in petroleum exploration. Vail et al. (1991) suggested the following sequence of procedures:

1. Determine the physical chronostratigraphic framework by interpreting sequences, systems tracts and parasequences and/or simple sequences on outcrops, well logs and seismic data and age date with high resolution biostratigraphy.
2. Construct geohistory, total subsidence and tectonic subsidence curves based on sequence boundary ages.
3. Complete a tectono-stratigraphic analysis including:

A. Relate major transgressive-regressive facies cycles to tectonic events.
B. Relate changes in rates on tectonic subsidence curves to plate-tectonic events.
C. Assign a cause to tectonically enhanced unconformities.
D. Relate magmatism to the tectonic subsidence curve.
E. Map tectono-stratigraphic units.
F. Determine style and orientation of structures within tectono-stratigraphic units.
G. Simulate geological history.
4. Define depositional systems and lithofacies tracts within systems tracts and parasequences or simple sequences.
5. Interpret paleogeography, geological history, and stratigraphic signatures from resulting cross sections, maps and chronostratigraphic charts.
6. Locate potential reservoirs and source rocks for possible sites of exploration.

The purpose of this section is to discuss the application of this methodology to practical problems of petroleum geology. As noted in Section 2.4, the main focus of this book is on steps 1, 3, 4 and 5.

17.2.2 The Development of an Allostratigraphic Framework

The basis for all sequence-stratigraphic studies is a tight network of lithostratigraphic correlation. This may be built from correlation of well logs and cores, or it may be constructed by analysis of a grid of seismic-reflection lines with few wells, or none at all. The first step therefore is to carry out a thorough data analysis. In mature basins this may involve the examination of hundreds or even thousands of well records. All available biostratigraphic data should be utilized. Principles of allostratigraphic terminology and classification should be employed, in which the geologist attempts to map and correlate unconformity-bounded successions (Sects. 1.4, 15.2).

Techniques of seismic-stratigraphic analysis are not dealt with in detail in this book. Some remarks on the subject are included in Chapters 2 and 4, and many excellent textbooks are available. Papers by Brown and Fisher (1977) and many of the papers by the Exxon research group, are also useful. Reading (1986), Miall (1990), and Walker and James (1992) discussed some of the broad principles of lithostratigraphic and facies analysis.

Some types of lithology are particularly easy to correlate in well logs because of their distinctive geophysical signature. Marker beds associated with condensed sections or marine hiatuses are a good example, because of the concentrations of authigenic minerals or shell hash that are commonly present. Many are "hot," in the sense that they yield high gamma-ray signatures. Sequence boundary unconformities may also have distinctive lithological or geophysical signatures, such as thin gravel lags. Another important practical hint is that shales are commonly more widespread than other lithologies in a section, and therefore easier to correlate. They commonly represent transgressive deposits, which may reach far into coastal areas during the maximum flooding stage. Some geologists (e.g., Galloway 1989a; Underhill and Partington 1993) have found that mapping flooding surfaces is a much more practical method of stratigraphic correlation that attempting to trace sequence boundaries. Correlations may have to be attempted in the absence of much sample or core material. In this case the ability of the geologist to recognize distinctive patterns and shapes in the logs becomes an important research skill.

Well-log correlations need to account for facies variations and the details of the architecture of sequences and facies successions. The emphasis on the presence of tabular parasequences may tend to encourage stratigraphically horizontal correlations, but more complex architectures are common, particularly where depositional systems contain channels or have evolved by lateral growth. Channels may be readily recognized by sudden thickness changes. Clinoform surfaces, indicating lateral accretion of deltas, beaches, continental margins, and submarine fans, may be readily identified in seismic-reflection records, but require careful correlation of subtle markers in well logs, and may not be detectable in areas of sparse well control. Dipmeter and formation microscanner logs may provide much useful information on the attitude of such dipping surfaces.

Lithostratigraphic correlation is nominally an empirical process, and in the case of mature basins with a dense well network the correlations of individual beds and sequences may develop with little need for geological insights. However, the exploration geologist or academic researcher needs to be aware of the possible stratigraphic architectures in the rocks under study. The work of Plint et al. (1986) on the Cardium Sandstone of Alberta was an important breakthrough because Plint was attuned to the possibility of the presence of basin-wide erosion surfaces, and tested his correlations of well logs with this concept in mind. Correct results may not be achievable without some guiding concepts of what is possible or likely. This is particularly the case where data are sparse. The geologist must have one or more architectural models in mind to guide correlations. These may include clinoform units, channels, stepped erosion surfaces, or other features that might be predicted based on reconnaissance analysis or comparison with other areas. The researcher also needs to be aware of the appropriate sequence models. A stratigraphic basin analysis proceeds by constant interplay between the application of the practical, empirical steps discussed in this section and the interpretive ideas, based on sedimentological (facies, sequence) models discussed in the next section, the latter providing architectural guidelines for the correlation process.

17.2.3 Choice of Sequence-Stratigraphic Models

Many examples of evolving concepts in sequence architecture are referred to in Chapter 15. For example, Sections 15.3.3 and 15.3.4 refer to the recent realization that many stratigraphic units that had been interpreted as shelf deposits, such as many "offshore bars" are coastal deposits generated during falling relative sea levels, at lowstands, or during trangression. Posamentier et al. (1992) made the point that correct interpretation of these beds is facilitated by choosing the right datum on which to "hang" lithostratigraphic correlations (Fig. 15.16). Another example is the improvement in our understanding of carbonate deposits and their response to sea-level change. Some aspects of the Exxon sequence models for carbonates are unrealistically similar to those developed for clastics. In Section 15.4.2 reference is made to revised models for carbonate slope environments, and the reader may wish to compare the systems-tract architecture for lowstand (carbonate slope) deposits in the two sequence models (Figs. 4.5 and 15.32).

Basin analyses cannot be completed successfully unless the geologist has in mind a suite of possible appropriate depositional-systems and sequence models and tests these as the construction of the correlation network proceeds.

Construction of the lithostratigraphy and sequence framework, complete with facies analysis and interpretation of depositional environments permits stratigraphic predictions which may be of considerable use in subsurface exploration. Such a framework also allows the subdivision of reservoir rocks into "flow units" for the purpose of calculating recoverabilities and designing production models.

17.2.4 The Search for Mechanisms

The Exxon approach has been to use their global cycle chart as a template for correlation, and all regional studies published by their staff and co-workers indicate correlations with this chart. In Part III of this book we discuss alternative mechanisms for the generation of regional suites of stratigraphic sequences; in Part IV of this book we examine the reliability of global correlations, and conclude that the global cycle chart should be abandoned. The practice of petroleum geology does not, of course, require the availability of a global cycle chart. The imprecision of the global data base and the questionable value of the existing sea-level curves (not just that of Exxon; see Sect. 14.3) means that *such curves are of no practical value in basin exploration and development.*

It is recommended, however, that effort be expended to develop a *regional* sequence framework, as this may permit tight correlations with regional tectonic events and other processes of basin-forming importance. Studies of unconformities and stratigraphic offsets along faults may yield considerable insight into the timing of tectonic events, and this may provide virtually the entire story, certainly as far as the fine details of stratigraphic architecture are concerned – the details of importance in petroleum exploration and production. The discussion of modern data sets in the North Atlantic-North Sea region in Section 11.2.2 reveals the importance of this approach. When such analyses are unbiased by forced references to the Exxon global cycle chart a wealth of tectonic detail may

emerge (as in the case of this area), which can provide far more useful insights into the basin evolution of the entire area. *In practice this is as far as most petrolem-exploration sequence analyses need to go.*

17.2.5 Reservoir Characterization

The application of sequence-stratigraphic concepts may have considerable predictive value in the work done by production geologists to define the distribution and internal heterogeneity of reservoir bodies. The definition of the sequence framework (Sect. 17.2.2) and the application of the appropriate sequence model (Sect. 17.2.3) set the stage for a detailed analysis of the internal characteristics of the reservoir. The natural subdivision of the reservoir into separate flow compartments, their relationship to each other, and their internal facies composition, are determined by the dynamic balance that existed during sedimentation between sediment supply and accommodation. Cross et al. (1993) referred to this process as volumetric partitioning. Examples of this dependency include the stacking pattern and internal geometry of reservoir units, such as channels, beach lenses, delta and submarine fan lobes, and reef cores. To a considerable extent sequences and their components ("parasequences") are self-similar on several scales (Cross et al. 1993). This provides the geometric or architectural basis for prediction, and can enable such predictions to be attempted from a few initial exploration holes.

There is a long tradition in petroleum geology of using outcrop analogs of reservoir rocks as a basis for subsurface description and prediction. In addition to basic measurements of the scale of potential flow units, outcrops may be analyzed by the use of portable minipermeameters to develop detailed descriptions of the porosity-permeability architecture of these flow units. Eschard and Doligez (1993) provided a useful compilation of theoretical analyses and case studies, illustrating the use of sequence stratigraphic principles in outcrop description as an aid in reservoir characterization. Data of this type are now widely used in numerical and graphical simulation of fluid flow through reservoir media (Flint and Bryant 1993).

17.3 Controversies in Practical Sequence Analysis

Two basin-analysis problems are discussed briefly in this section, to illustrate the points made above. The first deals with a detailed analysis of a relatively mature basin, within which, nonetheless, the application of careful lithostratigraphic analysis and modern sequence concepts has brought about a significant improvement in sequence interpretation. The second example illustrates the loss of detail that occurs when inappropriate global models are applied to a regional analysis.

17.3.1 The Case of the Tocito Sandstone, New Mexico

The Tocito Sandstone is a transgressive deposit constituting part of the Gallup Sandstone third-order sequence in the San Juan Basin, New Mexico (Fig. 7.23). It consists of several separate sandstone lenses containing evidence of tidal sedimentation. Nummedal et al. (1989), Nummedal (1990), and Nummedal and Riley (in Van Wagoner et al. 1991) interpreted the sandstone as a suite of shallow-marine, tide-dominated sandbodies formed as deltas and shelf sand ridges. This interpretation fits a general "third-order" model which places the Tocito within a trangressive systems tract (Figs. 7.24, 7.25B,C). However, a high-resolution sequence-stratigraphic analysis of these beds by Jennette et al. (in Van Wagoner et al. 1991) reached a different conclusion. They documented four separate erosion surfaces underlying Tocito lenses, and showed that the Tocito in its entirety cuts down deeply into the Gallup Sandstone. They reinterpreted the Tocito lenses as incised-valley-fills deposited following four separate phases of valley erosion. The regional

correlation diagrams erected by the two sets of authors are shown in Figs. 17.1 and 17.2.

The interpretation of Nummedal and his co-workers is based largely on biostratigraphic correlation of unconformities and the lack of evidence for subaerial exposure of the unconformities in the deeper part of the basin. This led them to interpret the surfaces as marine in origin, including formation by ravinement erosion (Fig. 7.25C).

Jennette et al. (in Van Wagoner et al. 1991) based their reinterpretation on detailed lithostratigraphic correlation, using the large subsurface data base. Although agreeing that there is no evidence for subaerial exposure of the unconformities, they pointed to other studies of interpreted subaerially eroded sequence boundaries, such as those in the Cardium Sandstone, where evidence of subaerial exposure was likewise removed before the next trangression (a recent paper on this topic by Arnott 1992, is discussed in Sect. 15.3.1). The correlation and mapping by Jennette and coworkers, a small sample of which is reproduced in Fig. 17.3, suggested the incised-valley interpretation of the basal Tocito erosion surfaces largely on the basis of the paleogeomorphology of the surfaces. The evidence includes the sinuosity of the valleys, their locally steep, erosional sides, and the presence of fluvial-like tributary junctions.

17.3.2 The Case of Gippsland Basin, Australia

Gippsland Basin constituted one of the first basins examined by Vail and his coworkers in their study of global eustasy. For example, it appears in their illustration of global correlation (Fig. 5.1). The main beds of interest are the petroleum-bearing Latrobe Group of uppermost Cretaceous to earliest Oligocene age. The succession consists of unconformity-

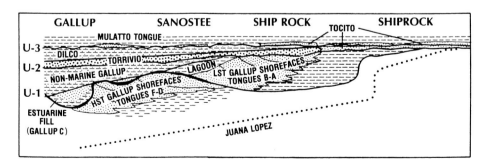

Fig. 17.1. Schematic correlation diagram of Gallup Sandstone, showing Tocito Sandstone lenses resting on three intra-Gallup erosion surfaces *U-1* to *U-3*. (Nummedal and Riley in Van Wagoner et al. 1991, reproduced by permission)

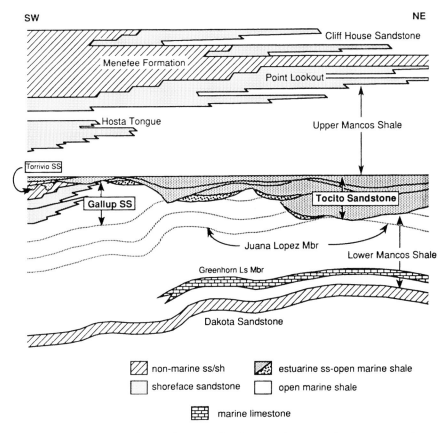

SW

NE

Fig. 17.2. Schematic correlation diagram of Gallup Sandstone showing Tocito Sandstone as marine-estuarine valley fills formed in four phases of cut-and-fill. *ss* Sandston; *ls* Limestone; *mbr* member. (Jennette et al. in Van Wagoner et al. 1991, reproduced by permission)

bounded marine-nonmarine packages. The biostratigraphic framework used in this analysis was first developed in the early 1970s and consists of spore-pollen and dinoflagellate assemblages, which were integrated with each other where the marine and nonmarine strata interfinger. Partridge (1976), who summarized the basin zonal schemes, provided a table correlating these with the planktonic foraminiferal zones. At this time the analysis did not extend below the Maastrichtian.

The zonal schemes established in the 1970s are shown along with the stratigraphic sequences in a Wheeler diagam (Fig. 17.4). The zones were clearly robust, because they formed the basis of later analyses of the Gippsland Basin by Duff et al. (1991), as shown in Fig. 17.5. Zonal correlations of the succession of 10^6-year sequences have been changed very little, but absolute ages have undergone some minor revisions in this later diagram. Lithostratigraphic correlation has also been revised, with the base of the Latrobe Group extended downward from the *Tricolpites longus* zone (Maas-

trichtian) in Partridge's (1976) analysis to the *N. senectus* zone (Early Campanian) in the paper of Duff et al. (1991). Earlier deposits, of the Golden Beach Group, of Cenomanian-Campanian age, were included in the Latrobe Group by Rahmanian et al. (1990).

The biostratigraphic work of Partridge (1976) formed part of the basis of the Exxon sequence analysis, a summary of which was published by Rahmanian et al. (1990). However, the sequence analysis underwent substantial revision between the mid-1970s and 1990. Partridge's (1976) analysis (Fig. 17.4) subdivided the Latrobe Group into eight sequences of Maastrichtian-Oligocene age, whereas Rahmanian et al. (1990) stated that they had identified 25 sequences spanning the Cenomanian–Early Oligocene. In the global cycle chart of Haq et al. (1987, 1988a) there are 37 sequence boundaries indicated between the Cenomanian and the earliest Oligocene (95–36 Ma). This contrasts with the nine sequences recognized in the same interval by Duff et al. (1991; Fig. 17.5).

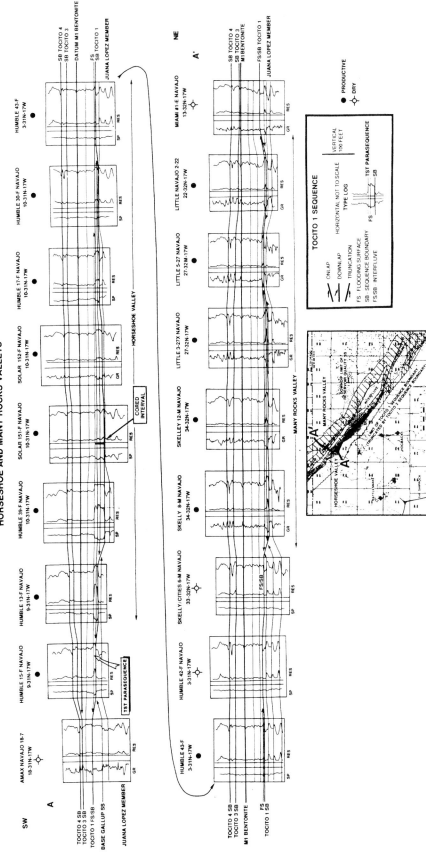

Fig. 17.3. Example of lithostratigraphic correlation carried out to develop a high-resolution sequence stratigraphy. In this case the correlations indicate the presence of an incised valley, filled with the first phase of Tocito estuarine deposits. (Jennette et al. in Van Wagoner et al. 1991, reproduced by permission)

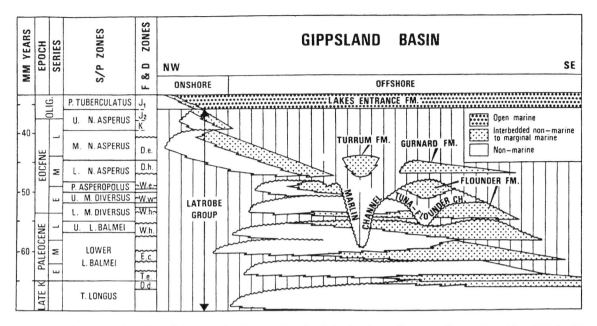

Fig. 17.4. Chronostratigraphy of the Latrobe Group, Gippsland Basin, Australia, according to Partridge (1976). *S/P* Spore/pollen; *F* and *D* foraminifera and dinoflagellates

Two examples of the sequence analysis of Rahmanian et al. (1990) are illustrated in Figs. 17.6 and 17.7. All the examples of sequence analysis published in this paper show all the sequence boundaries of the Exxon global cycle chart for the interval illustrated. This is in keeping with the principle of Exxon sequence analysis, which is to attempt to identify every eustatic event in every basin. It may be presumed that the 25 sequences referred to by Rahmanian et al. (1990) each correspond to the sequences indicated in the global cycle chart. If this is so, it is not made clear where the missing twelve are.

The differences between the Rahmanian et al. (1990) analysis and that of Duff et al. (1991) are considerable. The lithofacies and seismic facies analysis of the first paper (Figs. 17.6, 17.7) imply a sequence-boundary spacing averaging 1.6 m.y., contrasting with the 6.6 m.y. spacing of Duff et al. (1991). Without access to the details of the two analyses it is impossible to evaluate what are clearly two different approaches to sequence analysis. However, some comments of Duff et al. (1991) are relevant. They stated that they can trace sequence boundaries from depocenters, where they are virtually conformable, into adjacent wrench-fault zones or normal-fault complexes, where they are angular unconformities. "The increase in structural deformation within these zones with successively deeper sequence boundaries (with the usual onlap/

subcrop features) shows them to have been caused by recurrent deformational (mainly compressional) phases." Duff et al. (1991) concluded, after outlining the global-eustasy model:

The sequence model proposed in this paper envisages instead that first-order tectonic hiatuses (regional plate margin reorganization) are registered in the sequence record as the corresponding first-order unconformities or megasequence boundaries; that second-order tectonic pulses (reflecting the vagaries of local intra-basin basement-cover response to deformation) are registered as the corresponding second-order sequence boundaries ..., and that far from being the principal or sole determinant of sequence architecture, eustatic effects are much more subtly 'written into' the sequence record as the progression of regressive or trangressive seismic facies.... Contrary to the Exxon school therefore, at least for tectonically active areas such as the Gippsland Basin, eustatic changes are long-term processes, registered in the sequence record during intervals of tectonic quiescence, and tectonism unique to individual basins is the shorter-term process. The logical corollary is that because sequence boundaries reflect the vagaries of local tectonic style internal to basins, they do not provide a tool for consistent, global chronostratigraphic correlation between basins.

This is reminiscent of the reanalysis of the Moray Firth data by Underhill (1991; see Sect. 14.2), and indicates a similar approach to regional seismic analysis to that of Hubbard (1988) and Tankard and Welsink (1987; Sect. 6.3.2), in which emphasis is placed on local and regional tectonic control. I

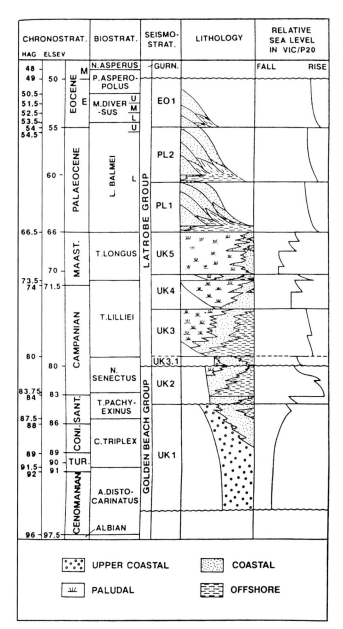

Fig. 17.5. Sequence analysis of the Latrobe Group and underlying strata, Gippsland Basin, according to Duff et al. (1991). Note the two different time scales at the *left*. That of Haq refers to the global cycle chart, but does not show all the sequence boundaries identified by Haq et al. (1987, 1988a)

suggest that this is the more realistic and useful approach.

17.3.3 Conclusions: A Modified Approach to Sequence Analysis for Practicing Petroleum Geologists and Geophysicists

The following is a summary of the recommended approach to sequence analysis, modified from that of Vail et al. (1991), which is given in Section 17.2.1:

1. Develop an allostratigraphic framework based on detailed lithostratigraphic well correlation and/or seismic facies analysis.
2. Develop a suite of possible stratigraphic models that conform with available stratigraphic and facies data.
3. Establish a regional sequence framework with the use of all available chronostratigraphic information.
4. Determine the relationships between sequence boundaries and tectonic events by tracing se-

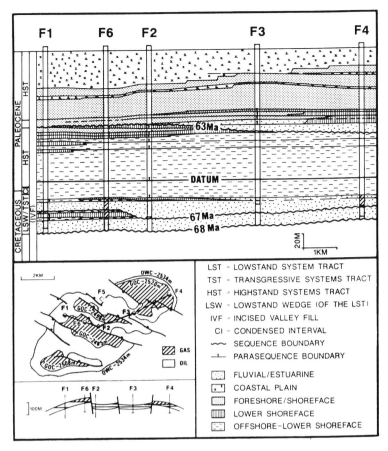

Fig. 17.6. Example of Exxon-style sequence analysis of part of the Latrobe Group. Flounder field, Gippsland Basin, Australia. (Rahmanian et al. 1990, reproduced by permission of The Geological Society, London)

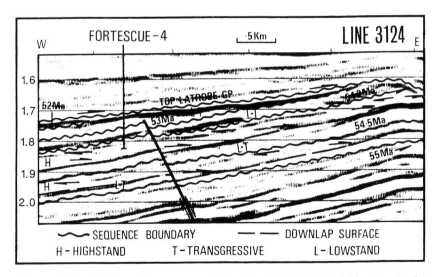

Fig. 17.7. Seismic facies and sequence analysis of the Cobia-Halibut fields, Gippsland Basin, Australia. (Rahmanian et al. 1990, reproduced by permission of The Geological Society, London)

quence boundaries into areas of structural deformation, and documenting the architecture of onlap/offlap relationships, fault offsets, unconformable discordancies, etc.

5. Establish the relationship between sequence boundaries and regional tectonic history, based on plate-kinematic reconstructions.
6. Refine the sequence-stratigraphic model. Subdivide sequences into depositional systems tracts and interpret facies.
7. Construct regional structural, isopach, and facies maps, interpret paleogeographic evolution, and develop plays and prospects based on this analysis.
8. Develop detailed subsidence and thermal-maturation history by backstripping/geohistory analysis.

It is suggested that subsidence and maturation analysis (step 8) be carried out at the conclusion of the detailed analysis, rather than near the beginning, as proposed in the Exxon approach. The reason is that a complete, thorough analysis requires the input of a considerable amount of stratigraphic data. Corrections for changing water depths and for porosity/lithification characteristics,

which are an integral part of such analysis, all require a detailed knowledge of the stratigraphic and paleogeographic evolution of the basin.

17.4 Main Conclusions

1. Sequence analysis should be based on the careful construction of an allostratigraphic framework.
2. The use of an appropriate sequence model is crucial to the understanding of basin architecture, and is therefore a necessary prerequisite to the mapping and prediction work carried out by petroleum geologists.
3. Establishing the relationship of the sequence architecture to regional tectonic events is a critical component of basin analysis.
4. The global cycle chart is of no practical use in basin analysis and exploration.
5. The subsidence and maturation history of a basin is best investigated at the conclusion of the tectonic-stratigraphic analysis because of the wealth of stratigraphic data required in the analysis.

18 Conclusions and Recommendations

18.1 Sequences in the Stratigraphic Record

Rank-order classifications of sequences (first-, second-order, etc.) are no longer recommended, in view of the continuous range of cycle frequencies that have now been demonstrated from the rock record, and in view of the overlapping frequencies of several of the major cycle-generating mechanisms.

18.1.1 Long-Term Stratigraphic Cycles

Many of the broad characteristics of the global stratigraphic record can be related to the changes in continental scale, climate, latitudinal position and eustatic sea level that result from the assembly and dispersal of supercontinents over a 200–500 m.y. episodicity. Cratonic sequences of about 10–100 m.y. duration can be traced and correlated between several of the earth's major continental interiors, including the interior of North America, Russia, and Brazil.

Other sequences of comparable time duration are the result of regional tectonism, including the effects of long-term changes in plate-kinematic patterns. These sequences may be correlatable with each other within the area, perhaps as large as two or three adjacent tectonic plates, the tectonics of which were dominated by the same major events, such as a large-scale plate rifting or collision event. Beyond the effects of these tectonic events correlatable sequences are not formed.

18.1.2 Cycles with Million-Year Episodicities

Stratigraphic sequences with episodicities in the 10^6-year range are common in rifted and extensional-margin basins. Their architecture constitutes repeated transgressive-regressive packages of siliciclastic or carbonate deposits, of tabular form on continental shelves, or comprises prograding clinoform slope wedges. Mixed carbonate-siliciclastic sequences, formed by the process of reciprocal sedimentation, are common. These sequences can readily be interpreted using the Exxon systems-tract models or modified versions of these models. In many of these basins close correlation of the sequence framework with the Exxon global cycle chart is claimed.

Sequences with similar episodicities are also common in ramp-type foreland basins and can also be interpreted using the systems-tract approach. Siliciclastic deposits are dominant, particularly thick alluvial deposits in proximal settings. Transgressive carbonate successions may occur on the distal ramp and forebulge of foreland basins. Evidence for tectonic control is common, for example, stratigraphic thinning and facies changes related to the position of blind thrusts and the forebulge. However, many authors emphasize correlations with the Exxon global cycle chart.

In forearc basins stratigraphic evolution is dominated by tectonic subsidence and faulting events and by volcanism. Few stratigraphic events can be unequivocally related to eustasy, and the correlation of these events with the Exxon global cycle chart is questionable. Backarc basins, especially on their cratonic, hinterland sides, have tectonic and stratigraphic histories comparable to those of extensional continental margins, and commonly contain well-developed records of carbonate or clastic third-order stratigraphic sequences.

Some evidence exists of million-year cyclicity in deep-marine sediments, as studied in DSDP cores. Fluctuations in cycle thickness reflect varying sediment supply, which may be controlled by eustasy or long-term climatic rhythms.

A unique type of cyclicity occurs in Upper Paleozoic rocks of the northern hemisphere. The cycles are of glacioeustatic origin, resulting from the great glaciation of the Gondwana supercontinent. high-frequency cycles, called cyclothems, are the most prominent type of stratigraphic sequence, but groupings of these into "mesothems" or "major cycles" with 10^6-year episodicities can be recognized, and intercontinental correlation of these has been proposed. Recent stratigraphic studies have thrown doubt on the mesothem concept as applied to the Upper Paleozoic record of northwestern Europe, whereas there is increasing evidence for such cycles within the United States Midcontinent.

18.1.3 Cycles with Episodicities of Less Than One Million Years

The Neogene stratigraphic records of continental margins, including the continental shelf, slope, and deep basin, have been intensively studied by reflection-seismic surveying and offshore drilling, including DSDP surveys. Most stratigraphic sections in both carbonate- and clastic-dominated successions are characterized by high-frequency cycles. These can be correlated with the ocean temperature cycles defined by the oxygen-isotope chronostratigraphic record, and are interpreted to be of glacioeustatic origin.

Many stratigraphic sections in the pre-Neogene record also contain prominent high-frequency cycles. They are particularly prominent, and have been well-described, in various carbonate-platform successions, notably the Triassic section of the Dolomites, in northern Italy. However, cycles with comparable episodicity are also present in many other types of succession. Examples of lacustrine, marginal-marine evaporite, pelagic-marine and nonmarine suites are described here. There is controversy regarding the origins of many of these cycles. Orbital forcing ("Milankovitch mechanisms") is the accepted mechanism for many.

A particularly well-known type of Milankovitch cycle is the Upper Paleozoic record of cyclothems in northwestern Europe and the Midcontinent region of North America. These cycles contain much of the economic coal deposits of the northern hemisphere and have been widely studied since the 1930s. It is now generally agreed that these cycles were driven by the Late Paleozoic glaciation of the Gondwana supercontinent, although specific correlations between cyclothems and glacial episodes have yet to be achieved.

Many examples of high-frequency clastic cycles have been described from the Cretaceous sedimentary record of the Western Interior of North America, notably in Alberta, Canada, and the Colorado Plateau area of the United States. These cycles were all deposited within a tectonically active foreland-basin setting. Tectonism is known to have been a significant control in the development of the large-scale architecture of the basin fill, and it remains a controversial question whether tectonism also controlled the high-frequency cyclicity. Some chemical cycles of the deep basin may be correlated with clastic cycles of the basin margin, and Milankovitch mechanisms have been suggested as a contributing or controlling factor. A possible comparison may be made with the Upper Paleozoic cyclothems of the Alleghenian foreland basin of the eastern United States.

18.2 Mechanisms

18.2.1 Long-Term Eustasy and Epeirogeny

Long-term (10^8-year) cycles of sea-level change result from the assembly and breakup of supercontinents on the earth's surface. Sea level is affected by eustatic processes resulting from global changes in sea-floor spreading rate, and changes in the average age of the oceanic crust.

Cycles of eustatic sea-level changes on a 10^7-year time scale are caused primarily by variations in ocean-basin volume generated by episodic spreading, and by the variations in total length and age of the sea-floor spreading centers as supercontinents disassemble and disperse. Many other processes have smaller effects on global sea levels through their effects on the volume of the ocean basins. These include oceanic volcanism, sedimentation, ocean temperature changes, and the desiccation of small ocean basins.

The term dynamic topography is used for the epeirogenic movements that are driven by the thermal effects of mantle convection associated with the supercontinent cycle. This cycle has long-term consequences, including the generation of persistent geoid anomalies. Vertical continental movements are related to the thermal properties of large- and small-scale convection cells, and can involve continent-wide uplift, subsidence, and tilts, and cratonic basin formation. These movements do not correlate in sign or magnitude from continent to continent.

Mantle processes during continental breakup may be "pulsed" on a 10^7-year time scale, as suggested by correlations with changes in the character of the earth's magnetic field.

18.2.2 Milankovitch Processes

The evidence for Milankovitch control of climate and sedimentation in the Late Cenozoic is now overwhelming. Stratigraphic cyclicities in many Late Cenozoic sections have been correlated with the oxygen-isotope record that tracks ocean-water temperature changes, and it is now generally agreed that the ^{18}O record can be used as an analog recorder of sea level bearing in mind the lag between the eustatic high and the attainment of the highest temperatures. This body of knowledge permits chronostratigraphically precise studies of Late Cenozoic sequence stratigraphy. The methods of cyclostratigraphy, as it is now termed, include sophisticated time-series analysis of cyclic parameters, such as cycle thickness and carbon content.

The earth's sensitivity to climate change depends on a wide range of parameters and feedback mechanisms, which make reconstruction of past climate change and of Milankovitch controls difficult. Of particular importance, as demonstrated by the initiation of the Gondwana glaciation and the Early Cenozoic glaciation of Antarctica, is the plate-tectonic control of large continental masses relative to the poles, and the oceanic and atmospheric circulation around them. Major glacial episodes may be initiated by regional uplift caused by collisional uplift of plateaus and mountain ranges, and by thermal uplift during supercontinent fragmentation.

Controversy remains regarding the periodicity of Milankovitch parameters in the geological past. Major changes may have occurred during the Phanerozoic, because of changes in the orbital behavior of the earth. In addition, sedimentological studies indicate that various autogenic mechanisms can generate apparent episodicities that simulate Milankovitch effects. For these reasons, it is considered unwise to attempt to correlate suites of cycles to specific periodicities or to invoke Milankovitch mechanisms purely on the basis of interpreted 10^4–10^5-year cycle frequencies.

Glacioeustasy is markedly affected by isostatic and geoidal changes within hundreds to a few thousands of kilometers of the edge of a major continental ice cap. The timing, direction and magnitude of sea-level changes within this region may be entirely different from the signatures elsewhere. Regional and global correlation should therefore be carried out with caution.

The postulation of Milankovitch controls for any given cyclic sequence should only be attempted following rigorous spectral analysis of one or more cyclic parameters within the succession, because other autogenic and allogenic mechanisms can generate cycles with similar thickness and repetitiveness. Demonstration of the presence of a hierarchy of cycle types with characteristic periodic ratios, such as the 1:5 precession:eccentricity combination, is additional evidence of Milankovitch controls, although the cautions noted above need to be borne in mind.

It seems likely that the earth was entirely ice free for only short periods during the Phanerozoic, and orbital forcing of some type seems to have occurred throughout the Phanerozoic. However, the evidence for major continental glaciation and consequent significant glacioeustasy outside the periods occupied by the Late Paleozoic Gondwana glaciation and the Late Cenozoic glaciation is at present weak.

Glacioeustasy is not the only cycle-forming mechanism driven by orbital forcing. Variations in climate and oceanic circulation have significant effects on such parameters as sediment yield and organic productivity, which have generated many successions of meter-scale cycles throughout the geological past. These are especially prominent in the deposits of many carbonate shelves and in fine-grained clastic deposits formed in lakes and deep seas.

18.2.3 Tectonic Mechanisms

Regarding Extensional Basins and Continental Margins. The stratigraphic architectures that have been used by Vail and his coworkers to interpret 10^7-year eustatic sea-level changes may in fact be generated by other processes. Coastal onlap (relative rises in sea level) may be caused by flexural downwarp of extensional continental margins. Relative falls in sea level may be caused by thermal doming accompanying rifting.

The processes which lead to long-term eustatic sea-level changes, such as changing rates of sea-floor spreading also lead to tectonic adjustents of the continents (e.g., initiation of the rift-thermal subsidence cycle), so that relative changes in sea level may have multiple causes which are not necessarily simply interpretable in terms of eustasy.

The North Sea Basin, one of the major type areas for the Vail curves, was affected by almost continuous tectonism during the Mesozoic and Cenozoic, and many of the sequence-boundary events and stratigraphic architectures there can now be interpreted in terms of specific local or regional episodes of rift faulting and thermal uplift and subsidence.

Regarding Convergent Plate Margins and Collision Zones, and Their Associated Foreland Basins. Arc-related basins are characterized by active regional tectonism as a result of variations in subduction history, arc migration, and the filling and deflation of magma chambers. The rates and episodicity of the relative changes in sea level caused by this tectonism have been determined by detailed mapping in modern arcs and other regions of convergent tectonism, and are consistent with the sea-level changes interpreted to have generated stratigraphic sequences with 10^4- to 10^7-year frequencies that have been mapped in arc-related basins.

The various elements of convergent tectonism that generate foreland basins by flexural loading (continental collision, terrane accretion, nappe migration, thrust movement, imbricate fault propagation) can cause relative changes in sea level on a 10^4- to 10^7-year time scale.

Stratigraphic sequences of 10^6- to 10^7-year duration in foreland basins, such as the Western Interior of North America and Pyrenean basins, may commonly be correlated with specific tectonic episodes in the adjacent orogen, and there is increasing evidence that high-frequency sequences, of 10^4- to 10^5-year duration, may be correlated with more localized tectonic events. The correlation of these sequences with the global cycle chart may be fortuitous.

Regarding Intraplate Stress. Stresses generated by plate-margin extension or compression, and transmitted horizontally, have an important modifying effect on the flexural behavior of sedimentary basins, acting to amplify or subdue the flexural deformation caused by regional tectonics. Changes in paleostress fields are capable of generating regional upwarp and downwarp, resulting in offlap and onlap patterns comparable in stratigraphic duration and magnitude to the "third-order" (10^6-year) stratigraphic cyclic changes in the Exxon global cycle chart.

High-frequency sequences (of 10^4- to 10^5-year duration), of limited, regional extent, may be generated by the transmission of stresses induced by flexural loading of individual structures and their structural components, as thrust faults and their imbricates are propagated during contractional tectonism.

Stratigraphic patterns induced by intraplate stress are of opposite sign in basin margins and basin centers. The proposition of intraplate stress is therefore a testable hypothesis, requiring detailed correlation of unconformities and their correlative conformities across major basins.

Regarding the Importance of Sediment Supply. The sediment supply to a basin may not be governed by local tectonics but by distant tectonic events, with sediment transported across cratons and through orogens by major rivers. In such cases the distribution and age of fluvial, coastal, and submarine clastic wedges does not relate to the sea-level cycle in the basin, but to the tectonic evolution of the sediment source area.

Regarding the Importance of Environmental Change. Carbonate sedimentation may be interrupted by changes in nutrients, water chemistry or clastic content, which inhibit carbonate production. Carbonate platforms are also susceptible to erosion by shifting submarine currents. The result, in all these cases, may be breaks in sedimentation that appear similar to but are unrelated to sequence boundaries produced by sea-level fall.

General Conclusions Regarding Mechanisms. If tectonic subsidence varies in rate within and between basins, as is typically the case, but is comparable to the rate of eustatic sea-level change, sequence boundaries and flooding surfaces are markedly diachronous. The availability of tectonic mechanisms to explain stratigraphic cyclicity of all types removes the need for global eustasy as a primary mechanism for the generation of stratigraphic architectures. This being the case, the onus is on the supporters of eustasy to prove their case by quantitative documentation of eustatic processes, and by rigorous global correlation of supposed eustatic events.

Given the finite size of the earth, major plate-tectonic events (e.g., collisions, ridge reordering events, changes in rotation vectors), and their stratigraphic responses, may be globally synchronous. The potential therefore exists for global stratigraphic correlation. However, the structural and stratigraphic signature would vary from region to region, such that, for example, an unconformity in one location may be genetically related to a transgression elsewhere.

The invocation of tectonic mechanisms to explain stratigraphic architecture has led to the suggestion that Vail's charts be "inverted," to provide a source of data from which tectonic episodicity could be extracted. While this is an interesting idea, there remains the problem that the global accuracy and precision of the Vail curves is in question, and tectonic arguments are in as much danger as eustatic arguments of falling into the trap of false correlation and circular reasoning.

18.3 Chronostratigraphy and Correlation

18.3.1 Concepts of Time

Only a small fraction of elapsed time normally is represented by sediment preserved in the geological record (50% to as little as 3%).

Sequences and parasequences constitute a hierarchy within which time spans and sedimentation rates differ from each other by orders of magnitude. Accurate and precise estimates of time spans of the component sequences and hiatuses in a succession cannot be achieved by determining bracketing ages from stratigraphically widely spaced marker beds that enclose several or many sequences.

Correlation of sequences to test hypotheses such as global eustasy requires that they be dated with a precision, or potential error range, equal to a fraction of the range of the shortest sequence in the succession. For example, sequences with 10^6-year episodicities require a correlation precision of ± 0.5 m.y., or less.

18.3.2 Correlation Problems, and the Basis of the Global Cycle Chart

The proposed supremacy of stratigraphic sequences as chronostratigraphic indicators and their suitability to form the basis for a superior standard of time represents a new paradigm in geology. The Exxon "school," led by Peter Vail, based their development of the global cycle chart on this paradigm. *However, at present the paradigm remains unproven.* The recommendation that sequence boundaries be used for the redefinition of stage boundaries indicates a misunderstanding of

the purpose of the geological time scale, which is to provide a practical subdivision of continuous time *independent* of any geological process.

The main theoretical basis for the paradigm is the supposition that global stratigraphic architecture is controlled primarily by eustatic sea-level changes. Following the eustatic-sequence paradigm, additional or missing sequences in a succession, relative to the global cycle chart, are attributed to local tectonic processes.

Given the availability of many other processes for generating sequences, the main test of eustasy is global correlation of the sequence framework. However, there are two significant problems with such tests: (a) Currently available chronostratigraphic techniques (biostratigraphy, radiometric dating, magnetostratigraphy, chemostratigraphy) are inherently imprecise. In many cases the imprecision and potential error in sequence correlation exceeds interpreted sequence boundary spacing, which means that sequences cannot be distinguished on chronostratigraphic grounds. (b) If rates of eustatic sea-level change and tectonic subsidence are comparable, their integration results in a curve of relative sea-level change in which the timing of highstands and lowstands varies in response to regional variations in tectonism. Therefore a synchronous signature of eustasy is not generated in the stratigraphic record.

Higher order sequences in the Exxon global cycle chart (those with episodicities of 10^6 years and less) have spacings less than the range of error involved in their dating and correlation (except for Late Cenozoic glacioeustatic cycles). *The precise ages that form part of the Exxon global cycle chart cannot therefore be supported.*

Given current levels of chronostratigraphic imprecision and event spacing, almost any stratigraphic succession can be made to correlate with the global cycle chart.

Rigorous correlation exercises carried out where detailed data are available reveal complex networks of regional hiatuses of varying duration and areal extent. Clear eustatic signals may not necessarily be present.

Many published sequence analyses include "forced" correlations to the global cycle chart, even where individual stratigraphic events can be interpreted in terms of local or regional tectonism.

The Exxon global cycle chart should be abandoned.

18.3.3 Comparison of Sea-Level Curves

The margin of error in the assigned ages of sequence boundaries in the Exxon global cycle curves is, in most cases, greater than the spacing of these boundaries. Distinguishing these boundaries on the basis of chronostratigraphic data is therefore not possible. The Exxon curves and the supporting work published in SEPM Special Publication 42 and elsewhere, contain errors and revisions that are not acknowledged or explained.

If the Exxon global cycle charts are a measure of eustasy, as claimed, the positions of sequence boundaries should not vary from location to location, or from curve to curve, although, according to the model, the amplitude of the curve may vary in response to local tectonics. However, different versions of the Exxon curves are in some cases markedly different, and these differences have not been explained or discussed by the authors of the curves.

Many other sea-level curves have been prepared by other authors, based on outcrop and subsurface stratigraphic data. In only a few cases are there genuinely close similarities between these curves, indicating the likely occurrence of eustatic events that show through the different methods of data compilation and analysis. Broad similarities between some of the curves, such as the overall rise in sea level through the Jurassic, reveal long-term trends corresponding to 10^7- to 10^8-year stratigraphic cyclicity, which is probably of eustatic origin.

The almost total lack of correspondence between the high-frequency (1–5 m.y.) events ("third-order" cycles) revealed in the various sea-level curves confirms that global signals on this time scale are not globally correlatable. The high-frequency (1–5 m.y.) events that characterize the Exxon global cycle chart do not appear on most of the other charts. This is interpreted as an outcome of the application of the paradigm of stratigraphic eustasy followed by the Exxon group, which allows them to interpret virtually all sea-level events as eustatic in origin. On this basis, every new sea-level event that is discovered, can be added to the global chart. However, the lack of proof for a global eustatic signal on the 10^6-year ("third-order") time scale (the lack of correspondence between the various other curves at that time scale) indicates that either or both of the following statements is correct: (a) The imprecision and inaccuracy of current chronostratigraphic correlation methods do not permit a reliable test of global eustasy of this frequency; and/ or (b) If a 10^6-year eustatic signal exists, it is largely or entirely masked by local to regional variations in sediment supply and tectonically driven uplift and subsidence of comparable rate and magnitude.

In a few cases there is growing evidence for widespread sea-level events that are simultaneous (within the limits of chronostratigrahic precision) but also occur at times of tectonism, as indicated by changes in thickness, facies or dispersal patterns. These may be exmples of a newly recognized class of tectono-eustatic sea-level changes, with continental-scale changes in in-plane stress as the major driving force.

18.4 Modern Sequence Analysis

18.4.1 Elaboration of the Basic Sequence Model

Regarding General Principles. The application of sequence-stratigraphic concepts to higher-order sequences has raised problems of the hierarchical nature of sequence architectures, requiring revisions in the stratigraphic nomenclature used to name and describe sequences and parasequences.

Concepts of hierarchies of architectural elements and bounding surfaces already developed for fluvial and tidal deposits may be reconciled with the sequence-parasequence framework of the Exxon school, but this requires attention to the time scales and physical scales of the processes involved and the units generated.

Sequence concepts and nomenclature and the literature of sequence stratigraphy have downplayed the importance of autogenic processes in generating facies successions of the scale of parasequences in both carbonate and siliciclastic depositional systems.

Sequence classifications are in a state of flux. A rank-order classification based on duration – the classic Exxon approach – is still widely used, but is not recommended in this book because recent work provides no empirical or genetic justification. A classification by descriptive characteristics has been proposed, based on areal extent, the amount of facies change at the sequence boundary and the degreee of deformation at the boundary, but this classification cannot satisfy all empirical observations. A descriptive terminology referring to the duration or episodicity/periodicity of sequences is useful but does not constitute a formal classification.

Regarding Siliciclastic Depositional Systems. The Exxon model which purports to explain the development of fluvial coastal-plain depositional systems is seriously flawed. Accommodation space for fluvial progradation is not generated during late-highstand seaward shift of the shoreline, as had been proposed. Most fluvial deposits are generated during the trangressive and highstand phases of the sea-level cycle, and are also a response to tectonic uplift of source areas.

The Exxon model of the bayline also creates some contradictory depositional models that are not supported by observations of modern depositional systems.

A need has been expressed for an additional category of systems tracts, corresponding to the "forced regression" of barrier-strandplain systems during falls in sea level. The term forced regressive wedge has been proposed. This systems tract lies above a regressive surface of marine erosion, but below the sequence boundary.

Deltaic styles, which have traditionally been interpreted in terms of the balance between fluvial and marine processes, may also depend on a sea-level control. River-dominated deltas are common in highstand settings; wave-dominated deltas are common shelf-margin systems tracts; tidal deltas are part of the estuarine realm, and are typical of transgressive environments.

Many units previously interpreted as shelf deposits, such as many so-called offshore bars, can now be shown to be the product of coastal depositional processes during the falling stage, lowstand and transgression. These processes include the development of forced regressive wedges, the reworking of shoreface deposits during transgression, and the filling of incised valleys during trangression.

Coastal-shelf stratigraphic architecture depends on a balance between the rate of creation of accommodation space and the rate of introduction and transportation of sediment.

Condensed sections and marine hiatuses are a characteristic component of trangressive systems tracts, but can also be generated by submarine current erosion. Examples in the Cenozoic sections of the United States Atlantic margin have been mistaken for sequence boundaries.

Turbidite depositional systems are a characteristic component of lowstand systems tracts, but the presence of major submarine fan units does not invariably indicate low sea levels. Tectonism and volcanic-dominated sediment supply are other important controls that can lead to submarine fan

sedimentation being out-of-phase with and unrelated to the eustatic sea-level curve.

In the common situation of sediment supply to the deep sea reaching a maximum during the falling leg of the sea-level curve, because of subaerial erosion and sediment bypass, base-of-slope fans accumulate most rapidly during this falling stage. The sequence boundary, which defines the time of lowest sea level, should therefore be drawn above these deposits, not below them as in the original Exxon model.

Regarding Carbonate Depositional Systems. The original Exxon sequence models for carbonates are unrealistically similar to those developed for siliciclastics. They did not incorporate the differences in the response of the two types of system to sea-level change. One of the most important differences is the absence of thick lowstand wedges in carbonate systems. The carbonate factory is at a low level of activity during sea-level lowstands. Deep-water carbonates are more typical of highstand systems tracts, as a result of highstand shedding from carbonate platforms.

The styles of platform sedimentation reflect a sensitive balance between sedimentation rates and sea-level change. Rapid sea-level rise may lead to drowning and backstepping of platforms, with subsequent "catch-up" aggradation. Late stages of sea-level rise are characterized by "keep-up" sedimentation of thick tidal deposits, and may be followed by still-stand progradation.

Drowning of platforms during rises in sea level by siliciclastic detritus, or influxes of cold, warm, or unusually saline water, or an oversupply of nutrients, may generate "drowning unconformities" that are similar in stratigraphic character to lowstand sequence boundaries.

18.4.2 Numerical and Graphical Modeling of Stratigraphic Sequences

Stratigraphic architecture is a complex response to variations in three major variables, sediment flux, the rate and style of subsidence, and the duration and amplitude of the eustatic sea-level cycle.

Computer models of continental-margin stratigraphy cannot yet simulate the physics of sedimentation in all its complex detail, but they can approximate the development of continental margins in two dimensions, by making use of diffusion equations and transport coefficients to provide realistic simulations of sediment dispersal. Stratigraphic architectures can be modeled in various tectonic settings by varying the input of subsidence styles and rates.

It has been demonstrated that there can be a phase lag between the sea-level cycle and the stratigraphic response, and that this lag may vary within and between basins by up to 1/4 of a wavelength of the sea-level cycle. Sequence boundaries may therefore vary in age locally and regionally by an amount of time equivalent to 1/4 of the duration of the sequence. In the case of 10^6-year ("third-order") sequences this could amount to several million years.

18.5 Implications for Petroleum Geology

Sequence analysis should be based on the careful construction of an allostratigraphic framework.

The use of an appropriate sequence model is crucial to the understanding of basin architecture, and is therefore a necessary prerequisite to the mapping and prediction work carried out by petroleum geologists.

Establishing the relationship of the sequence architecture to regional tectonic events is a critical component of basin analysis, whereas the global cycle chart is of no practical use in basin analysis and exploration.

The subsidence and maturation history of a basin is best investigated at the conclusion of the tectonic-stratigraphic analysis because of the wealth of stratigraphic data required in the analysis.

18.6 The Global-Eustasy Paradigm: Working Backwards from the Answer?

18.6.1 The Exxon Factor

At the end of this book it is hard to avoid the conclusion that many of the Exxon sequence concepts and models, with their entire glossaries of new terminology, are based on ideas that have remained basically unchanged since they were developed by Peter Vail and his colleagues in the 1960s. Deep in the bowels of Exxon Corporation Peter Vail initiated a revolution which shaped an entire corporate exploration philosophy. The juggernaut, once started, became unstoppable – management structures and budgets in large corporations may come to be dominated by a prevailing paradigm, especially if it appears to offer a competitive advantage and can be kept secret. However, corporate science which remains unexposed to public peer review is like a hot-house plant. It flourishes in its enclosed, protected environment, but may not enjoy the chill winds of critical comment on the outside. And so it is with the Exxon work. Very few of the criticisms levelled at the global cycle model have been answered directly by members of the Exxon research group.

I have deliberately referred to the global-eustasy model as a paradigm, because it represents a new approach to the interpretation of stratigraphic data, which has been explicitly set out in many of the Exxon publications (see Sect. 13.2). This is the fundamental concept that was born in the late 1960s and early 1970s, grandfathered (but not now fully supported; see below) by Sloss, Vail's graduate supervisor (Sloss 1988a). According to this concept any succession of sequences anywhere represents part of the global eustatic tape recording. The complete eustatic record may therefore be assembled from any collection of sequences. Exxon's burgeoning offshore seismic data base represented an ideal starting point for such an analysis. Once started, all sequence successions could be assumed to fit the standard. The really interesting question is when Peter Vail concluded that he had arrived at the standard succession. It seems likely that this happened very early on, probably in the 1960s. Vail (1992) has recounted some of the features of his own early scientific development. He drew his first Phanerozoic global cycle chart in 1959 (four years before Sloss's classic 1963 publication, which only dealt with sequences within North America!). The principles of seismic facies and systems tracts began to evolve in the early half of the 1960s, and Vail was the supervisor of an interpretation group in Exxon that did a lot of regional work in the 1960s. This group was broken up in 1971. Vail's personal prestige really began to increase with outside presentations (according to his own accounting in the 1992 article), and this started with a Penrose conference talk on trench-slope deposits in the early 1970s, and the AAPG seismic stratigraphy school in the mid-1970s. The evolution of concepts in his papers is consistent with this timing. For example, see Underhill's (1991) reappraisal of the work done in the Moray Firth, off Scotland (Sect. 14.2). Vail and Todd's (1981) key paper, which set out the basis for much of the Jurassic part of the global cycle chart, was based on the analysis of one seismic line from this area. In a highly tectonically active area, crossed by numerous faults, this one line managed to miss all the faults. However, following the new paradigm, this should not matter, only one section representing each part of the Phanerozic succession being all that is really necessary. Every new sea-level event that is discovered, can simply be added to the global chart. The only other use for additional sections is to assist in calibration of any "tectonic enhancement" of each sequence-bounding unconformity, to determine the magnitude, or amplitude, of the sea-level deflection. Given the new paradigm, the only real purpose of recent publications on the global cycle chart by the Exxon group has been to better illustrate and enunciate the principles and results that were essentially completed by the mid-1980s.

The unrealistic treatment of accuracy and precision in the chronostratigraphic data used by the Exxon group is the perfect framework for the new paradigm. In practice it provides virtually no constraints on correlations with the supposed global standard. Even where available data demonstrate an unavoidable conflict with this global standard (extra or missing sequences in a regional sequence stratigraphy), this can readily be dismissed as a result of local tectonics. The 1977 and 1987/1988 versions of the global cycle chart are very different. The time scale has been revised in the later version, many sequence boundaries have been adjusted with respect to the age-stage framework, and many boundaries have been assigned different ages. The lack of any detailed discussion of these revisions enables the authors to avoid dealing with the imprecise nature of their data and confronting the problems of accuracy and precision. Haq et al. (1988a) discussed error and imprecision in some detail, but only that error associated with the *generation* of their time scale. Such error was not discussed at the time of *application* of the time scale to the dating of actual stratigraphic sequences.

Phrases such as *tectonically enhanced unconformity,* a favorite of the Exxon group, are enormously revealing. This phrase constitutes a gesture toward those who emphasize the importance of tectonism, while retaining the unshakable conviction that the unconformity would be there anyway because of eustasy. Diagrams appearing in the Exxon publications that purport to explain the integration of tectonism and eustasy always show these two processes as occurring at very different rates (e.g., Fig. 11.1A). Tectonism is normally shown as a rather slow, steady, back-

ground effect. This enables the favored mechanism, eustasy, to receive prominence. For the process of tectonic enhancement to operate it is necessary that tectonic and eustatic episodes be exactly in phase. Although this is possible, this necessary correlation has not beeen pointed out or discussed by the Exxon researchers. The overwhelming evidence in the literature, summarized in this book, that tectonism and eustasy may in fact have very similar rates and magnitudes of effect on relative sea levels, is not mentioned or dealt with in the Exxon publications.

None of the important work carried out on the tectonic mechanisms that drive sea-level change (Chap. 11) has been properly addressed by the Exxon group, for example, that of Hubbard (1988), who reported a sequence analysis of various basins bordering the Americas. Most of his major ("second-order") sequence boundaries do not correlate with the corresponding boundaries in the global cycle chart of Haq et al. (1987, 1988a; see Fig. 6.15 and discussion in Sect. 6.2.2). Van Wagoner et al. (1990) dismissed this as the result of regional "tectonic enhancement" of third-order eustatic cycle boundaries. Vail et al. (1991) discussed thermal and flexural subsidence of sedimentary basins, but did not address the important implications this process has for eustatic interpretation of coastal-onlap charts, implications that were pointed out by Watts in the early 1980s (Sect. 11.2).

The conservatism of the Exxon research group is illustrated by the history of the sawtooth curve. Although this curve, as a graph of sea-level change, was very quickly abandoned (compare Vail et al. 1977 with Vail and Todd 1981), the alternative interpretations that were offered for the sawtooth pattern seem to reveal an unwillingness to abandon a faulty model, in this case the use of coastal onlap as an invariant indicator of relative sea-level rise. The model of late-highstand coastal-plain fluvial onlap that became the basis for subsequent interpretations has been critiqued by Miall (1991a, 1996). Figure 4.9 is from this period of revision (Vail et al. 1984), and purports to explain types 1 and 2 unconformities. As noted by Miall (1996), this diagram manages the feat of demonstrating both sea-level rise and sea-level fall in the same stratigraphic units. Onlap is shown at the left in each panel, where the stratigraphic units terminate against a steeply dipping basement surface (this is not a fault). At the same time facies belts are shown migrating seaward, indicating regression and relative sea-level fall. The observations on small-scale depositional systems (lateral movement of river meanders, sedimentation in reservoirs) reported by Posamentier and Vail (1988) provided examples of simultaneous basin-margin onlap and regression, and fed into an early model of Vail et al. (1977) regarding late-highstand fluvial aggradation, despite the fact that none of these observations or early models actually show a *fall* in base level.

The Exxon group and their coworkers have in several instances insisted on the retention of concepts or terms that have outlived their usefulness, indicating a resistance to new data or ideas. In Section 14.2 a quote from Baum and Vail (1988) is given regarding the naming of sea-level events in the global cycle chart with reference to their numerical radiometric ages. They urged the retention of the same suite of numerically named events even where more recent research indicates that the radiometric age is incorrect. In Section 15.2.2 the suggestion by Van Wagoner (1995) is mentioned that systems tracts could be defined on the basis of "rock properties," such as parasequence stacking patterns, regardless of their relationship to the sea-level cycle. For example, this permits the designation of shelf-margin deltas in the Gulf of Mexico as highstand deposits because they resemble highstand deposits (they fit the Exxon definition that is based entirely on stratigraphic architecture), even though it has been demonstrated that they were formed during Pleistocene glacioeustatic sea-level low-stands. In the same section the debate regarding the placement of the sequence boundary within shelf and deep-sea sediments is summarized. It is now realized that significant shoreline deposition may take place during "forced regressions," and that basin-floor fans are formed during the early phase of a relative fall in sea level. Most workers now advocate placing the sequence boundary above these falling-stage deposits (Hunt and Tucker 1992; Helland-Hansen and Gjelberg 1994; Embry 1995). However, the Exxon group and their coworkers (Posamentier et al. 1992; Mitchum et al. 1993) insist on placing the sequence boundary below these falling-stage sediments, as in the original model of Vail et al. (1977) and Posamentier et al. (1988), because of their adherence to a theoretical model that relates accommodation changes and facies shifts to inflection points in the sea-level curve.

Why has the Exxon global cycle model become so popular? Almost every stratigrapher describing primary data refers to correlations with one or other version of the Exxon global cycle charts. The fact that the Exxon model is an all-embracing theory is one feature that makes it very attractive. However,

something I like to call the "*Exxon factor*" may also be at play. Exxon is the world's largest petroleum company; the publications and conference presentations by the Vail group are always accompanied by superb graphics; they tend to emerge as blockbuster packages, such as AAPG Memoir 26 and SEPM Special Publication 42, in which funding subsidies from Exxon may be acknowledged. This style of presentation suggests corporate approval, even sponsorship, and provides the research with automatic prestige, which tends to encourage an uncritical acceptance that would not be forthcoming for the work of a single individual. Another facet of the Exxon factor may have evolved from the assumption that a company such as Exxon, which is active in exploration on a worldwide basis, would own one of the world's best collections of seismic-reflection lines and well records, and would be able to call on the best available scientific brain power for analysis and interpretation. Reference to the initial data base was made by Vail et al. (1977, p. 88) but very little actual data from it has ever been published. The tradition of industry confidentiality in North America enhances this assumption of the existence of secret proofs of the Vail curves, by allowing us to imagine whatever impressive data collection we wish. A few remarks by Vail and his coworkers may have encouraged this trend. For example, Vail et al. (1977, p. 86), in presenting their first global cycle chart, noted that "Cretaceous [third-order] cycles have not been released for publication."

The basing of an important new paradigm on data that are largely unpublished does not constitute good science. One of the principles of the scientific method is replication – different workers should be able to repeat an analysis or experiment and arrive at the same conclusions. Future work will require the publication and exhaustive analysis of large, public, data sets. In the few instances where this has been done using rigorous scientific principles, such as the work of Aubry (1991, 1995), discussed in Section 13.4.2, the results cast serious doubt on the global-eustasy paradigm.

Even where data and ideas have been published, questions need to be raised regarding the rigor of the critical review and editing that were involved in the preparation of the reports. The two major works containing the principal Exxon theoretical contributions, AAPG Memoir 26 and SEPM Special Publication 42, are both almost "in-house" reports, composed largely of Exxon output and edited in large measure by Exxon personnel. A later AAPG book on the sequence stratigraphy of foreland basins (Van Wagoner and Bertram 1995) includes many papers by Exxon personnel, and one editor is also from this corporation. According to AAPG editor Kevin Biddle (personal communication 1996) reviewers of the papers in this book included geologists from within and outside Exxon. Very limited references to other work on sequence stratigraphy is made by the Exxon scientists, particularly to data or ideas that challenge those of the authors. Few of the major papers by the Exxon group have ever appeared in refereed journals. I suggest that the principle of critical, arms-length, peer review has not been consistently applied to this body of work. These points seem to suggest the workings of the Exxon factor in the editorial offices of major publishers, which is not a healthy development. It is my personal opinion that without the influence of the Exxon factor much of the Exxon output would not have been published in its present form.

It is only fair to add that at least some of the later publications by present or former members of the Exxon team and their immediate associates (e.g., G.P. Allen, D.P. James, R.M. Mitchum Jr., H.W. Posamentier, J.F. Sarg, P. Weimer) that focus on sequence architecture models show a much more appropriate level of flexibility and openness to discussion and new developments, and some of their work is now appearing in refereed journals. This aspect of sequence stratigraphy has become a highly dynamic area of research, and Weimer and Posamentier (in their introduction to Weimer and Posamentier 1993) and Posamentier and James (in their introduction to Posamentier et al. 1993) have complained, with some justification, that they and their colleagues now receive blame for the misapplication of sequence models and concepts by others with whom they have no working relationship. They emphasize that sequence models can only be just that, models, that may or may not be suitable for particular field case examples, and that the originators of the models should not be held responsible where they do not work. Exactly the same criticism of the "model builders" was levelled at those who were involved in developing autogenic facies models in the 1970s (including me!). Many recent publications by this group (the paper by Posamentier and Allen 1993b, is a good example) specifically set out to demonstrate the variability in sequence architectures that may arise in response to variations in such major controlling parameters as tectonic setting, sediment supply, and climate.

The testing of research paradigms, such as the paradigm of global eustasy, is not a priority for corporations, and it is unlikely that petroleum-industry scientists will be given the time or the free use of corporate data sets to pursue this agenda in the coming years. Nor will it be easy for anybody else, in view of the limited amount of raw data that can now be accepted for publication by research journals, and the decline in the role of government survey organizations in publishing definitive stratigraphic reports.

18.6.2 Conclusions

1. The Exxon methodologies were developed "in house" in the 1960s and early 1970s and were not subject to independent critical peer review.
2. Although seismic stratigraphy represents an approach that was radically new at the time of development, the original Exxon research group appears to be highly conservative, and there has been some resistance to revisions and modifications.
3. The primacy of the Exxon school may in part reflect the operation of an "Exxon factor," in which preference for the Exxon work is based on the prestige attached to a corporation that happens to be the world's largest oil company, and scientists' normal critical faculties are suspended.
4. Recent work in the field of sequence architecture by present and former members of the Exxon group reveals a greater degree of openness and inventiveness, and to some extent now the "shoe is on the other foot," in that their emphasis on the flexibility and variability in sequence models and concepts is not always appreciated by the "users" of these concepts.

18.7 Recommendations

It is time to become more discriminating. Peter Vail has started a legitimate revolution of great importance in stratigraphy, but we need now to separate the real advances and contributions from the ideas that have failed. The dual nature of sequence stratigraphy was commented on by Carter et al. (1991; see quote from this paper in the introduction to Part IV). They advised acceptance of the guiding principles of sequence stratigraphy, while being extremely cautious in application and use of the global eustasy model. The architectural models that are emerging from the new stratigraphic syntheses and from computer modeling (Part IV) are providing invaluable insights into the evolution of sedimentary basins, and are becoming a predictive tool of enormous importance. Much important work on generating mechanisms has resulted from this new impetus to stratigraphic research.

As recommended by Walker (1992a), the concepts and practices of sequence stratigraphy may be employed without making use of Exxon's genetic sequence terminology. The mapping of unconformity-bounded sequences and the use of allostratigraphic terms and methods provide the necessary framework for the definition, classification and naming of sequences. Within this frame-

work great advances in practical petroleum exploration and in the search for generating mechanisms may be made by employing the standard methods of facies analysis and mapping in basin evaluation.

Let the last "countervailing" words go to Sloss, whose career spans the birth, flowering and maturing of sequence stratigraphy:

While we await the ultimate clarification of the relationship between sea levels and tectonics, let us avoid the obscurantism that clouded my generation's formative decades, the blind agreement, stubbornly adhered to, on fixed continents and oceans in the absence of an "acceptable" mechanism for lithospheric mobility. In terms of the current issue, it is to be hoped that the cult of Neo-Neptunism, which demands that continents, their degree of freeboard, and their cratonic basins and arches are subject to domination by eustasy, can be set aside as a rather quaint oversimplification. (Sloss 1991, p. 6616)

References

Abbotts IL (ed) (1991) United Kingdom oil and gas fields 25 years commemorative volume. Geological Society, London, Memoir 14, 573 pp

Adams CG, Benson RH, Ridd RB, Ryan WBF, Wright RC (1977) The Messinian salinity crisis and evidence of late Miocene eustatic changes in the world ocean. Nature 269:383–386

Ager DV (1964) The British Mesozoic committee. Nature 203:1059

Ager DV (1981) The nature of the stratigraphical record (2nd edn). Wiley, New York, 122 pp

Agterberg FP (1990) Automated stratigraphic correlation. Elsevier, Amsterdam, 424 pp

Aharon P, Goldstein SL, Wheeler CW, Jacobsen G (1993) Sea-level events in the South Pacific linked with the Messinian salinity crisis. Geology 21:771–775

Aitken JD (1966) Middle Cambrian to Middle Ordovician cyclic sedimentation, southern Rocky Mountains of Alberta. Bull Can Pet Geol 14:405–441

Aitken JD (1978) Revised models for depositional grand cycles, Cambrian of the southern Rocky Mountains, Canada. Bull Can Pet Geol 26:515–542

Algeo TJ (1993) Quantifying stratigraphic completeness: a probabilistic approach using paleomagnetic data. J Geol 101:421–433

Algeo TJ, Seslavinsky KB (1995) The Paleozoic world: continental flooding, hypsometry, and sealevel: Am J Sci 295:787–822

Algeo TJ, Wilkinson BH (1988) Periodicity of mesoscale Phanerozoic sedimentary cycles and the role of Milankovitch orbital modulation. J Geol 96:313–322

Allen JRL (1983) Studies in fluviatile sedimentation: bars, bar complexes and sandstone sheets (low-sinuosity braided streams) in the Brownstones (L Devonian), Welsh borders. Sediment Geol 33:237–293

Allen PA, Allen JR (1990) Basin analysis: principles and applications. Blackwell, Oxford, 451 pp

Allen PA, Collinson JD (1986) Lakes. In: Reading HG (ed) Sedimentary environments and facies. Blackwell, Oxford, pp 63–94

Allen PA, Homewood P, Williams GD (1986) Foreland basins: an introduction. In: Allen PA, Homewood P (eds) Foreland basins. Int Assoc Sedimentol Spec Publ 8:3–12

Anadon P, Cabrera L, Colombo F, Marzo M, Riba O (1986) Syntectonic intraformational unconformities in alluvial fan deposits, eastern Ebro basin margins (NE Spain). In: Allen PA, Homewood P (eds) Foreland basins. Int Assoc Sedimentol Spec Publ 8:259–271

Anderson DL (1982), Hotspots, polar wander, Mesozoic convection, and the geoid. Nature 297:391–393

Anderson DL (1984) The earth as a planet: paradigms and paradoxes. Science 223:347–355

Anderson DL (1994) Superplumes or supercontinents? Geology 22:39–42

Anderson JB, Thomas MA (1991) Marine ice-sheet decoupling as a mechanism for rapid, episodic sea-level change: the record of such events and their influence on sedimentation. Sediment Geol 70:87–104

Anderson RY (1984) Orbital forcing of evaporite sedimentation. In: Berger A, Imbrie J, Hays J, Kukla G, Saltzman B (eds) Milankovitch and climate. NATO ASI Series. Part 1. Reidel, Dordrecht, pp 147–162

Armentrout JM (1981) Correlation and ages of Cenozoic chronostratigraphic units in Oregon and Washington. Geol Soc Am Spec Pap 184:137–148

Armentrout JM, Malecek SJ, Fearn LB, Sheppard CE, Naylor PH, Miles AW, Desmarais RJ, Dunay RE (1993) Log-motif analysis of Paleogene depositional systems tracts, central and northern North Sea: defined by sequence stratigraphic analysis. In: Parker JR (ed) Petroleum geology of northwest Europe. Proceedings of 4th Conference. Geological Society, London, pp 45–57

Arnott RWC (1992) The role of fluvial processes during deposition of the (Cardium) Carrott Creek/Cyn-Pem Conglomerates. Bull Can Pet Geol 40:356–362

Ashley GM, Wellner RW, Esker D, Sheridan RE (1991) Clastic sequences developed during late Quaternary glacio-eustatic sea-level fluctuations on a passive margin: example from the inner continental shelf near Barnegat Inlet, New Jersey. Geol Soc Am Bull 103:1607–1621

Aubry M-P (1991) Sequence stratigraphy: eustasy or tectonic imprint. J Geophys Res 96B:6641–6679

Aubry M-P (1995) From chronology to stratigraphy: interpreting the lower and middle eocene stratigraphic record in the Atlantic Ocean. In: Berggren WA, Kent DV, Aubry M-P, Hardenbol J (eds) Geochronology, time scales and global stratigraphic correlation. Soc Sediment Geol Spec Publ 54:213–274

Austin JA Jr, Schlager W et al. (1988) Leg 101 – an overview. Proc ODP, Scientific Results, vol 101, pp 455–472

Autin WJ (1992) Use of alloformations for definition of Holocene meander belts in the middle Amite River, southeastern Louisiana. Geol Soc Am Bull 104:233–241

Baars DL, Stevenson GM (1982) Subtle stratigraphic traps in Paleozoic rocks of Paradox Basin. In: Halbouty MT (ed) The deliberate search for the subtle trap. Am Assoc Pet Geol Mem 32:131–158

Baars DL, Watney WL 1991) Paleotectonic control of reservoir facies. In: Franseen EK, Watney WL, Kendal

CGStC (eds) Sedimentary modeling: computer simulations and methods for improved parameter definition. Kans Geol Surv Bull 233:253–262

Badley ME, Price JD, Dahl CR, Agdestein T (1988) The structural evolution of the northern Viking Graben and its bearing upon extensional modes of basin formation: J Geol Soc (Lond) 145:455–472

Bally AW (ed) (1987) Atlas of seismic stratigraphy. Am Assoc Pet Geol Stud Geol 27 (3 vols)

Bally AW, Snelson S (1980) Realms of subsidence. In: Miall AD (ed) Facts and principles of world petroleum occurrence. Can Soc Pet Geol Mem 6:9–94

Barron EJ (1983) A warm, equable Cretaceous: the nature of the problem. Earth-Sci Rev 19:305–338

Barron EJ, Thompson SL (1990) Sea level and climate change. In: Revelle R (ed) Sea-level change. National Research Council, Studies in Geophysics. National Academy Press, Washington, pp 185–192

Bartek LR, Vail PR, Anderson JB, Emmet PA, Wu S (1991) Effect of Cenozoic ice sheet fluctuations in Antarctica on the stratigraphic signature of the Neogene. J Geophys Res 96B:6753–6778

Bassett MG (1985) Towards a "common language" in stratigraphy. Episodes 8:87–92

Bates RL, Jackson JA (1987) Glossary of geology, 3rd edn. American Geological Institute, Alexandria, 788 pp

Baum GR, Vail PR (1988) Sequence stratigraphic concepts applied to Paleogene outcrops, Gulf and Atlantic basins. In: Wilgus CK, Hastings BS, Kendall CGStC, Posamentier HW, Ross CA, Van Wagoner JC (eds) Sea-level changes: an integrated approach. Soc Econ Paleontol Mineral Spec Publ 42:309–327

Beaumont C (1981) Foreland basins. Geophys J R Astron Soc 65:291–329

Beaumont C (1982) Platform sedimentation. International Association of Sedimentologists. 11th Int Sedimentol Congr, Hamilton, Ontario, Abstr, p 132

Berger A (1988) Milankovitch theory and climate. Rev Geophys 26:624–657

Berger A, Imbrie J, Hays J, Kukla G, Saltzman B (eds) (1984) Milankovitch and climate. NATO ASI Series. Reidel, Dordrecht, 2 vols, 895 pp

Berger AL, Loutre MF (1994) Astronomical forcing through geological time. In: de Boer PL, Smith DG (eds) Orbital forcing and cyclic sequences. Int Assoc Sedimentol Spec Publ 19:15–24

Berger AL, Loutre MF, Dehant V (1989) Influence of the changing lunar orbit on the astronomical frequencies of pre-Quaternary insolation patterns. Paleoceanography 4:555–564

Berger AL, Loutre MF, Laskar J (1992) Stability of the astronomical frequencies over the earth's history for paleoclimate studies. Science 255:560–566

Berger WH, Winterer EL (1974) Plate stratigraphy and the fluctuating carbonate line. In: Hsü KJ, Jenkyns HC (eds) Pelagic sediments: on land and under the sea. Int Assoc Sedimentol Spec Publ 1:11–48

Berggren WA, Van Couvering JA (1978) Biochronology. In: Cohee GV, Glaessner MF, Hedberg HD (eds) Contributions to the geologic time scale. Am Assoc Pet Geol Stud Geol 6:39–55

Berggren WA, Hilgen FJ, Langereis CG, Kent DV, Obradovich JD, Raffi I, Raymo ME, Shackleton NJ (1995a) Late Neogene chronology, new perspectives in high-resolution stratigraphy. Geol Soc Am Bull 107:1272–1287

Berggren WA, Kent DV, Aubry M-P, Hardenbol J (eds) (1995b) Geochronology, time scales and global stratigraphic correlation. Soc Sediment Geol Spec Publ 54:386

Bergman KM, Walker RG (1987) The importance of sea level fluctuations in the formation of linear conglomerate bodies: Carrot Creek Member of the Cardium Formation, Cretaceous Western Interior Seaway, Alberta, Canada. J Sediment Petrol 57:651–665

Beuthin JD (1994) A sub-Pennyslvanian paleovalley system in the central Appalachian basin and its implications for tectonic and eustatic controls on the origin of the regional Mississippian-Pennyslvanian unconformity. In: Dennison JM, Ettensohn FR (eds) Tectonic and eustatic controls on sedimentary cycles. Soc Sediment Geol, Concepts Sedimentol Paleontol 4:107–120

Bhattacharya J (1988) Autocyclic and allocyclic sequences in river- and wave-dominated deltaic sediments of the upper Cretaceous Dunvegan formation, Alberta: core examples. In: James DP, Leckie DA (eds) Sequences, stratigraphy, sedimentology: surface and subsurface. Can Soc Pet Geol Mem 15:25–32

Bhattacharya J (1991) Regional to sub-regional facies architecture of river-dominated deltas, upper Cretaceous Dunvegan formation, Alberta subsurface. In: Miall AD, Tyler N (eds) The three-dimensional facies architecture of terrigenous clastic sediments and its implications for hydrocarbon discovery and recovery. Soc Econ Paleontol Mineral, Concepts Sedimentol Paleontol 3:189–206

Bhattacharya JP (1993) The expression and interpretation of marine flooding surfaces and erosional surfaces in core; examples from the upper Cretaceous Dunvegan formation, Alberta foreland basin, Canada. In: Posamentier HW, Summerhayes CP, Haq BU, Allen GP (eds) Sequence stratigraphy and facies associations. Int Assoc Sediment Spec Publ 18:125–160

Bhattacharya J, Walker RG (1991) Allostratigraphic subdivision of the upper Cretaceous Dunvegan, Shaftesbury, and Kaskapau formations in the northwestern Alberta subsurface. Bull Can Pet Geol 39:145–164

Bhattacharya J, Walker RG (1992) Deltas. In: Walker RG, James NP (eds) Facies models: response to sea level change. Geological Association of Canada, St John's, Newfoundland, pp 157–177

Biddle KT, Schlager W (eds) (1991) The record of sea-level fluctuations. Sediment Geol 70(2/4):85–270

Biddle KT, Schlager W, Rudolph KW, Bush TL (1992) Seismic model of progradational carbonate platform, Picco di Vallandro, the Dolomites, northern Italy. Am Assoc Pet Geol Bull 76:14–30

Birkelund T, Hancock JM, Hart MB, Rawson PF, Remane J, Robaszynski F, Schmid F, Surlyk F (1984) Cretaceous stage boundaries-proposals. Bull Geol Soc Den 33:3–20

Bishop DG (1991) High-level marine terraces in western and southern New Zealand: indicators of the tectonic tempo of an active continental margin. In: Macdonald

DIM (ed) Sedimentation, tectonics and eustasy: sea-level changes at active margins. Int Assoc Sedimentol Spec Publ 12:69–78

Bjørlykke D, Elvsborg A, Hoy R (1976) Late PreCambrian sedimentation in the central Sparagmite basin of south Norway. Norsk Geol Tidskr 56:233–290

Blair TC, Bilodeau WL (1988) Development of tectonic cyclothems in rift, pull-apart, and foreland basins: sedimentary response to episodic tectonism. Geology 16:517–520

Blum MD (1994) Genesis and architecture of incised valley fill sequences: a late Quaternary example from the Colorado River, Gulf Coastal Plain of Texas. In: Weimer P, Posamentier HW (eds) Siliclastic sequence stratigraphy: recent developments and applications. Am Assoc Pet Geol Mem 58:259–283

Boardman DR, Heckel PH (1989) Glacial-eustatic sea-level curve for early late Pennsylvanian sequence in north-central Texas and biostratigraphic correlation with curve for midcontinent North America. Geology 17:802–805

Boardman DH, Heckel PH (1991) Reply to comment on "glacial-eustatic sea-level curve for early late Pennsylvanian sequence in north-central Texas and biostratigraphic correlation with curve for midcontinent North America". Geology 19:92–94

Boldy SAR, Brealey S (1990) Timing, nature and sedimentary result of Jurassic tectonism in the Outer Moray Firth. In: Hardman RFP, Brooks J (eds) Tectonic events responsible for Britains's oil and gas reserves. Geological Society, London, Spec Publ 55, pp 259–279

Bond G (1976) Evidence for continental subsidence in North America during the late Cretaceous global submergence. Geology 4:557–560

Bond G (1978) Speculations on real sea-level changes and vertical motions of continents at selected times in the Cretaceous and tertiary periods. Geology 6:247–250

Bond G (1979) Evidence for some uplifts of large magnitude in continental platforms. Tectonophysics 61:285–305

Bond GC, Kominz MA (1984) Construction of tectonic subsidence curves for the early Paleozoic miogeocline, southern Canadian Rocky Mountains: implications for subsidence mechanisms, age of breakup, and crustal thinning. Geol Soc Am Bull 95:155–173

Bond GC, Kominz MA (1991a) Disentangling middle Paleozoic sea level and tectonic events in cratonic margins and cratonic basins of North America. J Geophys Res 96B:6619–6639

Bond GC, Kominz MA (1991b) Some comments on the problem of using vertical facies changes to infer accommodation and eustatic sea-level histories with examples from Utah and the southern Canadian Rockies. In: Franseen EK, Watney WL, Kendall CGStC (eds) Sedimentary modeling: computer simulations and methods for improved parameter definition. Kans Geol Surv Bull 233:273–291

Bond GC, Kominz MA, Steckler MS, Grotzinger JP (1989) Role of thermal subsidence, flexure, and eustasy in the evolution of early Paleozoic passive margin cabonate platforms. In: Crevello P, Wilson J, Sarg R, Read F (eds) Controls on evolution of carbonate platforms

and basin development. Soc Econ Paleontol Mineral Spec Publ 44:39–61

Bond GC, Devlin WJ, Kominz M, Beavan J, McManus J (1993) Evidence of astronomical forcing of the earth's climate in Cretaceous and Cambrian times. Tectonophysics 222:295–315

Boreen T, Walker RG (1991) Definition of allomembers and their facies assemblages in the Viking formation, Willesden Green area, Alberta. Bull Can Pet Geol 39:123–144

Borer JM, Harris PM (1991) Lithofacies and cyclicity of the Yates formation, Permian Basin: implications for reservoir heterogeneity. Am Assoc Pet Geol Bull 75:726–779

Bosellini A (1984) Progradation geometries of carbonate platforms: examples from the Triassic of the Dolomites, northern Italy. Sedimentology 31:1–24

Boss SK, Rasmussen KA (1995) Misuse of Fischer plots as sea-level curves. Geology 23:221–224

Boyd R, Sute J, Penland S (1989) Sequence stratigraphy of the Mississippi delta. Trans Gulf Coast Assoc Geol Soc 39:331–340

Bradley WB (1929) The varves and climate of the Green River epoch. U S Geol Surv Prof Pap 158:87–110

Brenchley PJ, Marshall JD, Carden GAF, Robertson DBR, Long DGF, Meidla T, Hints L, Anderson TF (1994) Bathymetric and isotopic evidence for a short-lived late Ordovician glaciation in a greenhouse period. Geology 22:295–298

Brink GJ, Keenan JHG, Brown LF Jr (1993) Deposition of fourth-order, post-rift sequences and sequence sets, lower Cretaceous (lower Valanginian to lower Aptian), Pletmos Basin, southern offshore, South Africa. In: Weimer P, Posamentier HW (eds) Siliciclastic sequence stratigraphy. Am Assoc Pet Geol Mem 58:43–69

Brookfield ME (1977) The origin of bounding surfaces in ancient aeolian sandstones. Sedimentology 24:303–332

Brookfield ME (1992) The paleorivers of central Asia: the interrelationship of Cenozoic tectonism, erosion and sedimentation. 29th International Geological Congress, Kyoto, Japan, Abstracts, vol 2, p 292

Brookfield ME (1993) The interrelations of post-collision tectonism and sedimentation in Central Asia. In: Frostick LE, Steel RJ (eds) Tectonic controls and signatures in sedimentary successions. Int Assoc Sedimentol Spec Publ 20:13–35

Brown LF Jr, Fisher WL (1977) Seismic-stratigraphic interpretation of depositional systems: examples from Brazilian rift and pull-apart basins. In: Payton CE (ed) Seismic stratigraphy – applications to hydrocarbon exploration. Am Assoc Pet Geol Mem 26:213–248

Brunet MF, Le Pichon X (1982) Subsidence of the Paris Basin. J Geophys Res 87:8547–8560

Burbank DW, Raynolds RGH (1984) Sequential late Cenozoic structural disruption of the northern Himalayan foredeep. Nature 311:114–118

Burbank DW, Raynolds RGH (1988) Stratigraphic keys to the timing of thrusting in terrestrial foreland basins: applications to the northwestern Himalaya. In: Kleinspehn KL, Paola C (eds) New perspectives in basin analysis. Springer-Verlag, Berlin Heidelberg New York, pp 331–351

Burbank DW, Raynolds RGH, Johnson GD (1986) Late Cenozoic tectonics and sedimentation in the north-western Himalayan foredeep. II Eastern limb of the Northwest Syntaxis and regional synthesis. In: Allen PA, Homewood P (eds) Foreland basins. Int Assoc Sedimentol Spec Publ 8:293–306

Burbank DW, Beck RA, Raynolds RGH, Hobbs R, Tahirkheli RAK (1988) Thrusting and gravel progradation in foreland basins: a test of post-thrusting gravel dispersal. Geology 16:1143–1146

Burbank DW, Puigdefàbregas C, Muñoz JA (1992) The chronology of the eocene tectonic and stratigraphic development of the eastern Pyrenean foreland basin, northeast Spain. Geol Soc Am Bull 104:1101–1120

Burgess PM, Gurnis M (1995) Mechanisms for the formation of cratonic stratigraphic sequences. Earth Planet Sci Lett 136:647–663

Burton R, Kendall CGStC, Lerche I (1987) Out of our depth: on the impossibility of fathoming eustasy from the stratigraphic record. Earth-Sci Rev 24:237–277

Busch RM, Rollins HB (1984) Correlation of Carboniferous strata using a hierarchy of transgressive-regressive units. Geology 12:471–474

Butcher SW (1990) The nickpoint concept and its implications regarding onlap to the stratigraphic record. In: Cross TA (ed) Quantitative dynamic stratigraphy. Prentice-Hall, Englewood Cliffs, pp 375–385

Butterworth PJ (1991) The role of eustasy in the development of a regional shallowing event in a tectonically active basin. Fossil Bluff Group (Jurassic-Cretaceous), Alexander Island, Antarctica. In: Macdonald DIM (ed) Sedimentation, tectonics and eustasy: sea-level changes at active margins. Int Assoc Sedimentol Spec Pub 12:307–329

Campbell CV (1967) Lamina, laminaset, bed and bedset. Sedimentology 8:7–26

Cant DJ (1979) Storm-dominated shallow marine sediments of the Arisaig Group (Silurian-Devonian) of Nova Scotia. Can J Earth Sci 17:120–131

Cant DJ (1984) Development of shoreline-shelf sand bodies in a Cretaceous epeiric sea deposit: J Sediment Petrol 54:541–556

Cant DJ (1989a) Zuni sequence: the foreland basin Lower Zuni sequence: middle Jurassic to middle Cretaceous. In: Ricketts BD (ed) Western Canada sedimentary basin: a case history. Canadian Society of Petroleum Geologists, Calgary, Alberta, pp 251–267

Cant DJ (1989b) Simple equations of sedimentation: applications to sequence stratigraphy. Basin Res 2:73–81

Cant DJ (1991) Geometric modelling of facies migration: theoretical development of facies successions and local unconformities. Basin Res 3:51–62

Cant DJ (1992) Subsurface facies analysis. In: Walker RG, James NP (eds) Facies models: response to sea level change. Geological Association of Canada, St John's, Newfoundland, pp 27–45

Cant DJ (1995) Sequence stratigraphic analysis of individual depositional successions: effects of marine/nonmarine sediment partitioning and longitudinal sediment transport, Mannville Group, Alberta foreland basin, Canada. Am Assoc Pet Geol Bull 79:749–762

Cant DJ, Stockmal GS (1989) The Alberta foreland basin: relationship between stratigraphy and terrane-accretion events. Can J Earth Sci 26:1964–1975

Caputo MV, Crowell JC (1985) Migration of glacial centers across Gondwana during Paleozoic era. Geol Soc Am Bull 96:1020–1036

Carmichael SMM (1988) Linear estuarine conglomerate bodies formed during a mid-Albian marine transgression; "Upper Gates" formation, Rocky Mountain foothills of northeastern British Columbia. In: James DP, Leckie DA (eds) Sequences, stratigraphy, sedimentology: surface and subsurface. Can Soc Pet Geol Mem 15:49–62

Carter RM, Abbott ST, Fulthorpe CS, Haywick DW, Henderson RA (1991) Application of global sea-level and sequence-stratigraphic models in southern hemisphere Neogene strata from New Zealand. In: Macdonald DIM (ed) Sedimentation, tectonics and eustasy: sea-level changes at active margins. Int Assoc Sedimentol Spec Publ 12:41–65

Cartwright JA, Haddock RC, Pinheiro LM (1993) The lateral extent of sequence boundaries. In: Williams GD, Dobb A (eds) Tectonics and seismic sequence stratigraphy. Geological Society, London, Spec Publ 71, pp 15–34

Cathless LM, Hallam A (1991) Stress-induced changes in plate density, Vail sequences, epeirogeny, and short-lived global sea level fluctuations. Tectonics 10:659–671

Catuneanu O, Sweet AR, Miall AD (in press) Reciprocal architecture of Bearpaw sequences, uppermost Cretaceous, Western Canada Basin. Bulletin of Canadian Petroleum Geology

Cecil CB (1990) Paleoclimate controls on stratigraphic repetition of chemical and siliciclastic rocks. Geology 18:533–536

Chang KH (1975) Unconformity-bounded stratigraphic units. Geol Soc Am Bull 86:1544–1552

Chappell J, Shackleton NJ (1986) Oxygen isotopes and sea level. Nature 324:137–140

Chesnut DR Jr (1994) Eustatic and tectonic control of deposition of the lower and middle Pennsylvanian strata of the central Appalachian Basin. In: Dennison JM, Ettensohn FR (eds) Tectonic and eustatic controls on sedimentary cycles. Soc Sediment Geol, Concepts Sedimentol Paleontol 4:51–64

Chow N, James NP (1987) Cambrian grand cycles: a northern Appalachian perspective. Geol Soc Am Bull 98:418–429

Christie-Blick N (1991) Onlap, offlap, and the origin of unconformity-bounded depositional sequences. Mar Geol 97:35–56

Christie-Blick N, Grotzinger JP, Von der Borch CC (1988) Sequence stratigraphy in Proterozoic successions. Geology 16:100–104

Christie-Blick N, Mountain GS, Miller KG (1990) Seismic stratigraphy: record of sea-level change. In: Revelle R (ed) Sea-level change. National Research Council, Studies in Geophysics. National Academy Press, Washington, pp 116–140

Cloetingh S (1986) Intraplate stresses: a new tectonic mechanism for fluctuations of relative sea-level. Geology 14:617–621

Cloetingh S (1988) Intraplate stresses: a new element in basin analysis. In: Kleinspehn K, Paola C (eds) New perspectives in basin analysis. Springer-Verlag, Berlin Heidelberg New York, pp 205–230

Cloetingh S (ed) (1991) Measurement, causes, and consequences of long-term sea-level changes. J Geophys Res 96B4:6584–6949

Cloetingh S, Kooi H (1990) Intraplate stresses: a new perspective on QDS and Vail's third-order cycles. In: Cross TA (ed) Quantitative dynamic stratigraphy. Prentice-Hall, Englewood Cliffs, pp 127–148

Cloetingh S, McQueen H, Lambeck K (1985) On a tectonic mechanism for regional sea-level variations. Earth Planet Sci Lett 75:157–166

Cloetingh S, Gradstein FM, Kooi H, Grant AC, Kaminski M (1990) Plate reorganization: a cause of rapid late Neogene subsidence and sedimentation around the North Atlantic? J Geol Soc (Lond) 147:495–506

Coleman JM, Gagliano SM (1965) Sedimentary structures, Mississippi River deltaic plain. In: Middleton GV (ed) Primary sedimentary structures and their hydrodynamic interpretation. Soc Econ Paleontol Mineral Spec Publ 12:133–148

Coleman JM, Wright LD (1975) Process framework for describing the morphologic and stratigraphic evolution of deltaic depositional systems. In: Broussard ML (ed) Deltas, models for exploration. Houston Geological Society, Houston, pp 99–149

Collinson JD (ed) (1989) Correlation in hydrocarbon exploration. Graham and Trotman, London, 381 pp

Collinson JD, Holdsworth BK, Jones CM, Martinsen OJ, Read WA (1992) The Millstone Grit (Namurian) of the southern Pennines viewed in the light of eustatically controlled sequence stratigraphy; discussion and reply. Geol J 27:173–180

Coniglio M, Dix GR (1992) Carbonate slopes. In: Walker RG, James NP (eds) Facies models: response to sea-level change. Geological Association of Canada, St John's, Newfoundland, pp 349–373

Cope JCW (1993) High resolution biostratigraphy. In: Hailwood EA, Kidd RB (eds) High resolution stratigraphy. Geological Society, London, Spec Publ 70, pp 257–265

Cotillon P (1987) Bed-scale cyclicity of pelagic Cretaceous successions as a result of world-wide control. Mar Geol 78:109–123

Cowan CA, James NP (1993) The interactions of sea-level change, terrigenous-sediment influx, and carbonate productivity as controls on Upper Cambrian grand cycles of western Newfoundland, Canada. Geol Soc Am Bull 105:1576–1590

Cox BM (1990) A review of Jurassic chronostratigraphy and age indicators for the UK. In: Hardman RFP, Brooks J (eds) Tectonic events responsible for Britain's oil and gas reserves. Geological Society, London, Spec Publ 55, pp 169–190

Crampton SL, Allen PA (1995) Recognition of forebulge unconformities associated with early stage foreland basin development: example from the North Alpine Foreland Basin. Am Assoc Pet Geol Bull 79:1495–1514

Croll J (1864) On the physical cause of the change of climate during geological epochs. Philos Mag 28:435–436

Cross TA (1986) Tectonic controls of foreland basin subsidence and Laramide style deformation, western United States. In: Allen PA, Homewood P (eds) Foreland basins. Int Assoc Sedimentol Spec Publ 8:5–39

Cross TA (1988) Controls on coal distribution in transgressive-regressive cycles, Upper Cretaceous, Western Interior, USA. In: Wilgus CK, Hastings BS, Kendall CGStC, Posamentier HW, Ross CA, Van Wagoner JC (eds) Sea-level changes: an integrated approach. Soc Econ Paleontol Mineral Spec Publ 42:371–380

Cross TA (ed) (1990) Quantitative dynamic stratigraphy. Prentice-Hall, Englewood Cliffs, 625 pp

Cross TA, Lessenger MA (1988) Seismic stratigraphy. Annu Rev Earth Planet Sci 16:319–354

Cross TA, Baker MR, Chapin MA, Clark MS, Gardner MH, Hanson MS, Lessenger MA, Little LD, McDonough K-J, Sonnenfeld MD, Valasek DW, Williams MR, Witter DN (1993) Applications of high-resolution sequence stratigraphy to reservoir analysis. In: Eschard R, Doligez B (eds) Subsurface reservoir characterization from outcrop observations. Institut Français du Petrole, Éditions Technip, Paris, pp 11–33

Crough ST, Jurdy DM (1980) Subducted lithosphere, hotspots, and the geoid. Earth Planet Sci Lett 48:15–22

Crowell JC (1978) Gondwanan glaciation, cyclothems, continental positioning, and climate change. Am J Sci 278:1345–1372

Crowley KD (1984) Filtering of depositional events and the completeness of sedimentary sequences. J Sediment Petrol 54:127–136

Crowley TJ, Baum SK (1991) Estimating Carboniferous sea-level fluctuations from Gondwana ice extent. Geology 19:975–977

Crumeyrolle P, Rubino J-L, Clauzon G (1991) Miocene depositional sequences within a tectonically controlled transgressive-regressive cycle. In: Macdonald DIM (ed) Sedimentation, tectonics and eustasy: sea-level changes at active margins. Int Assoc Sedimentol Spec Publ 12:373–390

Curray JR (1964) Transgressions and regressions. In: Miller RL (ed) Papers in marine geology, Shepard commemorative volume. MacMillan Press, New York, pp 175–203

Dahlstrom CDA (1970) Structural geology in the eastern margin of the Canadian Rocky Mountains. Bull Can Pet Geol 18:332–406

Dalziel IWD (1991) Pacific margins of Laurentia and East Antarctica-Australia as a conjugate rift pair: evidence and implications for an EoCambrian supercontinent. Geology 19:598–601

Dalrymple RW (1992) Tidal depositional systems. In: Walker RG, James NP (eds) Facies models: response to sea-level change. Geological Association of Canada, St John's, Newfoundland, pp 195–218

Dalrymple RW, Zaitlin BA (1994) High-resolution sequence stratigraphy of a complex, incised valley succession, Cobequid Bay-Salmon River estuary, Bay of Fundy, Canada. Sedimentology 41:1069–1091

Dalrymple RW, Boyd R, Zaitlin BA (eds) (1994) Incised-valley systems: origin and sedimentary sequences. Soc Sediment Geol Spec Publ 51:391

de Boer PL, Smith DG (eds) (1994a) Orbital forcing and cyclic sequences. Int Assoc Sedimentol Spec Publ 19:559

de Boer PL, Smith DG (eds) (1994b) Orbital forcing and cyclic sequences. Int Assoc Sedimentol Spec Publ 19:1–14

DeCelles PG, Lawton TF, Mitra G (1995) Thrust timing, growth of structural culminations, and synorogenic sedimentation in the type Sevier orogenic belt, western United States. Geology 23:699–702

Demarest JM II, Kraft JC (1987) Stratigraphic record of Quaternary sea levels: implications for more ancient strata. In: Nummedal D, Pilkey OH, Howard JD (eds) Sea-level fluctuation and coastal evolution. Soc Econ Paleontol Mineral Spec Publ 41:223–239

Dennison JM, Ettensohn FR (eds) (1994) Tectonic and eustatic controls on sedimentary cycles. Soc Sediment Geol, Concepts Sedimentol Paleontol 4:264

Deramond J, Souquet P, Fondecave-Wallez M-J, Specht M (1993) Relationships between thrust tectonics and sequence stratigraphy surfaces in foredeeps: model and examples from the Pyrenees (Cretaceous-Eocene, France, Spain). In: Williams GD, Dobb A (eds) Tectonics and seismic sequence stratigraphy. Geological Society, London, Spec Publ 71, pp 193–219

Devine PE (1991) Transgressive origin of channeled estuarine deposits in the Point Lookout Sandstone, northwestern New Mexico: a model for upper Cretaceous, cyclic regressive parasequences of the U S Western Interior. Am Assoc Pet Geol Bull 75:1039–1063

De Visser JP, Ebbing JHJ, Gudjonsson L, Hilgen FJ, Jorissen FJ, Verhallen PJJM, Zevenboom D (1989) The origin of rhythmic bedding in the Pliocene Trubi Formation of Sicily, southern Italy. Palaeogeogr Palaeoclimatol Palaeoecol 69:45–66

Dewey JF (1980) Episodicity, sequence, and style at convergent plate boundaries. In: Strangway DW (ed) The continental crust and its mineral deposits. Geol Assoc Can Spec Pap 20:553–573

Dewey JF (1982) Plate tectonics and the evolution of the British Isles. J Geol Soc (Lond) 139:371–414

Dickinson WR (1993) The Exxon global cycle chart: an event for every occasion? Discussion. Geology 21:282–283

Dickinson WR, Soreghan GS, Giles KA (1994) Glacioeustatic origin of Permo-Carboniferous stratigraphic cycles: evidence from the southern Cordilleran foreland region. In: Dennison JM, Ettensohn FR (eds) Tectonic and eustatic controls on sedimentary cycles. Soc Sediment Geol, Concepts Sediment Paleontol 4:25–34

Dixon J (1993) Regional unconformities in the Cretaceous of north-west Canada. Cretac Res 14:17–38

Dixon J, Dietrich JR (1988) The nature of depositional and seismic sequence boundaries in Cretaceous-Tertiary strata of the Beaufort-Mackenzie basin. In: James DP, Leckie DA (eds) Sequences, stratigraphy, sedimentology: surface and subsurface. Canadian Society of Petroleum Geologists Memoir 15, Calgary, Alberta, pp 63–72

Dixon J, Dietrich JR (1990) Canadian Beaufort Sea and adjacent land areas. In: Grantz A, Johnson L, Sweeney JF (eds) The Arctic Ocean region. The geology of North America, vol L. Geological Society of America, Boulder, pp 239–256

Dockal JA, Worsley TR (1991) Modeling sea level changes as the Atlantic replaces the Pacific: submergent versus emergent observers. J Geophys Res 96B:6805–6810

Dolan JF (1989) Eustatic and tectonic controls on deposition of hybrid siliciclastic/carbonate basinal cycles. Am Assoc Pet Geol Bull 73:1233–1246

Dominguez JML, Bittencourt ACSP, Martin L (1992) Controls on Quaternary coastal evolution of the east-northeastern coast of Brazil: roles of sea-level history, trade winds and climate. Sediment Geol 80:213–232

Donovan AD, Baum GR, Blechschmidt GL, Loutit TS, Pflum CE, Vail PR (1988) Sequence stratigraphic setting of the Cretaceous-tertiary boundary in central Alabama. In: Wilgus CK, Hastings BS, Kendall CGStC, Posamentier HW, Ross CA, Van Wagoner JC (eds) Sea-level changes: an integrated approach. Soc Econ Paleontol Mineral Spec Publ 42:299–307

Donovan DT, Jones EJW (1979) Causes of world-wide changes in sea level. J Geol Soc (Lond) 136:187–192

Donovan RN (1975) Devonian lacustrine limestones at the margin of the Orcadian Basin, Scotland. J Geol Soc (Lond) 131:489–510

Donovan RN (1978) The Middle Old Red Sandstone of the Orcadian Basin. In: Friend PF (ed) The Devonian of Scotland: international symposium on the Devonian system, field guide. Palaeontological Association, London, pp 37–53

Donovan RN (1980) Lacustrine cycles, fish ecology, and stratigraphic zonation in the middle Devonian of Caithness. Scott J Geol 16:35–50

Dott RH Jr (1992) An introduction to the ups and downs of eustasy. In: Dott RH Jr (ed) Eustasy: the historical ups and downs of a major geological concept. Geol Soc Am Mem 180:1–16

Dott RH Jr, Bourgeois J (1982) Hummocky stratification: significance of its variable bedding sequences. Geol Soc Am Bull 93:663–680

Douglas RJW, Gabrielse H, Wheeler JO, Stott DF, Belyea HR (1970) Geology of Western Canada. In: Douglas RJW (ed) Geology and economic minerals of Canada. Geol Surv Can Econ Geol Rep 1:489–488

Doyle JA (1977) Spores and pollen: the Potomac group (Cretaceous) angiosperm sequence. In: Kauffman EG, Hazel JE (eds) Concepts and methods of biostratigraphy. Dowden, Hutchinson and Ross, Stroudsburg, Pennsylvania, pp 339–364

Driscoll NW (1992) Stratigraphic response of a carbonate platform to relative sea level changes: Broken Ridge, southeast Indian Ocean. Reply. Am Assoc Pet Geol Bull 76:1037–1041

Drummond CN, Wilkinson BH (1993a) On the use of cycle thickness diagrams as records of long-term sealevel change during accumulation of carbonate sequences. J Geol 101:687–702

Drummond CN, Wilkinson BH (1993b) Carbonate cycle stacking patterns and hierarchies of orbitally forced eustatic sealevel change. J Sediment Petrol 63:369–377

Drummond CN, Wilkinson BH (1996) Stratal thickness frequencies and the prevalence of orderedness in stratigraphic sequences. J Geol 104:1–18

Duff BA, Grollman NG, Mason DJ, Questiaux JM, Ormerod DS, Lays P (1991) Tectonostratigraphic evolution of the south-east Gippsland Basin. Aust Pet Expl Assoc J 31:116–130

Duff PMD, Hallam A, Walton EK (1967) Cyclic sedimentation: developments in sedimentology, vol 10. Elsevier, Amsterdam, 280 pp

Eberli G, Ginsburg RN (1988) Aggrading and prograding infill of buried Cenozoic seaways, northwestern Great Bahama Bank. In: Bally AW (ed) Atlas of seismic stratigraphy. Am Assoc Pet Geol Stud Geol 27(2):97–103

Eberli G, Ginsburg RN (1989) Cenozoic progradation of northwestern Great Bahama Bank, a record of lateral platform growth and sea-level fluctuations. In: Crevello PD, Wilson JL, Sarg JF, Read JF (eds) Controls on carbonate platform and basin development. Soc Econ Paleontol Mineral Spec Publ 44:339–351

Edwards LE (1984) Insights on why graphic correlation (Shaw's method) works. J Geol 92:583–597

Edwards LE (1985) Insights on why graphic correlation (Shaw's method) works: a reply (to discussion). J Geol 93:507–509

Edwards LE (1989) Supplemented graphic correlation: a powerful tool for paleontologists and nonpaleontologists. Palaios 4:127–143

Edwards MB (1981) Upper Wilcox Rosita delta system of South Texas: growth-faulted shelf-edge deltas. Am Assoc Pet Geol Bull 65:54–73

Einsele G, Ricken W (1991) Larger cycles and sequences: introductory remarks. In: Einsele G, Ricken W, Seilacher A (eds) Cycles and events in stratigraphy. Springer-Verlag, Berlin Heidelberg New York, pp 611–616

Einsele G, Seilacher A (eds) (1982) Cyclic and event stratification. Springer-Verlag, Berlin Heidelberg New York, 536 pp

Einsele G, Ricken W, Seilacher A (eds) (1991a) Cycles and events in stratigraphy. Springer-Verlag, Berlin Heidelberg New York, 955 pp

Einsele G, Ricken W, Seilacher A (1991b) Cycles and events in stratigraphy – basic concepts and terms. In: Einsele G, Ricken W, Seilacher A (eds) Cycles and events in stratigraphy. Springer-Verlag, Berlin Heidelberg New York, pp 1–19

Elder WP, Gustason ER, Sageman BB (1994) Correlation of basinal carbonate cycles to nearshore parasequences in the late Cretaceous Greenhorn seaway, Western Interior, USA. Geol Soc Am Bull 106:892–902

Elrick M, Read JF (1991) Cyclic ramp-to-basin carbonate deposits, lower Mississippian, Wyoming and Montana: a combined field and computer modeling study. J Sediment Petrol 61:1194–1224

Embry AF (1988) Triassic sea-level changes: evidence from the Canadian Arctic Archipelago. In: Wilgus CK, Hastings BS, Kendall CGStC, Posamentier HW, Ross CA, Van Wagoner JC (eds) Sea-level research: an integrated approach. Soc Econ Paleontol Mineral Spec Publ 42:249–259

Embry AF (1990) A tectonic origin for third-order depositional sequences in extensional basins – implications for basin modelling. In: Cross TA (ed) Quantitative dynamic stratigraphy. Prentice-Hall, Englewood Cliffs, pp 491–501

Embry AF (1991) Mesozoic history of the Arctic Islands. In: Trettin HP (ed) Geology of the Innuitian Orogen and Arctic platform of Canada and Greenland. Geol Surv Can, Geol Can 3:369–433

Embry AF (1993) Transgressive-regressive (T-R) sequence analysis of the Jurassic succession of the Sverdrup Basin, Canadian Arctic Archipelago. Can J Earth Sci 30:301–320

Embry AF (1995) Sequence boundaries and sequence hierarchies: problems and proposals. In: Steel RJ, Felt VL, Johannessen EP, Mathieu C (eds) Sequence stratigraphy on the Northwest European margin. Norsk Petroleumsforening Spec Publ 5, Elsevier, Amsterdam, pp 1–11

Emiliani C (1955) Pleistocene temperatures. J Geol 63:538–578

Engel AEG, Engel CB (1964) Continental accretion and the evolution of North America. In: Subramaniam AP, Balakrishna S (eds) Advancing frontiers in geology and geophysics. Indian Geophysical Union, Hyderabad, pp 17–37

Epting M (1989) Miocene carbonate buildups of central Luconia, offshore Sarawak. In: Bally AW (ed) Atlas of seismic stratigraphy. Am Assoc Pet Geol Stud Geol 27(3):168–173

Erba E, Castradori D, Guasti G, Ripepe M (1992) Calcareous nannofossils and Milankovitch cycles: the example of the Albian Gault clay formation (southern England). Palaeogeogr Palaeoclimatol Palaeoecol 93:47–69

Erlich RN, Longo AP Jr, Hyare S (1993) Response of carbonate platform margins to drowning: evidence of environmental collapse. In: Loucks RG, Sarg JF (eds) Carbonate sequence stratigraphy. Am Assoc Pet Geol Mem 57:241–266

Eschard R, Doligez B (eds) (1993) Subsurface reservoir characterization from outcrop observations. Institut Français du Petrole, Éditions Technip, Paris, 189 pp

Ettensohn FR (1994) Tectonic control on formation and cyclicity of major Appalachian unconformities and associated stratigraphic sequences. In: Dennison JM, Ettensohn FR (eds) Tectonic and eustatic controls on sedimentary cycles. Soc Sediment Geol, Concepts Sedimentol Paleontol 4:217–242

Eugster HP, Hardie LA (1975) Sedimentation in an ancient playa-lake complex. The Wilkins Peak member of the Green River formation of Wyoming. Geol Soc Am Bull 86:319–334

Eyles N (1993) Earth's glacial record and its tectonic setting. Earth-Sci Rev 35:1–248

Fairbridge RW (1961) Eustatic changes in sea level. In: Ahrens LH, Press F, Rankama K, Runcorn SK (eds) Physics and chemistry of the Earth. Pergamon Press, London, pp 99–185

Feeley MH, Moore TC Jr, Loutit TS, Bryant WR (1990) Sequence stratigraphy of Mississippi fan related to oxygen isotope sea level index. Am Assoc Pet Geo Bull 74:407–424

Ferm JC (1975) Pennsylvanian cyclothems of the Appalachian plateau, a retrospective view. US Geol Surv Prof Pap 853:57–64

Filer JK (1994) High frequency eustatic and siliciclastic sedimentation cycles in a foreland basin, Upper

Devonian, Appalachian Basin. In: Dennison JM, Ettensohn FR (eds) Tectonic and eustatic controls on sedimentary cycles. Soc Sediment Geo Concepts Sedimentol Paleontol 4:133–145

Fischer AG (1964) The Lofer cyclothems of the Alpine Triassic. Kans Geol Surv Bull 169:107–149

Fischer AG (1981) Climatic oscillations in the biosphere. In: Nitecki MH (ed) Biotic crises in ecological and evolutionary time. Academic Press, New York, pp 102–131

Fischer AG (1984) The two Phanerozoic supercycles. In: Berggren WA, Van Couvering J (eds) Catastrophes and earth history. Princeton University Press, Princeton, pp 129–150

Fischer AG (1986) Climatic rhythms recorded in strata. Annu Rev Earth Planet Sci 14:351–376

Fischer AG, Arthur MA (1977) Secular variations in the pelagic realm. In: Cook HE, Enos P (eds) Deep water carbonate environments. Soc Econ Paleontol Mineral Spec Publ 25:19–50

Fischer AG, Bottjer DJ (1991) Orbital forcing and sedimentary sequences (introduction to special issue). J Sediment Petrol 61:1063–1069

Fischer AG, Roberts LT (1991) Cyclicity in the Green River formation (lacustrine Eocene) of Wyoming. J Sediment Petrol 61:1146–1154

Fischer AG, de Boer PL, Premoli Silva I (1990) Cyclostratigraphy. In: Ginsburg RN, Beaudoin B (eds) Cretaceous resources, events and rhythms: background and plans for research. Kluwer Academic Publishers, Dordrecht, pp 139–172

Fisher WL, McGowen JH (1967) Depositional systems in the Wilcox group of Texas and their relationship to occurrence of oil and gas. Trans Gulf Coast Assoc Geol Soc 17:105–125

Fisk HN (1939) Depositional terrace slopes in Louisiana. J Geomorphol 2:181–200

Fisk HN (1944) Geological investigations of the alluvial valley of the lower Mississippi River. U S Army Corps of Engineers, Mississippi River Commission, Vicksburg, Mississippi, 78 pp

Fjeldskaar W (1989) Rapid eustatic change – never globally uniform. In: Collinson JD (ed) Correlation in hydrocarbon exploration. Graham and Trotman, London, pp 13–19

Flemings PB, Jordan TE (1989) A synthetic stratigraphic model of foreland basin development. J Geophys Res 94:3851–3866

Flint SS, Bryant ID (eds) (1993) The geological modelling of hydrocarbon reservoirs and outcrop analogues. Int Assoc Sedimentol Spec Publ 15:269

Flint SS, Turner P, Jolley EJ (1991) Depositional architecture of Quaternary fan-delta deposits of the Andean fore-arc: relative sea-level changes as a response to aseismic ridge subduction. In: Macdonald DIM (ed) Sedimentation, tectonics and eustasy: sea-level changes at active margins. Int Assoc Sedimentol Spec Publ 12:91–103

Fortuin AR, de Smet MEM (1991) Rates and magnitudes of late Cenozoic vertical movements in the Indonesian Banda Arc and the distinction of eustatic effects. In: Macdonald DIM (ed) Sedimentation, tectonics and eustasy: sea-level changes at active margins. Int Assoc Sedimentol Spec Publ 12:79–89

Fouch TD, Lawton TF, Nichols DJ, Cashion WB, Cobban WA (1983) Patterns and timing of synorogenic sedimentation in upper Cretaceous rocks of central and northeast Utah. In: Reynolds M, Dolly E (eds) Mesozoic paleogeography of west-central United States. Soc Econ Paleontol Mineral, Rocky Mountain Section, Symposium, 2:305–334

Frakes LA (1979) Climates throughout geologic time. Elsevier, Amsterdam, 310 pp

Frakes LA (1986) Mesozoic-Cenozoic climatic history and causes of the glaciation. In: Hsü KJ (ed) Mesozoic and Cenozoic oceans. Geodynamics series. American Geophysical Union, Washington, vol 15, pp 33–48

Frakes LA, Francis JE (1988) A guide to Phanerozoic cold polar climates from high-latitude ice-rafting in the Cretaceous. Nature 333:547–549

Francis JE, Frakes LA (1993) Cretaceous climates. Sedimentol Rev 1:17–30

Franseen EK, Watney WL, Kendall CGStC (eds) (1991) Sedimentary modeling: computer simulations and methods for improved parameter definition. Kans Geol Surv Bull 233:524

Franseen EK, Goldstein RH, Whitesell TE (1993) Sequence stratigraphy of Miocene carbonate complexes, Las Negras area, southeastern Spain: implications for quantification of changes in relative sea level. In: Loucks RG, Sarg JF (eds) Carbonate sequence stratigraphy. Am Assoc Pet Geol Mem 57:409–434

Frazier DE (1974) Depositional episodes: their relationship to the Quaternary stratigraphic framework in the northwestern portion of the Gulf Basin. Bureau of Economic Geology, University of Texas, Geological Circular 74-1, 26 pp

Fulford MM, Busby CJ (1993) Tectonic controls on non-marine sedimentation in a Cretaceous fore-arc basin, Baja California, Mexico. In: Frostick LE, Steel RJ (eds) Tectonic controls and signatures in sedimentary successions. Int Assoc Sedimentol Spec Publ 20:301–333

Fulthorpe CS (1991) Geological controls on seismic sequence resolution. Geology 19:61–65

Fulthorpe CS, Carter RM (1989) Test of seismic sequence methodology on a southern hemisphere passive margin: the Canterbury Basin, New Zealand. Mar Pet Geol 6:348–359

Funnell BM (1981) Mechanisms of autocorrelation. J Geol Soc (Lond) 138:177–182

Gallagher K, Dumitru TA, Gleadow AJW (1994) Constraints on the vertical motion of eastern Australia during the Mesozoic. Basin Res 6:77–94

Galloway WE (1975) Process framework for describing the morphologic and stratigraphic evolution of the deltaic depositional systems. In: Broussard ML (ed) Deltas, models for exploration. Houston Geological Society, Houston, pp 87–98

Galloway WE (1989a) Genetic stratigraphic sequences in basin analysis I: architecture and genesis of flooding-surface bounded depositional units. Am Assoc Pet Geol Bull 73:125–142

Galloway WE (1989b) Genetic stratigraphic sequences in basin analysis II: application to northwest Gulf of Mexico Cenozoic basin. Am Assoc Pet Geol Bull 73:143–154

Galloway WE, Brown LF Jr (1973) Depositional systems and shelf-slope relations on cratonic basin margin, uppermost Pennsylvanian of north-central Texas. Am Assoc Pet Geol Bull 57:1185–1218

George TN (1978) Eustasy and tectonics: sedimentary rhythms and stratigraphical units in British Dinantian correlation. Proc Yorkshire Geol Soc 42:229–253

Gibling MR, Bird DJ (1994) Late Carboniferous cyclothems and alluvial paleovalleys in the Sydney Basin, Nova Scotia. Geol Soc Am Bull 106:105–117

Gilbert GK (1890) Lake Bonneville. U S Geological Survey Monograph 1, 438 pp

Gilbert GK (1895) Sedimentary measurement of geologic time. J Geol 3:121–127

Ginsburg RN, Beaudoin B (eds) (1990) Cretaceous resources, events and rhythms. NATO ASI Series C, vol 304. Kluwer Academic Publishers, Dordrecht, 352 pp

Glennie KW (ed) (1984) Introduction to the petroleum geology of the North Sea. Blackwell Scientific Publications, Oxford, 236 pp

Goldhammer RK, Harris MT (1989) Eustatic controls on the stratigraphy and geometry of the Latemar buildup (middle Triassic), the Dolomites of northern Italy. In: Crevello PD, Wilson JL, Sarg JF, Read JF (eds) Controls on carbonate platform and basin development. Soc Econ Paleontol Mineral Spec Publ 44:323–338

Goldhammer RK, Dunn PA, Hardie LA (1987) High-frequency glacio-eustatic sea level oscillations with Milankovitch characteristics recorded in middle Triassic platform carbonates in northern Italy. Am J Sci 287:853–892

Goldhammer RK, Dunn PA, Hardie LA (1990) Depositional cycles, composite sea-level changes, cycle stacking patterns, and the hierarchy of stratigraphic forcing: examples from Alpine Triassic platform carbonates. Geol Soc Am Bull 102:535–562

Goldhammer RK, Harris MT, Dunn PA, Hardie LA (1993) Sequence stratigraphy and systems tract development of the Latemar platform, middle Triassic of the Dolomites (northern Italy): outcrop calibration keyed by cycle stacking patterns. In: Loucks RG, Sarg JF (eds) Carbonate sequence stratigraphy. Am Assoc Pet Geol Mem 57:353–387

Goldstein RH, Franseen EK (1995) Pinning points: a method providing quantitative constraints on relative sea-level history. Sediment Geol 95:1–10

Goodwin PW, Anderson EJ (1985) Punctuated aggradational cycles: a general hypothesis of episodic stratigraphic accumulation. J Geol 93:515–533

Goodwin PW, Anderson EJ, Goodman WM, Saraka LJ (1986) Punctuated aggradational cycles: implications for stratigraphic analysis. Paleoceanography 1:417–429

Gould HR (1970) The Mississippi delta complex. In: Morgan JP (ed) Deltaic sedimentation modern and ancient. Soc Econ Paleontol Mineral Spec Publ 15:3–30

Gradstein FM, Agterberg FP, Aubry M-P, Berggren WA, Flynn JJ, Hewitt R, Kent DV, Klitgord KD, Miller KG, Obradovitch J, Ogg JG (1988) Sea level history. Science 241:599–601

Gradstein FM, Agterberg FP, D'Iorio MA (1990) Time in quantitative stratigraphy. In: Cross TA (ed) Quantitative dynamic stratigraphy. Prentice Hall, Englewood Cliffs, pp 519–542

Gradstein FM, Agterberg FP, Ogg JG, Hardenbol J, Van Veen P, Thierry J, Zehui Huang (1995) A Triassic, Jurassic and Cretaceous time scale. In: Berggren WA, Kent DV, Aubry M-P, Hardenbol J (eds) Geochronology, time scales and global stratigraphic correlation. Soc Sediment Geol Spec Publ 54:95–126

Grammer GM, Ginsburg RN, Harris PM (1993) Timing of deposition, diagenesis, and failure of steep carbonate slopes in response to a high-amplitude/high-frequency fluctuation in sea level, Tongue of the Ocean, Bahamas. In: Loucks RG, Sarg JF (eds) Carbonate sequence stratigraphy. Am Assoc Pet Geol Mem 57:107–131

Gray J, Boucot AJ (eds) (1979) Historical biogeography, plate tectonics, and the changing environment. Oregon State University Press, Corvallis, Oregon, 500 pp

Greenlee SM, Moore TC (1988) Recognition and interpretation of depositional sequences and calculation of sea-level changes from stratigraphic data – offshore New Jersey and Alabama tertiary. In: Wilgus CK, Hastings BS, Kendall CGStC, Posamentier HW, Ross CA, Van Wagoner JC (eds) Sea-level changes: an integrated approach. Soc Econ Paleontol Mineral Spec Publ 42:329–353

Guex J (1991) Biochronological correlations. Springer-Verlag, Berlin Heidelberg New York, 252 pp

Guidish TM, Lerche I, Kendall CGStC, O'Brien JJ (1984) Relationship between eustatic sea level changes and basement subsidence. Am Assoc Pet Geol Bull 68:164–177

Gurnis M (1988) Large-scale mantle convection and the aggregation and dispersal of supercontinents. Nature 332:695–699

Gurnis M (1990) Bounds on global dynamic topography from Phanerozoic flooding of continental platforms. Nature 344:754–756

Gurnis M (1992) Long-term controls on eustatic and epeirogenic motions by mantle convection. GSA Today 2:141–157

Hallam A (1963) Major epeirogenic and eustatic changes since the Cretaceous and their possible relationship to crustal structure. Am J Sci 261:397–423

Hallam A (1975) Jurassic environments. Cambridge University Press, London, 269 pp

Hallam A (1978) Eustatic cycles in the Jurassic. Palaeogeogr Palaeoclimatol Palaeoecol 23:1–32

Hallam A (1981) A revised sea-level curve for the early Jurassic. J Geol Soc (Lond) 138:735–743

Hallam A (1984) Pre-Quaternary sea-level changes. Annu Rev Earth Planet Sci 12:205–243

Hallam A (1986) Origin of minor limestone-shale cycles: climatically induced or diagenetic? Geology 14:609–612

Hallam A (1988) A reevaluation of Jurassic eustasy in the light of new data and the revised Exxon curve. In: Wilgus CK, Hastings BS, Kendall CGStC, Posamentier HW, Ross CA, Van Wagoner JC (eds) Sea-level changes: an integrated approach. Soc Econ Paleontol Mineral Spec Publ 42:261–273

Hallam A (1991) Relative importance of regional tectonics and eustasy for the Mesozoic of the Andes. In: Macdonald DIM (ed) Sedimentation, tectonics and eustasy: sea-level changes at active margins. Int Assoc Sedimentologists, Spec Publ 12, pp 189–200

Hallam A (1992) Phanerozoic sea-level changes. Columbia University Press, Irvington, New York, 224 pp

Hambrey MJ, Harland WB (eds) (1981) Earth's pre-Pleistocene glacial record. Cambridge University Press, Cambridge, 1004 pp

Hancock JM (1977) The historic development of biostratigraphic correlation. In: Kauffman EG, Hazel JE (eds) Concepts and methods of biostratigraphy. Dowden, Hutchinson and Ross, Stroudsburg, pp 3–22

Hancock JM (1984) Cretaceous. In: Glennie KW (ed) Introduction to the petroleum geology of the North Sea. Blackwell Scientific Publications, Oxford, pp 133–150

Hancock JM (1993a) Comments on the EXXON cycle chart for the Cretaceous system: Madrid. Cuadernos de Geología Ibérica, no 17, pp 3–24

Hancock JM (1993b) Transatlantic correlations in the Campanian-Maastrichtian stages by eustatic changes of sea-level. In: Hailwood EA, Kidd RB (eds) High resolution stratigraphy. Geological Society, London, Spec Publ 70, pp 241–256

Hancock JM, Kauffman EG (1979) The great transgressions of the late Cretaceous. J Geol Soc (Lond) 136:175–186

Haq BU (1991) Sequence stratigraphy, sea-level change, and significance for the deep sea. In: Macdonald DIM (ed) Sedimentation, tectonics and eustasy: sea-level changes at active margins. Int Assoc Sedimentol Spec Publ 12:3–39

Haq BU, Hardenbol J, Vail PR (1987) Chronology of fluctuating sea levels since the Triassic (250 million years ago to present). Science 235:1156–1167

Haq BU, Hardenbol J, Vail PR (1988a) Mesozoic and Cenozoic chronostratigraphy and cycles of sea-level change. In: Wilgus CK, Hastings BS, Kendall CGStC, Posamentier HW, Ross CA, Van Wagoner JC (eds) Sea-level changes: an integrated approach. Soc Econ Paleontol Mineral Spec Publ 42:71–108

Haq BU, Vail PR, Hardenbol J, Van Wagoner JC (1988b) Sea level history. Science 241:596–602

Hardenbol J, Vail PR, Ferrer J (1981) Interpreting paleoenvironments, subsidence history and sea-level changes of passive margins from seismic and biostratigraphy. Oceanol Acta Suppl 3:33–44

Hardman RFP, Brooks J (eds) (1990) Tectonic events responsible for Britain's oil and gas reserves. Geological Society, London, Spec Publ 55, 404 pp

Harland WB (1978) Geochronologic scales. In: Cohee GV, Glaessner MF, Hedberg HD (eds) Contributions to the geologic time scale. Am Assoc Pet Geol Stud Geol 6:9–32

Harland WB, Cox AV, Llewellyn PG, Pickton CAG, Smith AG, Walters R (1982) A geologic time scale. Cambridge Earth Science Series. Cambridge University Press, Cambridge, 131 pp

Harland WB, Armstrong RL, Cox AV, Craig LE, Smith AG, Smith DG (1990) A geologic time scale, 1989. Cambridge Earth Science Series. Cambridge University Press, Cambridge, 263 pp

Harper CW Jr, Crowley KD (1985) Insights on why graphic correlation (Shaw's method) works: a discussion. J Geol 93:503–506

Harris PM, Frost SH, Seiglie GA, Schneidermann N (1984) Regional unconformities and depositional cycles, Cretaceous of the Arabian Peninsula. In: Schlee JS (ed) Interregional unconformities and hydrocarbon accumulation. Am Assoc Pet Geol Mem 36:67–80

Harrison CGA (1990) Long-term eustasy and epeirogeny in continents. In: Revelle R (ed) Sea-level change. National Research Council, Studies in Geophysics. National Academy Press, Washington, pp 141–158

Hart BS, Plint AG (1993) Tectonic influence on deposition in a ramp setting: upper Cretaceous Cardium formation, Alberta foreland basin. Am Assoc Pet Geol Bull 77:2092–2107

Hartley RW, Allen PA (1994) Interior cratonic basins of Africa: relation to continental break-up and role of mantle convection. Basin Res 6:95–113

Hay WW, Leslie MA (1990) Could possible changes in global groundwater reservoir cause eustatic sea-level fluctuations? In: Revelle R (ed) Sea-level change. National Research Council, Studies in Geophysics. National Academy Press, Washington, pp 161–170

Hay WW, Southam JR (1978) Quantifying biostratigraphic correlation. Annu Rev Earth Planet Sci 6:353–375

Hays JD, Pitman WC III (1973) Lithospheric plate motion, sea level changes and climatic and ecological consequences. Nature 246:18–22

Hays JD, Imbrie J, Shackleton NJ (1976) Variations in the earth's orbit: pacemaker of the ice ages. Science 194:1121–1132

Heckel PH (1986) Sea-level curve for Pennsylvanian eustatic marine transgressive-regressive depositional cycles along midcontinent outcrop belt, North America. Geology 14:330–334

Heckel PH (1990) Evidence for global (glacio-eustatic) control over Upper Carboniferous (Pennsylvanian) cyclothems in midcontinent North America. In: Hardman RFP, Brooks J (eds) Tectonic events responsible for Britains's oil and gas reserves. Geological Society, London, Spec Publ 55, pp 35–47

Heckel PH (1994) Evaluation of evidence for glacio-eustatic control over marine Pennsylvanian cyclothems in North America and consideration of possible tectonic effects. In: Dennison JM, Ettensohn FR (eds) Tectonic and eustatic controls on sedimentary cycles. Soc Sediment Geol, Concepts Sedimentol Paleontol 4:65–87

Hedberg HD (ed) (1976) International stratigraphic guide. John Wiley, New York, 200 pp

Helland-Hansen W, Gjelberg JG (1994) Conceptual basis and variability in sequence stratigraphy: a different perspective. Sediment Geol 92:1–52

Helland-Hansen W, Kendall CGStC, Lerche I, Nakayama K (1988) A simulation of continental basin margin sedimentation in response to crustal movements, eustatic sea level change, and sediment accumulation rates. Math Geol 20:777–802

Heller PL, Angevine CL (1985) Sea-level cycles during the growth of Atlantic-type oceans. Earth Planet Sci Lett 75:417–426

Heller PL, Paola C (1992) The large-scale dynamics of grain-size variation in alluvial basins, 2: application to syntectonic conglomerate. Basin Res 4:91–102

Heller PL, Angevine CL, Winslow NS, Paola C (1988) Two-phase stratigraphic model of foreland-basin sequences. Geology 16:501–504

Heller PL, Beekman F, Angevine CL, Cloetingh SAPL (1993) Cause of tectonic reactivation and subtle uplifts in the Rocky Mountain region and its effect on the stratigraphic record. Geology 21:1003–1006

Hilgen FJ (1991) Extension of the astronomically calibrated (polarity) time scale to the Miocene/Pliocene boundary. Earth Planet Sci Lett 107:349–368

Hinnov LA, Goldhammer RK (1991) Spectral analysis of the middle Triassic Latemar limestone. J Sediment Petrol 61:1173–1193

Hiroki Y (1994) Quaternary crustal movements from facies distribution in the Atsumi and Hamana areas, central Japan. Sediment Geol 93:223–235

Hiscott RN, Wilson RCL, Gradstein FM, Pujalte V, García-Mondéjar J, Boudreau RR, Wishart HA (1990) Comparative stratigraphy and subsidence history of Mesozoic rift basins of North Atlantic. Am Assoc Pet Geol Bull 74:60–76

Hodell DA, Elmstrom KM, Kennett JP (1986) Latest Miocene benthic O^{18} changes, global ice volume, sea-level and "Messinian salinity crisis". Nature 320:411–414

Hoffman PF (1989) Speculations on Laurentia's first gigayear (20 to 10 Ga). Geology 17:135–138

Hoffman PF (1991) Did the breakout of Laurentia turn Gondwanaland inside-out? Science 252:1409–1412

Holdsworth BK, Collinson JD (1988) Millstone Grit cyclicity revisited. In: Besly BM, Kelling G (1988) Sedimentation in a synorogenic basin complex: the Upper Carboniferous of northwest Europe. Blackie, Glasgow, pp 132–152

House MR (1985) A new approach to an absolute timescale from measurements of orbital cycles and sedimentary microrhythms. Nature 315:721–725

House MR, Gale AS (eds) (1995) Orbital forcing timescales and cyclostratigraphy. Geological Society, London, Spec Publ 85, 210 pp

Hsü KJ, Cita MB, Ryan WBF (1973) The origin of the Mediterranean evaporites. In: Ryan WBF, Hsü KJ et al. (eds) Initial Reports of the Deep Sea Drilling Project, vol 13. US Government Printing Office, Washington, pp 1203–1231

Hubbard RJ (1988) Age and significance of sequence boundaries on Jurassic and early Cretaceous rifted continental margins. Am Assoc Pet Geol Bull 72:49–72

Hunt D, Tucker ME (1992) Stranded parasequences and the forced regressive wedge systems tract: deposition during base-level fall. Sediment Geol 81:1–9

Hunt D, Tucker ME (1995) Stranded parasequences and the forced regressive wedge systems tract: deposition during base-level fall – reply. Sediment Geol 95:147–160

Imbrie J (1985) A theoretical framework for the Pleistocene ice ages. J Geol Soc (Lond) 142:417–432

Imbrie J, Imbrie KP (1979) Ice ages: solving the mystery. Enslow, Hillside, New Jersey, 224 pp

International Subcommission on Stratigraphic Classification (1987) Unconformity-bounded stratigraphic units. Geol Soc Am Bull 98:232–237

Ito M (1992) High-frequency depositional sequences of the upper part of the Kazusa Group, a middle Pleistocene forearc basin fill on Boso Peninsula, Japan. Sediment Geol 76:155–175

Ito M (1995) Volcanic ash layers facilitate high-resolution sequence stratigraphy at convergent plate margins: an example from the Plio-Pleistocene forearc basin fill in the Boso Peninsula, Japan. Sediment Geol 95:187–206

Ito M, O'Hara S (1994) Diachronous evolution of systems tracts in a depositional sequence from the middle Pleistocene paleo-Tokyo Bay, Japan. Sedimentology 41:677–698

James DP, Leckie DA (eds) (1988) Sequences, stratigraphy, sedimentology: surface and subsurface. Can Soc Pet Geol Mem 15:586

James NP (1983) Reef environment. In: Scholle PA, Bebout DG, Moore CH (eds) Carbonate depositional environments. Am Assoc Pet Geol Mem 33:345–444

James NP, Bourque P-A (1992) Reefs and mounds. In: Walker RG, James NP (eds) Facies models: response to sea level change. Geological Association of Canada, St John's, Newfoundland, pp 323–347

James NP, Kendall AC (1992) Introduction to carbonate and evaporite facies models. In: Walker RG, James NP (eds) Facies models: response to sea level change. Geological Association of Canada, St John's, Newfoundland, pp 265–275

Janssen ME, Stephenson RA, Cloetingh S (1995) Temporal and spatial correlations between changes in plate motions and the evolution of rifted basins in Africa. Geol Soc Am Bull 107:1317–1332

Jenkins DG, Gamson P (1993) The late Cenozoic Globorotalia truncatulinoides datum plane in the Atlantic, Pacific and Indian Oceans. In: Hailwood EA, Kidd RB (eds) High resolution stratigraphy. Geological Society, London, Spec Publ 70, pp 127–130

Jervey MT (1988) Quantitative geological modeling of siliciclastic rock sequences and their seismic expression. In: Wilgus CK, Hastings BS, Kendall CGStC, Posamentier HW, Ross CA, Van Wagoner JC (eds) Sea level changes – an integrated approach. Soc Econ Paleontol Mineral Spec Publ 42:47–69

Johannessen EP, Embry AF (1989) Sequence correlation: upper Triassic to lower Jurassic succession, Canadian and Norwegian Arctic. In: Collinson JD (ed) Correlation in hydrocarbon exploration. Graham and Trotman, London, pp 155–170

Johnson DD, Beaumont C (1995) Preliminary results from a planform kinematic model of orogen evolution, surface processes and the development of clastic foreland basin stratigraphy. In: Dorobek SL, Ross GM (eds) Stratigraphic evolution of foreland basins. Soc Sediment Geol Spec Publ 52:3–24

Johnson JG (1971) Timing and coordination of orogenic, epeirogenic, and eustatic events. Geol Soc Am Bull 82:3263–3298

Johnson JG (1992) Belief and reality in biostratigraphic zonation. Newsl Stratigr 26:41–48

Johnson JG, Klapper G, Sandberg CA (1985) Devonian eustatic fluctuations in Euramerica. Geol Soc Am Bull 96:567–587

Johnson GD, Raynolds RGH, Burbank DW (1986) Late Cenozoic tectonics and sedimentation in the northwestern Himalayan foredeep. I Thrust ramping and associated deformation in the Potwar region. In: Allen PA, Homewood P (eds) Foreland basins. Int Assoc Sedimentol Spec Publ 8:273–291

Johnson NM, Sheikh KA, Dawson-Saunders E, McCrae LE (1988) The use of magnetic-reversal time lines in stratigraphic analysis: a case study in measuring variability in sedimentation rates. In: Kleinspehn K, Paola C (eds) New perspectives in basin analysis. Springer-Verlag, Berlin Heidelberg New York, pp 189–200

Jones B, Desrochers A (1992) Shallow platform carbonates. In: Walker RG, James NP (eds) Facies models: response to sea level change. Geological Association of Canada, St John's, Newfoundland, pp 277–301

Jones CE, Jenkyns HC, Coe AL, Hesselbo SP (1994a) Strontium isotope variations in Jurassic and Cretaceous seawater. Geochim Cosmochim Acta 58:3061–3074

Jones CE, Jenkyns HC, Hesselbo SP (1994b) Strontium isotopes in early Jurassic seawater. Geochim Cosmochim Acta 58:1285–1301

Jordan TE (1981) Thrust loads and foreland basin evolution, Cretaceous, Western United States. Am Assoc Pet Geol Bull 65:2506–2520

Jordan TE, Flemings PB (1990) From geodynamic models to basin fill – a stratigraphic perspective. In: Cross TA (ed) Quantitative dynamic stratigraphy. Prentice-Hall, Englewood Cliffs, pp 149–163

Jordan TE, Flemings PB (1991) Large-scale stratigraphic architecture, eustatic variation, and unsteady tectonism: a theoretical evaluation. J Geophys Res 96B:6681–6699

Jordan TE, Flemings PB, Beer JA (1988) Dating thrust-fault activity by use of foreland-basin strata. In: Kleinspehn KL, Paola C (eds) New perspectives in basin analysis. Springer Verlag, Berlin Heidelberg New York, pp 307–330

Kamola DL, Huntoon JE (1995) Repetitive stratal patterns in a foreland basin sandstone and their possible tectonic significance. Geology 23:177–180

Kamp PJJ, Turner GM (1990) Pleistocene unconformity-bounded shelf sequences (Wangnui Basin, New Zealand) correlated with global isotope record. Sediment Geol 68:155–161

Karner GD (1986) Effects of lithospheric in-plane stress on sedimentary basin stratigraphy. Tectonics 5:573–588

Kauffman EG (1977) Geological and biological overview, Western Interior Cretaceous basin. Mount Geol 14:75–99

Kauffman EG (1984) Paleobiogeography and evolutionary response dynamic in the Cretaceous Western Interior Seaway of North America. In: Westerman GE (ed) Jurassic-Cretaceous biochronology and paleogeography of North America. Geol Assoc Can Spec Pap 27:273–306

Kauffman EG, Hazel JE (eds) (1977a) Concepts and methods of biostratigraphy. Dowden, Hutchinson and Ross, Stroudsburg, 658 pp

Kauffman EG, Hazel JE (1977b) Preface. In: Kauffman EG, Hazel JE (eds) Concepts and methods of biostratigraphy. Dowden, Hutchinson and Ross, Stroudsburg, pp iii–v

Kauffman EG, Elder WP, Sageman BB (1991) High-resolution correlation: a new tool in chronostratigraphy. In: Einsele G, Ricken W, Seilacher A (eds) Cycles and events in stratigraphy Springer-Verlag, Berlin Heidelberg New York, pp 795–819

Kendall CGStC, Lerche I (1988) The rise and fall of eustasy. In: Wilgus CK, Hastings BS, Kendall CGStC, Posamentier HW, Ross CA, Van Wagoner JC (eds) Sea-level changes: an integrated approach. Soc Econ Paleontol Mineral Spec Publ 42:3–17

Kendall CGStC, Strobel J, Cannon R, Bezdek J, Biswas G (1991) The simulation of the sedimentary fill of basins. J Geophys Res 96B:6911–6929

Kendall CGStC, Moore P, Whittle G (1992) A challenge: is it possible to determine eustasy and does it matter? In: Dott RH Jr (ed) Eustasy: the historical ups and downs of a major geological concept. Geol Soc Am Mem 180:93–107

Kennedy WJ, Cobban WA (1977) The role of ammonites in biostratigraphy. In: Kauffman EG, Hazel JE (eds) Concepts and methods of biostratigraphy. Dowden, Hutchinson and Ross, Stroudsburg, pp 309–320

Kennett JP (1977) Cenozoic evolution of Antarctic glaciation, the circum-Antarctic Ocean, and their impact on global paleoceanography. J Geophys Res 82:3843–3860

Kent DV, Gradstein FM (1985) A Cretaceous and Jurassic geochronology. Geol Soc Am Bull 96:1419–1427

Kerr RA (1980) Changing global sea levels as a geologic index. Science 209:483–486

Kidd RB, Hailwood EA (1993) High resolution stratigraphy in modern and ancient marine sequences: ocean sediment cores to Paleozoic outcrop. In: Hailwood EA, Kidd RB (eds) High resolution stratigraphy. Geological Society, London, Spec Publ 70, pp 1–8

Kidwell SM (1984) Outcrop features and origin of basin margin unconformities in the Lower Chesapeake Group (Miocene), Atlantic coastal plain. In: Schlee JS (ed) Interregional unconformities and hydrocarbon exploration. Am Assoc Pet Geol Mem 36:37–58

Kidwell SM (1988) Reciprocal sedimentation and non-correlative hiatuses in marine-paralic siliciclastics: Miocene outcrop evidence. Geology 16:609–612

Kidwell SM (1989) Stratigraphic condensation of marine transgressive records: origin of major shell deposits in the Miocene of Maryland. J Geol 97:1–24

King PB (1977) The evolution of North America, revised edition. Princeton University Press, Princeton, 197 pp

Klein G deV (1990) Pennsylvanian time scales and cycle periods. Geology 18:455–457

Klein G deV (1992) Climatic and tectonic sea-level gauge for midcontinent Pennsylvanian cyclothems. Geology 20:363–366

Klein G deV (1994) Depth determination and quantitative distinction of the influence of tectonic subsidence and climate on changing sea level during deposition of midcontinent Pennsylvanian cyclothems. In: Dennison JM, Ettensohn FR (eds) Tectonic and eustatic controls on sedimentary cycles. Soc Sediment Geol, Concepts Sedimentol Paleontol 4:35–50

Klein G deV, Kupperman JB (1992) Pennsylvanian cyclothems: methods of distinguishing tectonically induced changes in sea level from climatically induced changes. Geol Soc Am Bull 104:166–175

Klein G deV, Willard DA (1989) Origin of the Pennsylvanian coal-bearing cyclothems of North America. Geology 17:152–155

Klitgord KD, Schouten H (1986) Plate kinematics of the central Atlantic. In: Vogt PR, Tucholke BE (eds) The Western North Atlantic region. The geology of North America: Boulder, Colorada. vol M. Geological Society of America Bulletin, pp 351–378

Klüpfel W (1917) Über die Sedimente der Flachsee im Lothringer Jura. Geolog Rundsch 7:98–109

Knight I, James NP, Lane TE (1991) The Ordovician St George unconformity, northern Appalachians: the relationship of plate convergence at the St Lawrence Promontory to the Sauk/Tippecanoe sequence boundary. Geol Soc Am Bull 103:1200–1225

Kocurek G (ed) (1988a) Late Paleozoic and Mesozoic eolian deposits of the Western Interior of the United States. Sediment Geol (special issue) 56:413

Kocurek G (1988b) First-order and super bounding surfaces in eolian sequences – bounding surfaces revisited. Sediment Geol 56:193–206

Kolb W, Schmidt H (1991) Depositional sequences associated with equilibrium coastlines in the Neogene of southwestern Nicaragua. In: Macdonald DIM (ed) Sedimentation, tectonics and eustasy: sea-level changes at active margins. Int Assoc Sediment Spec Publ 12:259–272

Kolla V, Macurda DB Jr (1988) Sea-level changes and timing of turbidity-current events in deep-sea fan systems. In: Wilgus CK, Hastings BS, Kendall CGStC, Posamentier HW, Ross CA, Van Wagoner JC (eds) Sea-level changes: an integrated approach. Soc Econ Paleontol Mineral Spec Publ 42:381–392

Kolla V, Posamentie HW, Eichenseer H (1995) Stranded parasequences and the forced regressive wedge systems tract: deposition during base-level fall – discussion. Sediment Geol 95:139–145

Kominz MA (1984) Ocean ridge volumes and sea-level change – an error analysis. In: Schlee JS (ed) Interregional unconformities and hydrocarbon accumulation. Am Assoc Pet Geol Mem 36:108–127

Kominz MA, Bond GC (1990) A new method of testing periodicity in cyclic sediments: application to the Newark Supergroup. Earth Planet Sci Lett 98:233–244

Kominz MA, Bond GC (1991) Unusually large subsidence and sea-level events during middle Paleozoic time: new evidence supporting mantle convection models for supercontinent assembly. Geology 19:56–60

Kooi H, Cloetingh S (1992a) Lithospheric necking and regional isostasy at extensional basins. 1 Subsidence and gravity modeling with an application to the Gulf of Lions margin (SE France). J Geophys Res 97B:17553–17571

Kooi H, Cloetingh S (1992b) Lithospheric necking and regional isostasy at extensional basins. 2 Stress-induced vertical motions and relative sea level changes. J Geophys Res 97B:17573–17591

Kreisa RD, Moiola RJ, Nottvedt A (1986) Tidal sand wave facies, Rancho Rojo Sandstone (Permian), Arizona. In: Knight RJ, McLean JR (eds) Shelf sands and sandstones. Can Soc Pet Geol Mem 11:277–291

Laferriere AP, Hattin DE, Archer AW (1987) Effects of climate, tectonics, and sea-level changes on rhythmic bedding patterns in the Niobrara Formation (upper Cretaceous), US Western Interior. Geology 15:233–236

Lambeck K (1980) The earth's variable rotation. Cambridge University Press, Cambridge, 449 pp

Lambeck K, Cloetingh S, McQueen H (1987) Intraplate stresses and apparent changes in sea level: the basins of northwestern Europe. In: Beaumont C, Tankard AJ (eds) Sedimentary basins and basin-forming mechanisms. Can Soc Pet Geol Mem 12:259–268

Langenheim RL Jr, Heckel PH, de Boer PL, Klein G deV (1991) Comments and replies on "Pennsylvanian time scales and cycle periods". Geology 19:405–410

Larson RL (1991) Geological consequences of superplumes. Geology 19:963–966

Larson RL, Pitman WC III (1972) World-wide correlation of Mesozoic magnetic anomalies and its implications. Geol Soc Am Bull 83:3645–3662

Larson RL, Golovchenko X, Pitman WC III (1982) Geomagnetic polarity time scale. American Association of Petroleum Geologists Plate-tectonic map, Circum-Pacific Region, Pacific Basin sheet, scale 1:20,000,000, Tulsa, Oklahoma

Laskar J (1989) A numerical experiment on the chaotic behaviour of the solar system. Nature 338:237–238

Lawrence DT (1993) Evaluation of eustasy, subsidence, and sediment input as controls on depositional sequence geometries and the synchroneity of sequence boundaries. In: Weimer P, Posamentie HW (eds) Siliciclastic sequence stratigraphy. Am Assoc Pet Geol Mem 58:337–367

Lawton TF (1986a) Compositional trends within a clastic wedge adjacent to a fold-thrust belt: Indianola Group, central Utah, USA. In: Allen PA, Homewood P (eds) Foreland basins. Int Assoc Sedimentol Spec Publ 8:411–423

Lawton TF (1986b) Fluvial systems of the upper Cretaceous Mesaverde group and Paleocene North Horn formation, central Utah: a record of transition from thin-skinned to thick-skinned deformation in the foreland region. In: Peterson JA (ed) Paleotectonics and sedimentation in the Rocky Mountain region, United States. Am Assoc Pet Geol Mem 41:423–442

Leckie DA (1986) Rates, controls, and sand-body geometries of transgressive-regressive cycles: Cretaceous Moosebar and Gates formations, British Columbia. Am Assoc Pet Geol Bull 70:516–535

Leckie DA, Krystinik LF (1993) Sequence stratigraphy: fact, fantasy, or work in progress (?) Can Soc Pet Geol, Calgary, Alberta, Reservoir 20(8):2–3

Leeder MR (1988) Recent developments in Carboniferous geology: a critical review with implications for the British Isles and NW Europe. Proc Geol Assoc 99:73–100

Lees GM, Falcon NL (1952) The geographical history of the Mesopotamian plains. Geogr J 118:24–39

Legarreta L (1991) Evolution of a Callovian-Oxfordian carbonate margin in the Neuquén Basin of west-central Argentina: facies, architecture, depositional sequences and global sea-level changes. Sediment Geol 70:209–240

Legarreta L, Uliana M (1991) Jurassic-Cretaceous marine oscillations and geometry of back-arc basin fill, central Argentine Andes. In: Macdonald DIM (ed) Sedimentation, tectonics and eustasy: sea-level changes at active margins. Int Assoc Sedimentol Spec Publ 12:429–450

Leggett JK, McKerrow WS, Cocks LRM, Rickards RB (1981) Periodicity in the lower Paleozoic marine realm. J Geol Soc (Lond) 138:167–176

Leggitt SM, Walker RG, Eyles CH (1990) Control of reservoir geometry and stratigraphic trapping by erosion surface E5 in the Pembina-Carrot Creek area, upper Cretaceous Cardium formation, Alberta, Canada. Am Assoc Pet Geol Bull 74:1165–1182

Leithold EL (1994) Stratigraphical architecture at the muddy margin of the Cretaceous Western Interior Seaway, southern Utah. Sedimentology 41:521–542

Leopold LB, Bull WB (1979) Base level, aggradation, and grade. Proc Am Philos Soc 123:168–202

Leopold LB, Wolman MG, Miller JP (1964) Fluvial processes in geomorphology. W.H Freeman, San Francisco, 522 pp

Lincoln JM, Schlanger SO (1991) Atoll stratigraphy as a record of sea level change: problems and prospects. J Geophys Res 96B:6727–6752

Linsley PN, Potter HC, McNab G, Racher D (1979) Beatrice field, Moray Firth, North Sea (abs). Am Assoc Pet Geol Bull 63:487

Long DGF (1993) Limits on late Ordovician eustatic sea-level change from carbonate shelf sequences: an example from Anticosti Island, Quebec. In: Posamentier HW, Summerhayes CP, Haq BU, Allen GP (eds) Sequence stratigraphy and facies associations. Int Assoc Sedimentol Spec Publ 18:487–499

Loucks RG, Sarg JF (eds) (1993) Carbonate sequence stratigraphy. Am Assoc Pet Geol Mem 57:545

Loup B, Wildi W (1994) Subsidence analysis in the Paris Basin: a key to Northwest European intracontinental basins. Basin Res 6:159–177

Loutit TS, Hardenbol J, Vail PR, Baum GR (1988) Condensed sections: the key to age dating and correlation of continental margin sequences. In: Wilgus CK, Hastings BS, Kendall CGStC, Posamentier HW, Ross CA, Van Wagoner JC (eds) Sea-level changes: an integrated approach. Soc Econ Paleontol Mineral Spec Publ 42:183–213

Ludvigsen R, Westrop SR, Pratt BR, Tuffnell PA, Young GA (1986) Paleoscene 3. Dual biostratigraphy: zones and biofacies. Geosci Can 13:139–154

Lumsden DN (1985) Secular variations in dolomite abundance in deep marine sediments. Geology 13:766–769

Luterbacher HP, Eichenseer H, Betzler C, Van den Hurk AM (1991) Carbonate-siliciclastic depositional systems in the Paleogene of the south Pyrenean foreland basin: a sequence-stratigraphic approach. In: Macdonald DIM (ed) Sedimentation, tectonics and eustasy: sea-level changes at active margins. Int Assoc Sedimentol Spec Publ 12:391–407

Macdonald DIM (ed) (1991) Sedimentation, tectonics and eustasy: sea-level changes at active margins. Int Assoc Sedimentol Spec Publ 12:518

Mackenzie FT, Pigott JD (1981) Tectonic controls of Phanerozoic sedimentary rock cycling. J Geol Soc (Lond) 138:183–196

MacLeod N, Keller G (1991) How complete are Cretaceous/Tertiary boundary sections? A chronostratigraphic estimate based on graphic correlation. Geol Soc Am Bull 103:1439–1457

Mancini EA, Tew BH (1993) Eustasy versus subsidence: lower Paleocene depositional sequences from southern Alabama, eastern Gulf coastal plain. Geol Soc Am Bull 105:3–17

Manley PL, Flood RD (1988) Cyclic sediment deposition within Amazon deep-sea fan. Am Assoc Pet Geol Bull 72:912–925

Mann KO, Lane HR (eds) (1995) Graphic correlation. Soc Sediment Geol Spec Publ 53:263

Martinsen OJ (1993) Namurian (late Carboniferous) depositional systems of the Craven-Askrigg area, northern England: implications for sequence-stratigraphic models. In: Posamentier HW, Summerhayes CP, Haq BU, Allen GP (eds) Sequence stratigraphy and facies associations. Int Assoc Sedimentol Spec Publ 18:247–281

Martinsen OJ, Helland-Hansen W (1994) Sequence stratigraphy and facies model of an incised valley fill: the Gironde Estuary, France – discussion. J Sediment Res B64:78–80

Martinsen OJ, Martinsen RS, Steidtmann JR (1993) Mesaverde group (upper Cretaceous), southeastern Wyoming: allostratigraphy versus sequence stratigraphy in a tectonically active area. Am Assoc Pet Geol Bull 77:1351–1373

Martire L (1992) Sequence stratigraphy and condensed pelagic sediments. An example from the Rosso Ammonitico Veronese, northeastern Italy. Palaeogeogr Palaeoclimatol Palaeoecol 94:169–191

Masuda F (1994) Onlap and downlap patterns in Plio-Pleistocene forearc and backarc basin-fill successions, Japan. Sediment Geol 93:237–246

Matthews RK (1984) Oxygen-isotope record of ice-volume history: 100 million years of glacio-isostatic sea-level fluctuation. In: Schlee JS (ed) Interregional unconformities and hydrocarbon accumulation. Am Assoc Pet Geol Mem 36:97–107

Matthews RK (1988) Sea level history. Science 241:597–599

Matthews RK, Frohlich C (1991) Orbital forcing of low-frequency glacioeustasy. J Geophys Res 96(B):6797–6803

Matthews R, Poore RA (1980) Tertiary ^{18}O record and glacio-eustatic sea-level fluctuations. Geology 8:501–504

Matthews SC, Cowie JW (1979) Early Cambrian transgression. J Geol Soc (Lond) 136:133–135

Maxwell JC (1984) What is the lithosphere? American Geophysical Union, Washington. Eos 65:321–325

May JA, Warme JE (1987) Synchronous depositional phases in west coast basins: eustasy or regional tectonics. In: Ingersoll RV, Ernst WG (eds) Cenozoic basin developmnet of coastal California. Prentice-Hall, Englewood Cliffs, pp 25–46

Mayer L (1987) Subsidence analysis of the Los Angeles Basin. In: Ingersoll RV, Ernst WG (eds) Cenozoic basin development of coastal California. Rubey vol VI. Prentice-Hall, Englewood Cliffs, pp 299–320

Maynard JR, Leeder MR (1992) On the periodicity and magnitude of late Carboniferous glacio-eustatic sea-level changes. J Geol Soc (Lond) 149:303–311

McCrossan RG, Glaister RP (eds) (1964) Geological history of Western Canada. Alberta Society of Petroleum Geologists, Calgary, Alberta, 232 pp

McDonough KJ, Cross TA (1991) Late Cretaceous sea level from a paleoshoreline. J Geophys Res 96B:6591–6607

McGinnis JP, Driscoll NW, Karner GD, Brumbaugh WD, Cameron N (1993) Flexural response of passive margins to deep-sea erosion and slope retreat: implications for relative sea-level change. Geology 21:893–896

McKenzie DP (1978) Some remarks on the development of sedimentary basins. Earth Planet Sci Lett 40:25–32

McKenzie DP, Sclater JG (1971) The evolution of the Indian Ocean since the late Cretaceous. Geophys J R Astron Soc 25:437–528

McKerrow WS (1979) Ordovician and Silurian changes in sea level. J Geol Soc (Lond) 136:137–146

McKinney ML (1986) Biostratigraphic gap analysis. Geology 14:36–38

McLaren DJ (1970) Presidential address: time, life and boundaries. J Paleontol 44:801–815

McLaren DJ (1973) The Silurian-Devonian boundary. Geol Mag 110:302–303

McMillan NJ (1973) Shelves of Labrador Sea and Baffin Bay, Canada. In: McCrossan RG (ed) The future petroleum provinces of Canada – their geology and potential. Can Soc Pet Geol Mem 1:473–517

McNeil DH, Dietrich JR, Dixon J (1990) Foraminiferal biostratigraphy and seismic sequences: examples from the Cenozoic of the Beaufort-Mackenzie Basin, Arctic Canada. In: Hemleben C, Kaminski MA, Kuhnt W, Scott DB (eds) Paleoecology, biostratigraphy, paleoceanography and taxononomy of agglutinated foraminifera. Kluwer, Dordrecht, pp 859–882

McShea DW, Raup DM (1986) Completeness of the geological record. J Geol 94:569–574

Melnyk DH, Smith DG, Amiri-Garroussi K (1994) Filtering and frequency mapping as tools in subsurface cyclostratigraphy, with examples from the Wessex Basin, UK. In: de Boer PL, Smith DG (eds) Orbital forcing and cyclic sequences. Int Assoc Sedimentol Spec Publ 19:35–46

Menning M (1989) A synopsis of numerical time scales 1917–1986. Episodes 12(1):3–5

Merriam DF (ed) (1964) Symposium on cyclic sedimentation. Kans Geoll Surv Bull 169:636

Miall AD (1978) Tectonic setting and syndepositional deformation of molasse and other nonmarine-paralic sedimentary basins. Can J Earth Sci 15:1613–1632

Miall AD (1981) Alluvial sedimentary basins: tectonic setting and basin architecture. In: Miall AD (ed) Sedimentation and tectonics in alluvial basins. Geol Assoc Can Spec Pap 23:1–33

Miall AD (1986) Eustatic sea-level change interpreted from seismic stratigraphy: a critique of the methodology with particular reference to the North Sea Jurassic record. Am Assoc Pet Geol Bull 70:131–137

Miall AD (1987) Epeirogeny: is it really orogeny or theology? Geosci Can 14:126–127

Miall AD (1988) Reservoir heterogeneities in fluvial sandstones: lessons from outcrop studies. Am Assoc Pet Geol Bull 72:682–697

Miall AD (1990) Principles of sedimentary basin analysis, 2nd edn. Springer, Berlin Heidelberg New York, 668 pp

Miall AD (1991a) Stratigraphic sequences and their chronostratigraphic correlation. J Sediment Petrol 61:497–505

Miall AD (1991b) Hierarchies of architectural units in terrigenous clastic rocks, and their relationship to sedimentation rate. In: Miall AD, Tyler N (eds) The three-dimensional facies architecture of terrigenous clastic sediments and its implications for hydrocarbon discovery and recovery. Soc Econ Paleontol Mineral, Concepts Sedimentol Paleontol 3:6–12

Miall AD (1992) The Exxon global cycle chart: an event for every occasion? Geology 20:787–790

Miall AD (1993) The architecture of fluvial-deltaic sequences in the Upper Mesaverde Group (Upper Cretaceous), Book Cliffs, Utah. In: Best JL, Bristow CS (eds) Braided rivers. Geological Society, London, Spec Publ 75, pp 305–332

Miall AD (1994) Sequence stratigraphy and chronostratigraphy: problems of definition and precision in correlation, and their implications for global eustasy. Geosci Can 21:1–26

Miall AD (1995) Whither stratigraphy? Sediment Geol 100:5–20

Miall AD (1996) The geology of fluvial deposits: sedimentary facies, basin analysis and petroleum geology. Springer-Verlag, Berlin Heidelberg New York, 582 pp

Milankovitch M (1930) Mathematische Klimalehre und astronomische Theorie der Klimaschwankungen. In: Koppen W, Geiger R (eds) Handbuch der Klimatologie, I (A). Gebruder Borntraeger, Berlin

Milankovitch M (1941) Kanon der Erdbestrahlung und seine Anwendung auf das Eiszeitenproblem. Akad Royale Serbe 133:633

Millan H, Aurell M, Melendez A (1994) Synchronous detachment folds and coeval sedimentation in the Prepyrenean External Sierras (Spain): a case study for a tectonic origin of sequences and systems tracts. Sedimentology 41:1001–1024

Millendorf SA, Heffner T (1978) Fortran program for lateral tracing of time-stratigraphic units based on faunal assemblage zones. Comput Geosc 4:313–318

Miller FX (1977) The graphic correlation method in biostratigraphy. In: Kauffman EG, Hazel JE (eds) Concepts and methods in biostratigraphy. Dowden, Hutchinson and Ross, Stroudsburg, pp 165–186

Miller KB, McCahon TJ, West RR (1996) Lower Permian (Wolfcampian) paleosol-bearing cycles of the U.S. midcontinent: evidence of climatic cyclicity. J Sediment Res 66:71–84

Miller KG (1990) Recent advances in Cenozoic marine stratigraphic resolution. Palaios 5:301–302

Miller KG, Kent DV (1987) Testing Cenozoic eustatic changes: the critical role of stratigraphic resolution. In: Ross CA, Haman D (eds) Timing and depositional history of eustatic sequences: constraints on seismic stratigraphy. Cushman Found Foraminiferal Res Spec Publ 24:51–56

Miller KG, Fairbanks RG, Mountain GS (1987) Tertiary oxygen isotope synthesis, sea level history and continental margin erosion. Paleoceanography 2:1–19

Mitchum RM Jr (1985) Seismic stratigraphic expression of submarine fans. In: Berg OR, Woolverton DG (eds) Seismic stratigraphy II: an integrated approach to hydrocarbon exploration. Am Assoc Pet Geol Mem 39: 117–138

Mitchum RM Jr, Van Wagoner JC (1991) High-frequency sequences and their stacking patterns: sequence-stratigraphic evidence of high-frequency eustatic cycles. Sediment Geol 70:131–160

Mitchum RM Jr, Vail PR, Sangree JB (1977a) Seismic stratigraphy and global changes of sea level, part 6: stratigraphic interpretation of seismic reflection patterns in depositional sequences. In: Payton CE (ed) Seismic stratigraphy – applications to hydrocarbon exploration. Am Assoc Pet Geol Mem 26:117–133

Mitchum RM Jr, Vail PR, Thompson S III (1977b) Seismic stratigraphy and global changes of sea level, part 2, the depositional sequence as a basic unit for stratigraphic analysis. In: Payton CE (ed) Seismic stratigraphy – applications to hydrocarbon exploration. Am Assoc Pet Geol Mem 26:53–62

Mitchum RM Jr, Uliana MA (1988) Regional seismic stratigraphic analysis of upper Jurassic-lower Cretaceous carbonate depositional sequences, Neuquen Basin, Argentina. In: Bally AW (ed) Atlas of seismic stratigraphy. Am Assoc Pet Geol Stud Geol 27(2):206–212

Mitchum RM Jr, Sangree JB, Vail PR, Wornardt WW (1993) recognizing sequences and systems tracts from well logs, seismic data, and biostratigraphy: examples from the late Cretaceous. In: Weimer P, Posamentier HW (eds) Siliciclastic sequence stratigraphy. Am Assoc Pet Geol Mem 58:163–197

Mitrovica JX, Jarvis GT (1985) Surface deflections due to transient subduction in a convecting mantle. Tectonophysics 120:211–237

Mitrovica JX, Beaumont C, Jarvis GT (1989) Tilting of continental interiors by the dynamical effects of subduction. Tectonics 8:1079–1094

Molenaar CM (1983) Principle reference section and correlation of Gallup Sandstone, northwestern New Mexico. In: Hook S (ed) Contributions to mid-Cretaceous paleontology and stratigraphy of New Mexico – part II. New Mexico Bureau of Mines and Mineral Resources Circular 185:29–40

Molenaar CM, Rice DD (1988) Cretaceous rocks of the Western Interior Basin. In: Sloss LL (ed) Sedimentary cover – North American Craton: US. The Geology of North America, vol D-2. Geological Society of America, Boulder, pp 77–82

Molnar P, Tapponnier P (1975) Cenozoic tectonics of Asia: effects of a continental collision. Science 189:419–426

Montañez IP, Read JF (1992) Eustatic control on early dolomitization of cyclic peritidal carbonates: evidence from the early Ordovician Upper Knox Group, Appalachians. Geol Soc Am Bull 104:872–886

Moore PF (1989) The Kaskasia sequence: reefs, platforms and foredeeps. The lower Kaskasia sequence – Devonian. In: Ricketts BD (ed) Western Canada sedimentary basin: a case history. Can Soc Pet Geol, Calgary, Alberta, pp 139–164

Moore RC (1936) Stratigraphic classification of the Pennsylvanian rocks of Kansas. Kans Geol Surv Bull 22:256

Moore RC (1964) Paleoecological aspects of Kansas Pennsylvanian and Permian cyclothems. In: Merriam DF (ed) Symposium on cyclic sedimentation. Kans Geol Surv Bull 169:287–380

Moore TC Jr, Romine K (1981) In search of biostratigraphic resolution. In: Warme JE, Douglas RG, Winterer EL (eds) The deep sea drilling project: a decade of progress. Soc Econ Paleontol Mineral Spec Publ 32:317–334

Moore TC Jr, Pisias NG, Dunn DA (1982) Carbonate time series of the Quaternary and late Miocene sediments in the Pacific Ocean: a spectral comparison. Mar Geol 46:217–233

Mork A, Embry AF, Weitschat W (1989) Triassic transgressive-regressive cycles in the Sverdrup Basin, Svalbard and the Barents Shelf. In: Collinson JD (ed) Correlation in hydrocarbon exploration. Graham and Trotman, London, pp 113–130

Mörner N-A (1994) Internal response to orbital forcing and external cyclic sedimentary sequences. In: de Boer PL, Smith DG (eds) Orbital forcing and cyclic sequences. Int Assoc Sedimentol Spec Publ 19:25–33

Morton RA, Price WA (1987) Late Quaternary sea-level fluctuations and sedimentary phases of the Texas coastal plain and shelf. In: Nummedal D, Pilkey OH, Howard JD (eds) Sea-level fluctuation and coastal evolution. Soc Econ Paleontol Mineral Spec Publ 41:181–198

Moxon IW, Graham SA (1987) History and controls of subsidence in the late Cretaceous-Tertiary Great Valley forearc basin, California. Geology 15:626–629

Muntingh A, Brown LF Jr (1993) Sequence stratigraphy of petroleum plays, post-rift Cretaceous rocks (lower Aptian to upper Maastrichtian), Orange Basin, western offshore, South Africa. In: Weimer P, Posamentier HW (eds) Siliciclastic sequence stratigraphy. Am Assoc Pet Geol Mem 58:71–98

Murphy MA (1977) On time-stratigraphic units. J Paleontol 51:213–219

Murphy MA (1988) Unconformity-bounded stratigraphic units: discussion. Geol Soc Am Bull 100:155

Muto T, Steel RJ (1992) Retreat of the front in a prograding delta. Geology 20:967–970

Mutti E (1985) Turbidite systems and their relations to depositional sequences. In: Zuffa GG (ed) Provenance of Arenites. Reidel Publishing Company, Dordrecht, pp 65–93

Mutti E, Normark WR (1987) Comparing examples of modern and ancient turbidite systems: problems and concepts. In: Leggett JK, Zuffa GG (eds) Marine clastic sedimentology: concepts and case studies. Graham and Trotman, London, pp 1–38

Newell ND (1967) Revolutions in the history of life. Geol Soc Am Spec Pap 89:63–91

Nio SD, Yang CS (1991) Sea-level fluctuations and the geometric variability of tide-dominated sandbodies. Sediment Geol 70:161–193

North American Commission on Stratigraphic Nomenclature (1983) North American stratigraphic code. Am Assoc Pet Geol Bull 67:841–875

Nummedal D (1987) Preface. In: Nummedal D, Pilkey OH, Howard JD (eds) Sea-level fluctuation and coastal evolution. Soc Econ Paleontol Mineral Spec Publ 41 pp iii-iv

Nummedal D (1990) Sequence stratigraphic analysis of upper Turonian and Coniacian strata in the San Juan Basin, New Mexico, USA. In: Ginsburg RN, Beaudoin B (eds) Cretaceous resources, events and rhythms: background and plans for research. Kluwer Academic Publishers, Dordrecht, pp 33–46

Nummedal D, Swift DJP (1987) Transgressive stratigraphy at sequence-bounding unconformities: some principles derived from Holocene and Cretaceous examples. In: Nummedal D, Pilkey OH, Howard JD (eds) Sea-level fluctuation and coastal evolution. Soc Econ Paleontol Mineral Spec Publ 41:241–260

Nummedal D, Wright R (eds) (1989) Cretaceous shelf sandstones and shelf depositional sequences, Western Interior Basin, Utah, Colorado and New Mexico. Field trip guidebook T119. 28th Int Geol Congr, American Geophysical Union, Washington, DC, 87 pp

Nummedal D, Pilkey OH, Howard JD (eds) (1987) Sea-level fluctuation and coastal evolution. Soc Econ Paleontol Mineral Spec Publ 41:267

Nummedal D, Wright R, Swift DJP, Tillman RW, Wolter NR (1989) Depositional systems architecture of shallow marine sequences. In: Nummedal D, Wright R (eds) Cretaceous shelf sandstones and shelf depositional sequences, Western Interior Basin, Utah, Colorado and New Mexico. Field trip guidebook T119. 28th Int Geol Congr, American Geophysical Union, Washington, DC, pp 35–79

Obradovich JD, Cobban WA (1975) A time-scale for the Late Cretaceous of the Western Interior of North American. In: Caldwell WGE (ed) The Cretaceous system of the Western Interior of North America. Geological Association of Canada Spec Pap 13, St John's, Newfoundland, pp 31–54

Officer CB, Drake CL (1985) Epeirogeny on a short geological time scale. Tectonics 4:603–612

Olsen H (1990) Astronomical forcing of meandering river behaviour: Milankovitch cycles in Devonian of East Greenland. Palaeogeogr Palaeoclimatol Palaeoecol 79:99–115

Olsen PE (1984) Periodicity of lake-level cycles in the late Triassic Lockatong formation of the Newark Basin (Newark Supergroup, New Jersey and Pennsylvania). In: Berger A, Imbrie J, Hays J, Kukla G, Saltzman B (eds) Milankovitch and climate: NATO ASI Series part 1. Reidel Publishing Company, Dordrecht, pp 129–146

Olsen PE (1986) A 40-million year lake record of early Mesozoic orbital climatic forcing. Science 234:842–848

Olsen PE (1990) Tectonic, climatic, and biotic modulation of lacustrine ecosystems – examples from Newark Supergroup of eastern North America. In: Katz BJ (ed) Lacustrine basin exploration: case studies and modern analogs. Am Assoc Pet Geol Mem 50:209–224

Olsen T, Steel RJ, Høgseth K, Skar T, Røe S-L (1995) Sequential architecture in a fluvial succession: sequence stratigraphy in the upper Cretaceous Mesaverde group, Price Canyon, Utah. J Sediment Res B65:265–280

Olsson RK (1988) Foraminiferal modeling of sea-level change in the late Cretaceous of New Jersey. In: Wilgus CK, Hastings BS, Kendall CGStC, Posamentier HW, Ross CA, Van Wagoner JC (eds) Sea-level changes: an integrated approach. Soc Econ Paleontol Mineral Spec Publ 42:289–297

Olsson RK (1991) Cretaceous to Eocene sea-level fluctuations on the New Jersey margin. Sediment Geol 70:195–208

Osleger D, Read JF (1991) Relation of eustasy to stacking patterns of meter-scale carbonate cycles, late Cambrian, USA. J Sediment Petrol 61:1225–1252

Osleger DA, Read JF (1993) Comparative analysis of methods used to define eustatic variations in outcrop: late Cambrian interbasinal sequence development. Am J Sci 293:157–216

Palmer AR (ed) (1983) The Decade of North American Geology 1983 geologic time scale. Geology 11:503–504

Pang M, Nummedal D (1995) Flexural subsidence and basement tectonics of the Cretaceous Western Interior Basin, United States. Geology 23:173–176

Parkinson N, Summerhayes C (1985) Synchronous global sequence boundaries. Am Assoc Pet Geol Bull 69:685–687

Partridge AD (1976) The geological expression of eustacy in the early Tertiary of the Gippsland Basin. Austr Pet Expl Assoc J 16:73–79

Pashin JC (1994) Flexurally influenced eustatic cycles in the Pottsville Formation (Lower Pennsylvanian), Black Warrior Basin, Alabama. In: Dennison JM, Ettensohn FR (eds) Tectonic and eustatic controls on sedimentary cycles. Soc Sediment Geol, Concepts Sedimentol Paleontol 4:89–105

Payton CE (ed) (1977) Seismic stratigraphy – applications to hydrocarbon exploration. Am Assoc Pet Geol Mem 26:516

Pedley M, Grasso M (1991) Sea-level changes around the margins of the Catania-Gela Trough, and Hyblean Plateau, southeast Sicily (African-European plate convergence zone): a problem of Plio-Quaternary plate buoyancy? In: Macdonald DIM (ed) Sedimentation, tectonics and eustasy: sea-level changes at active margins. Int Assoc Sedimentol Spec Publ 12:451–464

Peper T (1994) Tectonic and eustatic control on late Albian shallowing (Viking and Paddy formations) in the Western Canada Foreland Basin. Geol Soc Am Bull 106:254–263

Peper T, Cloetingh S (1995) Autocyclic perturbations of orbitally forced signals in the sedimentary record. Geology 23:937–940

Peper T, Beekman F, Cloetingh S (1992) Consequences of thrusting and intraplate stress fluctuations for vertical motions in foreland basins and peripheral areas. Geophys J Int 111:104–126

Peper T, Van Balen R, Cloetingh S (1995) Implications of orogenic wedge growth, intraplate stress variations, and eustatic sea-level change for foreland basin stratigraphy – inferences for numerical modeling. In: Dorobek SL, Ross GM (eds) Stratigraphic evolution of foreland basins. Soc Sediment Geol Spec Publ 52:25–35

Perlmutter MA, Matthews MD (1990) Global cyclostratigraphy – a model. In: Cross TA (ed) Quantitative
dynamic stratigraphy. Prentice Hall, Englewood Cliffs,
pp 233–260

Petrobras Exploration Department (1988) Pará-Maranhao
Basin, Brazil. In: Bally AW (ed) Atlas of seismic stratigraphy. Am Assoc Pet Geol Stud Geol 27(2):179–183

Pienkowski G (1991) Eustatically-controlled sedimentation in the Hettangian-Sinemurian (early Jurassic) of
Poland and Sweden. Sedimentology 38:503–518

Pinet PR, Popenoe P (1982) Blake Plateau: control of
Miocene sedimentation patterns by large-scale shifts of
the Gulf Stream axis. Geology 10:257–259

Pitman WC III (1978) Relationship between eustacy and
stratigraphic sequences of passive margins. Geol Soc
Am Bull 89:1389–1403

Pitman WC III (1979) The effect of eustatic sea level
changes on stratigraphic sequences at Atlantic margins. In: Watkins JS, Montadert L, Dickerson PW (eds)
Geological and geophysical investigations of continental margins. Am Assoc Pet Geol Mem 29:453–460

Pitman WC III (1986) Effects of sea level change on basin
stratigraphy. Am Assoc Pet Geol Bull 70:1762

Pitman WC III, Golovchenko X (1983) The effect of sea-
level change on the shelfedge and slope of passive
margins. In: Stanley DJ, Moore GT (eds) The
shelfbreak: critical interface on continental margins.
Soc Econ Paleontol Mineral Spec Publ 33:41–58

Pitman WC III, Golovchenko X (1988) Sea-level changes
and their effect on the stratigraphy of Atlantic-type
margins. In: Sheridan RE, Grow JA (eds) The Atlantic
continental margin. The geology of North America,
United States, vol 1-2. Geological Society of America,
Boulder, pp 429–436

Pitman WC III, Golovchenko X (1991) The effect of sea
level changes on the morphology of mountain belts. J
Geophys Res 9B:6879–6891

Plint AG (1988a) Sharp-based shoreface sequences and
"offshore bars" in the Cardium formation of Alberta:
their relationship to relative changes in sea level. In:
Wilgus CK, Hastings BS, Kendall CGStC, Posamentier
HW, Ross CA, Van Wagoner JC (eds) Sea-level
changes: an integrated approach. Soc Econ Paleontol
Minera Spec Publ 42:357–370

Plint AG (1988b) Global eustasy and the Eocene sequence
in the Hampshire Basin, England. Basin Res 1:11–22

Plint AG (1990) An allostratigraphic correlation of the
Muskiki and Marshybank Formations (Coniacian-
Santonian) in the foothills and subsurface of the
Alberta Basin. Bull Can Pet Geol 38:288–306

Plint AG (1991) High-frequency relative sea-level oscillations in upper Cretaceous shelf clastics of the Alberta
foreland basin: possible evidence for a glacio-eustatic
control? In: Macdonald DIM (ed) Sedimentation,
tectonics and eustasy: sea-level changes at active
margins. Int Assoc Sedimentol Spec Publ 12:409–428

Plint AG, Walker RG, Bergman KM (1986) Cardium
formation 6. Stratigraphic framework of the Cardium
in subsurface. Bull Can Pet Geol 34:213–225

Plint AG, Eyles N, Eyles CH, Walker RG (1992) Control of
sealevel change. In: Walker RG, James NP (eds) Facies
models: response to sea level change. Geological Association of Canada, St John's, Newfoundland, pp 15–25

Plint AG, Hart BS, Donaldson WS (1993) Lithospheric
flexure as a control on stratal geometry and facies
distribution in upper Cretaceous rocks of the Alberta
foreland basin. Basin Res 5:69–77

Poag CW, Schlee JS (1984) Depositional sequences and
stratigraphic gaps on submerged United States Atlantic margin. In: Schlee JS (ed) Interregional unconformities and hydrocarbon accumulation. Am Assoc Pet
Geol Mem 36:165–182

Poag CW, Sevon WD (1989) A record of Appalachian
denudation in postrift Mesozoic and Cenozoic sedimentary deposits of the US Middle Atlantic continental margin. Geomorphology 2:119–157

Poag CW, Ward LW (1987) Cenozoic unconformities and
depositional supersequences of North Atlantic continental margins: testing the Vail model. Geology
15:159–162

Posamentier HW, Allen GP (1993a) Siliciclastic sequence
stratigraphic patterns in foreland ramp-type basins.
Geology 21:455–458

Posamentier HW, Allen GP (1993b) Variability of the
sequence stratigraphic model: effects of local basin
factors. Sediment Geol 86:91–109

Posamentier HW, Vail PR (1988) Eustatic controls on
clastic deposition II – sequence and systems tract
models. In: Wilgus CK, Hastings BS, Kendall CGStC,
Posamentier HW, Ross CA, Van Wagoner JC (eds) Sea
level changes – an integrated approach. Soc Econ
Paleontol Mineral Spec Publ 42:125–154

Posamentier HW, Weimer P (1993) Siliciclastic sequence
stratigraphy and petroleum geology – where to from
here? Am Assoc Pet Geol Bull 77:731–742

Posamentier HW, Jervey MT, Vail PR (1988) Eustatic
controls on clastic deposition I – conceptual framework. In: Wilgus CK, Hastings BS, Kendall CGStC,
Posamentier HW, Ross CA, Van Wagoner JC (eds) Sea
level changes – an integrated approach. Soc Econ
Paleontol Mineral Spec Publ 42:109–124

Posamentier HW, Allen GP, James DP, Tesson M (1992)
Forced regressions in a sequence stratigraphic framework: concepts, examples, and exploration significance. Am Assoc Pet Geol Bull 76:1687–1709

Posamentier HW, Summerhayes CP, Haq BU, Allen GP
(eds) (1993) Sequence stratigraphy and facies associations. Int Assoc Sedimentol Spec Publ 18:644

Postma G (1995) Sea-level-related architectural trends in
coarse-grained delta complexes. Sediment Geol 98:3–
12

Potter PE (1978) Significance and origin of big rivers. J
Geol 86:13–33

Poulton TP (1988) Major interregionally correlatable
events in the Jurassic of Western Interior, Arctic,
and eastern offshore Canada. In: James DP, Leckie DA
(eds) Sequences, stratigraphy, sedimentology: surface
and subsurface. Can Soc Pet Geol Mem 15:195–205

Pratt BR, James NP (1986) The St George Group (lower
ordovician) of western Newfoundland: tidal flat island
model for carbonate sedimentation in shallow epeiric
seas. Sedimentology 33:313–343

Pratt BR, Smewing JD (1993) Early Cretaceous platform
margin, Oman, eastern Arabian Peninsula. In: Simo
JA, Scott RW, Masse J-P (eds) Cretaceous carbonate
platforms. Am Assoc Pet Geol Mem 56:201–212

Pratt BR, James NP, Cowan CA (1992) Peritidal carbonates. In: Walker RG, James NP (eds) Facies models: response to sea-level change. Geological Association of Canada, St John's, Newfoundland, pp 303–322

Prosser S (1993) Rift-related linked depositional systems and their seismic expression. In: Williams GD, Dobb A (eds) Tectonics and seismic sequence stratigraphy. Geological Society, London, Spec Publ 71, pp 35–66

Puigdefabregas C, Munoz JA, Marzo M (1986) Thrust belt development in the eastern Pyrenees and related depositional sequences in the southern foreland basin. In: Allen PA, Homewood P (eds) Foreland basins. Int Assoc Sedimentol Spec Publ 8:229–246

Quinlan GM (1987) Models of subsidence mechanisms in intracratonic basins, and their applicability to North American examples. In: Beaumont C, Tankard AJ (eds) Sedimentary basins and basin-forming mechanisms. Can Soc Pet Geol Mem 12:463–481

Quinlan GM, Beaumont C (1984) Appalachian thrusting, lithospheric flexure, and the Paleozoic stratigraphy of the eastern interior of North America. Can J Earth Sci 21:973–996

Rahmanian VD, Moore PS, Mudge WJ, Spring DE (1990) Sequence stratigraphy and the habitat of hydrocarbons, Gippsland Basin, Australia. In: Brooks J (ed) Classic petroleum provinces. Geological Society of London, Spec Publ 50, pp 525–541

Ramsbottom WHC (1979) Rates of transgression and regression in the Carboniferous of NW Europe. J Geol Soc (Lond) 136:147–153

Ravenne C, Coumes F, Esteve JP (1988) Relative sea level changes and depositional modes of the shelf and deep-sea fan of the Indus. In: Bally AW (ed) Atlas of seismic stratigraphy. Am Assoc Pet Geol Stud Geol 27(2):270–276

Rawson PF, Riley LA (1982) Latest Jurassic – early Cretaceous events and the 'late Cimmerian unconformity' in the North Sea. Am Assoc Pet Geol Bull 66:2628–2648

Ray RR (1982) Seismic stratigraphic interpretation of the Fort Union formation, western Wind River Basin: example of subtle trap exploration in a nonmarine sequence. In: Halbouty MT (ed) The deliberate search for the subtle trap. Am Assoc Pet Geol Mem 32:169–180

Read JF, Goldhammer RK (1988) Use of Fischer plots to define third-order sea-level curves in Ordovician peritidal cyclic carbonates, Appalachians. Geology 16:895–899

Read WA (1991) The Millstone Grit (Namurian) of the southern Pennines viewed in the light of eustatically controlled sequence stratigraphy Geol J 26:157–165

Reading HG (ed) (1986) Sedimentary environments and facies, 2nd edn. Blackwell Scientific Publications, Oxford, 615 pp

Reinson GE (1992) Transgressive barrier island and estuarine systems. In: Walker RG, James NP (eds) Facies models: response to sea-level change. Geological Association of Canada, St John's, Newfoundland, pp 179–194

Revelle R (ed) (1990) Sea-level change. National Research Council, Studies in Geophysics. National Academy Press, Washington, 234 pp

Reynolds DJ, Steckler MS, Coakley BJ (1991) The role of the sediment load in sequence stratigraphy: the influence of flexural isostasy and compaction. J Geophys Res 96B:6931–6949

Riba O (1976) Syntectonic unconformities of the Alto Cardener, Spanish Pyrenees, a genetic interpretation. Sediment Geol 15:213–233

Richardson RM (1992) Ridge forces, absolute plate motions, and the intraplate stress field. J Geophys Res 97B:11739–11748

Ricken W (1991) Time span assessment – an overview. In: Einsele G, Ricken W, Seilacher A (eds) Cycles and events in stratigraphy. Springer-Verlag, Berlin Heidelberg New York, pp 773–794

Riedel WR (1981) DSDP biostratigraphy in retrospect and prospect. In: Warme JE, Douglas RG, Winterer EL (eds) The deep sea drilling project: a decade of progress. Soc Econ Paleontol Mineral Spec Publ 32:253–315

Rine JM, Helmold KP, Bartlett GA, Hayes BJR, Smith DG, Plint AG, Walker RG, Bergman KM (1987) Cardium ormation 6. Stratigraphic framework of the Cardium in subsurface: discussions and reply. Bull Can Pet Geol 35:362–374

Rine JM, Tillman RW, Culver SJ, Swift DJP (1991) Generation of late Holocene sand ridges on the middle continental shelf of New Jersey, USA – evidence for formation in a mid-shelf setting based on comparisons with a nearshore ridge. In: Swift DJP, Oertel GF, Tillman RW, Thorne JA (eds) Shelf sand and sandstone bodies. Int Assoc Sedimentol Spec Publ 14:395–423

Rivenaes JC (1992) Application of a dual-lithology, depth dependent diffusion equation in stratigraphic simulation. Basin Res 4:133–146

Roberts AM, Yielding G, Badley ME (1993) Tectonic and bathymetric controls on stratigraphic sequences within evolving half-grabens. In: Williams GD, Dobb A (eds) Tectonics and seismic sequence stratigraphy. Geological Society, London, Spec Publ 71, pp 87–121

Robertson AHF, Eaton S, Follows EJ, McCallum JE (1991) The role of tectonics versus global sea-level change in the Neogene evolution of the Cyprus active margin. In: Macdonald DIM (ed) Sedimentation, tectonics and eustasy: sea-level changes at active margins. Int Assoc Sedimentol Spec Publ 12:331–369

Rogers JJW (1996) A history of continents in the past three billion years. J Geol 104:91–107

Rona PA (1973) Relations between rates of sediment accumulation on continental shelves, sea-floor spreading and eustacy inferred from the central North Atlantic. Geol Soc Am Bull 84:2851–2872

Rona PA (1982) Evaporites at passive margins. In: Scrutton RA (ed) Dynamics of passive margins. Geodynamics series 6. American Geophysical Union, Washington, and Geol Soc Am, Boulder, pp 116–132

Ronov AB (1994) Phanerozoic transgressions and regressions on the continents: a quantitative approach based on areas flooded by the sea and areas of marine and continental deposition. Am J Sci 294:777–801

Ross CA, Haman D (eds) (1987) Timing and depositional history of eustatic sequences: constraints on seismic stratigraphy. Cushman Found Foraminiferal Res Spec Publ 24:227

Ross CA, Ross JRP (1988) Late Paleozoic transgressive-regressive deposition. In: Wilgus CK, Hastings BS, Kendall CGStC, Posamentier HW, Ross CA, Van Wagoner JC (eds) Sea-level changes: an integrated approach. Soc Econ Paleontol Mineral Spec Publ 42:227–247

Ross WC (1991) Cyclic stratigraphy, sequence stratigraphy, and stratigraphic modeling from 1964 to 1989: twenty-five years of progress? In: Franseen EK, Watney WL, Kendall CGStC (eds) Sedimentary modeling: computer simulations and methods for improved parameter definition. Kans Geol Surv Bull 233:3–8

Roth PH, Bowdler JL (1981) Middle Cretaceous calcareous nannoplankton biogeography and oceanography of the Atlantic Ocean. In: Warme JE, Douglas RG, Winterer EL (eds) The deep sea drilling project: a decade of progress. Soc Econ Paleontol Mineral Spec Publ 32:517–546

Rowley DB, Markwick PJ (1992) Haq et al eustatic sea level curve: implications for sequestered water volumes. J Geol 100:703–715

Ruddiman WF, McIntyre A, Niebler-Hun V, Durazzi JT (1980) Oceanic evidence for the mechanism of rapid Northern Hemisphere glaciation. Quat Res 13:33–64

Ruffell AH, Rawson PF (1994) Palaeoclimate control on sequence stratigraphic patterns in the late Jurassic to mid-Cretaceous, with a case study from eastern England. Palaeogeogr Palaeoclimatol Palaeoecol 110:43–54

Russell LK (1968) Oceanic ridges and eustatic changes in sea level. Nature 218:861–862

Russell M, Gurnis M (1994) The planform of epeirogeny: vertical motions of Australia during the Cretaceous. Basin Res 6:63–76

Rust BR, Koster EH (1984) Coarse alluvial deposits. In: Walker RG (ed) Facies models, 2nd edn. Geosci Can (Reprint series) 1:53–69

Ryer TA (1977) Patterns of Cretaceous shallow-marine sedimentation, Coalville and Rockport areas, Utah. Geol Soc Am Bull 88:177–188

Ryer TA (1983) Transgressive-regressive cycles and the occurrence of coal in some upper Cretaceous strata of Utah. Geology 11:201–210

Ryer TA (1984) Transgressive-regressive cycles and the occurrence of coal in some upper Cretaceous strata of Utah, USA. In: Rahmani RA, Flores RM (eds) Sedimentology of coal and coal-bearing sequences. Int Assoc Sedimentol Spec Publ 7:217–227

Sadler PM (1981) Sedimentation rates and the completeness of stratigraphic sections. J Geol 89:569–584

Sadler PM, Osleger DA, Montañez IP (1993) On the labeling, length, and objective basis of Fischer plots. J Sediment Petrol 63:360–368

Sahagian DL (1987) Epeirogeny and eustatic sea level changes as inferred from Cretaceous shoreline deposits: applications to the central and western United States. J Geophys Res 92:4895–4904

Sahagian DL (1988) Epeirogenic movements of Africa as inferred from Cretaceous shoreline deposits. Tectonics 7:125–138

Sahagian DL, Holland SM (1991) Eustatic sea-level curve based on a stable frame of reference: preliminary results. Geology 19:1209–1212

Sahagian DL, Watts AB (1991) Introduction to the special section on measurement, causes, and consequences of long-term sea level changes. J Geophys Res 96B:6585–6589

Saito Y (1991) Sequence stratigraphy on the shelf and upper slope in response to the latest Pleistocene-Holocene sea-level changes off Sendai, northeast Japan. In: Macdonald DIM (ed) Sedimentation, tectonics and eustasy: sea-level changes at active margins. Int Assoc Sedimentol Spec Publ 12:133–150

Salvador A (ed) (1994) International stratigraphic guide. International Union of Geological Sciences, Trondheim, Norway, and Geological Society of America, Boulder, 214 pp

Sandberg PA (1983) An oscillating trend in Phanerozoic non-skeletal carbonate mineralogy. Nature 305:19–22

Sanford BV, Thompson FJ, McFall FJ (1985) Plate tectonics – a possible controlling mechanism in the development of hydrocarbon traps in southwestern Ontario. Bull Can Pet Geol 33:52–71

Sangree JB, Widmier JM (1977) Seismic stratigraphy and global changes of sea level, part 9. Seismic interpretation of clastic depositional facies. In: Payton CE (ed) Seismic stratigraphy – applications to hydrocarbon exploration. Am Assoc Pet Geol Mem 26:165–184

Sarg JF (1988) Carbonate sequence stratigraphy. In: Wilgus CK, Hastings BS, Kendall CGStC, Posamentier HW, Ross CA, Van Wagoner JC (eds) Sea level changes – an integrated approach. Soc Econ Paleontol Mineral Spec Publ 42:155–181

Savin SM (1977) The history of the Earth's surface temperature during the past 100 million years. Annu Rev Earth Planet Sci 5:319–355

Schlager W (1989) Drowning unconformities on carbonate platforms. In: Crevello PD, Wilson JL, Sarg JF, Read JF (eds) Controls on carbonate platforms and basin development. Soc Econ Paleontol Mineral Spec Publ 44:15–25

Schlager W (1991) Depositional bias and environmental change – important factors in sequence stratigraphy. Sediment Geol 70:109–130

Schlager W (1992a) Sedimentology and sequence stratigraphy of reefs and carbonate platforms. Continuing education course notes series. Am Assoc Pet Geol 34:71

Schlager W (1992b) Stratigraphic response of a carbonate platform to relative sea level changes: Broken Ridge, southeast Indian Ocean: discussion. Am Assoc Pet Geol Bull 76:1034–1036

Schlager W (1993) Accommodation and supply – a dual control on stratigraphic sequences. Sediment Geol 86:111–136

Schlee JS (ed) (1984) Interregional unconformities and hydrocarbon accumulation. Am Assoc Pet Geol Mem 36:184

Schmidt H, Seyfried H (1991) Depositional sequences and sequence boundaries in fore-arc coastal embayments: case histories from Central America. In: Macdonald DIM (ed) Sedimentation, tectonics and eustasy: sea-level changes at active margins. Int Assoc Sedimentol Spec Publ 12:241–258

Schopf TJM (1974) Permo-Triassic extinctions: relation to sea-floor spreading. J Geol 82:129–143

Schumm SA (1993) River response to baselevel change: implications for sequence stratigraphy. J Geol 101:279–294

Schwan W (1980) Geodynamic peaks in alpinotype orogenies and changes in ocean-floor spreading during late Jurassic-late Tertiary time. Am Assoc Pet Geol Bull 64:359–373

Schwarzacher W (1993) Cyclostratigraphy and the Milankovitch theory. Developments in sedimentology 52. Elsevier, Amsterdam, 225 pp

Sclater JG, Christie PAF (1980) Continental stretching: an explanation of the post-mid-Cretaceous subsidence of the central North Sea Basin. J Geophys Res 85B7:3711–3739

Sclater JG, Anderson RN, Bell ML (1971) The elevation of ridges and the evolution of the central eastern Pacific. J Geophys Res 76:7888–7915

Scott RW, Frost SH, Shaffer BL (1988) Early Cretaceous sea-level curves, Gulf Coast and southeastern Arabia. In: Wilgus CK, Hastings BS, Kendall CGStC, Posamentier HW, Ross CA, Van Wagoner JC (eds) Sea-level changes: an integrated approach. Soc Econ Paleontol Mineral Spec Publ 42:275–284

Scott RW, Evetts MJ, Stein JA (1993) Are seismic/depositional sequences time units? Testing by SHADS cores and graphic correlation. Offshore Technology Conference, Houston, Paper OTC 7110, pp 269–276

Sengör AMC (1992) Unconformities in mountain belts: does the sequence stratigraphy make sense? 29th International Geological Congress, Kyoto, Japan, Abstracts, 294 pp

Seyfried H, Astorga A, Amann H, Calvo C, Kolb W, Schmidt H, Winsemann J (1991) Anatomy of an evolving island arc: tectonic control in the south Central American fore-arc area. In: Macdonald DIM (ed) Sedimentation, tectonics and eustasy: sea-level changes at active margins. Int Assoc Sedimentol Spec Publ 12:217–240

Shackleton NJ, Hall MA (1984) Oxygen and carbon isotope stratigraphy of Deep Sea Drilling Project Hole 552A: Plio-Pleistocene glacial history. Initial Reports of Deep Sea Drilling Project, vol 85. US Government Printing Office, Washington, pp 599–609

Shanley KW, McCabe PJ (1991) Predicting facies architecture through sequence stratigraphy – an example from the Kaiparowits Plateau, Utah. Geology 19:742–745

Shanley KW, McCabe PJ (1994) Perspectives on the sequence stratigraphy of continental strata. Am Assoc Petrol Geol Bull 78:544–568

Shanley KW, McCabe PJ, Hettinger RD (1992) Tidal influences in Cretaceous fluvial strata from Utah, USA: a key to sequence stratigraphic interpretation. Sedimentology 39:905–930

Shanmugam G, Moiola RJ (1982) Eustatic control of turbidites and winnowed turbidites. Geology 10:231–235

Shanmugam G, Bloch RB, Mitchell SM, Beamish GWJ, Hodgkinson RJ, Damuth JE, Straume T, Syvertsen SE, Shields KE (1995) Basin floor fans in the North Sea: sequence stratigaphic models vs sedimentary facies. Am Assoc Pet Geol Bull 79:477–512

Shaub EJ, Buffler RT, Parsons JG (1984) Seismic stratigraphic framework of deep central Gulf of Mexico Basin. Am Assoc Pet Geol Bull 68:1790–1802

Shaw AB (1964) Time in stratigraphy. McGraw-Hill, New York, 365 pp

Sheridan RE (1987) Pulsation tectonics as the control of continental breakup. Tectonophysics 143:59–73

Sheridan RE (1989) The Atlantic passive margin. In: Bally AW, Palmer AR (eds) The geology of North America – an overview. The Geology of North America, vol A. Geological Society of America, Boulder, pp 81–96

Shurr GW (1984) Geometry of shelf-sandstone bodies in the Shannon sandstone of southeastern Montana. In: Tillman RW, Siemers CT (eds) Siliciclastic shelf sediments. Soc Econ Paleontol Mineral Spec Publ 34:63–83

Sinclair HD, Coakley BJ, Allen PA, Watts AB (1991) Simulation of foreland basin stratigraphy using a diffusion model of mountain belt uplift and erosion: an example from the central Alps, Switzerland. Tectonics 10:599–620

Sleep NH (1971) Thermal effects of the formation of Atlantic continental margins by continental breakup. Geophys J R Astron Soc 24:325–350

Sleep NH (1976) Platform subsidence mechanisms and "eustatic" sea level changes. Tectonophysics 36:45–56

Sloan RJ, Williams BPJ (1991) Volcano-tectonic control of offshore to tidal-flat regressive cycles from the Dunquin Group (Silurian) of southwest Ireland. In: Macdonald DIM (ed) Sedimentation, tectonics and eustasy: sea-level changes at active margins. Int Assoc Sedimentol Spec Publ 12:105–119

Sloss LL (1962) Stratigraphic models in exploration. Am Assoc Pet Geol Bull 46:1050–1057

Sloss LL (1963) Sequences in the cratonic interior of North America. Geol Soc Am Bull 74:93–113

Sloss LL (1972) Synchrony of Phanerozoic sedimentary-tectonic events of the North American craton and the Russian platform. 24th International Geological Congress, Montreal, Section 6, pp 24–32

Sloss LL (1979) Global sea level changes: a view from the craton. In: Watkins JS, Montadert L, Dickerson PW (eds) Geological and geophysical investigations of continental margins. Am Assoc Pet Geol Mem 29:461–468

Sloss LL (1982) The Midcontinent province: United States. In: Palmer AR (ed) Perspectives in regional geological syntheses. Decade of North American Geology Spec Publ 1, Geological Society of America, Boulder, pp 27–39

Sloss LL (1984) Comparative anatomy of cratonic unconformities. In: Schlee JS (ed) Interregional unconformities and hydrocarbon accumulation. Am Assoc Pet Geol Mem 36:1–6

Sloss LL (1988a) Forty years of sequence stratigraphy. Geol Soc Am Bull 100:1661–1665

Sloss LL (1988b) Tectonic evolution of the craton in Phanerozoic time. In: Sloss LL (ed) Sedimentary cover – North American craton: US. The Geology of North America, vol D-2. Geological Society of America, Boulder, pp 25–51

Sloss LL (1991) The tectonic factor in sea level change: a countervailing view. J Geophys Res 96B:6609–6617

Sloss LL, Speed RC (1974) Relationships of cratonic and continental margin episodes. In: Dickinson WR (ed) Tectonics and sedimentation. Soc Econ Paleontol Mineral Spec Publ 22:98–119

Sloss LL, Krumbein WC, Dapples EC (1949) Integrated facies analysis. In: Longwell CR (ed) Sedimentary facies in geologic history. Geol Soc Am Mem 39:91–124

Smith GA (1994) Climatic influences on continental deposition during late-stage filling of an extensional basin, southeastern Arizona. Geol Soc Am Bull 106:1212–1228

Smith PL (1988) Paleoscene 11: paleobiogeography and plate tectonics. Geosci Can 15:261–279

Soares PC, Landim PMB, Fulfaro VJ (1978) Tectonic cycles and sedimentary sequences in the Brazilian intracratonic basins. Geol Soc Am Bull 89:181–191

Southgate PN, Kennard JM, Jackson MJ, O'Brien PE, Sexton MJ (1993) Reciprocal lowstand clastic and highstand carbonate sedimentation, subsurface Devonian reef complex, Canning Basin, Western Australia. In: Loucks RG, Sarg JF (eds) Carbonate sequence stratigraphy. Am Assoc Pet Geol Mem 57:157–179

Srinivasan MS, Kennett JP (1981) A review of planktonic foraminiferal biostratigraphy: applications in the equatorial and South Pacific. In: Warme JE, Douglas RG, Winterer EL (eds) The deep sea drilling project: a decade of progress. Soc Econ Paleontol Mineral Spec Publ 32:395–432

Srivastava SP, Tapscott CR (1986) Plate kinematics of the North Atlantic. In: Vogt PR, Tucholke BE (eds) The Western North Atlantic region. The Geology of North America, vol M. Geological Society of America, Boulder, pp 379–404

Steckler MS, Watts AB (1978) Subsidence of the Atlantic-type continental margin off New York. Earth Planet Sci Lett 41:1–13

Steckler MS, Reynolds DJ, Coakley BJ, Swift BA, Jarrard R (1993) Modelling passive margin sequence stratigraphy. In: Posamentier HW, Summerhayes CP, Haq BU, Allen GP (eds) Sequence stratigraphy and facies associations Int Assoc Sediment Spec Publ 18:19–41

Stockmal GS, Cant DJ, Bell JS (1992) Relationship of the stratigraphy of the Western Canada foreland basin to Cordilleran tectonics: insights from geodynamic models. In: Macqueen RW, Leckie DA (eds) Foreland basin and fold belts. Am Assoc Pet Geol Mem 55:107–124

Suess E (1885–1890) Das Antliz der Erde. F Tempsky, Vienna, 3 vol

Suess E (1906) The face of the earth, vol 2. Clarendon Press, Oxford, 759 pp

Summerhayes CP (1986) Sea level curves based on seismic stratigraphy: their chronostratigraphic significance. Palaeogeogr Palaeoclimatol Palaeoecol 57:27–42

Surlyk F (1989) Mid-Mesozoic syn-rift turbidite systems: controls and predictions. In: Collinson JD (ed) Correlation in hydrocarbon exploration. Graham and Trotman, London, pp 231–241

Surlyk F (1990) A Jurassic sea-level curve for East Greenland. Palaeogeogr Palaeoclimatol Palaeoecol 78:71–85

Surlyk F (1991) Sequence stratigraphy of the Jurassic-lowermost Cretaceous of east Greenland. Am Assoc Pet Geol Bull 75:1468–1488

Suter JR, Berryhill HL Jr (1985) Late Quaternary shelf-margin deltas, northwest Gulf of Mexico. Am Assoc Pet Geol Bull 69:77–91

Suter JR, Berryhill HL Jr, Penland S (1987) Late Quaternary sea-level fluctuations and depositional sequences, southwest Louisiana continental shelf. In: Nummedal D, Pilkey OH, Howard JD (eds) Sea-level fluctuation and coastal evolution. Soc Econ Paleontol Mineral Spec Publ 41:199–219

Swift DJP, Thorne JA (1991) Sedimentation on continental margins, I: a general model for shelf sedimentation. In: Swift DJP, Oertel GF, Tillman RW, Thorne JA (eds) Shelf sand and sandstone bodies: geometry, facies and sequence stratigraphy. Int Assoc Sedimentol Spec Publ 14:3–31

Swift DJP, Hudelson PM, Brenner RL, Thompson P (1987) Shelf construction in a foreland basin: storm beds, shelf sandbodies, and shelf-slope depositional sequences in the upper Cretaceous Mesaverde group, Book Cliffs, Utah. Sedimentology 34:423–457

Swift DJP, Oertel GF, Tillman RW, Thorne JA (eds) (1991a) Shelf sand and sandstone bodies. Int Assoc Sedimentol Spec Publ 14:532

Swift DJP, Phillips S, Thorne JA (1991b) Sedimentation on continental margins, V: parasequences. In: Swift DJP, Oertel GF, Tillman RW, Thorne JA (eds) Shelf sand and sandstone bodies: geometry, facies and sequence stratigraphy. Int Assoc Sedimentol Spec Publ 14:153–187

Tandon SK, Gibling MR (1994) Calcrete and coal in late Carboniferous cyclothems of Nova Scotia, Canada: climate and sea-level changes linked. Geology 22:755–758

Tankard AJ (1986) On the depositional response to thrusting and lithospheric flexure: examples from the Appalachian and Rocky Mountain basins. In: Allen PA, Homewood P (eds) Foreland basins. Int Assoc Sedimentol Spec Publ 8:369–392

Tankard AJ, Welsink H (1987) Extensional tectonics and stratigraphy of Hibernia oil field, Grand Banks, Newfoundland. Am Assoc Pet Geol Bull 71:1210–1232

Thierstein HR (1981) Late Cretaceous nannoplankton and the change at the Cretaceous-Tertiary boundary. In: Warme JE, Douglas RG, Winterer EL (eds) The deep sea drilling project: a decade of progress. Soc Econ Paleontol Mineral Spec Publ 32:355–394

Thunell R, Rio D, Sprovieri R, Raffi I (1991) Limestone-marl couplets: origin of the early pliocene Trubi Marls in Calabria, southern Italy. J Sediment Petrol 61:1109–1122

Tipper JC (1993) Do seismic reflections necessarily have chronostratigraphic significance? Geol Mag 130:47–55

Toksöz MN, Bird P (1977) Formation and evolution of marginal basins and continental plateaus. In: Talwani M, Pitman WC III (eds) Island arcs, deep-sea trenches and back-arc basins. Maurice Ewing Series 1, American Geophysical Union, Wahington, DC, pp 379–393

Tucholke BE, Embley RW (1984) Cenozoic regional erosion of the abyssal sea floor off South Africa. In: Schlee JS (ed) Interregional unconformities and hydrocarbon exploration. Am Assoc Pet Geol Mem 36:145–164

Uchupi E, Emery KO (1991) Pangaean divergent margins: historical perspective. Mar Geol 102:1–28

Underhill JR (1991) Controls on late Jurassic seismic sequences, Inner Moray Firth, UK North Sea: a critical test of a key segment of Exxon's original global cycle chart. Basin Res 3:79–98

Underhill JR, Partington MA (1993) Use of genetic sequence stratigraphy in defining and determining a regional tectonic control on the "Mid-Cimmerian unconformity" – implications for North Sea basin development and the global sea level chart. In: Weimer P, Posamentier HW (eds) Siliciclastic sequence stratigraphy Am Assoc Pet Geol Mem 58:449–484

Vail PR (1987) Seismic stratigraphy interpretation using sequence stratigraphy, part 1: seismic stratigraphy interpretation procedure. In: Bally AW (ed) Atlas of seismic stratigraphy. Am Assoc Pet Geol Stud Geol 27(1):1–10

Vail PR (1992) The evolution of seismic stratigraphy and the global sea-level curve. In: Dott RH Jr (ed) Eustasy: the historical ups and downs of a major geological concept. Geol Soc Am Mem 180:83–91

Vail PR, Todd RG (1981) Northern North Sea Jurassic unconformities, chronostratigraphy and sea-level changes from seismic stratigraphy. In: Illing LV, Hobson GD (eds) Petroleum geology of the continental shelf of northwest Europe. Institute of Petroleum, London, pp 216–235

Vail PR, Mitchum RM Jr, Todd RG, Widmier JM, Thompson S III, Sangree JB, Bubb JN, Hatlelid WG (1977) Seismic stratigraphy and global changes of sea-level. In: Payton CE (ed) Seismic stratigraphy – applications to hydrocarbon exploration. Am Assoc Pet Geol Mem 26:49–212

Vail PR, Hardenbol J, Todd RG (1984) Jurassic unconformities, chronostratigraphy and sea-level changes from seismic stratigraphy and biostratigraphy. In: Schlee JS (ed) Interregional unconformities and hydrocarbon exploration. Am Assoc Pet Geol Mem 36:129–144

Vail PR, Audemard F, Bowman SA, Eisner PN, Perez-Crus C (1991) The stratigraphic signatures of tectonics, eustasy and sedimentology – an overview. In: Einsele G, Ricken W, Seilacher A (eds) Cycles and events in stratigraphy. Springer-Verlag, Berlin Heidelberg New York, pp 617–659

Valentine JW, Moores E (1970) Plate tectonic regulation of faunal diversity and sea level. Nature 228:657–669

Valentine JW, Moores E (1972) Global tectonics and the fossil record. J Geol 80:167–184

Van Hinte JE (1976a) A Jurassic time scale. Am Assoc Pet Geol Bull 60:489–497

Van Hinte JE (1976b) A Cretaceous time scale. Am Assoc Pet Geol Bull 60:498–516

Van Houten FB (1964) Cyclic lacustrine sedimentation, upper Triassic Lockatong formation, central New Jersey and adjacent Pennsylvania. In: Merriam DF (ed) Symposium on cyclic sedimentation Kans Geol Surv Bull 169:495–531

Van Houten FB (1981) The odyssey of molasse. In: Miall AD (ed) Sedimentation and tectonics in alluvial basin.: Geol Assoc Can Spec Pap 23:35–48

Van Siclen DC (1958) Depositional topography – examples and theory. Am Assoc Pet Geols Bull 42:1897–1913

Van Tassell J (1994) Evidence for orbitally-driven sedimentary cycles in the Devonian Catskill Delta complex. In: Dennison JM, Ettensohn FR (eds) Tectonic and eustatic controls on sedimentary cycles. Soc Sediment Geol, Concepts Sedimentol Paleontol 4:121–131

van Veen PM, Simonsen BT (1991) Comment on "glacial-eustatic sea-level curve for early late Pennsylvanian sequence in north-central Texas and biostratigraphic correlation with curve for midcontinent North America". Geology 19:91–92

Van Wagoner JC (1995) Overview of sequence stratigraphy of foreland basin deposits: terminology, summary of papers, and glossary of sequence stratigraphy. In: Van Wagoner JC, Bertram GT (eds) Sequence stratigraphy of foreland basins. Am Assoc Pet Geol Mem 64:9–21

Van Wagoner JC, Bertram GT (eds) (1995) Sequence stratigraphy of foreland basins. Am Assoc Pet Geol Mem 64:487

Van Wagoner JC, Mitchum RM Jr, Posamentier HW, Vail PR (1987) Seismic stratigraphy interpretation using sequence stratigraphy, part 2: key definitions of sequence stratigraphy. In: Bally AW (ed) Atlas of seismic stratigraphy. Am Assoc Pet Geol Stud Geol 27(1):11–14

Van Wagoner JC, Mitchum RM, Campion KM, Rahmanian VD (1990) Siliciclastic sequence stratigraphy in well logs, cores, and outcrops. Am Assoc Pet Geol Methods in Exploration Series 7:55

Van Wagoner JC, Nummedal D, Jones CR, Taylor DR, Jennette DC, Riley GW (1991) Sequence stratigraphy applications to shelf sandstone reservoirs. Am Assoc Pet Geol Field Conference Guidebook, Am Assoc Pet Geologists, Tulsa, Oklahoma

Veevers JJ (1990) Tectonic-climatic supercycle in the billion-year plate-tectonic eon: Permian Pangean icehouse alternates with Cretaceous dispersed-continents greenhouse. Sediment Geol 68:1–16

Veevers JJ, Powell CMA (1987) Late Paleozoic glacial episodes in Gondwanaland reflected in transgressive-regressive depositional sequences in Euramerica. Geol Soc Am Bull 98:475–487

Villien A, Kligfield RM (1986) Thrusting and synorogenic sedimentation in central Utah. In: Peterson JA (ed) Paleotectonics and sedimentation in the Rocky Mountain region, United States. Am Assoc Pet Geol Mem 41:281–307

Vinogradov AP, Nalivkin VD (eds) (1960) Atlas of lithopaleogeographical maps of the Russian Platform and its geosynclinal framing, part 1 – Late PreCambrian and Paleozoic. Acadademy of Science USSR, Moscow

Vinogradov AP, Ronov AB, Khain VE (eds) (1961) Atlas of lithopaleogeographical maps of the Russian Platform and its geosynclinal framing, part II – Mesozoic and Cenozoic. Academy of Science USSR, Moscow

Wagner HC (1964) Pennsylvanian megacyclothems of Wilson County, Kansas, and speculations concerning their depositional environments. Kans Geol Surv Bull 169:565–591

Walker JCG, Zahnle KJ (1986) Lunar nodal tide and distance to the moon during the PreCambrian. Nature 320:600–602

Walker RG (ed) (1984) Facies models, 2nd edn. Geosci Can Reprint Series 1:317

Walker RG (1992a) Facies, facies models and modern stratigraphic concepts. In: Walker RG, James NP (eds) Facies models: response to sea-level change. Geological Association of Canada, St John's, Newfoundland, pp 1–14

Walker RG (1992b) Turbidites and submarine fans. In: Walker RG, James NP (eds) Facies models: response to sea-level change. Geological Association of Canada, pp 239–263

Walker RG, Eyles CH (1988) Geometry and facies of stacked shallow-marine sandier upward sequences dissected by an erosion surface, Cardium Formation, Willesden Green, Alberta. Am Assoc Pet Geol Bull 72:1469–1494

Walker RG, James NP (eds) (1992) Facies models: response to sea-level change. Geological Association of Canada, St John's, Newfoundland, 409 pp

Walker RG, Plint G (1992) Wave- and storm-dominated shallow marine systems. In: Walker RG, James NP (eds) Facies models: response to sea-level change. Geological Association of Canada, pp 219–238

Wanless HR (1950) Late Paleozoic cycles of sedimentation in the United States. 18th International Geological Congress, Algiers, part 4, pp 17–28

Wanless HR (1964) Local and regional factors in Pennsylvanian cyclic sedimentation. In: Merriam DF (ed) Symposium on cyclic sedimentation. Kans Geol Surv Bull 169:593–606

Wanless HR (1972) Eustatic shifts in sea level during the deposition of Late Paleozoic sediments in the central United States. In: Elam JG, Chuber S (eds) Cyclic sedimentation in the Permian Basin. West Texas Geological Society Symposium, pp 41–54

Wanless HR (1991) Observational foundation for sequence modeling. In: Franseen EK, Watney WL, Kendall CGStC (eds) Sedimentary modeling: computer simulations and methods for improved parameter definition. Kans Geol Surv Bull 233:43–62

Wanless HR, Shepard EP (1936) Sea level and climatic changes related to Late Paleozoic cycles. Geol Soc Am Bull 47:1177–1206

Wanless HR, Weller JM (1932) Correlation and extent of Pennsylvanian cyclothems. Geol Soc Am Bull 43:1003–1016

Warme JE, Douglas RG, Winterer EL (eds) (1981) The deep sea drilling project: a decade of progress. Soc Econ Paleontol Mineral Spec Publ 32:564

Waschbusch PJ, Royden LH (1992) Episodicity in foredeep basins. Geology 20:915–918

Washington PA, Chisick SA (1994) Foundering of the Cambro-Ordovician shelf margin: onset of Taconian orogenesis or eustatic drowning. In: Dennison JM, Ettensohn FR (eds) Tectonic and eustatic controls on sedimentary cycles. Society for Sedimentary Geology, Tulsa Oklahoma, Concepts in Sedimentology and Paleontology, vol 4, pp 203–216

Watts AB (1981) The US Atlantic margin: subsidence history, crustal structure and thermal evolution. American Association of Petroleum Geologists, Tulsa, Oklahoma, Education Course Notes Series 19, Chap 2, 75 p

Watts AB (1989) Lithospheric flexure due to prograding sediment loads: implications for the origin of offlap/onlap patterns in sedimentary basins. Basin Res 2:133–144

Watts AB, Ryan WBF (1976) Flexure of the lithosphere and continental margin basins. Tectonophysics 36:24–44

Watts AB, Steckler MS(1979) Subsidence and eustacy at the continental margin of eastern North America. In: Talwani M, Hay W, Ryan WBF (eds) Deep drilling results in the Atlantic Ocean; continental margins and paleoenvironment. American Geophysical Union, Washington, DC, Maurice Ewing Series, No 3, pp 218–234

Watts AB, Karner GD, Steckler MS (1982) Lithospheric flexure and the evolution of sedimentary basins. In: Kent P, Bott MHP, McKenzie DP, Williams CA (eds) The evolution of sedimentary basins. Philos Trans R Soc Lond 305A:249–281

Weedon GP (1986) Hemipelagic shelf sedimentation and climatic cycles: the basal Jurassic (Blue Lias) of South Britain. Earth Planet Sci Lett 76:321–335

Weedon GP (1993) The recognition and stratigraphic implications of orbital-forcing of climate and sedimentary cycles. Sedimentol Rev 1:31–50

Weimer P (1990) Sequence stratigraphy, facies geometries, and depositional history of the Mississippi fan, Gulf of Mexico. Am Assoc Pet Geol Bull 74:425–453

Weimer P, Posamentier HW (eds) (1993) Siliciclastic sequence stratigraphy. Am Assoc Pet Geol Mem 58:492

Weimer RJ (1960) Upper Cretaceous stratigraphy, Rocky Mountain area. Am Assoc Pet Geol Bull 44:1–20

Weimer RJ (1970) Rates of deltaic sedimentation and intrabasin deformation, Upper Cretaceous of Rocky Mountain region. In: Morgan JP (ed) Deltaic sedimentation modern and ancient. SocEcon Paleontol Mineral Spec Publ 15:270–292

Weimer RJ (1986) Relationship of unconformities, tectonics, and sea level change in the Cretaceous of the Western Interior, United States. In: Peterson JA (ed) Paleotectonics and sedimentation in the Rocky Mountain region, United States. Am Assoc Pet Geol Mem 41:397–422

Welsink HJ, Tankard AJ (1988) Structural and stratigraphic framework of the Jeanne d'Arc Basin, Grand Banks. In: Bally AW (ed) Atlas of seismic stratigraphy. Am Assoc Pet Geol Stud Geol 27(2):14–21

Weltje G, de Boer PL (1993) Astronomically induced paleoclimatic oscillations reflected in Pliocene turbidite deposits on Corfu (Greece): implications for the interpretation of higher order cyclicity in ancient turbidite systems. Geology 21:307–310

Wernicke B (1985) Uniform-sense normal simple shear of the continental lithosphere. Can J Earth Sci 22:108–125

Wescott WA (1993) Geomorphic thresholds and complex response of fluvial systems – some implications for sequence stratigraphy. Am Assoc Pet Geol Bull 77:1208–1218

Wheeler HE (1958) Time-stratigraphy. Am Assoc Pet Geol Bull 42:1047–1063

Wheeler HE (1959a) Note 24 – unconformity-bounded units in stratigraphy. Am Assoc Pet Geol Bull 43:1975–1977

Wheeler HE (1959b) Stratigraphic units in time and space. Am J Sci 257:692–706

Wheeler HE (1963) Post-Sauk and pre-Absaroka Paleozoic stratigraphic patterns in North America. Am Assoc Pet Geol Bull 47:1497–1526

Wignall PB (1991) Ostracod and foraminifera micropaleontology and its bearing on biostratigraphy: a case study from the Kimmeridgian (Late Jurassic) of north west Europe. Palaios 5:219–226

Wilgus CK, Hastings BS, Kendall CGStC, Posamentier HW, Ross CA, Van Wagoner JC (eds) (1988) Sea level changes – an integrated approach. Soc Econ Paleontol Mineral Spec Publ 42:407

Williams DF (1988) Evidence for and against sea-level changes from the stable isotopic record of the Cenozoic. In: Wilgus CK, Hastings BS, Kendall CGSC, Posamentier HW, Ross CA, Van Wagoner JC (eds) Sea level changes – an integrated approach. Soc Econ Paleontol Mineral Spec Publ 42:31–36

Williams DF, Thunell RC, Mucciarone D (1984) Toward a new oxygen isotope chronostratigraphy of early to middle Pleistocene deep-sea sediments. Geological Society of America (Abstracts with Program), Boulder, vol 16, 694 pp

Williams GD, Dobb A (eds) (1993) Tectonics and seismic sequence stratigraphy. Geological Society, London, Spec Publ 71, 226 pp

Williams GE (1981) Megacycles, benchmark papers in geology, vol 57. Hutchinson Ross Publishing Company, Stroudsberg, 434 pp

Wilson JL (1975) Carbonate facies in geologic history. Springer-Verlag, Berlin Heidelberg New York, 471 pp

Winker CD (1979) Late Pleistocene fluvial-deltaic deposition, Texas coastal plain and shelf. Unpubl MA thesis, University of Texas, 187 pp

Winker CD, Edwards MB (1983) Unstable progradational clastic shelf margins. In: Stanley DJ, Moore GT (eds) The shelfbreak: critical interface on continental margins. Soc Econ Paleontol Mineral Spec Publ 33:427–448

Winsemann J, Seyfried H (1991) Response of deep-water fore-arc systems to sea-level changes, tectonic activity and volcaniclastic input in Central America. In: Macdonald DIM (ed) Sedimentation, tectonics and eustasy: sea-level changes at active margins. Int Assoc Sedimentol Spec Publ 12:273–292

Wise DU (1974) Continental margins, freeboard and the volumes of continents and oceans through time. In: Burk CA, Drake CL (eds) The geology of continental margins. Springer-Verlag, Berlin Heidelberg New York, pp 45–58

Worsley TR, Nance RD (1989) Carbon redox and climate control through earth history: a speculative reconstruction. Palaeogeogr Palaeoclimatol Palaeoecol 75:259–282

Worsley TR, Nance D, Moody JB (1984) Global tectonics and eustasy for the past 2 billion years. Mar Geol 58:373–400

Worsley TR, Nance D, Moody JB (1986) Tectonic cycles and the history of the earth's biogeochemical and paleoceanographic record. Paleoceanography 1:233–263

Worsley TR, Nance RD, Moody JB (1991) Tectonics, carbon, life, and climate for the last three billion years: a unified system? In: Schneider SH, Boston PJ (eds) Scientists on Gaia. MIT Press, Cambridge, Massachusetts, pp 200–210

Wright R (1986) Cycle stratigraphy as a paleogeographic tool: Point Lookout Sandstone, southeastern San Juan Basin, New Mexico. Geol Soc Am Bull 96:661–673

Wright VP (1992) Speculations on the controls on cyclic peritidal carbonates: ice-house versus greenhouse eustatic controls. Sediment Geol 76:1–5

Wright VP, Marriott SB (1993) The sequence stratigraphy of fluvial depositional systems: the role of floodplain sediment storage. Sediment Geol 86:203–210

Yang Chang-Shu, Nio S-D (1993) Application of high-resolution sequence stratigraphy to the Upper Rotliegend in the Netherlands offshore. In: Weimer P, Posamentier HW (eds) Siliciclastic sequence stratigraphy. Am Assoc Pet Geol Mem 58:285–316

Youle JC, Watney WL, Lambert LL (1994) Stratal hierarchy and sequence stratigraphy – Middle Pennsylvanian, southwestern Kansas, USA. In: Klein G deV (ed) Pangea: paleoclimate, tectonics, and sedimentation during accretion, zenith, and breakup of a supercontinent. Geol Soc Am Spec Pap 288:267–285

Young RG (1955) Sedimentary facies and intertonguing in the Upper Cretaceous of the Book Cliffs, Utah-Colorado. Geol Soc Am Bull 66:177–202

Yoshida S, Willis A, Miall AD (1996) Tectonic control of nested sequence architecture in the Castlegate Sandstone (Upper Cretaceous), Book Cliffs, Utah. J Sediment Res 66:737–748

Zeller EJ (1964) Cycles and psychology. Geol Surv Kans Bull 169:631–636

Ziegler AM (1965) Silurian marine communities and their environmental significance. Nature 207:270–272

Ziegler AM, Cocks LRM, McKerrow WS (1968) The Llandovery transgression of the Welsh borderland. Paleontology 11:736–782

Ziegler PA (1982) Geological atlas of western and central Europe. Shell Internationale Petroleum Maatschappij BV, The Hague

Zoback ML (1992) First- and second-order patterns of stress in the lithosphere: the world stress map project. J Geophys Res 97B:11703–11728

Author Index

Subject Index

Springer
and the
environment

At Springer we firmly believe that an international science publisher has a special obligation to the environment, and our corporate policies consistently reflect this conviction.

We also expect our business partners – paper mills, printers, packaging manufacturers, etc. – to commit themselves to using materials and production processes that do not harm the environment. The paper in this book is made from low- or no-chlorine pulp and is acid free, in conformance with international standards for paper permanency.

 Springer